T0192849

Lecture Notes of the Institute for Computer Sciences, Social Informatics and Telecommunications Engineering 549

The LNICST series publishes ICST's conferences, symposia and workshops.

LNICST reports state-of-the-art results in areas related to the scope of the Institute.

The type of material published includes

- Proceedings (published in time for the respective event)
- Other edited monographs (such as project reports or invited volumes)

LNICST topics span the following areas:

- General Computer Science
- E-Economy
- E-Medicine
- Knowledge Management
- Multimedia
- Operations, Management and Policy
- Social Informatics
- Systems

Lin Yun · Jiang Han · Yu Han
Editors

Advanced Hybrid Information Processing

7th EAI International Conference, ADHIP 2023
Harbin, China, September 22–24, 2023
Proceedings, Part III

Editors
Lin Yun
Harbin Engineering University
Harbin, China

Jiang Han
Harbin Engineering University
Harbin, China

Yu Han
Harbin Engineering University
Harbin, China

ISSN 1867-8211 ISSN 1867-822X (electronic)
Lecture Notes of the Institute for Computer Sciences, Social Informatics and Telecommunications Engineering
ISBN 978-3-031-50548-5 ISBN 978-3-031-50549-2 (eBook)
https://doi.org/10.1007/978-3-031-50549-2

This Springer imprint is published by the registered company Springer Nature Switzerland AG
The registered company address is: Gewerbestrasse 11, 6330 Cham, Switzerland

Paper in this product is recyclable.

Preface

We are delighted to introduce the proceedings of the 7th edition of the European Alliance for Innovation (EAI) International Conference on Advanced Hybrid Information Processing (ADHIP 2023). This conference brought together researchers, developers and practitioners around the world who are leveraging and developing advanced information processing technology. This conference aimed to provide an opportunity for researchers to publish their important theoretical and technological studies of advanced methods in social hybrid data processing, and their novel applications within this domain.

The technical program of ADHIP 2023 consisted of 108 full papers. The topics of the conference were novel technology for social information processing and real applications to social data. Aside from the high-quality technical paper presentations, the technical program also featured three keynote speeches. The three keynote speakers were Cesar Briso from Technical University of Madrid, Spain, Yong Wang from Harbin Institute of Technology, China, and Yun Lin from Harbin Engineering University, China.

Coordination with the steering chairs, Imrich Chlamtac, Shuai Liu and Yun Lin was essential for the success of the conference. We sincerely appreciate their constant support and guidance. It was also a great pleasure to work with such an excellent organizing committee team for their hard work in organizing and supporting the conference. In particular, the Technical Program Committee, led by our TPC Co-Chairs, Yun Lin, Ruizhi Liu and Shan Gao completed the peer-review process of technical papers and made a high-quality technical program. We are also grateful to the Conference Manager, Ivana Bujdakova, for her support and to all the authors who submitted their papers to the ADHIP 2023 conference.

We strongly believe that the ADHIP conference provides a good forum for all researchers, developers and practitioners to discuss all technology and application aspects that are relevant to information processing technology. We also expect that the future ADHIP conferences will be as successful and stimulating as indicated by the contributions presented in this volume.

Yun Lin

Organization

Organizing Committee

General Chair

Yun Lin	Harbin Engineering University, China

General Co-chairs

Zheng Dou	Harbin Engineering University, China
Yan Zhang	University of Oslo, Norway
Shui Yu	University of Technology Sydney, Australia
Joey Tianyi Zhou	Institute of High-Performance Computing, A*STAR, Singapore
Hikmet Sari	Nanjing University of Posts and Telecommunications, China
Bin Lin	Dalian Maritime University, China

TPC Chair and Co-chairs

Yun Lin	Harbin Engineering University, China
Guangjie Han	Hohai University, China
Ruolin Zhou	University of Massachusetts Dartmouth, USA
Chao Li	RIKEN-AIP, Japan
Guan Gui	Nanjing University of Posts and Telecommunications, China
Ruizhi Liu	Harbin Engineering University, China

Sponsorship and Exhibit Chairs

Yiming Yan	Harbin Engineering University, China
Ali Kashif	Manchester Metropolitan University, UK
Liang Zhao	Shenyang Aerospace University, China

Local Chairs

Jiang Hang	Harbin Engineering University, China
Yu Han	Harbin Engineering University, China
Haoran Zha	Harbin Engineering University, China

Workshops Chairs

Nan Su	Harbin Engineering University, China
Peihan Qi	Xidian University, China
Jianhua Tang	Nanyang Technological University, Singapore
Congan Xu	Naval Aviation University, China
Shan Gao	Harbin Engineering University, China

Publicity and Social Media Chairs

Jiangzhi Fu	Harbin Engineering University, China
Lei Chen	Georgia Southern University, USA
Zhenyu Na	Dalian Maritime University, China

Publications Chairs

Weina Fu	Hunan Normal University, China
Sicheng Zhang	Harbin Engineering University, China
Wenjia Li	New York Institute of Technology, USA

Web Chairs

Yiming Yan	Harbin Engineering University, China
Zheng Ma	University of Southern Denmark, Denmark
Jian Wang	Fudan University, China

Posters and PhD Track Chairs

Lingchao Li	Shanghai Dianji University, China
Jibo Shi	Harbin Engineering University, China
Yulong Ying	Shanghai University of Electric Power, China

Panels Chairs

Danda Rawat	Howard University, USA
Yuan Liu	Tongji University, China
Yan Sun	Harbin Engineering University, China

Demos Chairs

Ao Li	Harbin University of Science and Technology, China
Guyue Li	Southeast University, China
Changbo Hou	Harbin Engineering University, China

Tutorials Chairs

Yu Wang	Nanjing University of Posts and Telecommunications, China
Yi Zhao	Tsinghua University, China
Qi Lin	Harbin Engineering University, China

Technical Program Committee

Zheng Dou	Harbin Engineering University, China
Yan Zhang	University of Oslo, Norway
Shui Yu	University of Technology Sydney, Australia
Joey Tianyi Zhou	A*STAR, Singapore
Hikmet Sari	Nanjing University of Posts and Telecommunications, China
Bin Lin	Dalian Maritime University, China
Yun Lin	Harbin Engineering University, China
Guangjie Han	Hohai University, China
Ruolin Zhou	University of Massachusetts Dartmouth, USA
Chao Li	RIKEN-AIP, Japan
Guan Gui	Nanjing University of Posts and Telecommunications, China
Zheng Ma	University of Southern Denmark, Denmark
Jian Wang	Fudan University, China
Lei Chen	Georgia Southern University, USA
Zhenyu Na	Dalian Maritime University, China
Peihan Qi	Xidian University, China
Jianhua Tang	Nanyang Technological University, Singapore

Contents – Part III

Mobile Monitoring, Civilian Audio and Acoustic Signal Processing

Wireless Networks for Social Information Processing, Image Information Processing

Interactive Sharing Method of Digital Media Image Information Based on Differential Privacy Protection

Wei Li(✉) and Jieling Jiang

Guangzhou Academy of Fine Arts, Guangzhou 510006, China
iccu007@163.com

Abstract. Because digital media image information is vulnerable to external attacks during interactive sharing, resulting in poor interactive sharing effect, a digital media image information interactive sharing method based on differential privacy protection is proposed. According to the sensitivity of digital media images, calculate the amount of digital media image information to be protected, and build a defense model of differential privacy protection mechanism. With LAN as the main communication medium, the C/S structure is used to realize the interactive management of digital media image information, calculate the information flow injected by a node to the interaction center, and ensure the smooth interaction process. The analysis of information set is in danger of being violated, revealing that the information set has the diversified characteristics of entropy to resist external attacks. Calculate the energy and distance of non-uniform clustering nodes, obtain the template mapping radius, and determine the weighted minimum value according to the residual energy of cluster heads and the spacing between clusters. Combined with multi-source information fusion method, the information sharing process of single node and multiple nodes is designed. The experimental results show that this method can realize information interaction, and the information is saved completely. The maximum amount of shared information is 900 bits, which has a good sharing effect.

Keywords: Differential Privacy Protection · Digital Media Image · Information Interaction · Information Sharing

1 Introduction

In the process of the continuous development of modern digital technology and the Internet, digital media technology has improved the richness of information resources, especially in the process of the continuous application of digital media technology, computer technology and network communication technology, digital media has penetrated into all aspects of life, making human beings enter the digital era. The digital media environment has created a diverse and rich ocean of open information for people. In the interactive sharing of digital media image information, participants may need to share

L. Yun et al. (Eds.): ADHIP 2023, LNICST 549, pp. 3–19, 2024.
https://doi.org/10.1007/978-3-031-50549-2_1

personal image data, such as photos or videos. However, these data may contain sensitive personal information, such as facial features, geographic location, etc. In order to protect these personal privacy, researchers have begun to explore how to address security issues during the data sharing process.

Literature [1] proposed a sharing scheme based on blockchain, designed a digital media image information security sharing model in combination with the blockchain distributed architecture, accessed the information in the model to the information collector, encrypted the information using an improved ElGamal encryption algorithm, and exchanged information using points instead of tokens in the traditional negotiation mechanism. After all nodes reached an agreement, The information package is stored in the information database to realize information sharing in the form of query and submission; Literature [2] proposes a sharing method based on the whole network power supply topology model. This method achieves the standardization of digital media image graphics information by establishing a graphics information sharing architecture and interaction model, and decoupling and exchanging different types of information based on CIM. However, in the above methods, most of the information is stored in relational databases, making it difficult to effectively share unstructured information. The amount of shared information is relatively small, and there may be situations of information loss.

In order to solve the shortcomings of the above methods, a digital media image information interactive sharing method based on Differential privacy protection is proposed.

2 Design of Real-Time Interaction Scheme of Digital Media Image Information

2.1 Construction of Interactive Defense Model Based on Differential Privacy Protection

Because digital media images are vulnerable to external attacks in the process of information exchange, a differential privacy protection mechanism is proposed. Based on this, the built interactive defense model is shown in Fig. 1.

It can be seen from Fig. 1 that digital media images may have multiple attack modes in the information exchange process, so when the real system is attacked, the digital media image data in the Physical system is synchronized to the virtual system through virtual to real mapping [3]. When the virtual system is attacked, the digital media image data in the virtual system is synchronized to the Physical system through virtual to real mapping; By providing feedback on the system synchronization process through virtual dual redundancy information, an interactive defense mode is formed to resist external attacks on digital media images during information exchange, prevent network vulnerabilities from being invaded, and provide a secure environment for information exchange security identification [4].

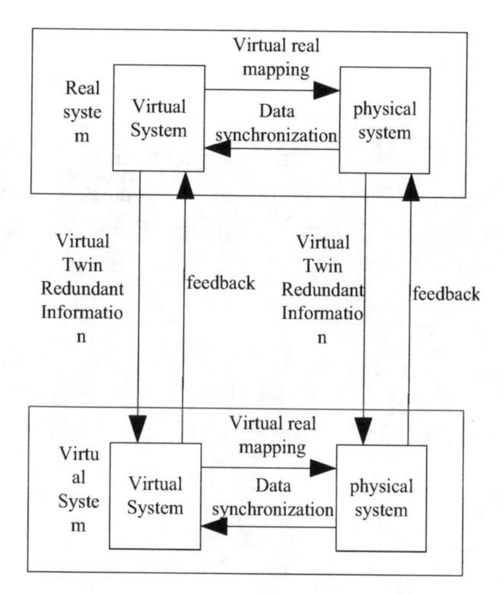

Fig. 1. Interactive Defense Model

In the defense link of differential privacy protection mechanism, the amount of digital media image information to be protected is calculated according to the sensitivity of digital media image, and the calculation formula is:

$$M = \left(M_x - \varepsilon \cdot \frac{|D_2 - D_1|^2}{\lambda} \right) \cdot t \tag{1}$$

In formula (1), M_x indicates the amount of input information; ε indicates the quantitative difference of sensitive information defense; λ indicates the level of information to be protected; D_1, D_2 indicates the attack intensity of any two messages; t indicates the defense duration.

In the process of malicious autonomous defense, due to the existence of the link layer, information at all levels is exchanged at will, resulting in tampering and collision of the link layer [5]. Therefore, in order to enhance the attack detection of the link layer, the defense capability of the link layer should be strengthened. Based on this, the defense model of differential privacy protection mechanism is constructed as follows:

$$W = \frac{\sqrt{\eta \cdot M}}{D_0 \times N^2} \tag{2}$$

In formula (2), η indicates information access rights; D_0 indicates the signal strength under attack; N Indicates that the information layer has been tampered with attack nodes. Differential privacy protection mechanism is the ability to provide a higher level of information security for the privacy of digital media images and better protect information from hacker intrusion by using differential private key technology on the premise that the attacker has mastered massive information [6]. By constructing this model, real-time detection and protection of digital media image information interaction can be realized.

2.2 Real Time Information Interaction

In order to realize the automatic collection, deep mining and feedback of digital media image information, the C/S structure is adopted to realize the interactive management of digital media image information with the whole process as the center, the computer as the core, and the LAN as the main communication medium. Its structure is shown in Fig. 2.

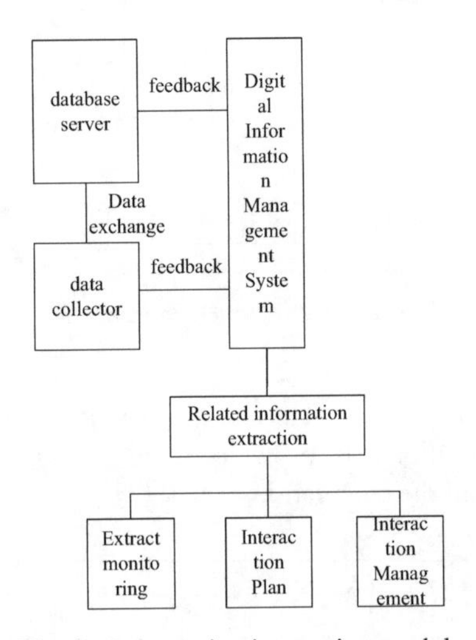

Fig. 2. Information interaction module

It can be seen from Fig. 2 that the information interaction module is mainly composed of the following three parts: the information acquisition part, which is responsible for providing the signal from the sensor to the computer for signal identification after effective processing; Information database service part, which stores the information of the information collection part in the information database, and can provide information support for information mining through information interaction with different information; In the application part of digital system, the digital information management system can deeply mine the information hidden in the production process and show it to the managers, making the managers' analysis of decision-making more intuitive and effective [7, 8].

The designed digital media image information interaction process is as follows:

On the information database server, establish an information directory for security detection of the interactive environment. In the information database server, the real-time information interaction mechanism is introduced to establish a group address composed of multiple users. In this way, each network element can establish a new information identification node [9] on a coordinator node. If the internal coordination node conflicts with the external coordination node during transmission, a real-time interaction technology is introduced according to the communication protocol between communication nodes.

Each interaction node acquires node information intelligently and autonomously through the Internet of Things access node. One node injects the calculation formula of information flow into the interaction center, which can be expressed as follows:

$$S = l_k^c \cdot \{s_k^c\} + l_k^d \cdot \{s_k^d\} \tag{3}$$

In formula (3): k indicates the number of packets; $\{s_k^c\}$ indicates the packet error correction retransmission information set, $\{s_k^d\}$ represents the packet control signaling set, l_k^c, l_k^d it respectively represents the length of error correction retransmission packet and the length of packet control signaling.

Each information is evenly distributed into the interaction defense model based on differential privacy protection to ensure smooth interaction process, so as to detect the non-zero kurtosis of each information. The spatial distribution vector based on differential privacy protection can be expressed as:

$$E_M = \frac{r_N^2}{T} \tag{4}$$

In formula (4), r_N represents the dangerous defense radius; T indicates the time-frequency distribution range of information. If the two kinds of interactive information are within the spatial distribution range of the above differential privacy protection, it indicates the information interaction security.

3 Design of Digital Media Image Information Security Sharing Scheme

3.1 Sharing Attack Resistance Based on Anonymous Differential Privacy Protection

The protection of differential privacy does not necessarily guarantee the security of digital media image information in the sharing process. In some special cases, the information set will still be in danger of being violated because it cannot completely avoid the attacker's prior knowledge and homogeneous attacks. The probability calculation formula is:

$$P = \beta(t) \cdot \frac{1}{\iota(\iota+1)} \tag{5}$$

In formula (5), $\beta(t)$ stay t the proportion of hazard information in the total information within the time; ι indicates the degree of adjacent information connection [10].

In order to overcome the shortcomings of anonymity and prevent different types of attacks, it is necessary to have different sensitivities in all information sets, and the following conditions must be met: if the anonymous tags in the disclosed information set conform to the above expression, the disclosed information set has the characteristics of entropy diversification:

$$\zeta = \sum_{a \in M} P(a, b) \cdot \log(P(a, b)) \tag{6}$$

In formula (6), $P(a, b)$ indicates that the sensitive information value is a the information of is on the label b probability in. The public information set meets the recursive diversity feature, if and only if any anonymous group label of the public information set meets the formula:

$$M_{\zeta} < \log(\zeta) \cdot (m_{x1} + m_{x2} + \ldots + m_{xn}) \tag{7}$$

In formula (7), $m_{x1}, m_{x2}, \ldots, m_{xn}$ an information set that represents the amount of input information. Recursive diversity can adjust the constant value to reduce the skew rate of the frequency of constant values given by different information sets of each anonymous group.

3.2 Information Sharing Process Design

In the process of information sharing, affected by the shared platform server, the location of information sharing nodes cannot be accurately determined, resulting in poor information sharing effect. To this end, the sharing process of non-uniform clustering information fusion is designed, which integrates decentralized information with non-uniform clustering information fusion technology. The interaction process designed is as follows:

Step 1: abstract the topology of the network and give the topology relationship;

Step 2: Obtain the topology in the network;

Step 3: Although the interactive platform server is the source of information, the accessed object can be used as the source server during configuration.

Step 4: Establish the fusion cluster based on this principle. In order to achieve non-uniform clustering, it is necessary to introduce different competition radii at the cluster head node to reduce the number of "hot zone" problems. It is necessary to calculate the energy and distance of non-uniform clustering nodes. The formula is:

$$Q = i \cdot \frac{Q_i}{Q_0} \tag{8}$$

$$L = \frac{L' - L_{\min}}{L_{\max} - L_{\min}} \tag{9}$$

In the above formula, i indicates the number of uneven clustering wheels; Q_i, Q_0 represent the initial and current energy of the node respectively; L' represents the distance between any two nodes; $L_{\min}$, $L_{\max}$ it respectively represents the nearest and farthest distance of any two nodes in all nodes.

To facilitate the rational use of information, the template mapping radius is designed in the information database mode, and the calculation formula is:

$$r' = 1 - Q \cdot L \tag{10}$$

In formula (10), the candidate node uses the existing information to construct a new node and uses it as the initial cluster head. After the initial cluster head is determined, broadcast the cluster head message. The unselected cluster head will no longer be in the sleep state, and select the cluster head with the lowest communication cost to complete the cluster creation.

Step 5: In the process of intra cluster and inter cluster information fusion, it can be divided into two stages: intra cluster communication and inter cluster communication. In the process of inter cluster communication, the intra cluster and cluster head direct communication is realized through the single hop technology. Between two adjacent secondary clusters, the weighted minimum value is determined according to the residual energy of the cluster head and the spacing between clusters. The formula is:

$$\omega_j = \min\left(\mu \cdot \frac{L_j}{Q_j}\right) \tag{11}$$

In formula (11), μ respectively represent the j adjustment coefficient of nodes. According to the results of the above calculation, when two information centers cannot support each other, they cannot be merged. In order to distinguish between trusted and untrusted information, multi-source information fusion method is adopted. The formula is:

$$A_{x_1x_2} = \begin{cases} 0 \\ 1 - \dfrac{\sum\limits_{x_1 \cap x_2 \in M} m(x_1)m(x_2) \cdot \omega_j}{\sum\limits_{x_1 \cap x_2 \notin M} m(x_1)m(x_2) \cdot \omega_j} \end{cases} \tag{12}$$

In formula (10), $m(x_1)$, $m(x_2)$ respectively x_1, x_2 The amount of trusted and untrusted information. Using the regularization method, the relative importance of the fusion order is determined, the untrusted information is eliminated, and the effective information fusion is completed.

Step 6: Considering the special requirements of the information demand side, share the information of a single node, as shown in Fig. 3.

Each cluster head node broadcasts information with the competition radius as the transmission radius. After the cluster head receives the broadcast information from the cluster head node, a single node requests information sharing. The digital media image dispatching center uses its private key and the public key provided by the information demander to generate a proxy re encryption key, and provides a proxy re encryption key and an information request to multiple signature notaries. The notary will input the information into the information management contract, input the information into the corresponding file, download the corresponding password information from the file, and use the trap gate to verify whether the digital media image information required by the information demander is consistent with the provider. Homomorphic encryption mechanism is adopted to integrate the downloaded heterogeneous information ciphertext according to the required information requirements. The proxy key is used to re encrypt the encrypted ciphertext, and the ciphertext information is transmitted to the information requester through the off chain channel. Information demander uses private key to decrypt heterogeneous information and realize information sharing of single node.

Step 7: Based on the single node sharing mode, combine it with secure multi-party computing technology to form a multi node sharing mode, as shown in Fig. 4.

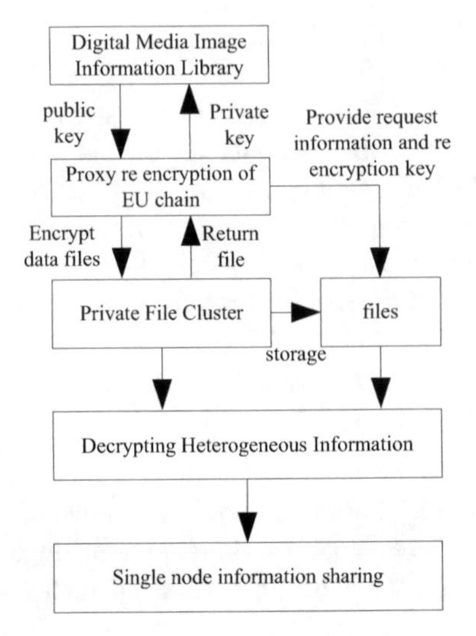

Fig. 3. Single node information sharing process

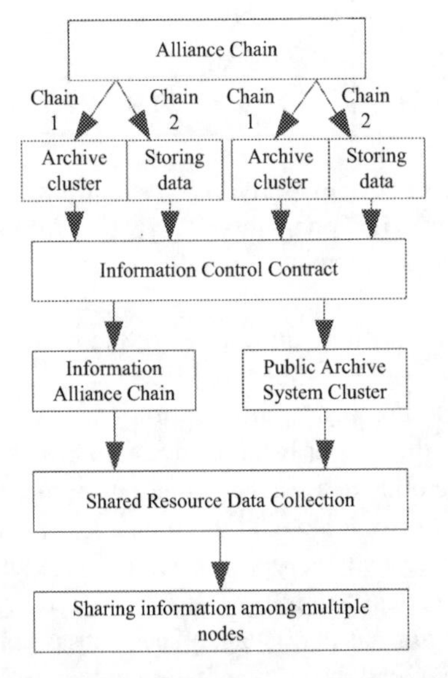

Fig. 4. Information sharing process of multiple nodes

As shown in Fig. 4, when the information demander generates a success rate information sharing request, the notary will query the file cluster according to the information demander's requirements, introduce the information control contract, and perform ciphertext operations on the information sets required by different nodes. The encrypted information and its information are divided into several blocks, and the proxy re encryption method is used to divide and re encrypt them, and they are transmitted on the link. The information demander reconstructs the encrypted text and generates the encrypted text according to the specific reconstruction algorithm. The obtained information collection ciphertext is uploaded to a common file cluster and information alliance chain. Each node will download the required information from the research information alliance chain and the public archive system cluster, thus realizing the information sharing of multiple nodes.

4 Experiment

In order to verify the effectiveness of the interactive sharing method for digital media image information based on differential privacy protection, the blockchain based sharing method proposed in reference [1] and the sharing method based on the full network power topology model proposed in reference [2] were used as comparative methods to jointly test the effectiveness of comparative experiments.

4.1 Experimental Scenario Setting

Network based virtual world refers to multiple users in different physical environment locations, or multiple virtual worlds connected with each other through the network, or multiple users participating in a virtual reality world at the same time, using the interaction between computers and other users. The experiment scenario based on virtual reality system is shown in Fig. 5.

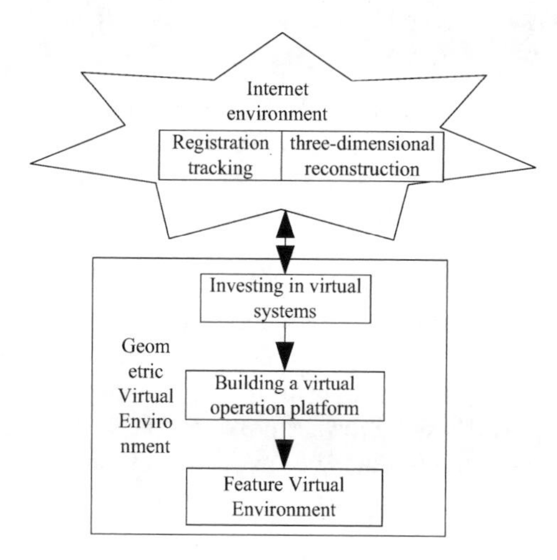

Fig. 5. Experimental scenario based on virtual reality system

In the application process, using the virtual environment of the Internet, you can digitize the information and then conduct interactive operations without any restrictions, and conduct a complete simulation process operation in a virtual environment, and record the whole process.

4.2 Simulation of Experiment Process

In the experimental scenario based on virtual reality system, establish a life cycle of digital media image information, and its model is shown in Fig. 6.

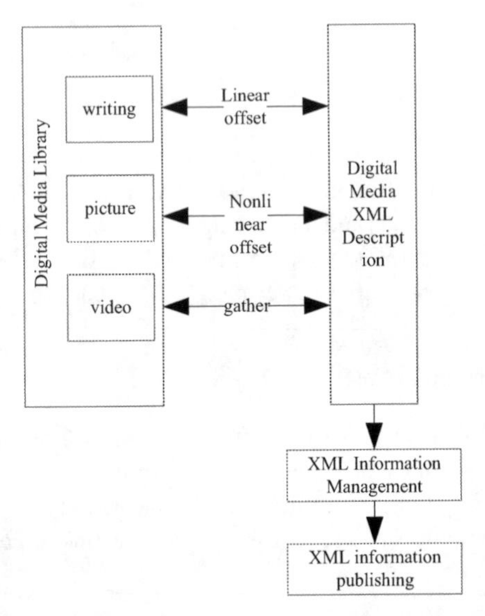

Fig. 6. Life cycle of digital media image information

Because XML is a structure oriented and content oriented description language, its inherent advantages make it easy to handle the access of various applications to resources as an exchange language, so XML schema is introduced into this model, and this technology is used to standardize the various stages of the digital life cycle.

(1) Generation phase

In this model, XML description is introduced in the generation phase, that is, the corresponding metadata description is generated at the same time as the material of digital media is generated, focusing on recording the relevant information during the generation of digital media, such as the author/owner of the media, creation time, keywords and instructions for use, and recording in the form of XML documents.

(2) Management phase

Different from traditional media, digital results enable digital media to be stored, managed and reused. In the current application requirements, the management of digital media mainly includes storage management, retrieval support (such as content-based query), copyright management, security management, etc. Taking content-based retrieval as an example, there are two main technical means at present: identification technology and MetaData retrieval technology. The recognition technology uses some special algorithms to search and match the original digital media files, such as matching the hue of the image; The database retrieval method is to store the MetaData information of each digital media in a readable language in the database, and then use the retrieval mechanism provided by the database to retrieve information. Although the technical principles used in various applications are similar, the structure of management information may be completely different in different applications, which causes difficulties in information sharing.

(3) Release phase

The main task of the publishing phase is to send the information to the place where users can reach, and the publishing phase will also generate an XML document for outward delivery, which contains various information of digital media required by users.

This process needs to use the local weighted fitting method to correct the image. First, take the control point as the center, control the control point in the circle, and perform local surface fitting on the control point, because the selected polynomial is quadratic. After the control points are locally fitted, the control points correspond to a fitting result.

If the local control points in the image are sparse, several auxiliary control points need to be added to ensure that the control circle contains enough control points. The specific operation steps are as follows:

Step 1: Select an inscribed square in the control circle, and use the points on each side of the square as new control points. Since the square can be selected under control, the new points can be controlled within the image range.

Step 2: In the standard space, there will be several control points near the newly added control points. Triangle these points to obtain several triangles and determine the triangle nearest the newly added control points.

Step 3: Use the control points formed by each vertex of the nearest triangle to build a correction model, and the collinear position of the new control points in the distorted image can be calculated through this model.

Step 4: After all control points complete the above local surface fitting, any point will correspond to a fitting result. The mapping relationship of each distortion point in the image can be obtained from nearby calculated points, so as to achieve distortion point correction.

4.3 Experimental Results and Analysis

Two scenarios are set, one is that there is a large amount of untrusted information in Scenario 1, and the other is that there are information leakage points in multi-channel in Scenario 2. In these two cases, the sharing scheme based on blockchain, the sharing method based on the whole network power supply topology model and the sharing method based on differential privacy protection are respectively used to compare and analyze the effect of information interaction and sharing.

4.3.1 Analysis of Information Interaction Results

For information interaction analysis, compare the information interaction results of the three methods.

(1) Situation 1

In this case, compare the information interaction of the three methods, as shown in Fig. 7.

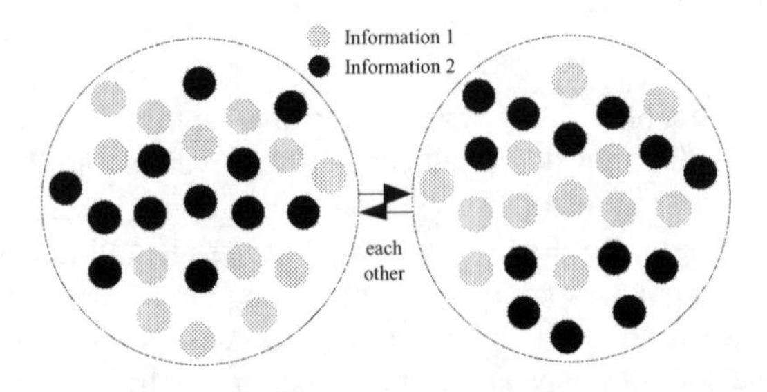

(a) Blockchain based sharing scheme

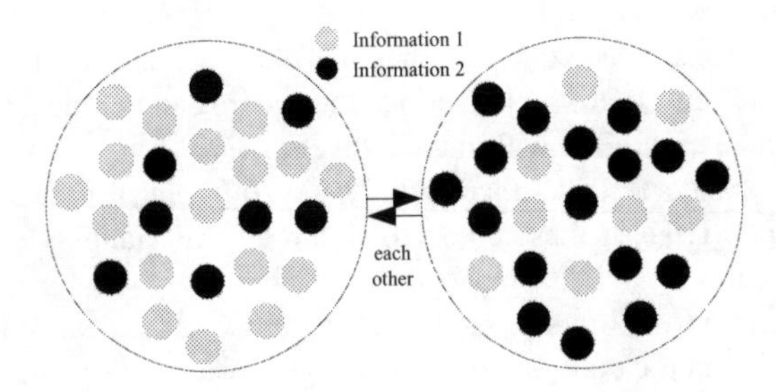

(b) Sharing method based on whole network power supply topology model

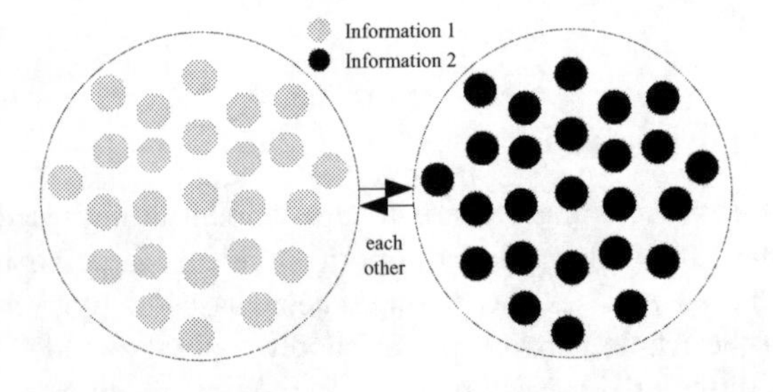

(c) Sharing method based on differential privacy protection

Fig. 7. Information interaction of three methods in case 1

It can be seen from Fig. 7(a) and Fig. 7(b) that in an environment with a large amount of untrusted information, the two comparison methods mix the two kinds of information together in the process of information interaction. After the interaction is completed, the two kinds of different information are not completely distinguished, and more information is lost, which cannot guarantee the integrity of the interaction information. From Fig. 7(c), it can be seen that the method proposed in this article fully distinguishes the two types of information after interacting with each other. Both types of information fully participate in the interaction process without any information loss, indicating that the information exchange effect is good and the security is high. This is because this method uses a trap gate to verify whether the digital media image information required by the information demander is consistent with the provider, and uses the Homomorphic encryption mechanism to integrate the downloaded heterogeneous information ciphertext according to the required information requirements, ensuring the integrity of the interactive information.

(2) Situation 2

In this case, compare the information interaction of the three methods, as shown in Fig. 8.

It can be seen from Fig. 8 that using the sharing scheme based on blockchain and the sharing method based on the whole network power supply topology model can not only fail to realize all information interaction, but also cause information loss; However, using the sharing method based on differential privacy protection, although individual information cannot interact, it does not affect the overall interaction effect, and the information is saved completely.

4.3.2 Analysis of Information Sharing Results

For information sharing analysis, compare the information sharing results of three methods.In the process of information sharing, if there is a large amount of untrusted information, it will cause some interference to the sharing process. In this case, compare the amount of information shared by the three methods, and the comparison results are shown in Fig. 9.

It can be seen from Fig. 9 that the maximum amount of shared information is 630 bit in case 1 and 505 bit in case 2 when using the blockchain based sharing scheme; Using the sharing method based on the whole network power supply topology model, the maximum amount of shared information in case 1 is 800 bits, and the maximum amount of shared information in case 2 is 500 bits; Using the sharing method based on differential privacy protection, the maximum amount of shared information in case 1 is 900 bits, and the maximum amount of shared information in case 2 is 780 bits. This is because the method in this article divides the encrypted information and its information into several blocks, uses proxy re encryption method to divide and re encrypt, and transmits it on the link, achieving maximum information sharing.

4.3.3 Analysis of Real-Time Results of Information Interaction and Sharing

In order to verify the real-time performance of the method proposed in this article in the process of information exchange and sharing, 5000 bit digital media image information was randomly selected as sample data, and three different methods were used for

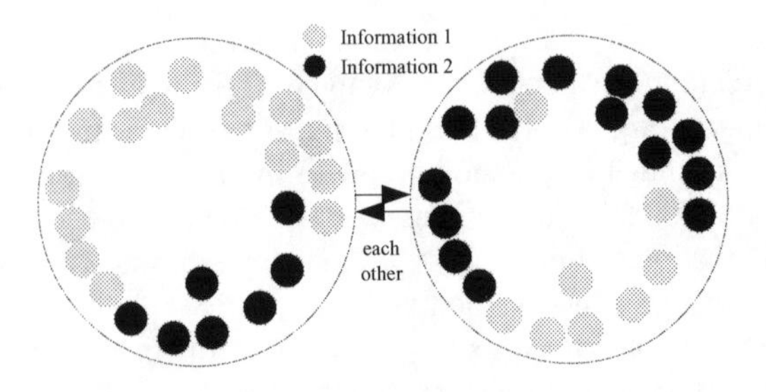

(a) Blockchain based sharing scheme

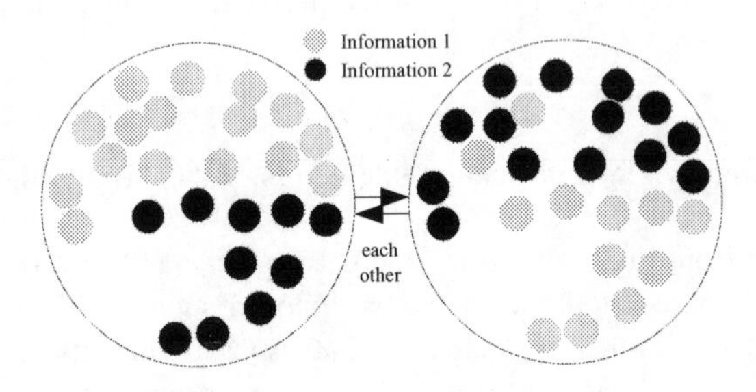

(b) Sharing method based on whole network power supply topology model

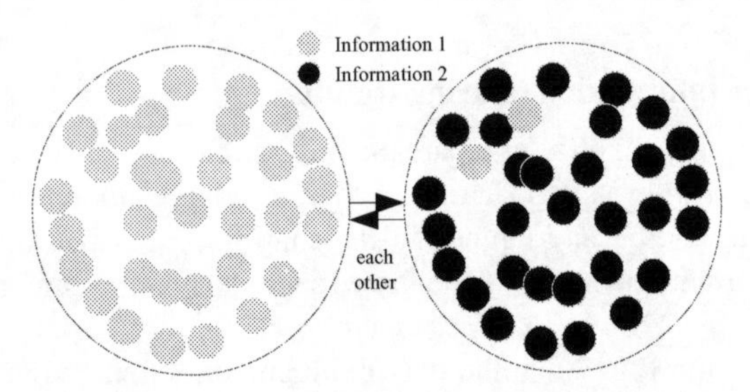

(c) Sharing method based on differential privacy protection

Fig. 8. Information interaction of three methods in case 2

interaction and sharing. The time spent completing interaction and sharing by different methods was counted. The less time it takes, the better the real-time performance of information exchange and sharing. The results are shown in Table 1.

According to Table 1, it can be seen that using a blockchain based sharing scheme to complete interactive sharing between 1000 bit and 5000 bit takes 1.45 ms to 5.21 ms; The sharing method based on the entire network power supply topology model takes 1.67 ms to 5.37 ms to complete the interactive sharing of 1000 bit to 5000 bit; Using the sharing method based on Differential privacy protection to complete 1000 bit–5000 bit interactive sharing takes 0.78 ms–4.25 ms; The relatively short time consumption indicates that the information exchange and sharing method proposed in this article has good real-time performance. This is because the proposed method in this article takes computers as the

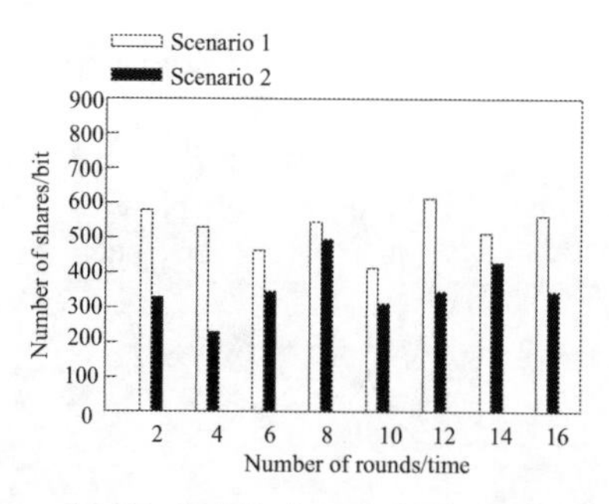

(a) Blockchain based sharing scheme

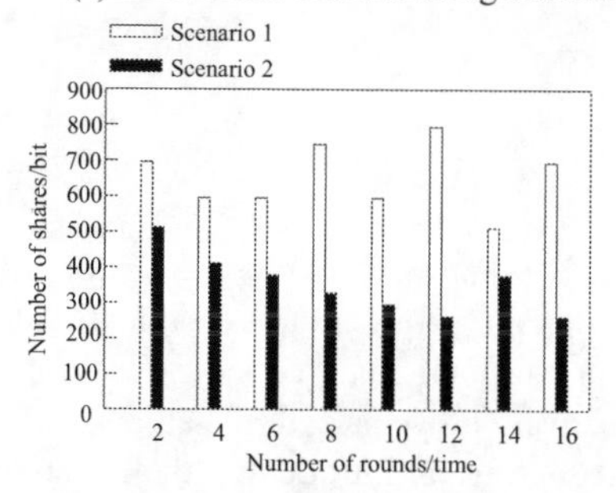

(b) Sharing method based on whole network power supply topology model

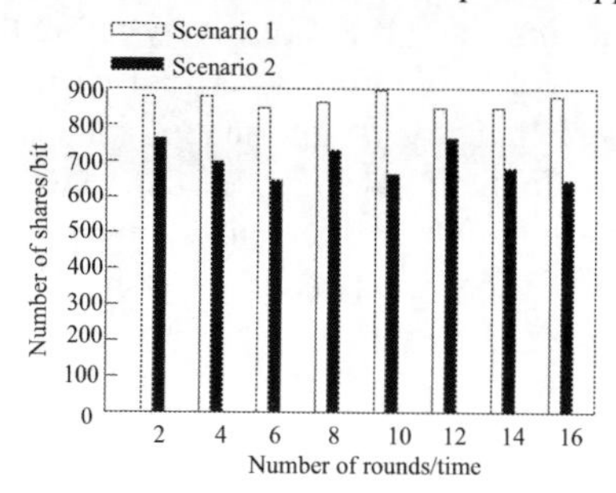

(c) Sharing method based on differential privacy protection

Fig. 9. Information sharing of three methods

core and local area networks as the main communication medium, and adopts a C/S structure to achieve interactive management of digital media image information. Each interactive node intelligently and independently obtains node information through the Internet of Things access node, and each cluster head node broadcasts information with a competitive radius as the transmission radius, saving time for information exchange and sharing.

From the above analysis results, it can be seen that the interaction effect of the research method is good, the amount of shared information is large, and it has good real-time performance.

Table 1. Time consumption/ms for interaction and sharing between different methods

Amount of information/bit	Blockchain based sharing scheme	Sharing method based on whole network power supply topology model	Sharing method based on differential privacy protection
1000	1.45	1.67	0.78
2000	2.66	2.98	1.88
3000	3.34	3.56	2.65
4000	4.08	4.32	3.48
5000	5.21	5.37	4.25

5 Conclusion

Based on the existing interactive sharing technology, an information interactive sharing method based on differential privacy protection is designed. This method combines the differential privacy protection mechanism to reduce the loss rate of information. At the same time, this method effectively prevents external attacks and ensures the security of information interaction and sharing by introducing the differential privacy protection mechanism into the process of information anonymization. The experimental results show that the proposed protection method can effectively ensure the availability of information while improving the security of information. However, due to time constraints, this article did not test the effectiveness of information interaction and sharing in the presence of noise interference. In the following research, we will focus on the impact of information noise on the sharing effect, continuously improve the design method, and provide technical support for the secure interaction and sharing of digital media image information.

References

1. Lihua, Z., Cao, Y., Ganzhe, Z., et al.: Blockchain-based secure data sharing scheme for microgrid. Comput. Eng. **48**(01), 43–50 (2022)
2. Guangyu, T., Xinkun, J.: Research and application of graphic model data sharing technology based on power supply topology model of the whole network. Electric. Measure. Instrument. **59**(06), 105–112 (2022)
3. Zhang, F., Gong, Z.: Supply chain inventory collaborative management and information sharing mechanism based on cloud computing and 5G Internet of Things. Math. Probl. Eng. **2021**, 1–12 (2021). https://doi.org/10.1155/2021/6670718
4. Kang, Y., Li, Q.: Design and implementation of data sharing traceability system based on blockchain smart contract. Sci. Program. **2021**, 1–14 (2021). https://doi.org/10.1155/2021/1455814
5. Yang, Y., Wang, B.: Information-sharing mechanism of synergistic incentive among EPC subjects of energy efficiency retrofitting of existing buildings against COVID-19. Int. J. Low-Carbon Technol. **3**, 3 (2021)
6. Gill, A.Q.: A theory of information trilogy: digital ecosystem information exchange architecture. Information (Switzerland) **12**(7), 283 (2021)

7. Cheng, T.: Information hiding mechanism based on QR code and information sharing algorithm. Int. J. Embedded Syst. **14**(1) (2021)
8. Zhu, L., Li, F.: Agricultural data sharing and sustainable development of ecosystem based on block chain. J. Clean. Product. **315**(Sep.15), 127869.1–127869.9 (2021)
9. Defranco, J.F., Ferraiolo, D.F., Kuhn, D.R., et al.: A trusted federated system to share granular data among disparate database resources. Computer **54**(3), 55–62 (2021)
10. Cheng, X., Niu, T., Wang, Y.: Information hiding mechanism based on QR code and information sharing algorithm. Int. J. Embedded Syst. **14**(1), 1–8 (2021)

A Hierarchical Smoothing Method for Animation Image Based on Scale Decomposition

Jieling Jiang(✉) and Wei Li

Guangzhou Academy of Fine Arts, Guangzhou 510006, China
icu007@qq.com

Abstract. In order to solve the problem of low quality of animated images affected by noise, a hierarchical smoothing method for animated images based on scale decomposition is proposed. Get the animation base image, and obtain the detail layer of the source image and target image. Use U-net convolutional neural network to select the decomposition box, select the results according to the remote sensing image segmentation box, and design the image decomposition process. Adjust the animation decomposition scale, focus on measuring multi-scale morphology, use the mean coordinate method to fuse the brightness of the target image, and retain rich details of the image. The fusion image smooth mosaic processing flow is designed, and the minimum variance standard is used to obtain the best matching combination. The gradient is used to represent the direction and size of the pixel changes in the animation image, and the details of the animation image are enhanced by means of superposition correction to achieve image edge smoothing. The experimental results show that the image details obtained by this method are consistent with the image samples, the signal-to-noise ratio is above 90 dB, and the longest smoothing processing time is 27 s, which can obtain high-quality animation images.

Keywords: Scale Decomposition · Animated Image · Hierarchy · Smoothing

1 Introduction

With the rapid development of China's animation industry in the context of modernization, the overall creation form has also undergone tremendous changes, and the corresponding image production method has also been upgraded [1, 2]. At present, the animation industry is developing rapidly, and the use of computer applications for animation creation has also become the main form of animation creation at this stage. Animation is the creation of animation by decomposing the actions of people or objects into many pictures, and then combining these scattered pictures in a certain way to give people a sense of continuous change in vision. The two-dimensional animation is to improve and innovate the traditional animation. Compared with the traditional image design form, the multi frame two-dimensional animation image design has higher flexibility and structural

L. Yun et al. (Eds.): ADHIP 2023, LNICST 549, pp. 20–34, 2024.
https://doi.org/10.1007/978-3-031-50549-2_2

integrity, and is more targeted to the details processing, which to some extent provides great convenience for the subsequent image processing, maintenance and adjustment work [3–5].

In the process of animation image processing, multi scale detail enhancement methods are an effective way to improve the quality of animation. The research on image detail enhancement at home and abroad mainly focuses on enhancing the resolution of animation images in order to improve the design quality of animation images [6]. For this reason, reference [7] proposes a detail blur enhancement method for 3D animation images based on optical parametric amplification. This method first constructs the transmission Relational model of 3D animation image detail features, and then carries out the detail color attenuation processing of 3D animation image through the optical parametric amplification method. The details of the input image are filtered by the filter holding method, the Sobel operator is used to detect the weak edge information of the animation image, and the adaptive interpolation method is used in the edge area to realize the details enhancement of the 3D animation image and the optical parameter detection amplification, and improve the image's fuzzy enhancement and identification ability; A multi-level encoding and decoding image description model based on attention mechanism was proposed in reference [8]. This model uses Faster R-CNN to extract image features, and then uses Transformer to extract three high-level features of the image. The features are effectively fused using a pyramid shaped fusion method. Finally, three long-term and short-term memory (LSTM) networks are constructed to hierarchically decode the features at different levels. In the decoding part, the soft attention mechanism is used to enhance the model to process image edges; Reference [9] proposed an image reconstruction method based on Gaussian smooth Compressed sensing fractional full variation algorithm. This method not only processes the low-frequency components of the fractional order differential loss image, but also increases the high-frequency components of the image, achieving the goal of enhancing image details. The Gaussian smoothing filter operator updates the Lagrange gradient operator to filter out the increase of the high-frequency component of the additive white Gaussian noise caused by the differential operator.

At present, although the traditional methods mentioned above can enhance the details of animated images, they are highly susceptible to external and environmental factors, resulting in unsatisfactory smoothing effects. Therefore, this study proposes a hierarchical smoothing method for animated images based on scale decomposition. This article conducts research from two perspectives: hierarchical scale decomposition and hierarchical smoothing processing of animated images. Based on the results of hierarchical scale decomposition, the image smoothing processing effect and efficiency are improved through scale adjustment, hierarchical fusion, morphological focusing, mean fusion, and smooth stitching processing.

2 Hierarchical Scale Decomposition of Animation Image

2.1 Animation Base Image Extraction

The base signal of the image is called the base image, and the process of preserving the edge image decomposition is the process of finding the base image [10].

For the extraction of the base image, the problem of preserving edges in the base image is considered first. For one-dimensional signals, the existence of edges can generally be determined by the four adjacent extreme points [11, 12].

The workflow on the brightness layer is shown in Fig. 1.

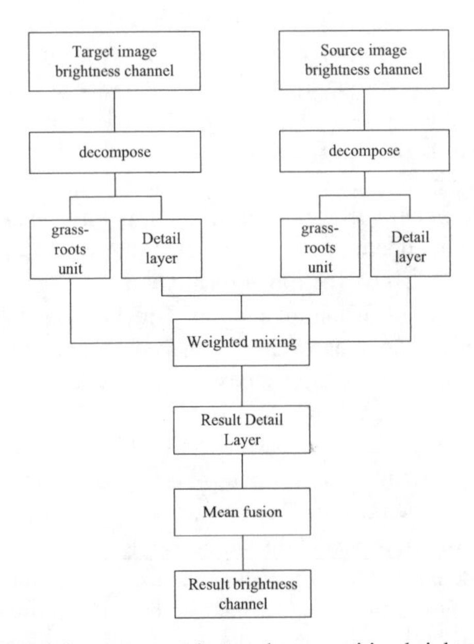

Fig. 1. Workflow on target image decomposition brightness layer

First, the source image and target image are converted to CIELAB color space, and then the brightness channel is decomposed into base layer and detail layer. The base layer is a structural layer containing image structure information. Edge smoothing methods such as bilateral filtering and weighted least squares filtering can realize the decomposition problem of the base layer and detail layer [13, 14]. For this reason, the weighted least squares method is used, because it has excellent performance in solving the blur level near the edge. Given a brightness channel, the base layer in the channel can be obtained by solving the following minimum values:

$$\min S = \sum_{m} (H(m) - O(m))^2 + \alpha \left(\frac{|H_x(m)|^2}{|O_x(m)|^\lambda} + \frac{|H_y(m)|^2}{|O_y(m)|^\lambda} \right) \tag{1}$$

In formula (1), m represents each pixel in the fusion area, H represents the smoothed base layer, O indicates the image brightness channel, $O_x(m)$, $H_x(m)$ means on x gradient in direction, $O_y(m)$, $H_y(m)$ represents the gradient in the Y direction [5]. The first is to make H try to connect with O keep similar, the second item is to minimize H yo smooth it. coefficient α it is used to balance the two: α the higher the value of, the smoother the base layer will be subscript x, y the partial derivatives are calculated on the abscissa and ordinate respectively coefficient λ decided O sensitivity of gradient.

Based on this, the algorithm uses subtraction to obtain the detail layers of the source image and the target image respectively:

$$C = O - H \tag{2}$$

In formula (2), C represents the detail layer of the image. The brightness balance can be achieved by transferring the brightness histogram of the target image base layer to the source image base layer, and the hole repair method is used to modify the detail layer of the target image in the fusion area [15, 16].

2.2 Image Hierarchical Scale Decomposition

Based on the obtained animation base image, the U-net convolutional neural network selection method is used in the image scale decomposition process, but the size of the decomposition box is not suitable for image decomposition. If only the original anchor box size is used for learning, the decomposition block will be misjudged, and its average identification rate is high, so the anchor box size needs to be corrected. By optimizing the given decomposition framework, the decomposition framework can cover the target in the image well [17, 18].

Determine the length, width, height and rotation angle of the center point of the bounding box according to the calculated position of the animation base image decomposition box, that is $box\{l, c, h, \theta\}$ after adding the rotation angle to the original decomposition box, calculate the decomposition box information, which is expressed as:

$$\begin{cases} \gamma_l = \Delta l \cdot \iota_l + \iota_l \\ \gamma_c = \Delta c \cdot \iota_c + \iota_c \\ \gamma_h = \Delta h \cdot \iota_h + \iota_h \\ \gamma_\theta = \Delta \theta \cdot \pi + \iota_\theta \end{cases} \tag{3}$$

In formula (3), $\Delta l, \ \Delta c, \ \Delta h, \ \Delta \theta$ respectively represent the displacement between the four vectors and their upper left positions; $\iota_l, \ \iota_c, \ \iota_h, \ \iota_\theta$ represent the minimum decomposition frame information of the surrounding target of the above four vectors respectively. Based on this, the candidate deviation value of the decomposition box is obtained, and the decomposition box is precisely determined according to this value [19].

Select the results according to the remote sensing image segmentation box, and design the image decomposition process:

In the decomposition area, the distance between the decomposed image blocks is represented by the color variance information:

$$d = \sum_{n=1}^{m} \left(\varepsilon_{n,m} - \left(u_1 \sigma_{n,m} + u_2 \varepsilon_{n,m} \right) \right) \tag{4}$$

In formula (4), n represents the set of adjacent points in the decomposed region image; $\varepsilon_{n,m}$ represents the mean square deviation of each parameter of the image; u_1,

u_2 represent the number of pixels of two images respectively [20, 21]. Using this equation, it can ensure that after decomposition, regions with small color variance can be decomposed first, while maintaining the same characteristics of colors, and the histogram characteristics of regions can be reflected in a sense.

3 Hierarchical Image Smoothing Based on Scale Decomposition

3.1 Animation Decomposition Scale Adjustment

According to the hierarchical fusion results of the scale decomposition image, the basic enhancement nodes are laid out by integrating the image detail enhancement requirements and standards, and corresponding associations are formed within a reasonable range.

First, use the test animation image of professional device and equipment atlas, and use the basic color of details to replace the dark primary color to set the image gray scale unchanged. Secondly, calculate the relative value of pixel depth of field, and set the image detail enhancement spacing as the basis. Finally, the relevant enhancement nodes are set according to the change of the initial pixel transmission intensity. When setting nodes, it is necessary to ensure that nodes are interrelated and can form a circular detail enhancement program [22, 23]. The recognition efficiency of the image can be further enhanced after the overlapping correlation is made for the change of the frame number of the animation image and the positioning of the enhancement details. Set marks at the location of enhancement defects to complete initial directional enhancement inspection and enhancement node setting and control.

The image recognition degree and frame number of animation are fixed, and animation basic decomposition is required for changes in requirements and standards. First, the intercepted animation is described and identified in a basic way, then the animation is separated, and finally the separation step is calculated. The specific formula is:

$$\phi = (B + E)^2 \times \frac{\delta \times \sum_{i=1} (e_i + 1)}{B} \tag{5}$$

In formula (5), ϕ is the animation separation step size; B is the preset total range; E is the stacking range; δ Is the resolution deviation. According to formula (5), the animation separation step size can be measured and set as the basic animation classification standard. The difference between the separation step size of two orthogonal processing directions and the scale of the original image is 1/4. After the design and adjustment of the principle of identifiable animation features are completed, the preset animation decomposition scale is constantly adjusted according to the needs, which can create a stable environment for the details enhancement of subsequent images.

3.2 Scale Decomposition Image Hierarchical Fusion

Due to the limited depth of field of the camera's optical lens, the optical lens can only capture images focused on local scenes. Therefore, only objects within the depth of field are focused and clear, while objects outside the depth of field are blurred. However, the information transmitted by partially focused images is incomplete, because not all meaningful objects are focused in one image. When the boundary between the image focus area and the defocus area is complex, image fusion processing needs to be carried out.

This study completes the hierarchical fusion processing of scale decomposed images from the perspectives of multi-scale morphological focusing measurement and parallel mean fusion.

3.2.1 Multi Scale Morphological Focusing Measurement

In image processing, gradient represents the sharpness information of the image, and morphological gradient operator can well extract the gradient information of the image, and can expand to multi-scale morphological gradient operator by changing the size of structural elements, and carry out filtering and other operations. Similarly, the multi-scale morphological gradient operator is used to extract the gradient information of the image at different scales, and then these information gradients are integrated to form an effective focus measurement, which is called multi-scale morphological focus measurement [24]. Build many scale structure elements, which can be expressed as:

$$A_n = a_1 \oplus a_2 \oplus \cdots \oplus a_n \tag{6}$$

In formula (6), $\oplus$ denotes the expansion operator in the morphological gradient operator; $\{a_1, a_2, \cdots, a_n\}$ express n basic elements a a collection of components. In mathematical morphology, structural elements are virtual tools used to extract image features, and structural elements of different shapes are used to extract different types of image features. In addition, changing the size of structural elements can expand to multi-scale, and these multi-scale structural elements can be used to extract comprehensive gradient features in the image.

Using morphological gradient operator to calculate image mesoscale as n the gradient characteristics of are:

$$D_n(x, y) = f(x, y) \oplus A_n - f(x, y) \otimes A_n \tag{7}$$

In formula (7), $f(x, y)$ image; $\otimes$ represents the erosion operator in the morphological gradient operator. In morphological operations, morphological gradient is equal to expansion operator minus erosion operator.

Morphological gradients at different scales are integrated into multi-scale morphological gradients, and multi-scale morphological gradients are constructed by integrating morphological gradients at different scales with weighted sum. Different weighting values are allocated under different scales. The larger the scale, the smaller the weighting value. On the contrary, the smaller the scale, the larger the weighting value. The integrated weighted gradient map can express the gradient information well, and can also clearly and effectively transfer the focus information of the source image.

The multi-scale morphological gradient in the region is summed to construct the multi-scale morphological focus measurement of the region. Since the sum of gradients helps to measure the sharpness of the area and suppress noise, the focus measurement of the area is described by the sum of multi-scale morphological gradients as follows:

$$L_n(x, y) = \sum_{n=1}^{k} \omega_k D_n(x, y) \tag{8}$$

In formula (8), ω_k express k weighted values. In the whole process of multi-scale morphological gradient measurement, it is also necessary to set the shape and scale values of structural elements. The shape of the structural element determines the shape information of the signal extracted by the operation. Different results can be obtained by processing images with different structural elements.

3.2.2 Parallel Mean Fusion

Once the base layer and detail layer information of the source image is obtained, the new source image brightness channel can be calculated from the channel. Then the new brightness channel is fused to the brightness layer of the target image using the mean coordinate method. Based on this, build a hierarchical image fusion model, as shown in Fig. 2.

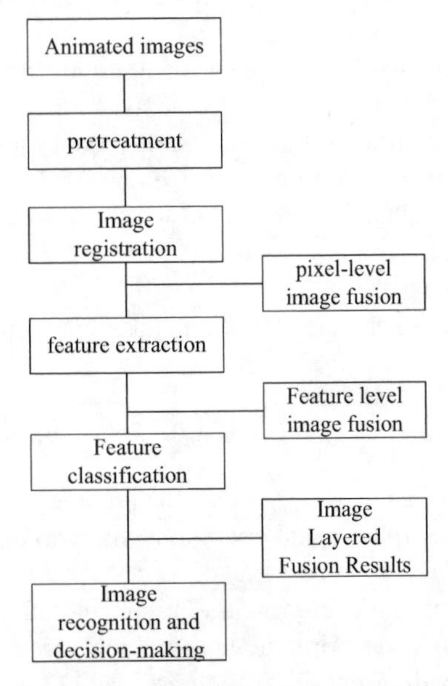

Fig. 2. Hierarchical image fusion model

Given a region to be fused and its chain of boundary pixels given in counterclockwise order, the mean coordinate of each pixel is the weight, and is calculated by the following formula:

$$\omega_i = \frac{\tan(\beta_{i-1}/2) + \tan(\beta_i/2)}{\|v_i - \vartheta\|} \tag{9}$$

In formula (9), β_i indicate the process i angle obtained by the second cycle; ϑ represents pixels. Using the mean coordinates of these pixels, the mean interpolation can be defined as follows:

$$\varsigma(\vartheta) = \sum_{i=1}^{m} \omega_i(\vartheta)\left(O_i - O_i'\right) \tag{10}$$

In formula (10), O_i' represents the new source image brightness channel.

Because the mean interpolation of each pixel in the fusion area is only related to the boundary coordinates, the interpolation of each pixel can be processed separately. It takes time to implement mean interpolation on CPU. In order to solve this performance bottleneck, CUDA technology is used to accelerate interpolation. The results obtained through the new source image brightness channel may exceed the brightness range, which may result in the loss of the visual effect of the image, so these values should be converted to the display range. The algorithm migrates the intensity histogram of the target brightness to the current brightness, and successfully retains the rich details of the image.

3.3 Smooth Mosaic Processing of Fused Images

First, create a matching relationship for the extracted image contour features. The image contour features taken include point features, edge features, contour features and area features. Match the image contour features with the boundary correlation constraint method. After matching the matched contour feature points, many matching pairs will appear near the boundary of the point. Therefore, it is necessary to detect the matching process by introducing the information of contour feature vector. The specific process is shown in Fig. 3.

After extracting and processing the image contour features, the image contours to be spliced have corresponding contour descriptors, and the minimum variance standard is used to obtain the best matching combination.

3.4 Image Edge Smoothing

After completing the histogram equalization of the background layer image, multi-scale animation images are fused by bilateral filtering, which mainly uses the edge stability of the bilateral filtering image and the translation invariance of the direction filtering to capture the feature structure of the image and avoid the loss of edge details or excessive noise after image fusion. The gradient weight factor is introduced to apply the gradient estimation result to the bilateral filtering. The gradient quoted here represents the direction and size of the pixel changes in the animation image.

The multi-scale edge of the image is perpendicular to the gradient direction, and the pixel with a smaller angle between the two included angles will eventually output a smaller weight. The size and direction of the pixel value change in the above are mainly obtained by derivation calculation, which can also reflect the contrast and change trend of image edge pixels. Calculate the horizontal partial derivative array and the vertical partial

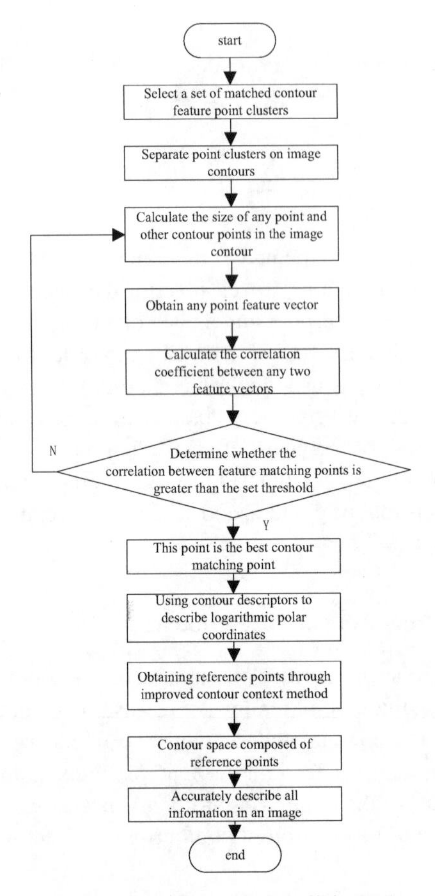

Fig. 3. Fusion image smooth splicing process

derivative array respectively. After obtaining the calculation results, use the second order norm to calculate the pixel gradient size and direction.

After the image gradient calculation is completed, considering the local direction information of the image edge contained in the gradient vector and the information perpendicular to the gradient direction, the gradient function is used to process the image edge pixels. In the process of multi-scale decomposition, the pixels in the vector direction occupy a high weight, so a gradient kernel function is constructed, and the gradient value corresponding to the image edge pixel points is calculated by using the first order partial derivative. The calculated gradient value is substituted into the gradient kernel function, and the pixel angle included in the gradient vector is calculated.

Then, the details of the animated image are enhanced by means of superposition correction. First, the scanning animation image is recognized by light shadow mapping, and the key frame number position and recognition features are marked. Secondly, set the optical parameter correction structure to analyze the change of optical parameters through the change of image spectrum. Finally, in case of any abnormality in the image detail enhancement process, the full coverage positioning shall be carried out immediately, and the correction amount of changes shall be adjusted evenly to ensure the correction effect of the detail enhancement, and the image recognition degree and frame number shall be set. The above process needs to set a directional correction edge value, and the calculation formula is:

$$\mu = \sum_{i=1}^{m} \left[(\upsilon \times 0.5b)^2 - \phi \right] \tag{11}$$

In formula (11), υ represents the pixel transmission intensity; b indicates stack offset. The corrected edge value obtained from the test is taken as the measurement standard of animation image correction processing, and the visual feature recognition model is comprehensively constructed to synchronously enhance the application quality of animation images, so as to achieve image edge smoothing.

4 Experiment

In order to verify the practical application effect of the hierarchical smoothing processing method of animation image based on scale decomposition, and considering the rigor and scientificity of the experiment, the comparative experiment is designed. The experimental methods are the fuzzy enhancement method of 3D animation image details based on optical parametric amplification, the image description model of multi-level encoding and decoding based on attention mechanism. The image reconstruction method based on Gaussian smooth compression perception fractional order total variation algorithm and the method proposed in this paper. In the experiment, the smoothness of image details is taken as the experimental item, and the image signal-to-noise ratio and processing time are taken as the comparison experimental criteria. The performance of the smoothing methods is compared and analyzed from this aspect.

4.1 Experimental Conditions

The image data used in the experiment is mainly from the animated image Standard library ImageNet, which contains animated images of people, landscapes, animals, plants, etc. This is a very large image dataset that contains over one million high-resolution images and covers thousands of categories. The ImageNet dataset is widely used in research on computer vision tasks such as image classification, object detection, and image segmentation. In the experiment, color images with unclear details in various categories are mainly selected. The selection of animation image samples mainly includes the following characteristics. From the aspect of image elements, image samples must include natural scenes or artificial scenes; From the aspect of image illumination, the selection of image samples should consider two situations: sufficient illumination and insufficient illumination; In terms of image color, image samples need to change from color rich scenes to color free scenes.

According to the above analysis, 600 animated images that meet the above requirements are used for testing, including 100 building images, people images, tree images,

Fig. 4. Image sample

dim light images, single color images, and rich color images. One image sample is shown in Fig. 4.

In the experiment, the i5-3570K computer was used as the platform, and the experimental results were obtained through third-party software.

4.2 Experiment and Analysis of Image Detail Smoothing Performance

Image detail smoothing is the basic indicator for effective verification of image hierarchical smoothing processing. Therefore, during the experiment, the detail smoothing performance of four different processing methods is compared and analyzed, as shown in Fig. 5.

It can be seen from Fig. 5 that the image detail information obtained by using the studied method is consistent with the image sample, while the details of the other three methods are blurred, indicating that the image details are not smooth.

4.3 Experiment and Analysis of Image Signal-to-Noise Ratio

For noisy animation images, the signal-to-noise ratio is a standard evaluation index. The larger the signal-to-noise ratio, the better the image quality is, and the better the hierarchical smoothing effect of the image is. Based on the above experimental results, the signal-to-noise ratio of image samples processed by different methods is calculated to measure the denoising level of different methods. The signal-to-noise ratio calculation results of different methods are shown in Table 1.

The results in Table 1 show that under the experimental conditions of animation images with different calculation times, the signal-to-noise ratio of the three traditional methods is below 75 dB, which indicates that the image contains more noise; In contrast, the signal-to-noise ratio of the designed scale decomposition method is more than 90 dB, indicating that the image quality is better and there is almost no noise interference. According to the experimental results of image detail smoothness, the designed method has smoother detail processing, better denoising effect, and higher quality and better hierarchy of animation images.

(a) A Method of 3D Animation Image Detail Fuzzy Enhancement Based on Optical Parametric Amplification

(b) An Image Description Model Based on Attention Mechanism and Multilevel Coding and Decoding

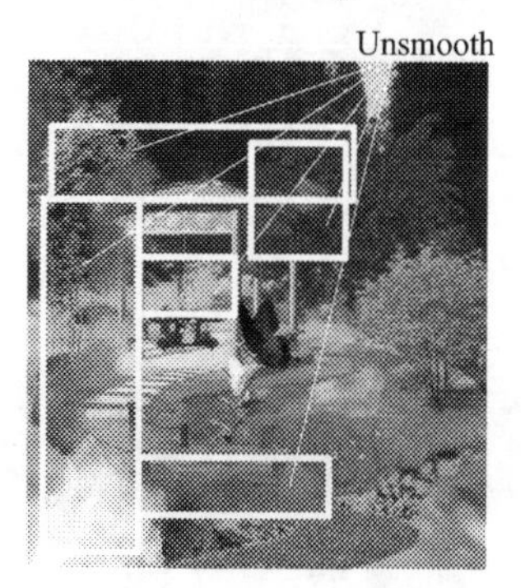

(c) Image reconstruction method based on Gaussian smooth compressed sensing fractional order total variation algorithm

Fig. 5. Comparison and Analysis of Image Detail Smoothing Performance of Different Methods

(d) A Hierarchical Smoothing Method for Animation Image Based on Scale Decomposition

Fig. 5. (*continued*)

Table 1. Signal to noise ratio calculation results of different methods/dB

Calculation times /Times	Optical parametric amplification method	Attention mechanism method	Gaussian smooth compressed sensing method	Scale decomposition method
2	70	60	52	91
4	68	58	50	92
6	67	59	48	92
8	68	58	49	91
10	69	57	50	90
12	68	59	51	91
14	70	60	52	92
16	68	55	52	93
18	69	56	53	92
20	68	58	50	93

4.4 Experiment and Analysis of Image Hierarchical Smoothing Processing Time

Compare the image hierarchical smoothing processing time of different methods, and the comparison results are shown in Fig. 6.

It can be seen from Fig. 6 that the longest smoothing processing time of the 3D animation image detail blur enhancement method based on optical parametric amplification, the image description model based on attention mechanism multi-level coding and decoding, and the image reconstruction method based on Gaussian smooth compression perception fractional order full variation algorithm are respectively 140 s, 175 s, 200 s, and the longest smoothing processing time of the studied method is 27 s.

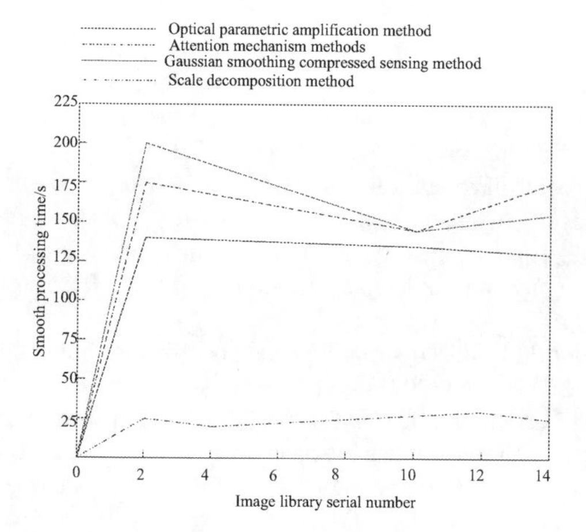

Fig. 6. Hierarchical smoothing processing time of images by different methods

5 Conclusion

A hierarchical smoothing processing method of animation image based on scale decomposition is proposed. This method uses scale decomposition method to produce visually realistic fused image details, integrates highly parallel mean fusion technology, and realizes hierarchical smoothing processing of animation image. In a complex animation processing environment, it can quickly identify the problems existing in details, and further improve the overall smoothing effect. Ensure the detail expression ability of images and create a more stable innovation environment. Experimental results show that the proposed method can match the target image well in detail.

Although this algorithm has the above advantages, it still has some limitations. For example, the current implementation is only partially parallel and region filling consumes a lot of CPU time. Therefore, in the future work, we should try to use GPU to handle the whole process, so as to improve the computational efficiency of the algorithm.

References

1. Gao, C., Song, C., Zhang, Y., et al.: Improving the performance of infrared and visible image fusion based on latent low-rank representation nested with rolling guided image filtering. IEEE Access **16**(5), 1–14 (2021)
2. Chen, H., Deng, L., Qu, Z., et al.: Tensor train decomposition for solving large-scale linear equations. Neurocomputing **464**, 203–217 (2021)
3. Chew, A., Ji, A., Zhang, L.: Large-scale 3D point-cloud semantic segmentation of urban and rural scenes using data volume decomposition coupled with pipeline parallelism. Autom. Constr. **133**(01), 1–19 (2022)
4. Zhang, X., Yan, H.: Medical image fusion and noise suppression with fractional-order total variation and multi-scale decomposition. IET Image Proc. **15**(8), 1688–1701 (2021)
5. Ren, L., Pan, Z., Cao, J., et al.: Infrared and visible image fusion based on edge-preserving guided filter and infrared feature decomposition. Signal Process. **186**(2), 108–116 (2021)

6. Chen, J., Wu, K., Cheng, Z., et al.: A saliency-based multiscale approach for infrared and visible image fusion. Signal Process. **182**(4), 107–118 (2021)
7. Yanxia, Y.: Detail blur enhancement method of 3D animation image based on optical parametric magnification. Laser J. **43**(09), 114–118 (2022)
8. Li, K., Zhang, J.: Multi-layer encoding and decoding model for image captioning based on attention mechanism. J. Comput. Appl. **41**(09), 2504–2509 (2021)
9. Yali, Q., Jicai, M., Hongliang, R., et al.: Image reconstruction based on Gaussian smooth compressed sensing fractional order total variation algorithm. J. Electron. Inf. Technol. **43**(07), 2105–2112 (2021)
10. Madureira, A.: Hybrid localized spectral decomposition for multiscale problems. SIAM J. Numer. Anal. **59**(2), 829–863 (2021)
11. Wang, K., Wang, S., Sun, Q., Liu, C., Chen, S.: Point cloud segmentation matching for 3d reconstruction using multi-layer lidar. J. Changchun Univ. Sci. Technol. **43**(04), 49–56 (2020)
12. Li, Y., Wang, J.: Edge feature extraction method of Brillouin scattering spectral image. Opt. Commun. Technol. **45**(03), 37–41 (2021)
13. He, L., Su, L., Zhou, G., Yuan, P., Lu, B., Yu, J.: Image super-resolution reconstruction based on multi-scale residual aggregation feature network. Laser Optoelectron. Progress **58**(24), 250–259 (2021)
14. Li, C. Research on optimization of 3D image enhancement based on adaptive. Comput. Simul. **37**(12), 358–361 (2020)
15. Liu, W., et al.: Research on intelligent image processing based on deep learning. Autom. Instrument. **12**(08), 60–63 (2020)
16. Zhao, S., Yang, T.: A coherent coefficient based filter of complex number images. Comput. Technol. Develop. **30**(02), 7–11 (2020)
17. Dr. Neetu, A.: Image recognition through human eyes, computers and artificial intelligence. J. Res. Sci. Eng. **3**(3), 132–141 (2021)
18. Duan, H., Wang, Z., Wang, Y.: Two-channel saliency object recognition algorithm based on improved YOLO network. Laser Infrared **50**(11), 1370–1378 (2020)
19. Ajay, R., et al.: Computer vision and machine learning for image recognition: A review of the convolutional neural network (CNN) model. Asian J. Multidimen. Res. **10**(10), 1023–1029 (2021)
20. Jin, H., Cao, T., Xiao, C., Xiao, Z.: Video summary generation based on multi-feature image and visual saliency. J. Beijing Univ. Aeron. Astron. **47**(03), 441–450 (2021)
21. Jeya, C.A., Dhanalakshmi, K.: Content-based image recognition and tagging by deep learning methods. Wireless Pers. Commun. **123**(1), 813–838 (2021)
22. Xiang, J., Xv, H.: Research on image semantic segmentation algorithm based on deep learning. Appl. Res. Comput. **37**(S2), 316–317+3 (2020)
23. Li, M., Li, L., Lei, S.: Application of unsupervised fuzzy clustering algorithm in image recognition. Techn. Autom. Appl. **39**(01), 121–124+159 (2020)
24. Li, P., Li, J., Wu, L., Hu, J.: Image recognition algorithm based on threshold segmentation method and convolutional neural network. J. Jilin Univ.(Science Edition) **58**(06), 1436–1442 (2020)

Video Image Based Monitoring Method for Operation Status of Internet of Things Network Equipment

Liang Yuan(✉)

State Grid Huitong Jincai (Beijing) Information Technology Co., Ltd., Beijing 100031, China
lihongsheng51@163.com

Abstract. Conventional monitoring methods for the operation status of Internet of Things network equipment mainly use CBM (condition based maintenance) to obtain the early symptom characteristics of equipment, which is vulnerable to the impact of dynamic operation instructions, resulting in abnormal monitoring and early warning. Therefore, a video image-based IoT network device operational status monitoring method is designed. That is to say, the video image technology is used to process the monitoring information of equipment operation status. Combined with fuzzy logic, the monitoring framework of equipment operation status of the Internet of Things is constructed, and the monitoring algorithm of equipment operation status of the Internet of Things is designed, thus realizing the monitoring of equipment operation status of the Internet of Things. The experimental results show that the designed method for monitoring the operation status of Internet of Things network equipment based on video images has a good monitoring effect, can effectively warn, has reliability, and has certain application value, and has made certain contributions to improving the operation security of Internet of Things network equipment.

Keywords: Video Image · Internet of Things · Network Equipment · Operation Status · Monitoring Methods

1 Introduction

The Internet of things (IoT) concept was formally proposed and released by the International Telecommunication Union at the Information Society Summit held in Tunis in November 2005 [1, 2]. The Internet of Things refers to the realization of the interconnection of things and the formation of a highly intelligent information network based on the Internet and computer related technologies [3], which can realize intelligent identification and management, provide safe, controllable and personalized real-time online monitoring, positioning and tracing, alarm linkage, dispatching command, plan management, remote control, security prevention, remote maintenance, online upgrade statistical report, decision support and other management and service functions [4]. The Internet of Things realizes the Internet of Things and information exchange on the basis of the Internet. With the arrival of the 5G era, the application prospect is extremely broad [5, 6].

L. Yun et al. (Eds.): ADHIP 2023, LNICST 549, pp. 35–50, 2024.
https://doi.org/10.1007/978-3-031-50549-2_3

In terms of equipment management, information technology is applied to implement equipment visualization and whole process management. Use information technology and means to strengthen the management of equipment [7]. The Internet of Things technology will play an irreplaceable role in data visualization, process control, real-time information collection and feedback and other management. At present, the existing equipment operation environment monitoring methods (regular manual inspection records) of each organization can only cover the normal working hours. At night and during holidays [8], the environment of the machine room may be out of control and can not be found in time, with great security risks and hidden dangers. In combination with the current situation and shortcomings of management, dynamic monitoring of the operating environment and real-time mastering of the data related to the equipment room environment will help to identify potential hazards as soon as possible [9], intervene in a timely manner, eliminate the crisis in the bud, ensure the smooth operation of equipment, ensure safety and reliability, and reduce the cost of maintenance support [10].

Now some large equipment manufacturers in the market are also monitoring the operating environment and status of equipment, but the data they collect are not open to customers [11]. Only by purchasing their warranty service will they open up some data sharing, which is a great economic burden for most organizations; Moreover, when manufacturers collect and transmit data, they may also involve information security issues in the process of information exchange between internal and external networks of institutions. In addition, the operation monitoring systems of different manufacturers are incompatible with each other. Therefore, it is difficult to achieve the two requirements of full coverage and low cost required by various institutions by using the environment and operation status monitoring systems of the original manufacturers of large equipment. Therefore, it is very necessary to design a set of general, low-cost and reliable equipment operation environment and condition monitoring methods on the basis of fully understanding and absorbing the existing mature technical solutions of various manufacturers and Internet of Things related technologies [12], and combining the suggestions and needs of large equipment users and engineering technology management personnel. Under the current background, this paper designs an effective monitoring method for the running state of Internet of Things network equipment based on video image technology, which has made some contributions to improving the running reliability of equipment.

2 Design of Monitoring Method for Operation Status of Internet of Things Network Equipment Based on Video Image

The overall framework for monitoring the operational status of IoT network devices based on video images is shown in Fig. 1.

Based on the above framework, specific monitoring methods are designed as follows.

2.1 Operation Status Monitoring Information Based on Video Image Processing

Generally speaking, the research foundation of video image recognition technology is a large number of sample video images, whether it is traditional image processing methods or image recognition methods based on deep learning [13]. Especially for the mouth tag

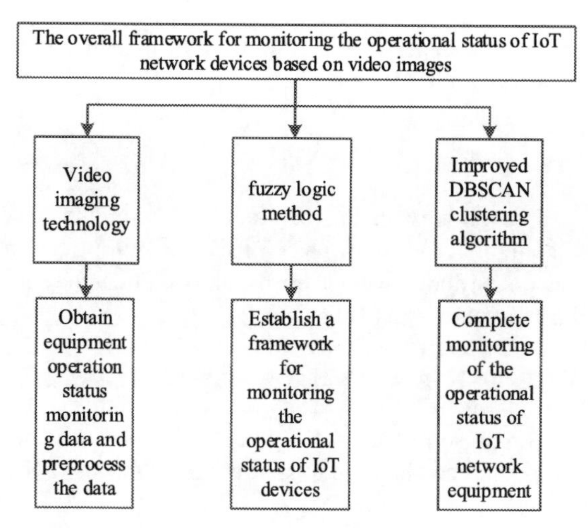

Fig. 1. Overall framework diagram for monitoring the operational status of IoT network devices based on video images

detection model, to achieve a good detection effect, there must be a large number of parameters as the support, and the premise to ensure the accuracy of these parameters is that they can be obtained only through a large number of data sample training, so the amount of data in the training set must be increased. If noise data is added to the original data set. It can also improve the robustness of the detection model.

Data enhancement is a method often used in the absence of data samples. It uses existing images to flip, shift, add noise and other operations to generate more images, so as to ensure that the detection model has a better generalization effect. There are two ways to enhance data: offline enhancement and online enhancement. The offline enhancement method is to operate directly on the original data set, which is often used for the lack of sample data set before training. The online enhancement method is to operate the batch data after obtaining it, which is generally used on larger data sets. Now, many machine learning frameworks can support online enhancement, and GPU can also be used to improve the calculation puzzle and effect. The method designed in this paper has carried out gray-scale image conversion according to the operating characteristics of the device $F(i,j)$ as shown in (1) below.

$$F(i,j) = \max(R(i,j) + G(i,j) + B(i,j)) \tag{1}$$

In formula (1), $R(i,j)$, $G(i,j)$, $B(i,j)$ they represent different color components. Because the manually set threshold does not combine with the specific distribution of pixel gray values on the image, it is arbitrary, so it will lead to the integration of mouth tags and background. Therefore, this topic also uses OTSU method for segmentation, which can find a more reasonable value as the threshold between the two peaks of the gray distribution histogram. The core idea is that the threshold value found meets the following conditions: all pixels in the image are divided into two categories, one is all pixels less than or equal to the threshold value, and the other is pixels greater than the threshold value. When the variance between the two categories of pixel values reaches the maximum, then the most appropriate threshold value is found, and the average pixel value at this time m as shown in (2) below.

$$m = m_0 \times \frac{n_0}{n} \tag{2}$$

In formula (2), m_0, n_0 represents the maximum and minimum pixel values respectively, n it represents the inter class variance of the monitoring and recognition pixel category combined with the above average pixel value to extract the equipment operating status characteristics g as shown in (3) below.

$$g = m \times (m_0 + n_0) + p_a \tag{3}$$

In formula (3), p_a represents the total number of pixels. Combined with the above formula, the OTSU method can be used to split the three devices. The effect picture is shown in Fig. 2 below.

It can be seen from Fig. 2 that the vertical encryption authentication gateway panel and background are both divided into black, and this level of adhesion cannot be separated using basic morphological operations (erosion and inflation).The main reason why the single threshold segmentation method can not effectively separate the mouth tag from the background is that the difference between the mouth tag and the background after graying is too small.

When there are multiple targets in the image, the single threshold segmentation method will cause the phenomenon that different mouthmarks stick to each other. The reason is that different mouthmarks may be in the same gray range. At this time, multiple thresholds must be calculated to distinguish each mouthmark from the background. Multi threshold segmentation is also called adaptive threshold segmentation, which is simply understood as calculating the threshold in each local area of the image, comparing the size relationship between the point in the area and the regional threshold to determine whether the pixel is classified as black or white, and the size of the area can be adjusted according to experience, so that multiple thresholds are calculated on an image, which is called multi threshold segmentation. This topic uses adaptive average threshold method and adaptive Gaussian threshold method to calculate the threshold value of each pixel, which means that the threshold value to be compared for each pixel is different. The basic idea is to calculate the average value of all pixels in the NXN region where the pixel is located, and then subtract the parameter C to obtain the threshold value of the point. The size of the NXN area can be adjusted to a range with the best effect after multiple comparisons. At this time, the adaptive threshold segmentation image of network equipment operation status monitoring is shown in Fig. 3 below.

It can be seen from Fig. 3 that although the adaptive threshold segmentation can have a good segmentation effect for the image with uneven lighting, there are two devices in the switch image. After the morphological operation, the two devices will stick together and cannot be separated. In this case, if you continue to use contour lookup to locate the device panel of the switch, two device panels will be selected by one contour. Therefore, the threshold based segmentation method is not ideal for separating devices from complex backgrounds.

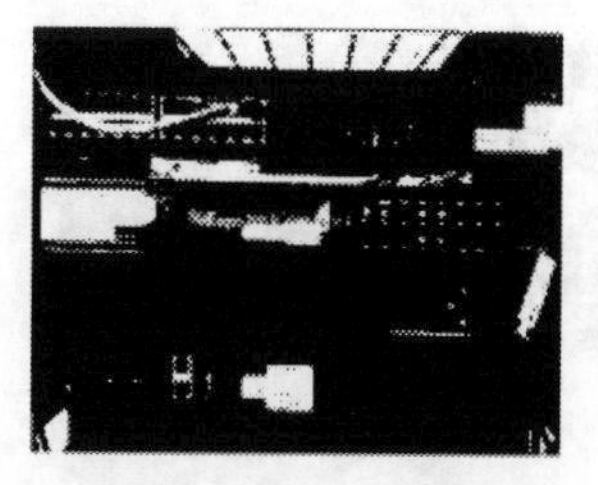

Vertical encryption
authentication gateway

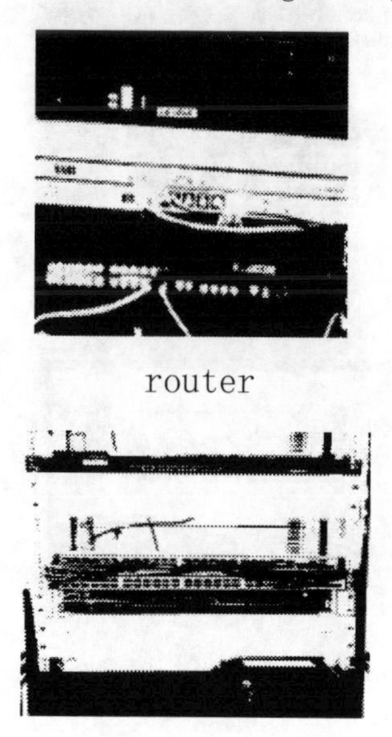

exchange board

Fig. 2. Effect Diagram of Equipment Operation Status Monitoring

The basic idea of the region based segmentation method is to directly find the region of the mouth marking device. Starting from a group of original pixels, these original pixels grow from different regions, merge the pixels that meet certain conditions in their neighborhood into the region they represent, and repeat the merging process with the newly added pixels as new original pixels, The growth process is ended until no new eligible pixels can be merged. From this process, we can determine the appropriate original pixels of the monitoring image and determine the monitoring status. At this time, the processing flow of the operation status data of the video image device is shown in Fig. 4 below.

It can be seen from Fig. 4 that the gray level of the current area can be specified by marking on the image. The purpose of this operation is to make the "lake water" in this area rise from a certain height instead of starting from the "bottom of the lake"

Vertical encryption
authentication gateway

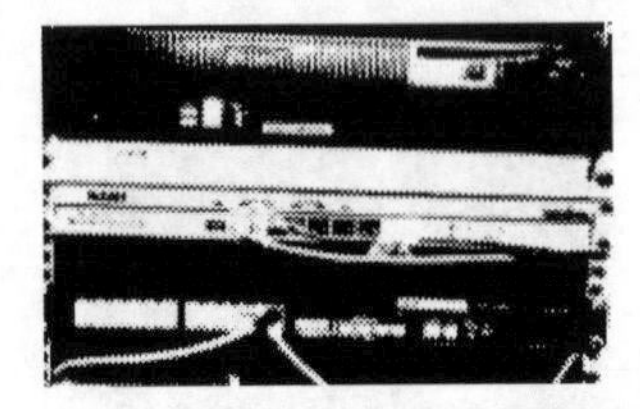

router

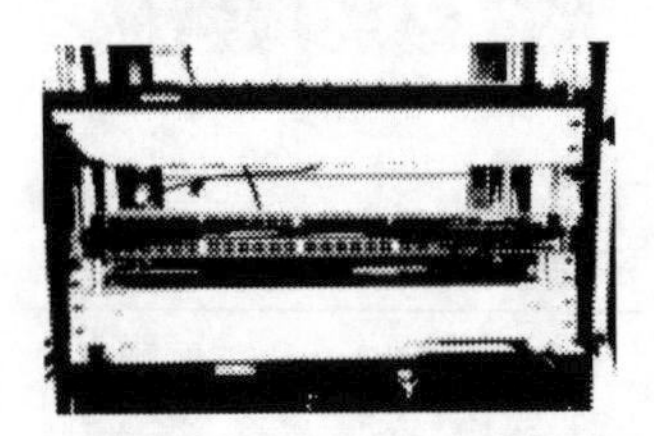

exchange board

Fig. 3. Monitoring adaptive threshold segmentation image

in the process of submerging, so as to avoid the interference of some very small noise extremes.

This topic uses the watershed algorithm function provided in OpenCV to build a mouse interactive segmentation algorithm in the python environment. The three devices are divided by manual marking, which is the segmentation effect of the vertical encryption authentication gateway, router, and switch. The vertical encryption authentication gateway used the mouse to mark the panel before splitting, which was beyond the scope of the panel, so the panel and the environment were integrated. After comparison, it is found that in the case of accurate marking, the image segmentation effect based on the region is better than the threshold segmentation method. The mouth mark is roughly separated from the background, and the noise is reduced a lot. However, the manual marking has uncertainty in guiding the rules of region growth, so the segmentation effect cannot be guaranteed.

The device recognition method based on template matching is simply to compare the existing device template with the image to be detected and find the matching mouth tag in the image. The device template starts from the upper left corner of the image. Each time it moves from left to right and from top to bottom in the unit of the upper left corner

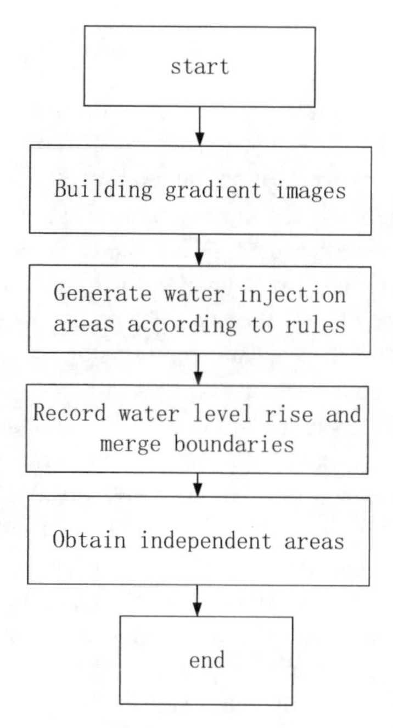

Fig. 4. Data processing flow of video image running status

pixel of the template, each time it reaches a pixel, it will take this pixel as the upper left corner vertex to cut an image of the same size as the template from the image to be detected and compare it with the template. At this time, the monitoring image matching degree $r(x, y)$ as shown in (4) below.

$$r(x, y) = \sum_{x,y} T(x, y) - I \tag{4}$$

In formula (4), $T(x, y)$ represents the abnormal state monitoring point, I it represents the calculation parameters of image matching matrix. The category prediction is multi label classification. The network structure replaces the Softmax layer originally used for single label multi classification with a classifier used for multi label multi classification. The category probability of object conditions is an array of probabilities. The length of the array is the number of categories detected by the current model. It means that when the prediction box considers that the object is currently included, the probability of each category in all categories should be detected. The probability of each category of POLO v3 algorithm is calculated separately by using the logical regression function (Sigmaid), so each category does not have to be mutually exclusive. Therefore, an object can be predicted to have multiple categories. After the above steps are processed, the monitoring data of the operation status of Internet of Things network equipment can be effectively obtained, which serves as the basis for the subsequent construction of the monitoring framework.

2.2 Build a Monitoring Framework for the Operation Status of IoT Equipment

The status of IOT node equipment has an important impact on the reliability of the IOT system and the safety of the monitoring environment. To solve this problem, this chapter will elaborate the overall framework of online monitoring methods for abnormal status of IOT node equipment, and study the theoretical basis respectively.

Compared with big data on the Internet, IoT data has its distinctive characteristics. Through the analysis of different IoT scenarios, this paper summarizes the characteristics of IoT data as follows: (1) IoT data is time series data with time stamps, usually including location information. These data are often numerical structured data, which is different from unstructured data such as text and pictures. (2) IoT data has massive high dimensionality. (3) The IoT data is streaming data. Streaming data is a group of sequential, large, fast, and continuous data sequences. Unlike static data, streaming data is an infinite dynamic data set that continues to grow over time. (4) The spatial distribution of IoT data sets has different characteristics from that of general data sets. Its spatial distribution is compact and dense, with a large amount of data closely overlapping, and only a small amount of data is distributed dispersedly, and the number is far less than that of densely distributed data. (5) IoT data also has the characteristics of multi-source heterogeneity, but at present this paper only analyzes the single observation characteristics of IoT sensors. (6) Finally, IoT data has very important relevance, that is, temporal correlation and spatial correlation.

Due to the above characteristics of IoT data, the processing of IoT data also has different processing requirements: (1) For IoT data, certain frequency reduction processing is required, and data users analyze the trend of data over a period of time rather than the data value at a specific time. (2) For Internet big data processing applications such as personalized recommendation system and user profile generation, only batch processing of data sets is required. (3) In the real-time processing of IoT data, it is usually calculated based on the time window and comprehensively analyzed by the time series of multiple node devices. (4) In the process of data processing of the Internet of Things, it is necessary to combine the real-time data collected at present with the historical data for processing to improve the data utilization. In view of the above characteristics, the monitoring framework of the operating status of the Internet of Things equipment built in this paper is shown in Fig. 5 below.

It can be seen from Fig. 5 that, on the whole, it can be divided into two key parts. The first part is the research on the online detection method of IOT node anomaly based on clustering, and the second part is the identification method of IOT node anomaly source based on fuzzy logic system. In the part of outlier detection, we first study the clustering method of outlier detection, propose a composite time series similarity measurement criterion, and propose an improved density based clustering algorithm. Secondly, the method of abnormal node detection based on clustering is studied. Using the time correlation of the IoT data, the method is divided into three steps: the division of the time dimension of the node data, the training phase, and the detection phase, to realize the detection of the abnormal status of each sensor node in the IoT application. In the part of anomaly source identification, firstly, the extraction method of node spatial correlation features is studied. By calculating the number of node spatial correlations, the geometric features of nodes are selected as the spatial correlation features. Secondly,

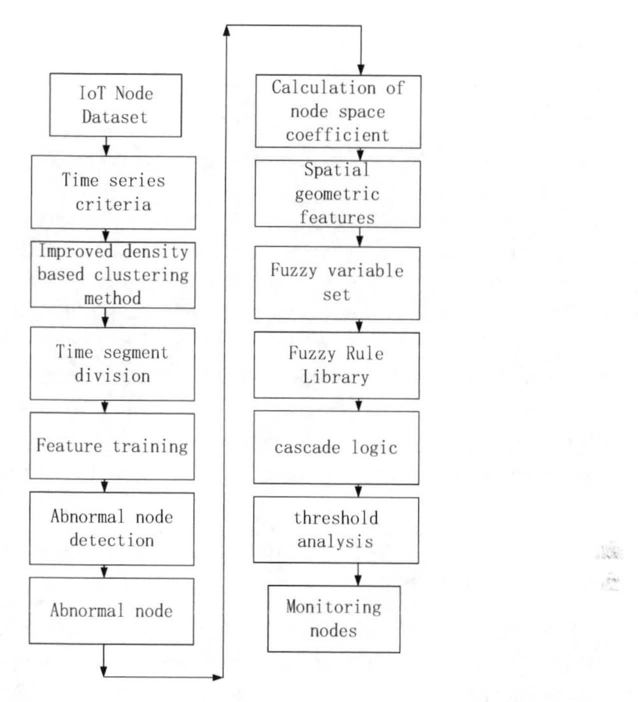

Fig. 5. Monitoring framework for operation status of IoT equipment

the fuzzy logic system that can evaluate the spatial correlation degree of nodes is studied, including the construction of fuzzy language variable set and fuzzy membership function, the establishment of temporal and spatial fuzzy rule base and the design of fuzzy logic system. Finally, the threshold analysis of the spatial correlation index of abnormal nodes output by the fuzzy logic system is carried out, and the identification of abnormal node types is realized by using the spatial correlation principle of Internet of Things nodes.

Unlike the lock step metric, the elasticity metric can compare two time series one to many. The most typical is the dynamic time adjustment DTW, which calculates the similarity between two time series by stretching or shrinking the time series to match. It is a similarity measurement method based on shape. It allows the points in the time series to be matched with equal length after self replication, which overcomes the problem that Euclidean distance cannot be matched due to the distortion of the time series; The time warping of dynamic window is introduced to improve the calculation efficiency and the accuracy of similarity measurement. However, DTW is very sensitive to noise and cannot be a measurement function. Therefore, Minkowski distance is introduced in this paper $Minkowsi\,Dis\tan ce : D$, the calculation formula is as follows (5).

$$Minkowsi\,Dis\tan ce : D = \left(\sum_{i=1}^{n} \left| s_{ij} - s_{io} \right| \right)^{2} \tag{5}$$

In formula (5), n represents the device operation status features extracted by monitoring and identifying pixel categories combined with average pixel values, s_{ij} represents the monitoring metric Euclidean distance, s_{io} represents similarity parameters. At this time, we can assume the corresponding monitoring positions of different monitoring nodes, and the generated monitoring set $N_{EPS}(P)$ as shown in (6) below.

$$N_{EPS}(P) = dist(p, q) \tag{6}$$

In formula (6), $dist(p, q)$ represents the monitoring measurement object. DBSCAN introduces intuitive definitions of clusters and noise points in the dataset. While clustering data sets with arbitrary shapes, it is also possible to identify "noise data". Select a point from the data set at random, start from the point, search all data points that meet the global density parameters and conditions and can reach the density of each other, and build them into a cluster. Objects that do not belong to any cluster are marked as noise points or outliers, Therefore, DBSCAN can be applied to abnormal data processing, and each monitoring node can be obtained by using the above designed operating state monitoring framework, effectively improving the reliability of monitoring.

2.3 Design the Operation Status Monitoring Algorithm of Internet of Things Network Equipment

Through the research on the clustering strategy for anomaly detection of IoT data, this chapter adopts the density based clustering method DBSCAN as the clustering method for anomaly detection, and further proposes an improved DBSCAN clustering algorithm against the defect of traditional DBSCAN that is sensitive to parameter values, and proposes a composite distance metric to comprehensively measure the similarity between IoT time series. After studying the clustering method of abnormal data detection of the Internet of Things, in order to study the online detection method of abnormal nodes of the Internet of Things, this chapter uses the time correlation of the data of the Internet of Things. Finally, based on the above work, this chapter proposes an online anomaly detection method for IoT nodes based on clustering. The method is divided into three parts: the division of the time dimension of IoT data, the training phase, and the detection phase. For each sensor node in the IoT application, online anomaly detection is carried out. The content block diagram of the design algorithm is shown in Fig. 6 below.

It can be seen from Fig. 6 that the principle of cluster analysis is to divide the sample set into several similar subsets according to the similarity of samples in the population. The sample items in the same subset are highly similar, but the sample items between subsets are not. Therefore, for the clustering task of time series mining, computing the similarity between time series is the focus of clustering algorithm, and reasonable time series distance measurement criteria have a direct and critical impact on the accuracy of clustering. IoT data is often a multimodal data stream, which is a multivariable time series composed of multiple time related variables. The time series generated by a single sensor node is generally a single variable time series, and different sensor nodes monitoring the same environmental variable generally collect data at a fixed frequency between nodes, so the time series of time stamps are of the same length. Only the single variable time series of sensor nodes are analyzed, and the observation time series of the same length can be designed at this time F_a, as shown in (7) below.

$$F_a = \{f_1, f_2 .. f_n\} \tag{7}$$

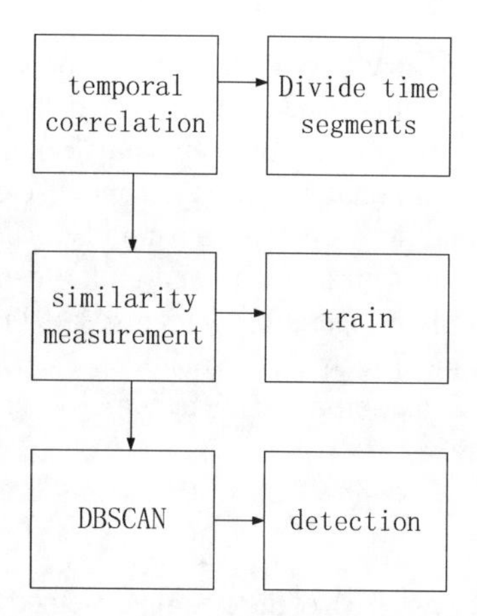

Fig. 6. Block Diagram of Operation Status Monitoring Algorithm

In formula (7), $f_1, f_2 .. f_n$ represent the observation node, and the difference amplitude of sequence status can be calculated according to the similarity relationship between time series d_{ij}, as shown in (8) below.

$$d_{ij} = \frac{1}{2} F_a + d \tag{8}$$

In formula (8), d represents the composite distance. The function can be estimated by combining the above formula, and the probability density obtained at this time $f(a)$ as shown in (9) below.

$$f(a) = \frac{d_{ij}}{p(f)} \tag{9}$$

In formula (9), $p(f)$ represent the dissimilarity function, which can generate effective equipment operation status monitoring algorithm by combining the above formula N, as shown in (10) below.

$$N = \| f(a) - d_{ij} \| * {}^1/_2 (P_C + P_R) \tag{10}$$

In formula (10), P_C, P_R they respectively represent the correct and wrong operation status monitoring samples in the data stream. The clustering based online detection method of IoT node anomalies studied in this paper combines the time correlation of IoT data to detect abnormal data points in the sensor time series through the clustering method. According to the number of consecutive abnormal sample points in the time sequence, whether the corresponding sensor node is an abnormal node is judged. Therefore, the research of efficient clustering strategy to achieve anomaly detection of time series data points of the Internet of Things is the key to the implementation of anomaly detection methods of Internet of Things nodes.

The value of the global density parameter is related to the distribution of data points and the size of data. It requires scientific calculation algorithms and cannot be selected only based on experience. In view of this, this paper proposes an improved DBSCAN clustering algorithm that adapts to the selection of global density parameters. This algorithm analyzes the distance distribution of each sample point in the data set. Aiming at the characteristics of small amount of noise data and discrete distribution, combined with statistical thinking, it finds the best density parameter value applicable to the global cluster, Thus, a large amount of normal data and outlier data can be accurately separated. The algorithm is simple in operation, which solves the problem that traditional algorithms are sensitive to the selection of global density parameter values, makes up for the defects of DBSCAN, and improves the accuracy of clustering results.

3 Experiment

In order to verify the monitoring effect of the designed method for monitoring the running state of Internet of Things network equipment based on video images, this paper built an experimental platform, and compared it with the conventional method for monitoring the running state of Internet of Things network equipment based on digital twin technology and big data technology, and carried out experiments as follows.

3.1 Experiment Preparation

Combined with the experimental requirements, this paper selects the experimental sample data set. In practical applications, anomaly detection is usually an unsupervised learning task. Labeling abnormal nodes of the Internet of Things to obtain tagged data sets requires expert verification of the status of equipment of the Internet of Things nodes in turn, which is costly and difficult to achieve. Therefore, this paper proposes an experimental verification of online monitoring method for abnormal status of IOT node equipment by manually simulating the abnormal status of nodes using unlabeled real IOT data sets.

In the experimental step, the time and space correlation analysis of the data set is first carried out, and then the artificial simulation of the abnormal mode of the node when a fault or event occurs is injected into the node data to generate the fault node and event node. Then it is carried out according to the two aspects of node anomaly detection and anomaly source identification of the online monitoring method of node abnormal state, in which node anomaly detection includes training stage and detection stage. At the same time of node anomaly detection, the clustering results of the improved DBSCAN algorithm and the classical algorithm in the training phase are compared. Finally, the experiment verifies the anomaly detection ability and anomaly recognition effect of the online anomaly monitoring method.

The online monitoring method of abnormal state of IOT node equipment proposed in this paper is a general method, which is applicable to different IOT application scenarios. Here, the application scenario of the method is verified by the traffic monitoring system example. According to the traffic flow data set monitored by the traffic monitoring system PeMS of the California Department of Transportation through more than 39000 sensor stations deployed in the highway system of major urban areas in California. This paper

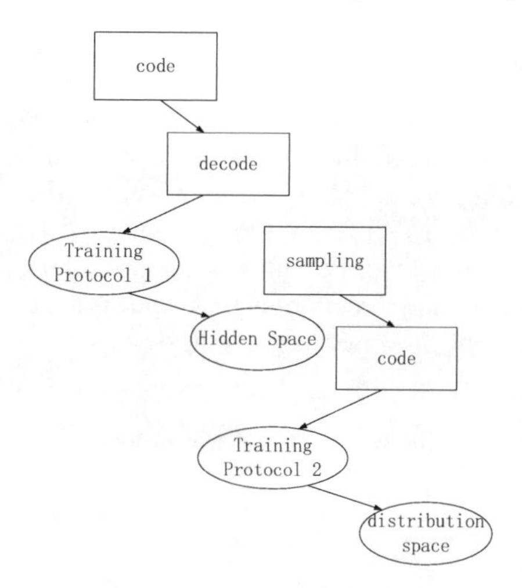

Fig. 7. Experimental data set training

selects a part of the monitoring data from PeMSD7 as the experimental data set, and then trains the data in this data set, as shown in Fig. 7 below.

It can be seen from Fig. 7 that data pre-processing is an essential and very important part of the anomaly detection process. Due to the instability of the automatic collection equipment and the variability of the environment, the selected traffic flow data set needs pre-processing work to clean, so that the data is neat and clean for subsequent data analysis.

First of all, this paper uses the Z-Core standardized method to standardize data and convert data of different orders of magnitude to the same order of magnitude. Secondly, the traffic flow data set is cleaned. Because it is difficult for sensor devices to synchronize, this paper uses linear interpolation to solve the problem of misaligned time points of data collected by different devices and missing data in the data set; For the wrong data in the data set, such as the data that obviously exceeds the normal range, such as the traffic speed is greater than 200 or negative, the threshold method is used to identify and the nearest neighbor interpolation method is used to repair.

For the noise data in the data set, the local weighted regression LOESS method is used to smooth the noise in the traffic time series. Then, the data set is aggregated and preprocessed. The data interval of metadata is set from 30s to 5min. Therefore, the time series length of each node in the data set is 12672, including 288 data points every day. Finally, because the training phase of the anomaly node detection method in this article requires clustering of reasonable data features, the time correlation between daily traffic flow data analyzed in Sect. 5.3.1 was used in the experiment, 10 nodes with very large variance between daily data were artificially pre screened as abnormal nodes, and the data set of the remaining 218 nodes was taken as reasonable node observation data, After the experimental data processing is completed, the subsequent equipment operation status monitoring experiment can be carried out.

3.2 Experimental Results and Discussion

On the basis of the above experimental preparations, we can monitor the operation status of IoT equipment, that is, we can use the video image based IoT network equipment operation status monitoring method designed in this paper, the digital twin based IoT network equipment operation status monitoring method, and the big data based IoT network equipment operation status monitoring method to monitor, It is known that early warning is required when the monitoring risk value is higher than 0.5. At this time, the monitoring results of the three methods are shown in Table 1 below.

Table 1. Experimental Results

Equipment	Monitoring risk value	Early warning status of this method	Early warning status of monitoring method based on digital twin technology	Early warning status of monitoring method based on big data technology
Switch	0.4	No early warning	No early warning	No early warning
Router	0.6	Early warning	No early warning	No early warning
Firewall	0.6	Early warning	No early warning	No early warning
Bridge	0.7	Early warning	Early warning	No early warning
Hub	0.3	No early warning	No early warning	No early warning
Gateway	0.2	No early warning	No early warning	No early warning
VPN	0.8	Early warning	Early warning	Early warning
Network interface card	0.7	Early warning	No early warning	No early warning
WAP	0.4	No early warning	No early warning	No early warning
Modem	0.2	No early warning	Early warning	No early warning

From Table 1, it can be seen that in switches, routers, firewalls, bridges, hubs, gateways, VPNs, network interface cards, WAP, and modem devices, the video image based IoT network device operation status monitoring method designed in this article can effectively provide early warning, while the digital twin technology based IoT network device operation status monitoring method is effective in routers, firewalls, network interface cards, an early warning error occurs in the modem equipment, and the Big data based monitoring method for the running state of the Internet of Things network equipment has an early warning error in the router, firewall, bridge, and network interface card equipment. The above two methods cannot trigger the early warning when the risk value of the early warning is low. The above experimental results prove that the detection and early warning effect of the video image based monitoring and early warning method for the running state of Internet of Things network equipment designed in this paper is good, reliable, and has certain application value.

4 Conclusion

The emergence of the Internet of Things has provided a new mode and idea for the health monitoring and diagnosis of equipment. With the Internet of Things, information perception technology, network technology and intelligent computing technology can be integrated to complete real-time collaborative collection, intelligent processing, timely feedback and other functions of equipment health status information; Build a three-layer system framework of the perception layer, network layer and application layer, which can realize an intelligent and efficient monitoring and diagnosis mode integrating fault prediction, remote monitoring, remote diagnosis, online diagnosis and artificial intelligence. The real-time status information of the equipment is obtained through sensors, and the working status and operating environment information of the monitoring equipment are analyzed using advanced analysis technology, so as to obtain the health status of the equipment. Through the data resource sharing service of Ethernet distributed database, effective monitoring, diagnosis and prediction can be carried out according to the synergy and complementarity of multi-sensor data, which can process the monitoring data quickly and effectively. The model with superior classification performance is established, and the fault feature extraction technology is used to carry out fault diagnosis with the method of fuzzy recognition to evaluate the health status of equipment operation. According to the monitoring advantages of the Internet of Things, this paper uses video images to design a new monitoring method for the operating status of the Internet of Things equipment. In response to the problem of poor monitoring performance caused by low data sample size, video image processing technology was used to obtain equipment operation status monitoring data, and the data was preprocessed. Based on fuzzy logic, a monitoring framework for the operation status of IoT devices was established to improve the accuracy of monitoring results. The improved DBSCAN clustering algorithm was used to design the monitoring method for the operation status of IoT network devices. The experimental results show that the proposed method can provide effective early warning in different device states, which has made certain contributions to improving the operating reliability of network equipment.

References

1. Cheng, Q., Jiang, L.: Simulation of charge state monitoring for lithium ion power batteries. Comput. Simul. (008), 039 (2022)
2. Liu, S., Liu, X., Wang, S., et al.: Fuzzy-aided solution for out-of-view challenge in visual tracking under IoT assisted complex environment. Neural Comput. Appl. **33**(4), 1055–1065 (2021)

3. Kim, J., Lee, H., Jeong, S., et al.: Sound-based remote real-time multi-device operational monitoring system using a Convolutional Neural Network (CNN). J. Manuf. Syst. **58**(2), 431–441 (2021)
4. Bandurin, M.A., Yurchenko, I.F., Bandurina, I.P., et al.: Problems and prospects of using mobile devices when performing operational monitoring of water supply structures. IOP Conf. Ser. Mater. Sci. Eng. **1111**(1), 012006 (6pp) (2021)
5. Mao, J., Yang, C., Wang, H., et al.: Bayesian operational modal analysis with genetic optimization for structural health monitoring of the long-span bridge. Int. J. Struct. Stabil. Dyn. **22**(05) (2022)
6. Tan, Y.L., Farhani, Z.N., Rizal, K.S., et al.: Hybrid statistical and numerical analysis in structural optimization of silicon-based RF detector in 5G network. Mathematics **10**(3), 326 (2022)
7. Abdollahnejad, H., Ochbelagh, D.R., Azadi, M.: An investigation on the 133Xe global network coverage for the International Monitoring System of the Comprehensive Nuclear-Test-Ban Treaty. J. Environ. Radioact. **237**(9), 106701 (2021)
8. Mahajan, G.R., Das, B., Murgaokar, D., et al.: Monitoring the foliar nutrients status of mango using spectroscopy-based spectral indices and PLSR-combined machine learning models. Remote Sens. **13**(4), 641 (2021)
9. Mieszek, M., Mateichyk, V., Tsiuman, M., et al.: Information system for remote monitoring the vehicle operational efficiency. IOP Conf. Ser. Mater. Sci. Eng. **1199**(1), 012081 (2021)
10. Gao, F., Tan, S., Shi, H., et al.: A status-relevant blocks fusion approach for operational status monitoring. Eng. Appl. Artific. Intell. **106**,104455 (2021)
11. Aqueveque, P., Radrigan, L., Pastene, F., et al.: Data-driven condition monitoring of mining mobile machinery in non-stationary operations using wireless accelerometer sensor modules. IEEE Access PP(99), 1–1 (2021)
12. Ferraz, G., Damio, L., Capelini, R.M., et al.: Assessment of the insulation conditions of power transformers through online monitoring of partial discharges. In: International Conference on Electrical Materials and Power Equipment. IEEE (2021)
13. Toldinas, J., Venčkauskas, A., Damaševičius, R., et al.: A novel approach for network intrusion detection using multistage deep learning image recognition. Electronics **10**(15), 1854 (2021)

Design of Power System Remote Video Monitoring System Based on RTP Technology

Liang Yuan(✉)

State Grid Huitong Jincai (Beijing) Information Technology Co., Ltd., Beijing 100031, China
lihongsheng51@163.com

Abstract. Most of the conventional power system remote video monitoring systems are designed based on the SIP principle. In the actual monitoring operation process, there are problems such as poor real-time monitoring and high packet loss rate. In order to solve this problem, RTP technology is introduced and a remote video monitoring system for power system based on RTP technology is designed. Based on the optimized design of the monitoring system hardware, the monitoring system software and software functions are designed. First, the remote video surveillance image of power system is filtered to remove most of the signal noise in the remote video surveillance image. Secondly, the background of power system remote video monitoring is updated to achieve the ideal segmentation effect of video monitoring image. The remote video monitoring module is designed to carry out remote video real-time monitoring of the operating conditions of the power system. The real-time transmission function of remote video image in power system is designed by using RTP technology. According to the system test results, after the proposed monitoring system is applied, with the increase of system running time, its packet loss rate is below 1%, and the real-time performance of remote video monitoring is high.

Keywords: RTP Technology · Power System · Long-Range · Video · Monitor · System

1 Introduction

Nowadays, with the continuous development of energy related technologies, the economic level has also been gradually improved. Electric power energy can be used as the secondary energy converted from primary energy, which is clean and renewable, in line with the concept of sustainable development proposed by the country, and also has a far-reaching impact on building a harmonious society [1]. However, in the power supply process, for example, if the power supply fails to provide stable electric energy, there is a problem in the power operation, or the power system fails to operate, the economic loss caused by the power failure in a large range is very huge [2]. At the current stage, the power development in China is very rapid, and has reached the working stage of ultra-high voltage, large power grid and large units. In addition, the application fields

L. Yun et al. (Eds.): ADHIP 2023, LNICST 549, pp. 51–65, 2024.
https://doi.org/10.1007/978-3-031-50549-2_4

of power technology are increasing, and the scale of the system formed by it is also increasing. This situation puts forward very high requirements for the safe and stable operation of power [3]. Therefore, we should take security and stability as the first consideration for all problems and solutions, and reflect them in the layout, operation and maintenance of the power system [4]. With the continuous improvement of technology, the hardware facilities and software performance of the power system remote video monitoring system we used earlier can no longer be applied to the current large-scale power system. Therefore, optimizing and improving the traditional power system remote video monitoring system has become our current focus of work. In order to realize the true remote video monitoring, it is necessary to add the functions of substation image monitoring and image transmission on the basis of the original telemetering, remote signaling, remote adjustment and remote control of the substation, and then realize the remote viewing function. At present, the image monitoring systems applied in power system include industrial television system and computer managed image monitoring system [5]. The industrial television system is that the camera installed on the site transmits the video signal through the video cable to the monitor through the video switcher. Because the video cable will cause loss of analog video signal, the transmitted image is limited to 500 m and can only be used for local monitoring: the image monitoring system managed by computer is to digitally process and re compress the analog or digital video signal collected by the camera through the computer video capture card, and then carry out local or remote monitoring. When the number of acquisition points increases, the workload of the computer is huge, forming a bottleneck of information transmission, image distortion, and system instability.

RTP (Real time Transport Protocol) is a real-time network transport protocol for multimedia data streams. In modern life, RTP plays an important role [6] when TCP/IP, UDP and other network protocols in the field of streaming media transmission cannot meet people's needs. For example, RTP protocol has been applied to video surveillance, video conference, voice phone and network transmission of audio and video.

Based on this, on the basis of the traditional power system remote video monitoring system, this paper makes an optimization design, introduces RTP technology, and carries out an all-round research on the power system remote video monitoring system based on RTP technology. The hardware system is designed through EWS, bus controller, video capture card, and high-speed intelligent ball camera, providing high-quality images for the remote video monitoring system of the power system. Real time transmission of remote video images is achieved based on RTP technology, improving the stability of the system.

2 Hardware Design of Power System Remote Video Monitoring System

2.1 EWS Hardware Design

EWS hardware includes microcontroller, memory, peripheral devices and IO ports, and its core is embedded microcontroller. In order to meet the demand of Internet access, embedded microcontroller should not only have the traditional control function, but also have the function of connecting to the Internet. Through comprehensive comparison of performance requirements and implementation costs, Ubicom's SX52BD is used as the EWS microcontroller: as for the transmission medium, the Internet is connected through Ethernet, and the data link layer protocol is implemented using Realtek's NF2000 compatible chip RTL8019AS. The EWS hardware structure block diagram is shown in Fig. 1.

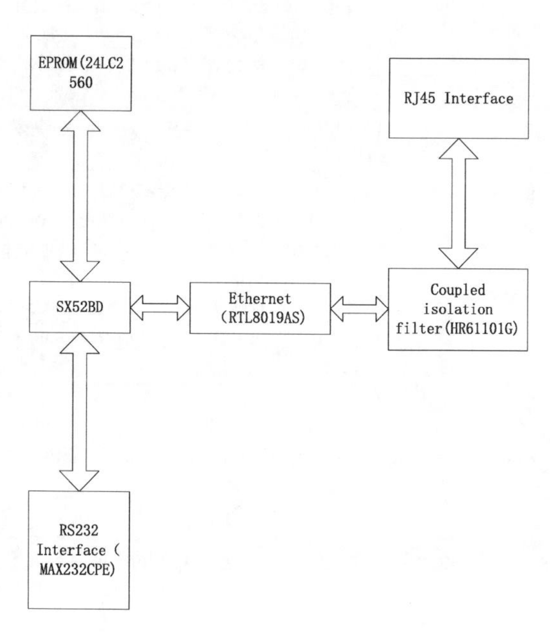

Fig. 1. EWS Hardware Structure Diagram

As shown in Fig. 1, the data flow direction is: request and control information comes from the network and is sent to RTL8019AS through RJ45. RTL8019AS is responsible for stripping the header and trailer information of the Ethernet frame, sending the processed data packet to the TCP/IP protocol stack of SX52BD, and the protocol stack parses the data packet to obtain the original request and control information [7]. The request and control information is encapsulated in protocol format by MAX232CPE, and then communicated with the on-site equipment. The process of requesting and controlling reply information to the network is just the opposite. First, set the technical parameters of microcontroller and serial memory in EWS hardware, as shown in Table 1.

Table 1 shows the technical parameters of the microcontroller and serial memory in the EWS hardware of the power system remote video monitoring system designed in this paper. Secondly, set the technical parameters of other sub hardware in EWS hardware, as shown in Table 2.

As shown in Table 2, it is the technical parameters of other sub hardware in EWS hardware. In the protection or measurement and control equipment of the remote video monitoring system, EWS is mainly used to realize the remote control, adjustment and maintenance of the equipment through IE on the client side, and browse the real-time

Table 1. Technical Parameters of EWS Microcontroller and Serial Memory

Number	Hardware	Parameter
1	SX52BD microcontroller	The instruction execution speed is 100 Mbps;It is internally integrated with fast on-chip program memory, data memory, analog comparator, timer, internal R/C oscillator and other functional devices; It supports the "virtual peripheral" function, that is, it can flexibly configure the five I/O ports of the SX52BD microcontroller, and simulate the function of hardware peripherals by executing virtual software modules to drive I/O ports
2	Serial memory	Produced by Microchip, 24LC256 chip uses 32 KB serial EPROM memory to store webpage resources of embedded web server. The stored web page resources include web pages, image files, PDF documents and other files in various formats.24LC256 is compatible with rc bus interface to communicate with external controllers

Table 2. EWS Sub hardware Technical Parameter Settings

Number	Hardware	Parameter
1	Ethernet control interface	It is implemented by RTL8019AS, a highly integrated full duplex Ethernet controller chip produced by Realte. It has a 16 bit data bus and a 24 bit address bus, which are used in the 10 megabit ISA (Industrial Standard Architecture) interface network adapter
2	Coupling isolation adapter	HR61101G chip, produced by Hanren Company, is used to convert and filter the pulse on the network.RJ45 connector is also integrated in the chip, so that it can be directly connected to twisted pair to access Ethernet
3	Serial communication level conversion interface	The MAX232CPE chip is selected, which is a special serial level conversion chip produced by Maxim Company and conforms to all EARS-22C interface standards. The chip contains a voltage multiplier circuit and a conversion circuit. Only+5 V power supply is needed to realize the conversion of TIL level to RS-232C level

information of the equipment. In order to make the operation and maintenance personnel more convenient to control and operate the equipment and monitor the equipment status more intuitively, the realization forms of its basic functions should be diversified. In addition, considering the limited resources of embedded devices, some unimportant functions should be restricted.

2.2 Bus Controller

In the power monitoring system, the monitored equipment is scattered and the distance between the equipment is relatively far. In such application systems, stable, reliable, convenient and fast data communication is the basis and guarantee for realizing the functions of the application system. Therefore, it is very important to select appropriate communication interfaces and protocols, and reasonably design communication software and hardware control circuits according to the actual working environment conditions of the system. When EWS is applied in power system, it is necessary to connect several relatively scattered equipment. Due to the short transmission distance of RS232 bus, other communication protocols [8] are required to enable communication in a wide range.

RS485 is the most widely used two-way and balanced transmission standard interface in the industry. It communicates in half duplex mode, supports multipoint connection, allows the creation of a network of up to 32 nodes (some driver modules can be increased to 128), has the advantages of long transmission distance (the maximum transmission distance is 1200 m), fast transmission speed (100 kbits/s when 1200 m), and is easy to set up when used for multi station interconnection Bus network with high reliability and wide distribution range [9]. Therefore, the bottom part of the designed power monitoring system adopts RS485 bus structure.

2.3 Video Capture Card

Video Capture Card, also known as video card, can be used to digitize video information, that is, convert it into digital data that can be recognized by the computer, and input the digitized information into the computer for editing, processing, storage or playback. In addition, many video capture cards also have the function of hardware compression, and the acquisition speed is very fast. Using the video capture card, the original videotape video can be converted into digital information that can be recognized by the computer and saved, and then made into VCD, DVD and other video situations. Video capture cards can be divided into broadcast video capture cards, professional video capture cards and civil video capture cards according to the quality level of captured images. The quality of captured images can be divided into high, medium and low levels. The technical parameters of the video capture card are shown in Table 3.

According to the technical parameters shown in Table 3, considering the functional requirements of the design system, the performance of the acquisition card, the system cost and other factors, we decided to use the 10MOONSSDK-2000 video acquisition card of Tianmin. It is compatible with Windows, including Windows VFW software architecture and WDM mode, plug and play, and supports one computer with multiple cards; The video capture speed can reach 30 frames per second, and the images obtained

Table 3. Technical Parameters of Video Capture Card

Number	Project	Parameter
1	Acquisition speed	25 fps
2	Maximum image resolution	1024 * 768
3	Acquisition card	Vision RGB PRO card
4	Graphics card	AGP card above 32 M
5	Video conference software	Comply with rrUH.324 video conference standard

are fluent and smooth; Support to provide a variety of save formats to complete dynamic image capture.

2.4 High Speed Intelligent Spherical Camera

In the remote video surveillance system, the selection of camera hardware is very important, which directly determines the clarity of the system output video surveillance image [10]. After comprehensive consideration of various performance and models of camera equipment, this paper selects high-speed intelligent spherical camera as the remote video surveillance image acquisition hardware equipment of the monitoring system. Table 4 shows the basic configuration parameter settings of the high-speed intelligent spherical camera.

Table 4. Basic configuration parameter settings of high-speed intelligent spherical camera

Number	Project	To configure
1	Optical Zoom	Not less than 22 times
2	Electronic zoom multiple	Not less than 10 times
3	Focal length	3.6–82.8 mm, variable
4	Minimum illumination	Not less than 0.01lx, with automatic color conversion function of day and night images
5	Signal system	PAL
6	Horizontal resolution	Not less than 480 TV lines, with backlight compensation and automatic white balance function
7	Signal-to-noise ratio	50 db
8	Working temperature	−35~+60 °C
9	Rotation angle	Horizontal continuous 360°;Vertical 0–90°
10	Rotation speed	Horizontal 0.5–300°/s manual;300°/s preset cruise
11	Programmable preset position	No less than 64

As shown in Table 4, it is the basic configuration parameters of the high-speed intelligent spherical camera used in the power system remote video monitoring system set in this paper. The camera has automatic scanning, automatic cruise and alarm linkage output functions, which can collect remote video images of power system in real time and ensure the quality of collected video images.

3 Software Design of Remote Video Monitoring System for Power System

3.1 Power System Remote Video Monitoring Image Filtering Processing

In the power system remote video monitoring system, the monitoring image acquisition and filtering processing are very important, which directly affects the quality of video monitoring.

Whether we can directly obtain the gray-scale image or the gray-scale image obtained by color conversion, there will be some noise in these images. Noise will have a great impact on the quality of the image, so we usually take the method of removing noise to correlate the image.

There are common methods to remove noise. We usually use the median filtering method. Its working principle is to replace each point in the picture with the median value of each point in its field, so that the pixel values around it can be close to each other, reducing the impact of noise on the picture. At first, this method is suitable for single dimension digital signal processing, and then it can be applied to two-dimensional images. The processing method of median filtering can reduce the effect of unclear picture caused by linear filtering, and can also alleviate the interference of filtering pulse and the scanning noise of picture [11].

Median filtering can not only deal with the characteristics of isolated point noise, but also maintain the edge characteristics of the image, which is very suitable for the relevant experiments of face images. During the operation of median filtering, a single sliding window containing several points is often selected, and the median value of the gray value of each point in the window replaces the gray value of the specified point. Since the image is a two-dimensional distribution, the size and shape of the selected window will greatly affect the filter effect. The median value is defined as follows: if N the observed values are $X1, X2, XN$, the input expression of the median filter is:

$$med(x_i) = \begin{cases} x_{k+1}, n = 2k+1 \\ \frac{1}{2}\left[x_k + x_{k+1}\right], n = 2k \end{cases} \tag{1}$$

among them, x_k in the window $2k+1$ or $2k$ median of data.set up $\{x_i\}(1 \leq x \leq N)$ is the image signal containing noise, m_i for with x_i is the center and the length is $n = 2k+1$ the median value of all data in the sliding window of, that is:

$$f_i = \begin{cases} 1, |x_i - m_i| \geq T_d \\ 0 \end{cases} \tag{2}$$

Among them, T_d represents the noise threshold of the power system remote video surveillance image; $f_i = 1$ indicates that this point is the noise point to be removed; $f_i = 0$ it indicates that this point is a normal signal point. Through constant adjustment T_d most of the signal noise in the remote video surveillance image can be removed.

3.2 Power System Remote Video Monitoring Background Update

After the system remote video monitoring image filtering processing is completed, the next step is to update the design of the power system remote video monitoring background to achieve the ideal segmentation effect of the video monitoring image. Avoid the problem of undesirable segmentation and wrong segmentation when directly subtracting the current frame from the original background image for image segmentation.

Because the change of light is a kind of interference information, the change of pixel gray value caused by this change will be far less than the change of pixel gray value caused by moving objects when they pass, so the background update can only update those pixels whose gray change is less than the threshold value through threshold comparison, that is, the area where no moving objects appear. The selection update method is expressed by the formula:

$$D_{p1} \succ T \tag{3}$$

$$B_{pt1} = B_{pt-1} \tag{4}$$

$$B_{pt1} = C_{pt-1} \tag{5}$$

Among them, D_{p1} indicates the gray difference between the current remote video surveillance image and the background image; T represents threshold pixels; B_{pt1} represents the updated background image pixel value; B_{pt-1} indicates the pixel value of the background image that has not been updated; C_{pt-1} indicates the pixel value of the field input image of the power system in the previous frame. In the designed power system remote video monitoring system, threshold pixels should be strictly selected. If the threshold is not properly selected, it is difficult to distinguish moving objects from background, resulting in a large gap between the updated background image and the actual background image, and losing the value of tracking the actual background.

3.3 Remote Video Monitoring of Power System Operation Conditions

After the above power system remote video monitoring background is updated, on this basis, a remote video monitoring module is designed to carry out remote video real-time monitoring of power system operating conditions. The functional composition structure of the remote video monitoring module designed in this paper is shown in Fig. 2.

As shown in Fig. 2, the power system remote video monitoring module designed in this paper includes a total of 6 functions. The video to be monitored can be selected in the tree form according to the region > regional monitoring center > substations of various voltage levels > monitoring area > cameras (which can be multiple cameras,

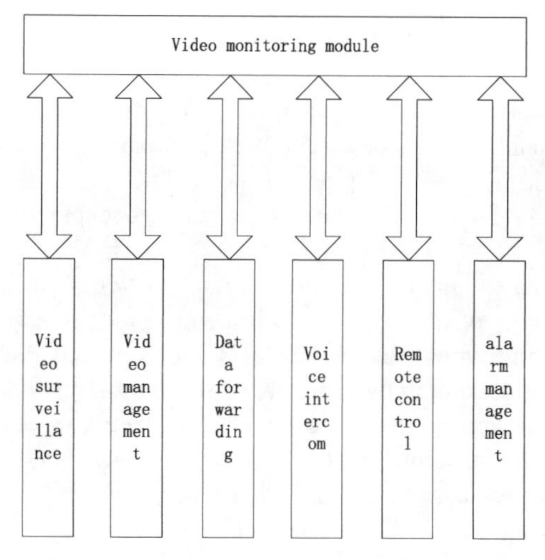

Fig. 2. Functional composition of remote video monitoring module

and can be divided into one main camera and multiple auxiliary cameras according to the monitoring target). The video information of substations under the jurisdiction of the region can be monitored in real time, and multiple (1, 4, 9, 16) real-time videos of the same substation can be monitored in real time, and the simultaneous monitoring of one machine on the same screen can be realized; It can simultaneously monitor single real-time video of multiple substations (1, 4, 9) in real time; Multi angle video of the device can be monitored at the same time. Different regional inspection centers can only monitor the video of their substations. You can click the multi angle real-time video (referred to as multi angle video for short) of multiple cameras of the primary equipment monitored at the same time by the primary equipment in the substation layout plan or primary equipment connection diagram carried by electronic map.

Relevant videos can be directly viewed according to the deployment, withdrawal and alarm status on the layout plan of the substation. On this basis, the automatic patrol function of the remote video monitoring system is set to conduct video patrol on the monitoring points of the system. The objects participating in the round patrol can be set arbitrarily, including the videos of different substations, different cameras of the same substation, different preset positions of the same camera, etc. The interval time of the round patrol can be set, and the cameras that complete the round patrol task can be automatically reset.

3.4 Real Time Transmission of Remote Video Image Based on RTP Technology

Next, RTP technology is used to design the real-time transmission function of remote video image in power system. RTP is a transmission protocol designed by IETF for real-time transmission of video and audio. RTP protocol is generally located above UDP protocol and functionally independent of the underlying transport layer and network layer, but it cannot exist as a single layer. Usually, it uses the low-level UDP protocol to multicast or unicast real-time video and audio data, thus realizing the transmission of multi-point or single point video and audio data.

The connection between RTP transmission service users is called RTP session. For each session participant, the session is identified by a pair of transport layer addresses (that is, one network layer address plus two port addresses, one port is occupied by sending/receiving RTP messages, and the other port is occupied by sending/receiving RTCP messages).

In multimedia data communication, different media types of data are transmitted by different RTP sessions, so RTP sessions correspond to the logical channels contained in the multimedia connection. With the help of RTCP, session participants can not only monitor the quality of data communication, but also perform some basic session control functions, such as participating in/exiting the session, identifying other participants, and controlling other communications. In the RTP protocol, there is no specific limit on the type of transmission services that the underlying network can provide. Generally, the RTP protocol is combined with the IP protocol stack and runs on UDP, as shown in Fig. 3.

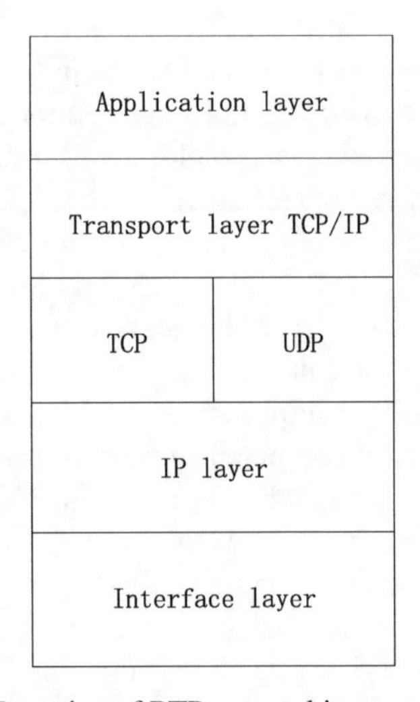

Fig. 3. Location of RTP protocol in protocol stack

As shown in Fig. 3, the RTP protocol itself does not guarantee the reliability of any data transmission. It is not responsible for resource reservation, does not guarantee the quality of service of the transmitted services, and does not guarantee the orderliness of data packets. These tasks are completed by other protocols or the underlying network. This is also the price paid to ensure real-time performance. At the sending end, the upper application program transmits the encoded media data to the RTP communication module in the form of packets. As the payload of the RTP message, the RTP communication module will add the RTP header in front of the payload according to the parameters provided by the upper application to form the RTP message, which will be sent through the socket interface to select the protocol.

The channel from the substation to the main station is optical cable. The optical cable is directly placed to the optical distribution frame in the communication room of the substation. Through the pigtail, it is directly jumpered to the network switch without passing through SDH. The network switch needs to be equipped with optical fiber connection or equipped with optical fiber transceiver. The data signals output by the RPU are directly uploaded to the regional master station or provincial master station through the network switch. The master station is equipped with SDH that provides Ethernet interface, output signal to network switch of master station system through Ethernet interface.

The direct optical fiber transmission mode can be used to transmit video lesson data, which is suitable for occasions with high requirements for image quality. The raw data is not compressed after analog-to-digital conversion, achieving the maximum degree of non damage during transmission, and obtaining the image effect with the lowest distortion when displayed at the back end. After the collected video data is compressed, it is transmitted to the receiving end of the system through the network protocol, thus realizing the goal of real-time image transmission of the remote video monitoring system of the power system.

4 System Test

In order to objectively analyze the feasibility and application effect of the power system remote video monitoring system proposed in this paper, the system test is carried out as shown below. The system is verified in terms of integration stability and effectiveness by means of actual operation, verification test cases and other test means, and the possible errors, causes and distribution are summarized to ensure the normal operation of the system. The system can be put into use in the power system only after the function and performance of the system meet the relevant requirements.

4.1 Test Preparation

For the test of this system, the operating system WINDOWS7 or WINDOWS SERVER 2000 browser E10. According to the above discussion, the remote video monitoring system of power system is built, and the system test environment is built according to the requirements of system operation. The system test environment information is shown in Table 5.

According to the configuration performance parameters shown in Table 5, the system test environment was built to provide a good guarantee for the system test. The system tests the network topology, as shown in Fig. 4.

According to the network topology shown in Fig. 4, the LAN network type is adopted, and the network bandwidth is set to 100 Mbps. Users can connect to the server through the remote login software of the client to view the system operation, which can minimize the impact on the server; Users can use all kinds of mainstream browsers to view the Web server. The types of browsers are unlimited, and there is no need to add any plug-ins; The server needs to install and set various software resources needed to start the system. The

Table 5. System Test Environment Information Description

Model	Configuration/performance parameters
Server compatible computer	CPU: Intel Xeon E5-2620 v4 Main frequency: 10 GHz Memory: 32 GB Hard disk: 100 GB
Client compatible machine	CPU: Intel Core i7-7700 Primary frequency: 3.60 GHz Memory: 16 GB Hard disk: 128 GB+2 TB
Browser	Mainly IE10, IE11 and Google browser
Database system	PostgreSQL 1.0

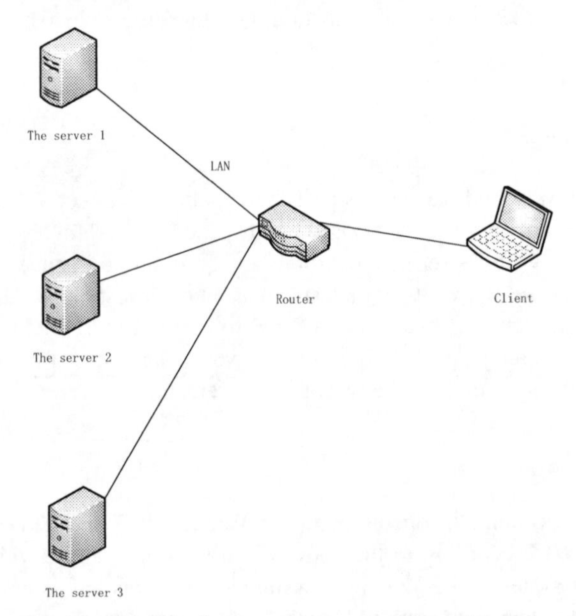

Fig. 4. Schematic diagram of system test network topology

mainstream PostgreSQL database is selected for the database, and it is easy to conduct secondary development later.

The system has unlimited requirements for client hardware. The system server does not need cluster. In the server hardware, the intermediate proxy server needs to receive and process a large number of RTP data packets. The Web end and database are selected to be placed on the same server. Both servers need 16 core CPUs and large hard disk space, while the AES server is a server that connects the external network and the internal network to exchange commands with lower requirements; There is no requirement for mobile phones. You can only receive WeChat enterprise account messages online.

4.2 Result Analysis

After completing the above system test preparation and function test, the next step is to test the operation performance of the power system remote video monitoring system. In order to make the system test results more convincing and comparative, the method and principle of comparative analysis are introduced. The power system remote video monitoring system based on RTP technology proposed in this paper is set as the experimental group, and the an intelligent and cost-effective remote underwater video device for fish size monitoring (The system of reference [4]), Design of the Power System and Remote Control System for the Unmanned Ship (The system of reference [5]) are set as the control group to make an objective comparison and analysis of the operation performance of the 3 systems. The packet loss rate of power system remote video monitoring system is selected as the evaluation index of this test. The packet loss rate (Loss Tolerance or Packet Loss Rate) refers to the ratio of the number of packets lost in the test to the number of data groups sent. The calculation formula of packet loss rate is:

$$Q = [(M_a - M_c)/M_a] \times 100\% \tag{6}$$

Among them, Q indicates packet loss rate of power system remote video monitoring system; M_a indicates system input message; M_c Indicates system output message. The packet loss rate is an important factor that affects the real-time performance of video in the study of real-time performance of remote video monitoring transmission in power system. In the whole system test process, the network data packet loss rate is calculated by the number of video data packets lost during network transmission, so the number of network packet losses must be calculated. The specific process is described as: the video capture device end establishes communication with the network through 3G wireless network, the client software runs on the Windows platform, and is connected to the network in a wired form. During the test process, specify that the sending end of the device sends video packets every 10 min.

Finally, count the amount of video data sent by the sender, add variables in the receiver program to count the total size of the received data, and add timers for encoding start time and encoding end time at the shooting end to calculate encoding time. In addition, the Framecount variable is added to the program to count the number of frames encoded, which is used to calculate the frame rate. The so-called packet loss number is the difference between the sent data and the received data, and the packet loss rate is calculated based on the number of packet losses.

Set the operation time of the power system remote video monitoring system as 24 h, 48 h, 72 h, 96 h, 120 h and 144 h respectively, use MATLAB simulation analysis software to measure and calculate the corresponding packet loss rate of the system when the monitoring time gradually increases, make a comparison, and then judge the performance test effect of the proposed monitoring system. The comparison results are shown in Fig. 5.

From the comparison results in Fig. 5, it can be seen that the packet loss rates of all three monitoring systems increase with the increase of system operation time. Among them, the highest packet loss rates of the system of reference [4] and the system of reference [5] reached 4.7% and 3.2% respectively, while the corresponding packet loss rates of the designed system during each operating period were below 1%, which was

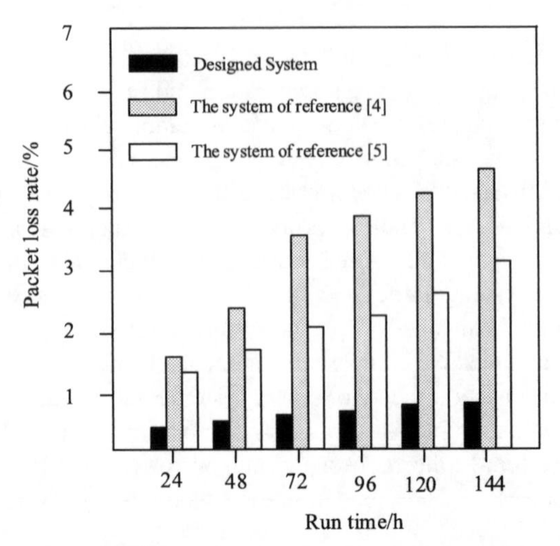

Fig. 5. Comparison Results of Packet Loss Rate of Power System Remote Video Monitoring System

smaller than the comparison system and did not experience significant fluctuations. From the comparison results, it is easy to see that the monitoring system proposed in this paper has a low packet loss rate, a high real-time remote video monitoring, good network status and system monitoring effect, and a high stability of the system operation, which meets the requirements of the power system remote video monitoring system and can meet the development needs of power enterprises in an all-round way.

5 Conclusion

With the in-depth development of digitalization and networking, the rapid development of video monitoring system has brought good opportunities, which has gradually expanded the application scope and application field of video monitoring system, and has an inseparable relationship with people's life. At present, video surveillance system has become a hot spot in the field of social security. With the development of network, the function and performance of remote video monitoring system are gradually improved. In order to improve the problems and shortcomings of the remote video monitoring system used in the current power system in terms of operation function and performance, this paper introduces RTP technology and proposes a new remote video monitoring system for power system. The hardware system of this article consists of EWS, bus controller, video capture card, and high-speed intelligent ball camera. Based on this, RTP technology effectively improves the quality and efficiency of the monitoring system operation, enhances the real-time performance of the video, and achieves the goal of real-time monitoring. It also lays the foundation for future system upgrades.

In addition to the basic real-time video monitoring function, future research can consider expanding the application scenarios of this design. For example, combining sensor data and Big data analysis, more comprehensive power system status monitoring and fault diagnosis functions can be realized.

References

1. Hidayat, R., Reza, F., Herawati, et al.: Remote monitoring application for automatic power supply system in telecommunication network. J. Phys. Conf. Series **1933**(1), 012103 (2021)
2. Wang, J., Zhao, Q., Ye, Q.: Design of remote monitoring system for substation DC power supply under the background of big data. J. Phys: Conf. Ser. **2037**(1), 012005 (2021)
3. Xu, S., Leng, X.: Design of real-time power quality monitoring system for active distribution network based on computer monitoring. J. Phys. Conf. Ser. **1992**(3), 032127 (2021)
4. Coro, G.: An intelligent and cost-effective remote underwater video device for fish size monitoring. Ecol. Inform. Int. J. Ecoinform. Comput. Ecol. **63**(1), 101311 (2021)
5. Zhang, H., Zhang, Z.: Design of the power system and remote control system for the unmanned ship. J. Phys. Conf. Ser. **2121**(1), 012021 (2021)
6. Cheragee, S.H., Hassan, N., Ahammed, S., et al.: A study of IoT based real-time solar power remote monitoring system. Int. J. Amb. Syst. Appl. **9**(2), 27–36 (2021)
7. Yang, L., Dai, M., Wang, H., et al.: A wireless, low-power, and miniaturized EIT system for remote and long-term monitoring of lung ventilation in the isolation ward of ICU. IEEE Trans. Instrum. Meas. **70**, 1–11 (2021)
8. Abdullah, H.M., Hariati, A.M., Fakhri, M., et al.: Raspberry Pi based IoT for aquaculture realtime remote monitoring system with self energy harvesting. IOP Conf. Series Mater. Sci. Eng. **1077**(1), 012062 (2021)
9. Golovan, A., Honcharuk, I., Deli, O., et al.: System of water vehicle power plant remote condition monitoring. IOP Conf. Series Mater. Sci. Eng. **1199**(1), 012049 (2021)
10. Liu, S., Wang, S., Liu, X., et al.: Human inertial thinking strategy: a novel fuzzy reasoning mechanism for IoT-assisted visual monitoring. IEEE Internet Things J. **10**(5), 3735–3748 (2023)
11. Ahmid, M., Kazar, O., Kahloul, L.: A secure and intelligent real-time health monitoring system for remote cardiac patients. Int. J. Med. Eng. Inform. **14**(2), 134–150 (2022)

RLE Algorithm Based Image Data Coding Method of Tujia Brocade Double Knitting Pattern

Yongchang Yao[1(✉)] and Zhuorong Li[2]

[1] Chongqing Vocational Institute of Tourism, Chongqing 409099, China
yaoyongchang0456@163.com

[2] Dalian University of Science and Technology, Dalian 116000, China

Abstract. In order to ensure the resolution of Tujia brocade double-sided woven pattern image and reduce the image storage space, a data encoding method of Tujia brocade double-sided woven pattern image based on RLE algorithm is proposed. A double-sided braided pattern image data encoder is designed as the executive component of the encoding method. Through filtering, image enhancement, illumination compensation and other steps, the pre-processing of pattern image is realized. Using the image data transformation of RLE algorithm, the image data coding of Tujia brocade double-sided woven patterns is realized. Through the performance test experiment, it is concluded that compared with the traditional coding method, the peak signal-to-noise ratio of the optimized design method is improved by about 5.7 db, the image compression ratio is improved by about 0.22, and the image data coding efficiency is significantly improved.

Keywords: RLE Algorithm · Tujia Woven Cotton · Double-Sided Woven Pattern Image · Image Data Coding

1 Introduction

Tujia brocade is also known as "Xilankapu" and "Dahua Bedding". According to the custom of Tujia nationality, when Tujia girls got married in the past, they had to cross stitch and embroider when weaving to make "Xilankapu" with beautiful patterns. After more than 4000 years of development, Tujia women have created countless brocade patterns, and have retained more than 220 patterns. The Tujia brocade has a wide range of patterns and themes, covering almost every aspect of Tujia people's life, including flowers, birds, fish and insects, scenery of mountains and rivers, auspicious characters, and the weaving of folk stories, fables and other pictures. The selection of themes is closely related to Tujia people's life and customs, and is a vivid portrayal of the relationship between Tujia people and nature [1]. Its patterns and patterns include natural image patterns, geometric patterns, and character patterns. They have the common characteristics of taking materials from life, abstract and mysterious patterns, balanced and symmetrical composition, and so on. The double-sided woven pattern of Tujia brocade belongs to China's intangible heritage and has high protection value.

L. Yun et al. (Eds.): ADHIP 2023, LNICST 549, pp. 66–81, 2024.
https://doi.org/10.1007/978-3-031-50549-2_5

Under the background of the rapid development of modern communication technology, computer technology, network technology and information processing technology, image information processing, storage and transmission play an increasingly important role in social life, and people are more and more urgent to accept image information. Image communication will be the biggest challenge in the development of communication industry, and will also be a hot market in the future communication field. Through a large amount of statistics and visual perception research on image data, people have shown that there is a strong correlation between adjacent pixels, between adjacent lines or between adjacent frames of images, that is, image signals have redundancy in space, time, structure, vision and knowledge, Using some coding method to eliminate these correlations or redundancies to a certain extent, we can achieve image data compression coding. In a word, large amount of image information will bring huge pressure on memory storage capacity, communication trunk channel broadband and computing speed. It is unrealistic to solve this problem simply by increasing storage capacity, channel bandwidth and computer processing speed. Therefore, in the transmission and storage of image data, compression coding is imperative.

Image coding is also called image compression, refers to the technology of representing the image or the information contained in the image with fewer bits under the condition that certain quality is met. At present, more mature image data coding methods mainly include: Reference [2] designs an embedded image coding algorithm based on discrete cosine transform. Firstly, the original image is subjected to discrete cosine transform, and then the cosine coefficients in the transform domain are scanned in a specific scanning order to accurately determine the positions of each coefficient. Secondly, in order to represent the cosine coefficients in the transform domain with fewer coding bits, the scanned coefficients need to be quantified to achieve image coding. Reference [3] proposes a sparse approximation image coding method that comprehensively applies Matrix decomposition and sparse representation. By comprehensively applying Matrix decomposition and sparse constraints, a matrix optimization model is constructed for image coding. However, the above image coding methods have a problem of poor coding performance in practical operation when dealing with large amounts of repetitive information, mainly reflected in the encoding quality and encoding space of the image. Therefore, this article proposes a data encoding method for Tujia brocade double-sided pattern images based on the RLE algorithm. RLE algorithm, also known as Run-length encoding algorithm, is a compression method using spatial redundancy, and also a basic lossless compression coding method in statistical coding. The basic principle is to use corresponding symbols and string length to represent consecutive strings with the same value, thereby reducing the number of symbols and achieving compression. In the case of large amount of duplicate information, Run-length encoding stipulates that only when the data code in the row or column changes, can the specific value and continuous number of data be recorded, which plays a certain role in eliminating redundant information and realizing data compression. This article achieves preprocessing of pattern images through steps such as filtering, image enhancement, and lighting compensation. It innovatively applies RLE algorithm to eliminate redundant information, complete image data transformation, and achieve image data encoding of Tujia brocade double-sided patterns.

2 Design of Data Encoding Method for Double-Sided Woven Pattern Image

Image compression is a part of data compression, which has the basic characteristics of data compression. Data compression mainly studies the methods of data representation, transmission and conversion, and reduces data storage space and transmission time by representing the source signal with the smallest data. The classic data compression technology is based on information theory, called source coding. Information theory believes that sources contain more or less redundancy, that is, there is a great correlation between signals, or the probability distribution of sources is uneven. By removing the correlation or changing the non-uniformity of the probability distribution, the data can be effectively compressed. The basic operation process of optimizing the design of double-sided braided pattern image data coding method is shown in Fig. 1.

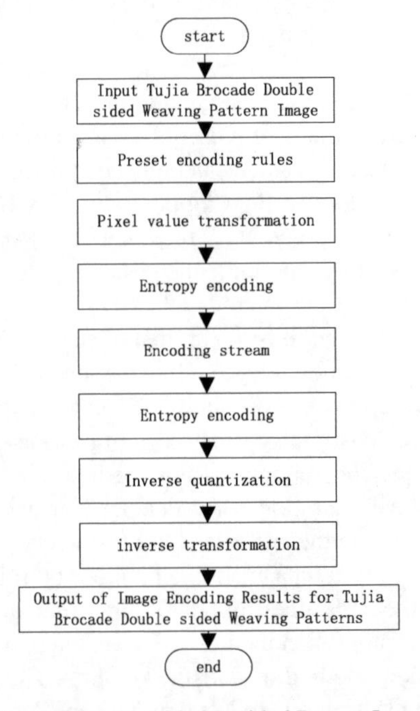

Fig. 1. Flow Chart of Double sided Braided Pattern Image Data Coding

The basic steps of image compression coding are described as follows: Step 1: First, perform some transformation on the original image. There are many transformation methods available here, which can be frequency domain transformation or spatial domain transformation, such as discrete cosine transform or wavelet transform. After transformation, a series of transformation coefficients are obtained. Step 2: Quantize the obtained transformation coefficient. The quantization method is divided into scalar quantization and vector quantization. Select the appropriate quantization method according to different needs. Step 3: The quantized data is processed by entropy coding, and then they are organized into different stream data. For this stream data, it can then be transmitted through the channel or stored using a storage device. The above is the basic encoding steps. If the image data after encoding is to be decoded, the decoding process is just the opposite of the encoding process, that is, entropy transformation is first followed by inverse quantization and inverse transformation.

2.1 Design of Double-Sided Braided Pattern Image Data Encoder

The double-sided braided pattern image data encoder is the running part of the image data coding method, including wavelet transform and zero tree coding. Its overall structure is shown in Fig. 2.

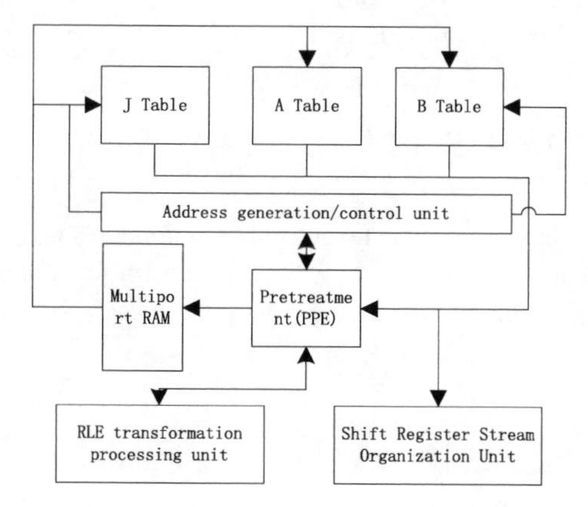

Fig. 2. Structure diagram of double-sided braided pattern image data encoder

As shown in Fig. 2, the original double-sided braided pattern image first enters the wavelet transform processing unit. After the wavelet transform is completed, the wavelet coefficients are sent to the preprocessing unit to calculate the highest bit level and the significance value of each coefficient. The main processing process carries out zero tree coding according to the wavelet coefficients and tables J, A and B, and finally outputs the compressed code stream. The pre-processing module mainly reads the wavelet coefficients generated by the wavelet transform processing unit, and generates the indexes of Table J, Table A and Table B for the main processing. Since each parent node has four direct child nodes after the wavelet transform, the preprocessing unit adopts the data packaging mode of four pixels in a group. The saliency processing can be completed by reading each pixel once. The main processing module mainly completes the sorting of wavelet zero tree and the significance judgment.

2.2 Preprocessing of Double-Sided Woven Pattern Image

2.2.1 Filtering and Denoising

Filtering and de-noising is the operation of double-sided woven pattern image, the purpose is to remove cross links, breakpoints and blurred parts in the image. Linear filtering is used to filter and denoise the double-sided woven pattern image. Linear filtering can remove some types of noise in the image. For example, the average filter using the neighborhood average method is very suitable for removing the particle noise in the scanned image. Neighborhood average method for given double-sided woven pattern image $f(x, y)$ each pixel in (i, j), take its neighborhood U. set up U contain M the average of pixels is taken as the image points after processing (i, j) the grayscale at. Replace the original gray level of a pixel with the average gray level of each pixel in the neighborhood of the pixel, that is, use the neighborhood average technology. Neighborhood U the shape and size of are determined according to the image characteristics [4]. Generally, the shapes are square, rectangle and cross, U the shape and size of can remain unchanged during the whole image processing, and can also change according to the local statistical characteristics of the image (i, j) generally located in U center, Assumption U is 3 * 3 neighborhood, then:

$$\overline{f}(x, y) = \frac{1}{3 \times 3} \sum_{i=-1}^{1} \sum_{j=-1}^{1} f(i + x, j + y) \tag{1}$$

Hypothetical noise N it is additive noise, and each point in space is uncorrelated, and the expectation is 0, and the variance is χ, $g(x, y)$ it represents the unpolluted image. After neighborhood averaging of the image containing noise, the filtering and denoising results are as follows:

$$\overline{f}(x, y)' = \frac{\sum g(x, y) + \sum n(x, y)}{U} \tag{2}$$

In the formula, $n(x, y)$ represents the original gray scale of the pixels. It can be seen from Formula 2 that after neighborhood averaging, the mean value of noise remains unchanged, and the noise variance is:

$$\chi_{new} = \frac{\chi}{U} \tag{3}$$

That is, the noise variance becomes smaller, indicating that the noise intensity is weakened, that is, the noise is suppressed. At the same time, it can be seen from the formula that the neighborhood average method also smoothes the image signal, especially the boundary of the image target area may become blurred.

2.2.2 Image Enhancement

The purpose of image enhancement is achieved by expanding the dynamic range of the image. Define the dynamic range of the color Tujia brocade double-sided woven pattern image as:

$$W(f) = \mathrm{Min}_{x \in D}\{f_R(x), f_G(x), f_B(x)\} - \mathrm{Max}_{x \in D}\{f_R(x), f_G(x), f_B(x)\} \tag{4}$$

In the above formula D represents the pixel x field, $f_R(x)$, $f_G(x)$ and $f_B(x)$ is the R, G, B channels of the image. The image enhancement operation is realized by histogram equalization. In the histogram, if the gray level is concentrated in the high gray area, the low gray level of the image is not easy to distinguish. If the gray level is concentrated in the low gray area, the high gray level is not easy to distinguish [5]. In order to make the high and low gray levels easy to distinguish, the best way is to convert the image so that the distribution probability of gray levels is the same. This is the purpose of histogram equalization. Suppose the image is transformed as follows:

$$s = T(r), 0 \leq r \leq L - 1 \tag{5}$$

In the formula, $T(r)$ represents the gray distribution probability and L represents the gray rank. The purpose of image enhancement is to make the probability distribution of gray s level equal:

$$p(s) = \frac{1}{L - 1} \tag{6}$$

Variables in Formula 5 and Formula 6 L is grayscale. The relationship between the gray level distribution before transformation and the gray level distribution after transformation is as follows:

$$p(s) = p(r)\left|\frac{dr}{ds}\right| \tag{7}$$

Through simultaneous and integral processing of the above formula, we can get:

$$\begin{cases} T'(r) = (L - 1)T(r) \\ s = T(r) = (L - 1)\int_0^r p(r)wdw \end{cases} \tag{8}$$

In the formula, w represents the grayscale weights. Finally, the enhancement processing results of Tujia brocade double-sided woven pattern image are as follows:

$$s_k = (L - 1)\sum_{r=0}^{w} p(r) \tag{9}$$

According to the above process, all pixels in the Tujia brocade double-sided woven pattern image are processed to complete the image enhancement operation.

2.2.3 Light Compensation

Considering that the pixel values of the double-sided woven patterns of Tujia brocade will be significantly different when photographed under different light intensities, the transformed similar images will reflect the actual situation of the double-sided woven patterns of Tujia brocade through light compensation [6, 7]. Linear brightness transformation is carried out for the original Tujia brocade double-sided woven pattern image, and the process can be quantified as:

$$f_{\text{illumination compensation}}(x, y) = \alpha f(x, y) + \beta \tag{10}$$

where parameters α and β scale and offset light compensation parameters respectively. The goal of light compensation is to make the image $f_{\text{illumination compensation}}(x, y)$ as close as possible to the real image, this is expressed by the following formula:

$$q = \min \sum_{1 \leq i \leq M} \left(h(x, y) - f_{\text{illumination compensation}}(x, y)\right) \tag{11}$$

among M is the number of all matching feature points i, $h(x, y)$ indicates the position of the feature point in the current image (x, y) the pixel value at. According to the calculation result of Formula 11, the specific value of the light compensation parameter can be obtained. The calculation result of the compensation parameter can be substituted into Formula 10 to complete the light compensation processing of the Tujia brocade double-sided woven pattern image.

2.2.4 Image Binarization

In digital image processing, binary images occupy a very important position, especially in practical image processing systems, there are many systems with binary image processing as the core. To process and analyze binary images, first we need to binary gray images to obtain binary images [8, 9]. The advantage of this is that when the image is further processed, the geometric properties of the image are only related to the positions of 0 and 1, not the gray value of pixels, which makes the processing simple and the data compression amount is large. The so-called image binarization refers to transforming a grayscale image into a binary image that uses only two values to represent the target and background of the image by setting a threshold value. The image binarization can be performed according to the following threshold processing:

$$g(x, y) = \begin{cases} 1; f(x, y) \geq \phi \\ 0; f(x, y) < \phi \end{cases} \tag{12}$$

In the above formula ϕ the threshold set for. the part of the binary image whose pixel value is 1 represents the target sub image, and the part whose pixel value is 0 represents the background sub image.

2.3 Image Data Transformation Using RLE Algorithm

RLE algorithm is also called run length coding, and run length is a cursor on a linear unit; The cursor can also be set under the mark of run. In this way, the cursor at the upper level is called the parent run, and the cursor at the lower level is called the child run. Figure 3 shows the basic structure of run.

In Fig. 3, Y represents the compressed data character, C represents the number of times Y repeats, and B represents the character not used in the data character set Y, which is used to indicate that there is a string with the length of C at this position in the data stream. There are intra frame data redundancy and inter frame data redundancy in the storage unit of image data. We want to eliminate redundant data as much as possible, that is, fully compress the data redundancy in time domain and space domain, so as to reduce

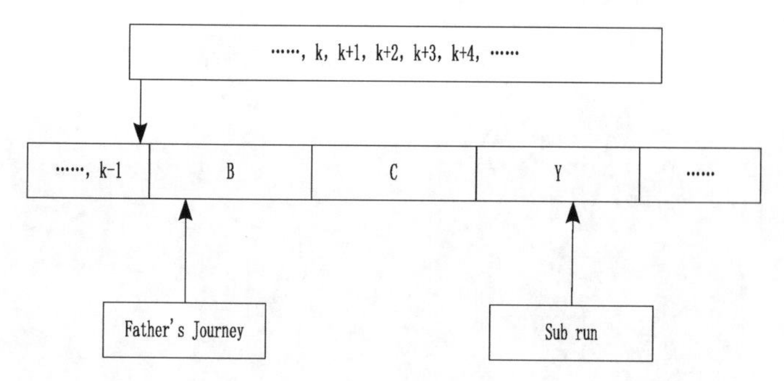

Fig. 3. Basic Structure of Run

the amount of data. Therefore, the simple run length coding principle is to form a new code [10] by moving the cursor and linearly scanning the records of consecutive cells, and saving the same cell name and number of cells. In general, the more consecutive identical records, the more redundant data removed, the smaller the amount of newly formed data, and the higher its compression ratio. In the same way, 2D or 3D data can also be encoded through multi run and linear processing. This is the basic principle of run length coding. The basic operation flow of RLE algorithm is shown in Fig. 4.

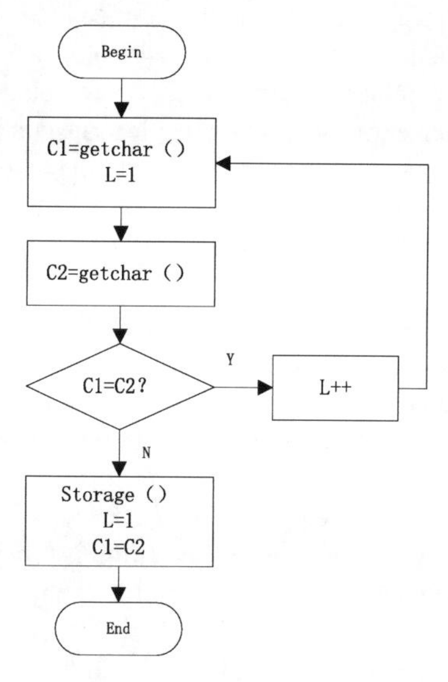

Fig. 4. RLE algorithm flow chart

According to the flow in Fig. 4, in the process of image data transformation of Tujia brocade double-sided woven patterns, set $v_1, v_1, \ldots, v_n$ if it is a row of pixels in the image, the data transformation principle of the row of pixels under the RLE algorithm is shown in Fig. 5.

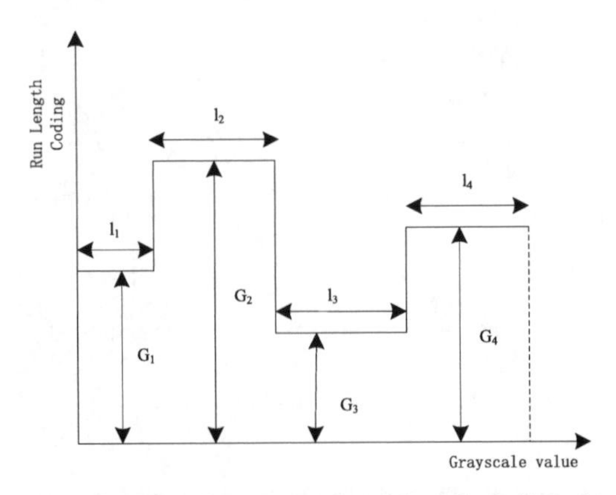

Fig. 5. Schematic diagram of RLE transformation of one line of image pixels

As can be seen from Fig. 5, each line of image is composed of k segment length is l_k, gray value is G_k the line image can be represented by even pairs:

$$(\mathrm{x}_1, \mathrm{x}_2, \cdots, \mathrm{x}_\mathrm{n}) \rightarrow (\mathrm{G}_1, l_1), (\mathrm{G}_2, l_2), \cdots, (\mathrm{G}_\mathrm{k}, l_k) \tag{13}$$

Each even symmetry is a grayscale run. Assume that the binary image line starts from the blank run. In this way, only the run length needs to be encoded. The resulting encoding is one-dimensional, and the one-dimensional transformation of image pixels can be completed [11, 12]. On this basis, the concept of wavelet transform is introduced to obtain two-dimensional image transformation results. The continuous wavelet transform process of Tujia brocade double-sided woven pattern image is as follows:

$$\varphi = \frac{1}{\sqrt{|\zeta|}} \int \psi\left(\frac{t-\delta}{\zeta}\right) dt \tag{14}$$

Among ζ the spectrum structure of the signal after wavelet analysis can be changed. By changing the size t and shape δ of the time window of the function, more detailed time-frequency analysis can be carried out on the signal. Parameters can be added $|\zeta|$ the value of makes the wavelet basis function $\psi()$ the frequency spectrum of $\psi()$ the width of is gradually increased. Conversely, by reducing $|\zeta|$ the value of makes the wavelet basis function $\psi()$ the frequency spectrum of $\psi()$ the width of decreases gradually. The multi-resolution analysis of wavelet analysis algorithm is reflected in the corresponding change of its time window when the frequency range changes. When the function time window is small, its frequency appears in the high frequency range. When the time window is large, the frequency of the signal appears in the low frequency range, so the wavelet transform algorithm has a higher resolution in the time domain or space domain. The original signal can be decomposed by Formula 14, and the image signal $f(t)$ The reconstruction is shown in Formula 15.

$$f(t) = \int_{-\infty}^{+\infty} \int_{-\infty}^{+\infty} \frac{d\zeta d\delta}{\zeta^2} W_f(\zeta, \delta) \psi(t) \tag{15}$$

In formula 15, $W_f(\zeta, \delta)$ represents the transform coefficient of the image. The processing result of image data transformation is obtained according to the above flow.

2.4 Realize Data Coding of Double-Sided Woven Pattern Image

Through the above description of the algorithm principle, take an 8 bit byte to record the run, the first bit of the byte as the inter frame and intra frame flag bit, and take a piece of Tujia brocade double-sided woven pattern image area for analysis; Keep two screen images in memory. When encoding, scan the pixels at the same position of the previous frame while scanning the current frame image, and count the number of the same pixels in the current frame image frame n_1 The number of pixels that are the same as the current frame and the previous frame n_2. When coding, if n_1 value greater than n_2, record n_1 and pixel values, if n_1 the value is less than n_2 only record n_2, because another buffer has saved the pixel data value of the previous frame, and the pixel value of the current frame is the same as the pixel value at the same position of the previous frame. To speed up, a byte record is used n_1 and n_2, in order to distinguish between encoded data streams n_1 and n_2, make n_1 and n_2 different data ranges in one byte, such as n_1 occupy 0–127, n_2 after the value is increased by 128, it will occupy 128–255.The final result of Tujia brocade double-sided woven pattern image data coding is:

$$f_{\text{code}}(x, y) = \kappa_{\text{code}} f(x, y) \tag{16}$$

In Formula 16 κ_{code} is the image data coding coefficient, which is determined by RLE algorithm. According to the above process, the data encoding result of double-sided braided pattern image is obtained.

3 Experimental Analysis of Coding Performance Test

In order to test the coding performance of the RLE algorithm based image data coding method for Tujia brocade double-sided woven patterns, the performance test experiment is designed by means of comparative testing. This experiment tests the coding quality and coding efficiency of Tujia brocade double-sided woven pattern image data. The contrast methods set in the experiment are reference [2] image coding method and reference [3] image coding method. To ensure the uniqueness of experimental variables, the optimization design method is the same as the coding object of the contrast method. And run in the same experimental environment.

3.1 Build Image Data Coding Performance Test Environment

This experiment selects MATLAB software as the development tool of the experiment. MATLAB software has powerful numerical calculation and data analysis capabilities. At the same time, MATLAB software can also be integrated with image processing technology and programming technology to solve scientific and practical engineering problems. MATLAB software also provides solutions to other professional problems, such as text and voice processing, symbol calculation, visual modeling, engineering and circuit simulation. MATLAB software mainly consists of two main parts: core function part and toolbox. The disciplinary toolbox mainly consists of professional tools such as signal processing toolbox and control toolbox. The functional toolbox is used to complete the functions of simulation, modeling and extended symbolic operations. The

7.0 version of MATLAB software is used in the performance test experiment. Compared with the older version, MATLAB 7.0 has made a greater degree of upgrading in the programming development environment, data processing, numerical operation and I/O interface. It mainly includes redefining the desktop environment, enhancing the functions of array editor and workspace browser, and adding the M-Lint code analysis function. In MATLAB7.0, users can define desktop windows and shortcut keys according to their own requirements, so that they can access and manage the document interface more conveniently. The directory browser adds coverage analysis tools, code efficiency analysis and M-Lint code analysis tools. M-Lint code analysis tool can help programmers complete the analysis of program efficiency, redundancy and other parameters, so as to effectively improve the efficiency of program execution. Install MATLAB7.0 software in the main test computer, and debug relevant functions.

3.2 Preparation of Double-Sided Woven Pattern Image Samples

This time, several double-sided woven patterns of Tujia brocade were taken as the research object, and Nikon D50 digital SLR camera was used to collect image samples of double-sided woven patterns of Tujia brocade. The sensitive element of the camera is CCD, the sensor size is about 24 * 16 mm, the total number of pixels is 610 pixels, and the resolution can reach more than 2000 dpi. Place the double-sided woven pattern of Tujia brocade at a place with sufficient light, set the camera focus and other image acquisition parameters, and obtain the preparation results of the double-sided woven pattern image samples. Figure 6 shows the preparation of some samples of Tujia brocade double-sided woven pattern images.

Fig. 6. Image sample of Tujia brocade double-sided woven pattern

Before the experiment, the image sample size was unified by image clipping, geometric transformation, etc. In order to ensure the credibility of the experimental results, the number of Tujia brocade double-sided woven pattern image samples prepared for this experiment is 150.

3.3 Describe the Experimental Process of Coding Performance Test

The RLE algorithm based image data coding method of Tujia brocade double-sided woven pattern is converted into computer code that can be run directly by using programming tools to realize the development of optimization design coding method. Input the prepared double-sided braided pattern image samples into the image data encoding program one by one to obtain the corresponding encoding results. Figure 7 shows the coding results of some Tujia brocade double-sided woven pattern image samples.

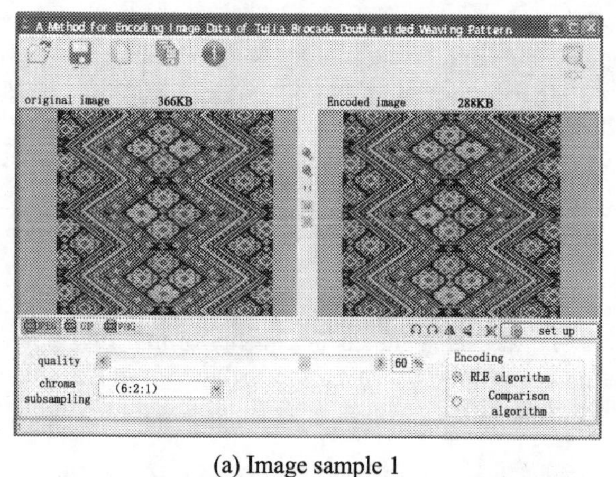

(a) Image sample 1

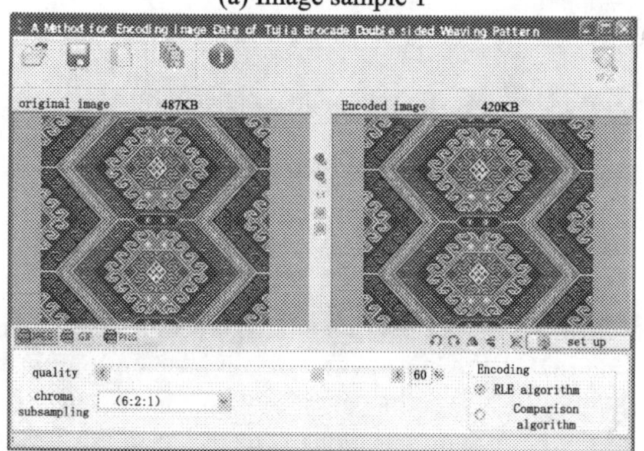

(b) Image sample 2

Fig. 7. Image coding results of Tujia brocade double-sided woven patterns

In the same way, the coding processing results of all image data samples in the experiment can be obtained. According to the above method, the development of two contrast encoding methods is realized, and the image samples are input into them to obtain the image data encoding results of the contrast method.

3.4 Setting Image Data Coding Performance Test Indicators

According to the purpose of the experiment, the test is carried out from two aspects: coding effect and operation efficiency. The test indicators set in the coding effect part are the peak signal to noise ratio and compression ratio, where the numerical results of the peak signal to noise ratio indicator are:

$$\varepsilon = 10\lg\left(\frac{G_{\max}\cdot MN}{\sum_{x=1}^{M}\sum_{y=1}^{N}\left[f(x,y)-f_{\text{code}}(x,y)\right]^2}\right) \tag{17}$$

Where parameters $G_{\max}$ is the maximum grayscale value of the image, M and N the length and width of the image respectively, and the peak signal to noise ratio are commonly used as the measurement method of signal reconstruction quality in image compression and other fields. The test result of compression ratio index is:

$$\sigma = \frac{\gamma(f(x,y))}{\gamma(f_{\text{code}}(x,y))} \tag{18}$$

Variables in Eq. 18 $\gamma(f(x,y))$ and $\gamma(f_{\text{code}}(x,y))$ is the data amount of the original image and the compressed image respectively. The higher the peak signal to noise ratio and compression ratio, the better the coding quality of the corresponding method. In addition, the test index of image data coding method operation efficiency is coding efficiency, and the test result is:

$$\eta = \frac{H(f(x,y))}{\overline{L}} \times 100\% \tag{19}$$

Among $H(f(x,y))$ is the information source entropy of the image, $\overline{L}$ is the average code length of image compression data, variable $\overline{L}$ the calculation formula of is as follows:

$$\overline{L} = \frac{\gamma(f_{\text{code}}(x,y))}{MN} \tag{20}$$

The coding efficiency reflects the closeness between the average code length after data compression and the source entropy η When it is equal to 1, it indicates that this compression algorithm is the best coding algorithm that reaches the theoretical limit and completely eliminates redundant information in image data. If the data of the same image is encoded with different encoding methods, the higher the efficiency of the algorithm, the better the algorithm selected.

3.5 Experimental Results and Analysis of Coding Performance Test

Count the maximum gray value data in the coding results of Tujia brocade double-sided woven pattern image, and obtain the test results of peak signal to noise ratio index through the calculation of Formula 17, as shown in Fig. 8.

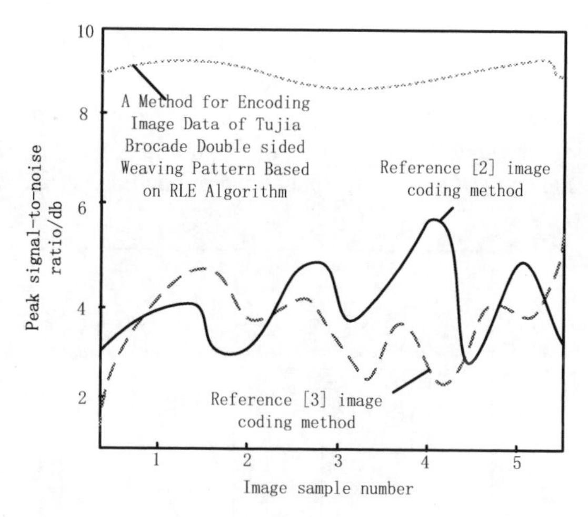

Fig. 8. Peak Signal to Noise Ratio Test Results of Image Data Coding

It can be seen intuitively from Fig. 8 that compared with the two comparison methods, the peak signal to noise ratio of the optimized design method is significantly improved by about 5.7 db. This is because this paper realizes the preprocessing of the pattern image through filtering, image enhancement, and illumination compensation, and removes most of the noise. The test results of compression ratio index are shown in Table 1.

Table 1. Test Data of Image Compression Ratio Index

Image sample number	Amount of original image data/KB	Image data amount/KB obtained by reference [2] image coding method	Image data amount/KB obtained by reference [3] image coding method	Tujia brocade double-sided woven pattern image data encoding method based on RLE algorithm
1	366	351	347	288
2	487	482	475	420
3	505	500	496	415
4	389	384	379	304
5	510	501	498	412
6	495	487	480	424
7	396	389	382	304
8	518	512	508	399

By substituting the data in Table 1 into Formula 18, the average image compression ratio of the two comparison methods is 1.02 and 1.03 respectively, while the average compression ratio of the optimized RLE algorithm based Tujia brocade double-sided woven pattern image data coding method is 1.24. This is because this paper innovatively applies RLE algorithm to eliminate redundant information, complete the image data transformation, and realize the high compression code of double-sided pattern of Tujia brocade. In addition, through the calculation of Formula 19, we can get the test results of the operation efficiency of the image data encoding method, as shown in Fig. 9.

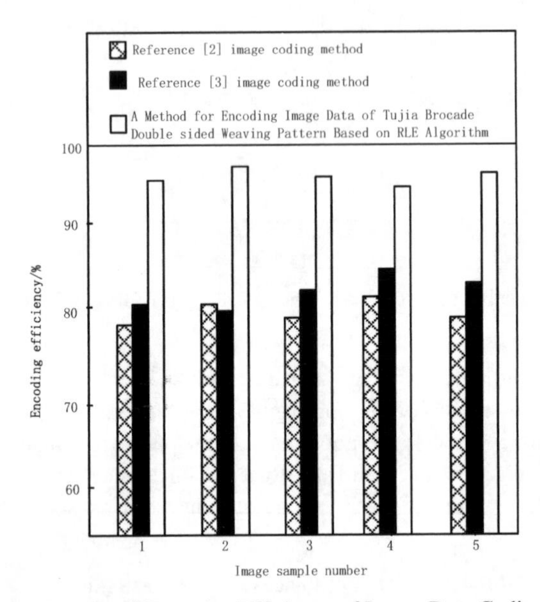

Fig. 9. Test Results of Operation Efficiency of Image Data Coding Method

The average coding efficiency of the two comparison methods is 78.4% and 80.2% respectively, and the average coding efficiency of the optimal design method is 95.5%. In conclusion, the optimization design method has obvious advantages in coding effect and operation efficiency. This is because this paper innovatively applies RLE algorithm to eliminate redundant information and improves the image data coding efficiency of double-sided pattern of Tujia brocade.

4 Conclusion

Tujia brocade double-sided woven pattern is the product of Chinese history, and has high application value. This paper innovatively uses the RLE algorithm to encode the Tujia brocade double-sided pattern image data, which effectively solves the problems of large space occupation and high distortion rate. From the experimental results, it can be seen that the optimization design method has high application value. In addition to the double-sided woven patterns of Tujia brocade, the optimization design method can also be applied to other complex image compression and coding work, providing auxiliary tools for image processing. However, due to the limited conditions, the basic coding data of this method is limited, and the coding patterns are not rich enough. The future research will improve the pattern range and improve the coding effect.

Acknowledgement. Science and technology research project of Chongqing Municipal Education Commission: Research on the popularization and application of new technology of double-sided weaving of Tujia nationality. (Project No.:KJQN202104603).

References

1. Xie, F., Wei, H.: Research on controllable deep learning of multi-channel image coding technology in Ferrographic Image fault classification. Tribol. Int. **173**, 107656 (2022). https://doi.org/10.1016/j.triboint.2022.107656
2. Chen, X., You, M., Cui, Y., et al.: An embedded image coding algorithm based on discrete cosine transform. Chin. J. Med. Phys. **37**(12), 1525–1528 (2020)
3. Yanlin, L., Yuting, T., Yan, Z.: Analysis and improvement of image coding algorithm based on sparse approximation. J. Jinling Inst. Technol. **36**(04), 18–21 (2020)
4. Hejda, M., Robertson, J., Bueno, J., et al.: Neuromorphic encoding of image pixel data into rate-coded optical spike trains with a photonic VCSEL-neuron. APL Photon. **6**(6), 060802 (2021)
5. Barannik, V., Sidchenko, S., Barannik, N., et al.: Development of the method for encoding service data in cryptocompression image representation systems. Eastern-Eur. J. Enterprise Technol. **3**(9(111)), 103–115 (2021)
6. Shakeel, M.S., Lam, K.-M.: Deep low-rank feature learning and encoding for cross-age face recognition. J. Visual Commun. Image Represent. **82**, 103423 (2022). https://doi.org/10.1016/j.jvcir.2021.103423
7. Hardi, S.M., Zarlis, M., Lubis, D., et al.: Text file compression using hybrid Run Length Encoding (Rle) algorithm With Even Rodeh Code (Erc) and Variable Length Binary Encoding (Vlbe) to save storage space. J. Phys. Conf. Series **1830**(1), 012022 (6pp) (2021)
8. Salih, H.M., Kadhim, A.M.: Medical image compression based on SPIHT-BAT algorithms. IOP Conf. Series Mater. Sci. Eng. **1076**(1), 012037 (2021). https://doi.org/10.1088/1757-899X/1076/1/012037
9. Swetha, V., Premjyoti Patil, G., Shantakumar Patil, B.: Lossless compression of satellite images using a versatile hybrid algorithm. IOP Conf. Series Mater. Sci. Eng. **1166**(1), 012048 (2021). https://doi.org/10.1088/1757-899X/1166/1/012048
10. Shan, Y., Chen, X., Qiu, C., et al.: Implementation of fast huffman encoding based on FPGA. J. Phys. Conf. Ser. **2189**(1), 012021 (2022)
11. Oswald, C., Sivaselvan, B.: Smart multimedia compressor—intelligent algorithms for text and image compression. Comput. J. **2**, 2 (2021)
12. Liu, S., Huang, S., Wang, S., Muhammad, K., Bellavista, P., Del Ser, J.: Visual tracking in complex scenes: a location fusion mechanism based on the combination of multiple visual cognition flows. Inform. Fusion **96**, 281–296 (2023)

A Blockchain Based Real-Time Sharing Method for Ideological and Political Mobile Education Resources

Peihua Zhang(✉), Jie Yang, and Zefeng Li

School of Marxism, Xi'an Eurasia University, Xi'an 710000, China
18829032042@163.com

Abstract. In order to improve the efficiency and effectiveness of real-time sharing of mobile education resources, this article proposes a blockchain based method for real-time sharing of ideological and political mobile education resources. Firstly, build a blockchain network for sharing ideological and political mobile education resources; Secondly, extract the features of video resource data, audio resource data, and text resource data from ideological and political mobile education resources, and fuse the resource features based on Tucker decomposition data fusion algorithm; Train the features based on the fuzzy neural network model again; Finally, resource sharing is achieved based on blockchain technology. Through experimental results, it has been proven that the proposed method has a resource request processing time of 5.20 s when the data volume is 200 GB, and can achieve 98% sharing of the total data volume. The sharing efficiency is high, the sharing effect is good, and it has good practical application performance. However, current blockchain technology still has some limitations when dealing with large-scale data and high concurrency access. Future research can explore how to improve the scalability and performance of blockchain systems to meet the needs of more users.

Keywords: Blockchain · Ideological And Political · Mobile Education · Resource Sharing · Resource Characteristics

1 Introduction

With the rapid development of mobile Internet technology, mobile phones, tablets and other mobile devices have become an important tool for people to learn. Especially under the support of cloud computing, artificial intelligence and other technologies, the online, intelligent, personalized and scenario-based mobile education resources show more and more advantages, which also provides good conditions for universities to build mobile teaching ecology [1]. With the continuous advancement of education reform in colleges and universities, more and more colleges and universities begin to build localized, open and shared independent intellectual property teaching resource libraries. The development, integration and sharing of these resources have become the necessary means for

L. Yun et al. (Eds.): ADHIP 2023, LNICST 549, pp. 82–95, 2024.
https://doi.org/10.1007/978-3-031-50549-2_6

colleges and universities to realize digital transformation and improve teaching effect [2]. The content of ideological and political courses includes political system, ideological and moral, laws and regulations, national security, world politics and other aspects. It has a strong crossover nature with other disciplines, involving philosophy, history, culture, economy and other fields. Therefore, in addition to pure knowledge, ideological and political courses emphasize the cultivation of students' political accomplishment, thinking ability, moral emotion and other aspects. The ideological and political curriculum has a very far-reaching influence on the national, cultural and ideological aspects. Therefore, actively promoting online education of ideological and political education can help improve the quality of ideological and political education, strengthen students' self-learning ability, and promote educational fairness and efficiency. Mobile education resource sharing can not only break geographical and time constraints, improve the efficiency and accuracy of educational resource utilization, but also meet the actual needs of personalized teaching, promoting the improvement of the quality and innovation level of university curriculum teaching [3]. Mobile education resource sharing can make educational resources more equitable, reduce the gap in educational resources between regions and schools, and promote educational equity. Through mobile education resource sharing, schools can access validated mobile education resources and teaching methods from other schools, which can not only improve teachers' teaching level but also improve students' learning outcomes. Fully utilize existing educational resources, avoid resource waste and duplicate construction, reduce educational investment costs, and improve resource utilization efficiency. Sharing educational resources can promote the exchange of information and knowledge among schools, and help schools improve their level of educational informatization and modernization [4].

At present, many scholars have carried out relevant research on the method of mobile educational resource sharing. Some researchers have designed a resource sharing model based on MB+ tree test method. The resource collection subsystem uses crawler technology to retrieve related resources and store them in the resource library. The MB+ tree test method is adopted to complete the test of shared resources, extract the correlation feature quantity of relevant resource sharing big data, obtain the cluster center of multi-module collaborative data mining, and adopt the method of regional networking design to obtain the resource sharing model and complete the resource sharing [5]. Although this method can effectively improve the resource utilization rate, with high resource integrity, and meet the purpose of mobile education resource sharing, in the actual process of mobile education resource sharing, when resource demanders initiate resource requests, they are prone to be affected by the surrounding environment, which increases the request processing time and the sharing process time, and affects the effect of real-time resource sharing. Some researchers put forward the sharing method of database resources based on alliance chain. Build a collaborative and shared mode for thematic database resources based on alliance chain technology, design the overall architecture of the mode, and conduct research according to the data layer, alliance chain system, and application layer of the architecture. Using practical Byzantine fault-tolerant consensus algorithm and three smart contracts for identity authentication, resource sharing, and copyright protection, effectively achieve the storage and sharing of thematic database resources [6]. Although the practical application performance of this method is good and can effectively share resources, as the number of uploaded resources continues to increase, the number of resources that this method can share begins to decrease, weakening a certain degree of resource sharing ability.

Blockchain technology can ensure the security, fairness, and transparency of data in data sharing, while improving the efficiency of data sharing. This article proposes a blockchain based real-time sharing method for ideological and political mobile education resources to address issues such as poor resource sharing capabilities and long resource request processing time in traditional methods. In order to improve the efficiency of resource sharing and improve the application performance of resource sharing through blockchain technology.

2 Blockchain Technology

Blockchain is a distributed database with basic features such as decentralization, tamper resistance, and high trust. Blockchain is a network composed of encryption technology and distributed nodes, which completes data exchange and value transfer without the need for centralized institutions [7]. Specifically, a blockchain is composed of blocks, each containing transaction data and hash values, which are encrypted and verified before being recorded on the blockchain and cannot be tampered with. Blockchain adopts a consensus mechanism to maintain the security and integrity of data, which means that it requires collaborative verification from multiple nodes to complete the confirmation and recording of a transaction, effectively avoiding the risk of hacker attacks and data tampering [8].

Blockchain technology can provide a secure, decentralized, traceable, and highly transparent solution when it comes to data sharing. Because every data point recorded by blockchain is immutable and can be updated in real time among all parties involved, blockchain can ensure data integrity, consistency and traceability, as well as data privacy protection [9].

The application of blockchain technology can make the sharing of data more efficient and secure. Using blockchain to share data can eliminate middlemen while also encouraging sharing through enhanced security. In this case, data is ultimately stored in a distributed network, rather than restricted to a specific business application or an organization's application, where sharing and collaboration is performed among all parties involved. In a blockchain-based data-sharing network, only certain parts of the data can be disclosed to certain parties; In this case, security permissions and permissions will be assigned to ensure that unauthorized access will never be authorized in order to maintain data privacy and security [10].

Therefore, applying blockchain technology to educational resource sharing can greatly improve the credibility and security of educational resource data by utilizing its immutability and decentralization characteristics, effectively ensuring the privacy and security of educational resource data. Assist in the decentralization of educational resources, promote information sharing and exchange among the education industry, government agencies, and academia, thereby improving the efficiency and value of the entire industry.

3 Design of Real-Time Sharing Method of Mobile Education Resources

3.1 Construction of Ideological and Political Mobile Education Resource Sharing Blockchain Network

In the process of sharing ideological and political mobile education resources, the primary goal is to achieve secure sharing of mobile education resources, ensuring to the greatest extent that the shared ideological and political mobile education resource data will not be tampered with. Therefore, in the process of data sharing, it is necessary to rely on a complete ideological and political mobile education resource sharing blockchain network. Utilize the characteristics of blockchain technology to obtain decentralized data and uniformly name the data. The essence of blockchain is a decentralized and distributed storage database, which can trace the source during the traceability process. Reasonable deployment of blockchain in the ideological and political mobile education resource sharing network, and the use of asymmetric encryption methods for encryption measures when using blockchain technology for decentralized storage, can improve the application performance of the ideological and political mobile education resource sharing blockchain network.

Blockchain technology can determine the time stamp of the shared resource data of ideological and political mobile education through the identification of resource data, sort the resource data according to the time stamp, and generate the pointing sequence of blockchain technology according to the sorting result, providing an effective verification process for the data sharing of ideological and political mobile education resources. To ensure the security of data sharing process of ideological and political mobile education resources.

Figure 1 shows the basic architecture of ideological and political mobile education resource data sharing based on blockchain technology.

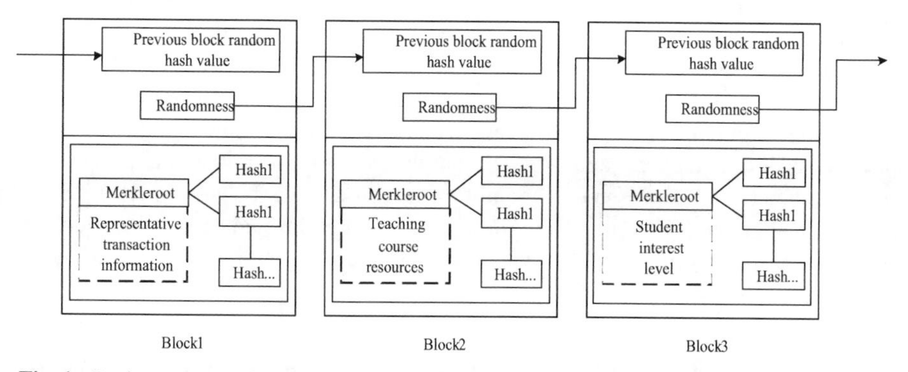

Fig. 1. Basic structure of data sharing of ideological and political mobile education resources

Observing Fig. 1, it can be seen that the basic architecture for resource data sharing built through blockchain technology can provide a secure sharing environment for resource data. In the basic architecture of ideological and political mobile education resource data sharing, additional resource data naming services, resource data authorization storage technology, and resource data distribution technology can be added to achieve equal information exchange behavior between the ideological and political mobile education resource data sharing end and the receiving end. The blockchain network architecture for ideological and political mobile education resource sharing is shown in Fig. 2.

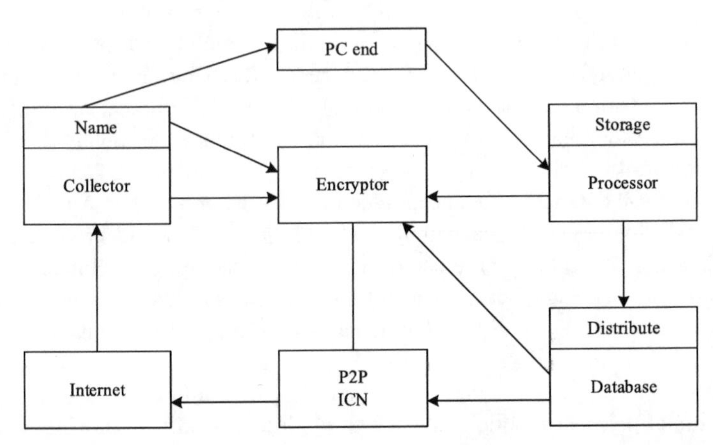

Fig. 2. Blockchain network architecture of ideological and political mobile education resource sharing

As can be seen from Fig. 2, the unified naming service of resource data is proposed on the basis of the resource sharing information model. This naming mechanism is centralized, open and independent. The purpose of this service is to process the naming of attributes that need to share resource data and reduce the calculation amount of resource data sharing. The implementation of the resource data authorization storage technology is based on the blockchain technology, which is mainly used to encrypt and store the data of ideological and political mobile education resources, so as to prevent the security of the data from being threatened due to the loopholes of the sharing network in the process of data sharing. The resource data distribution protocol is the restriction protocol of the ideological and political mobile education resource data peer-to-peer interaction network. This protocol simplifies the resource data sharing network into a point-to-point transmission network. Through data addressing and naming identification analysis, the security of ideological and political mobile education resource data sharing can be protected.

3.2 Feature Extraction of Ideological and Political Mobile Education Resources

In the ideological and political mobile education resources, the main collected resource data is video, audio, and text data. Therefore, this article extracts the features of video, audio, and text data from the collected multi-source data to obtain multi-source data features, making the fusion of multiple types of data more rapid and effective, and improving the efficiency of resource sharing. The specific feature extraction process is as follows:

(1) Video resource data: Using the FACET facial expression analysis framework, extract video features of ideological and political mobile education resources to form a video feature set;
(2) Audio resource data: In this paper, COVAREP acoustic analysis framework is used to extract audio features from ideological and political mobile education resources to form audio feature sets;
(3) Text resource data: Firstly, the spoken words in ideological and political mobile education resources are preprocessed through global word vector, and the spoken words are coded; Then, we use the short-short memory network to learn the spoken text with time correlation. Finally, the text after continuous learning is input into the neural network for convolution, and the convolution layer is used to extract finer grained text features. The feature extraction process of text resource data is shown in Fig. 3.

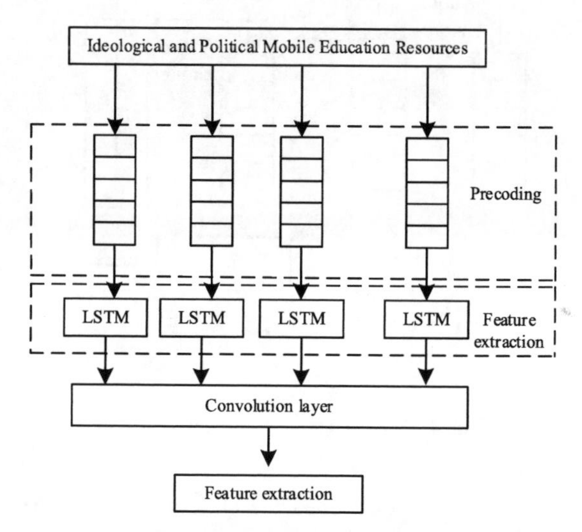

Fig. 3. Feature extraction process of text data for ideological and political mobile education resources

3.3 Feature Integration of Ideological and Political Mobile Education Resources

After completing the feature extraction of ideological and political mobile education resources, in order to improve the accuracy of resource sharing, Tucker decomposition algorithm is used. After feature fusion, the ideological and political mobile education resources are integrated together, so that the teaching resource features of different formats of ideological and political mobile education resources can be uniformly processed.

Set $z_v = \left(z_v^1, z_v^2, z_v^p\right)$ as the video data feature of the extracted ideological and political mobile education resources, $z_a = \left(z_a^1, z_a^2, z_a^q\right)$ as the audio data feature of the ideological and political mobile education resources, and $z_t = \left(z_t^1, z_t^2, z_t^m\right)$ as the text data feature of the ideological and political mobile education resources. Through the data fusion algorithm based on Tucker decomposition, three kinds of feature data sets are fused and the feature set Z after fusion is output. In the process of multi-source data fusion of ideological and political mobile education resources, a high-order tensor W is introduced, and the tensor W exists in the data feature space. Through the mode of the tensor W, the spatial mapping of data features of each ideological and political mobile education resources is realized. Therefore, by using the high-order tensor W, the characteristic modes of the data of ideological and political mobile education resources can be effectively corrected and the modal characteristics of the data of ideological and political mobile education resources can be memorized. The multi-source data fusion process of ideological and political mobile education resources is shown in Fig. 4.

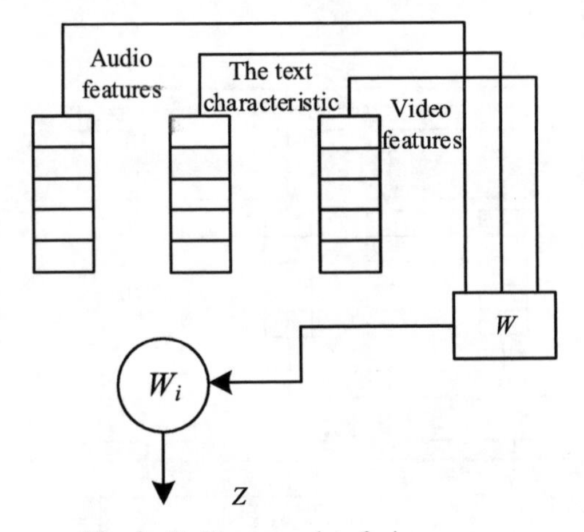

Fig. 4. Multi-source data fusion process

From Fig. 3, it can be seen that when the characteristics of the ideological and political mobile education resource data to be processed are z_a, z_t, and z_v in sequence, the higher-order tensor W is in a third-order form, that is, the three dimensions of tensor W correspond to the feature space of the teaching resource data features z_a, z_t, and z_v. At this time, by multiplying the teaching resource data features and corresponding feature spaces, the memory unit $\hat{W}$ of the current teaching resource data features can be obtained, Starting from this, fusion processing is carried out, and feature fusion is mainly achieved in three steps:

(1) Make modular multiplication of memory unit $\hat{W}$ and audio data feature z_a according to the first-order state to obtain a new memory unit $\hat{W}_1$ with feature z_a.
(2) Make modulo multiplication of memory unit $\hat{W}_1$ and video data feature z_v according to the second-order state to obtain a new memory unit $\hat{W}_2$ with features z_a and z_v.
(3) Perform modular multiplication of memory unit $\hat{W}_2$ and text data feature z_t according to the third order state to obtain the fusion tensor Z with three features, and represent the above steps through formula (1):

$$Z = ((W \times z_a) \times z_v) \times z_t \tag{1}$$

Through the operations of the above steps, data fusion of different types of teaching resources is realized, and the feature set Z after fusion is output.

3.4 Feature Training

In order to meet the demand for resource sharing and improve the sharing ability of data, a model based on fuzzy neural network is constructed by combining fuzzy theory with neural network according to feature categories. The network model consists of input layer, fuzzification layer, fuzzy rule layer, and output layer. In the fuzzy neural network model, the dataset of ideological and political mobile education resources is used as input samples. After processing at each layer, it is compared with the expected values, and the weights of each layer are adjusted. After continuous correction, the error of each training is minimized. This operation is repeated until all ideological and political mobile education resource features are trained, and the best dataset after training is obtained.

The number of input layers is determined by the number of parameters in the feature set Z, and then Z uses the fuzzy layer to obtain the membership function and fuzzy it. At the same time, the feature training is carried out in the fuzzy rule layer, so that the nodes with the same result are structured into a fuzzy quantity. The sum operation is adopted in the output layer to achieve the target fuzzy and output the result after decision level fusion. The training process is achieved through the following steps:

Suppose the first layer of the model's fuzzy input of ideological and political mobile education resources z_i in the following form:

$$z_i = [z_1, z_2, ..., z_n]^T \tag{2}$$

In the formula, n is a natural integer; T is the number of iterations.

In the first layer of the model, the fuzzy input is dimensionally reduced to generate the first layer network output. The specific output calculation is as follows:

$$I_i = z_i(i), i = 1, 2, ..., n \tag{3}$$

Where, I_i represents the data set of ideological and political mobile education resources after dimension reduction operation.

The processed I_i is transferred to the second layer of the model, where the fuzzy set of input feature quantity is analyzed, and the Gaussian function is adopted as the membership function. The specific output calculation is shown as follows:

$$\mu = \exp\left|\frac{-(z_i - c_{ij})^2}{2\sigma_{ij}^2}\right| \tag{4}$$

where, μ represents the fuzzy set of input characteristics; Among them, c_{ij} is the core of the Gaussian membership function of I_i; σ_{ij} is the width of the function.

Through the second layer of fuzzy processing, the fuzzy dataset h_i is output:

$$h_i = f\left(\sum_i^n W_{zy} \times I_i + \beta_i\right) \tag{5}$$

where, W_{zy} is the fuzzy weight coefficient; β_i represents fuzzy threshold; f is the weight coefficient of objective function.

In the third layer of the network model, fuzzy rules are designed, and each rule is described by nodes. In the output process of this layer, all fuzzy nodes are Defuzzification by summation, and fuzzy rule f_j is represented by formula (6):

$$f_j = \min(z_{1j}, z_{2j}, ..., z_{nj}, 1), z_{1j} = \alpha_1, \alpha_2, ..., \alpha_n \tag{6}$$

In the formula, z_{nj} represents the fuzzy feature set; α_n represents the applicability of fuzzy rules, and outputs the results of deblurring operations at this layer to obtain the optimal dataset after training, as shown in formula (7):

$$y_j = F\left(\sum_i^n W_{ij} \times g_i + \beta_j\right) \tag{7}$$

In the formula, y_j represents the best dataset after training; F represents the local weight of fuzzy rules; β_j represents the speed of deblurring; g_i represents the fuzzy median.

3.5 Real Time Sharing of Blockchain

After completing the feature training of ideological and political mobile education resources, the best data set was obtained. Based on this data set, real-time sharing of ideological and political mobile education resources is completed according to the resource sharing request issued by users.

Reg, short for registry regedit, is a registry script file. It is a command that adds, changes, imports, exports, and performs other operations on registry subkey information and registry key values. For a specific resource request Reg, the resource requester R will the request information as a transaction, stored in the blockchain system. The resource request record contains the ID of R, the request description, and the corresponding time stamp. The resource request description information includes R's description of the type of resources, data format, and size that need to be obtained for ideological and political mobile education resources. The purpose of this request transaction information is for resource retrieval to achieve real-time resource sharing. Specifically, based on the resource request information, a similarity matching algorithm is used to retrieve the data summary information of each node on the blockchain. The feature of the request information is matched with different types of teaching resource data to reduce request processing time and forward the request to meet the resource needs of the requester, achieving effective resource sharing. The specific sharing steps are as follows:

(1) Resource requester R publishes the sharing request Reg for ideological and political mobile education resources on the blockchain;

(2) Based on the resource request information, the blockchain system matches the data summary information of all nodes on the chain to query the resource characteristics related to the requested resource;
(3) Reg the resource sharing request to the queue of each resource provider;
(4) Each data provider iterates in response to the corresponding resource sharing request;
(5) The resource update record is used as the shared transaction information to generate the block, which is verified by the block broadcast network and reaches a consensus to upload the block to the chain;
(6) According to the consensus mechanism, the specific node is responsible for uploading the model file to the server for encrypted storage;
(7) Data requester R downloads the encrypted resource file, obtains the shared ideological and political mobile education resources after decryption, and the resource sharing task is finished.

4 Experimental Analysis

4.1 Experimental Setup

To verify the sharing ability of the method in practical application, this article selects the ideological and political mobile education resources of a local university as test data, and shares the teaching resources of ideological and political courses on the online education platform of the university to verify the resource sharing ability of the method in this article. Select the resource sharing method based on MB+tree test proposed in reference [5] as control group 1, and the resource sharing method based on alliance chain proposed in reference [6] as control group 2, and conduct experimental testing together with the method proposed in this article.

A large number of data of ideological and political course teaching resources are randomly selected and input into the self-built database. The data in the database is checked, and the invalid data less than 30 s in video resources and voice resources and less than 20 KB in text resources are eliminated. Finally, a total of 200 GB of ideological and political mobile education resources are retained in the database. The data sets of specific audio resources, video resources and text resources are shown in Table 1.

Table 1. Test data set

The data type	Data quantity/piece
Teaching video resource dataset	234220
Teaching Voice Resource Dataset	762342
Teaching Text Resource Dataset	324000
Teaching courseware dataset	354320
Course Dataset	213450
Course Extension Content Dataset	23500

On board inter@core In the Windows 10 system with i7 processor, this article runs the self built database using Matlab2019a simulation software, and uses three different methods to share and test the ideological and political mobile education resources in the database.

Taking the processing time of resource requests and the amount of shared data as the testing indicators for this experiment, the lower the processing time of resource requests, the higher the amount of shared data, the better the real-time sharing effect, and the higher the sharing efficiency. This proves that the application performance of the sharing method is better.

4.2 Resource Request Processing Time

When resource demanders initiate resource requests, the processing time of resource requests for ideological and political course teaching resources under different resource sharing methods is calculated, and the results are shown in Fig. 5.

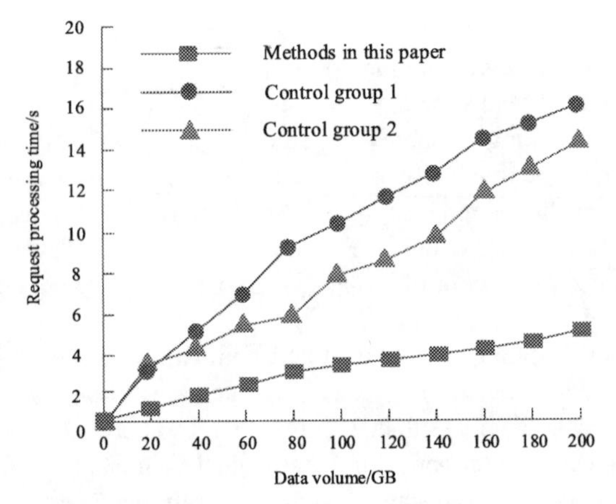

Fig. 5. Comparison of resource request processing time of ideological and political curriculum mobile education resources

Analyzing Fig. 5, it can be seen that the processing time of resource requests for mobile education resources in ideological and political courses will increase with the increase of data volume. The resource sharing method based on MB+tree test method from control group 1, when the data volume of mobile education resources in ideological and political courses is 100 GB, the processing time of resource requests is 10.25 s, and when the data volume is 200 GB, the processing time of resource requests is 16.33 s; The resource sharing method based on alliance chain from control group 2, when the data volume of mobile education resources in ideological and political courses is 100 GB, the processing time of resource requests is 8.02 s, and when the data volume is 200 GB, the processing time of resource requests is 14.25 s; The resource sharing method proposed in this article takes 3.22 s to process resource requests when the data volume of mobile education resources in ideological and political courses is 100 GB, and 5.20 s when the data volume is 200 GB, which is much lower than the two comparison methods and can always maintain within 6 s. This indicates that the blockchain based real-time resource sharing method proposed in this article has high sharing efficiency. This is because the sharing method designed in this paper adopts the similarity matching algorithm to retrieve the data summary information of each node on the blockchain according to the resource request information, and matches the characteristics of the requested information with the data characteristics of different types of teaching resources, which can reduce the request processing time to a certain extent and improve the efficiency of resource request processing. It can meet the requirements of real-time resource sharing.

4.3 Data Quantity of Shared Resources

When the number of uploaded mobile education resources for ideological and political courses continues to increase, the resource sharing ability of different methods is evaluated based on the resource sharing ability of different methods. The comparison results are shown in Table 2.

Table 2. Comparison of Shared Data Volume by Different Methods

Number of experiments/times	Amount of uploaded resource data/GB	Shared data volume/GB		
		Method in this paper	Control group 1	Control group 2
1	20	100.00	96.35	97.25
2	40	200.00	195.63	196.44
3	60	300.00	291.42	285.83
4	80	400.00	394.42	388.55
5	100	498.82	49.742	491.62
6	120	594.23	577.46	585.73
7	140	686.82	654.13	652.71
8	160	781.63	736.14	716.14
9	180	876.64	825.34	817.36
10	200	976.53	903.42	886.45

According to Table 2, under different experiment times, with the increasing amount of uploaded mobile education resources of ideological and political courses, the amount of shareable resources of different methods also decreases to a certain extent. Among them, when the amount of uploaded resources is less than 80 GB, effective and comprehensive sharing can be achieved by applying the proposed method. At this time, the amount of shareable data of the resource sharing method based on MB+tree test from control group 1 is 394.42 GB, and that of the resource sharing method based on alliance chain from control group 2 is 388.55 GB. Are significantly lower than the method in this paper. When the amount of uploaded resource data reaches 200 GB, the three literature methods compared have a significant decrease in the amount of shared data. Although the amount of shared data in this method has also shown a downward trend, the amount of shared data can still reach 976.53 GB, which can achieve 98% sharing of the total data. This indicates that this method has the ability to share a large amount of data and has a good sharing effect. This is because the method proposed in this article combines fuzzy theory with neural networks to construct a model based on fuzzy neural networks. Through feature training, the error of each training is minimized to obtain the optimal dataset after training, meeting the requirements of resource sharing and improving the ability to share data.

5 Conclusion

In view of the shortcomings of current resource sharing methods in the process of real-time sharing, this paper proposes a real-time sharing method of ideological and political mobile education resources based on blockchain.

(1) The ideological and political mobile education resource sharing blockchain network is constructed, the features of video resource data, audio resource data and text resource data in ideological and political mobile education resources are extracted, and the resource features are fused based on the data fusion algorithm of Tucker decomposition; Feature training is conducted based on fuzzy neural network model, and real-time sharing of ideological and political mobile education resources is completed based on blockchain technology.

(2) The experimental results show that when the data volume is 200 GB, the processing time of resource request is 5.20 s, and the method proposed in this paper can realize the sharing of 98% of the total data volume, with high sharing efficiency, good sharing effect and good practical application performance.

References

1. Qiancheng, D., Nan, Z.: Construction of an off campus practical teaching resource sharing platform for ideological and political courses. Teach. Ref. Middle School Politics **03**, 83–85 (2022)
2. Fu, J., Zhang, Z., Guo, Y., et al.: Research on sharing mechanism of practical teaching resources from value-added perspective. Experim. Technol. Manage. **37**(03), 222–225+229 (2020)
3. Juan, L., Yun, F.: Research on the sharing strategy of practical teaching resources for school enterprise collaborative construction. Chin. Vocat. Techn. Educ. **08**, 76–80 (2020)
4. Zhang, Y.: Research on public education resource sharing mechanism based on video live broadcasting technology. Electron. Compon. Inform. Technol. **5**(08), 231–232+237 (2021)
5. Tang, S.: National cultural resource sharing model based on MB+Tree test method. Inform. Sci. **39**(10), 95–100+117 (2021)
6. Li, S., Jiaqi, Y.: A research on the co-construction and sharing model of "the Belt and Road" thematic database resources based on consortium blockchain. Res. Lib. Sci. **04**, 40–46 (2022)
7. Jie, M., Qiuhui, X.: A design of archival information sharing platform based on blockchain. Beijing Arch. **06**, 26–28 (2022)

8. Chi, Z., Rui, W., Junchao, C., et al.: Application design of marine data resource sharing based on block chain. Sci. Technol. Rev. **38**(21), 69–74 (2020)
9. Dongpo, G., San, L., Qing, L., Dongbo, Z.: Research on crowdsourcing model of digital education resources from the perspective of social practice. China Audio Visual Educ. **18**(02), 51–60 (2021)
10. Fulian, X., Li, L., Dongfeng, M.: High quality education resource sharing path based on the "three classrooms." Ningxia Educ. **03**, 60–61 (2023)

Architecture Design of Employment Education Network Platform Based on Blockchain Technology

Weiwei Zhang(✉)

Changchun University of Finance and Economics, Changchun 130000, China
zhangww6565@163.com

Abstract. The Employment education network platform because of its convenience its business has been rapid development, too much platform integration and paid knowledge spread also breeds the platform and users and users and users with data sharing security issues. At the same time, the development of Internet technology also increases the education platform knowledge rights protection and fight against piracy difficulty. To this end, the employment education network platform architecture based on blockchain technology is designed. Based on blockchain technology, an employment and education network platform composed of network layer, infrastructure layer and business application layer was built. Intelligent contract was compiled by online editor Remix to design intelligent contract function of employment and education network. Based on the integral incentive mechanism contract, the teaching resource sharing function is designed. Design employment education data recommendation function based on residual neural network. The test results of the platform show that the fastest response time of the platform is 1.24 s, and the maximum amount of data that the platform can manage at the same time is 40000 Byte when the user visits are 5000 times. The platform can operate normally and has extremely high operating efficiency.

Keywords: Blockchain Technology · Employment Education · Network Platform · Platform Architecture · Residual Neural Network

1 Introduction

At present, with the continuous development and application of information technology, people's demand for knowledge and skills is also increasing. Especially when it comes to employment and career development, people are paying more attention to the quality and effect of education. For the job market, traditional recruitment methods also have some problems, such as cumbersome recruitment process, asymmetric information and high recruitment costs [1]. At the same time, with the popularization and development of Internet technology, many new Internet enterprises began to emerge. They have profound insight into and driving force to open up the Internet job market, and are in urgent

L. Yun et al. (Eds.): ADHIP 2023, LNICST 549, pp. 96–109, 2024.
https://doi.org/10.1007/978-3-031-50549-2_7

need of a new, efficient recruitment mode that can meet the recruitment needs of enterprises. Therefore, under the current social background, it has become an urgent issue to study the way of employment education based on the network platform and establish a more fair, just and transparent employment mechanism. The network platform can not only realize information openness and transparency, ensure consensus among students, enterprises and educational institutions on the authenticity and accuracy of information, but also provide students with more personalized and professional career development suggestions through big data analysis and other means, to help enterprises find the right talents more quickly and accurately. Therefore, the research and exploration of the architectural design of an employment education network platform has become one of the urgent problems to be solved.

Liu Xuejing [2] proposed the design of ideological and political education platform based on network multimedia technology. With the control of the computer, the compression technology is taken as the core technology, and the camera with adjustable position and the pickup with high sensitivity are installed to collect all kinds of video and audio materials, complete the compression and preservation processing in the hard disk video recorder, and use the network transmission to the server point-to-point connection and internal multipoint broadcasting to realize the on-demand and live broadcasting in the platform. Peng Xiaohua et al. [3] proposed the design of an online education platform for multi-concurrent high-speed communication. By analyzing the buffer and bandwidth at the receiving end, network delay and loss rate, the distribution shunt strategy of communication network link was designed. Then, combined with Greedy algorithm, Pareto distribution and queuing model theory, a multi-stream concurrent communication transmission control model in heterogeneous network environment is established. Finally, the firefly swarm algorithm is used to simulate the optimal solution target and complete the overall design of the system.

Blockchain technology provides a new way to solve the above problems. The characteristics of blockchain technology, such as decentralization, intamability and transparency, provide guarantee for information exchange and trust building among schools, enterprises and students, and also provide new opportunities and possibilities to realize information sharing and collaboration among students, schools and enterprises [4]. The employment education network platform based on blockchain technology can improve the traditional way of employment education by building trust, improving the reliability and security of information, and improving the effect of education and the quality of talents. Therefore, the research and exploration on the architecture design of the employment education network platform based on blockchain technology has both practical application value and theoretical academic value. A three-tier architecture pattern is adopted: Network layer, infrastructure layer and business application layer, through P2P network technology to provide services for the communication between nodes in the blockchain, establish the employment education network platform architecture, through the employment education network intelligent contract function, teaching resource sharing function, employment education data recommendation function, realize the employment education network platform based on blockchain technology; Furthermore, the education resources of students and employment education courses are embedded in the residual neural network model to predict the degree of students' preference for target resources. Experimental verification shows that the fastest response time of the platform designed in this paper is 1.24 s, which can effectively solve the problems of inaccurate information matching and inaccurate information often existing in traditional employment information platforms.

2 Employment Education Network Platform Architecture Design

The main purpose of the construction of the employment education network platform is to provide more convenient, efficient and comprehensive vocational skills training and employment services [5], help the majority of workers adapt to economic development and changes in employment situation, improve employment competitiveness and self-development ability, and promote employment stability and economic development. In addition, online platforms can also facilitate enterprises' recruitment and talent matching, speeding up industrial restructuring and talent cultivation. Therefore, this paper builds an employment education network platform based on blockchain technology. On the basis of making full use of high-quality teaching resources in various universities, it provides a convenient network platform for students and teachers, which can help users develop personalized learning plans more quickly and accurately according to their own learning conditions. At the same time, it can also recommend courses, so that users can obtain a wider range of educational opportunities [6]. In this paper, when using blockchain technology to design the overall framework of the employment education network platform, the idea of layered architecture should be followed, as shown in Fig. 1:

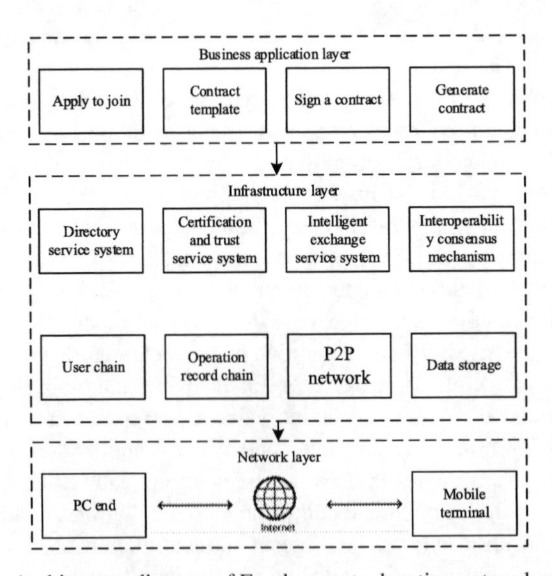

Fig. 1. Architecture diagram of Employment education network platform

As shown in Fig. 1, the employment education network platform designed in this paper mainly adopts three-level architecture mode: network layer, infrastructure layer and business application layer. The network layer mainly realizes information interaction and provides services for communication between nodes in the blockchain through P2P network technology, so as to achieve effective point-to-point communication. The network layer transmits the business information of each node and synchronizes the block information of adjacent nodes, which effectively improves the robustness of the system so that the system will not crash due to the attack of the information of a node. In this way, all nodes in the blockchain network can jointly maintain the system and manage the system data equally.

Infrastructure layer is the core layer of the whole model, which can provide support for basic services of mutual trust and interoperation of employment and education resources. It mainly covers three service support systems: blockchain directory service system, certification and trust service system and intelligent exchange service system. The blockchain directory service system provides overall management for all education resources on the blockchain network, carries out unified registration of employment education resources, provides organization and storage of information resources, as well as educational resource inquiry and jump services; The certification and trust service system provides unified trust services for all entities and educational resources on the blockchain network, and solves the mutual trust and trust between all nodes.

Sharing and other issues; The intelligent exchange service system provides a consensus mechanism to realize the consistency of information of all nodes.

The business application layer encapsulates the application scenario of the employment and education network platform, and provides corresponding programs and interfaces for the needs of users. Users interact with the blockchain through the protocols and contracts built into the application. To sum up, this paper uses blockchain technology architecture to design a three-tier employment and education network platform, which can improve user data security, data sharing efficiency, education certificate credibility, institutional fairness, and create a better learning experience.

3 Function Realization of Employment Education Network Platform

3.1 Intelligent Contract Function of Employment and Education Network

Smart contracts, the most important feature on the platform, are automated programs that can safely interact with the blockchain, avoiding the problem of human intervention. Smart contracts used on the platform can facilitate information exchange and transactions between educational institutions and students, ensuring the security and fairness of the platform. Strictly reviewing the indicators, automatically issue the qualification level certificate formulated by the authorities, through the pre-set certification standards and systems, without manual intervention, the system is on duty all day long, under the supervision of enterprises, schools, society and other parties to complete the issuance of certificates. The e-certificate based on block chain is the only certificate for authentication of the authenticity of the certificate. By tracing the block information attached to the e-certificate, any educational institution and recruitment unit can track and query the online education of students learning and certificate authenticity, in order to ensure the credibility and effectiveness of the e-certificate, to solve the problem of certificate fraud.The implementation of intelligent contract function of employment and education network is shown in Fig. 2.

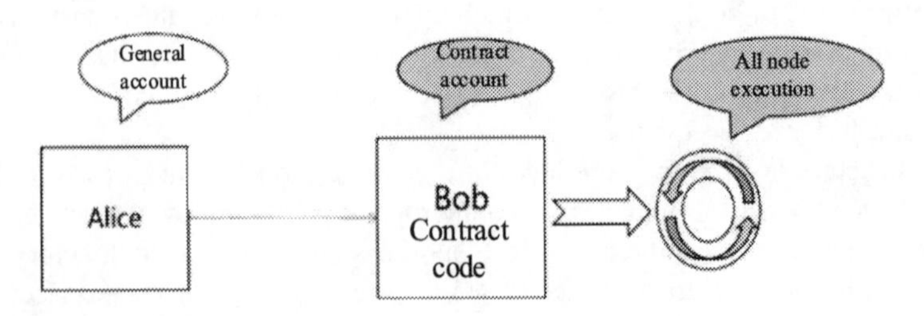

Fig. 2. Employment and Education Network smart contract function

The smart contract is compiled through the online editor Remix to obtain the corresponding binary coding file and application binary interface (ABI), and submit the deployment contract request on any blockchain node. The ABI interface defines the name of the function that the smart contract can call, input/output data format, gas consumption, and other parameters. Then, after the request is recorded on the chain, the address and block hash value containing the intelligence contract are returned to the request originator. The nodes in the synchronous state will update the block height in real time to obtain the successful deployment of smart contract information. Platform users can invoke smart contracts using Python's Web3 library to initiate platform login requirements. The specific operations are as follows: 1) Check node interconnection status; 2) Call the inter-process communication (IPC) file when the check status is synchronous; 3) Unlock the account and check the address. If the check is valid, continue; if it is invalid, unlock the account again; 4) Define ABI interface of intelligent energy contract; 5) Define the deployment address of the intelligent energy contract; 6) A real example of the Contract between ABI interface and deployment address; 7) Use Web3 library Contract class function to call smart contract.

3.2 Teaching Resource Sharing Function

Teaching resource sharing functions include personal data sharing, teaching resources sharing, teaching results sharing and evaluation system sharing. Specifically, students' personal information, class schedule, transcript and other data can be stored on the blockchain of the platform. These resources can be independently authorized by students and shared with recruitment companies or other educational institutions to demonstrate their learning process and results. Educational institutions can upload their courses, courseware, handouts and other teaching resources to the blockchain of the platform and share them with other educational institutions. Students' graduation works, project experience and other teaching results can be stored on the platform's blockchain and shared with recruitment companies or educational institutions. This can enhance the competitiveness of students in employment and facilitate the translation of learning results into employability and work experience. The platform can store students' evaluation records and evaluation results on the blockchain for reference by other educational institutions and facilitate recruitment enterprises to evaluate and screen students.

Therefore, in the employment education network platform designed in this paper, a point incentive mechanism is set up to ensure that the rights of uploaders and downloaders of educational resources of employment education courses can be guaranteed. The point incentive mechanism mainly takes the number of educational resource uploads of employment education courses as the evaluation index, and provides corresponding points rewards to users who upload resources. Therefore, the course resource sharing function of the platform is set up in this paper, as shown in Fig. 3:

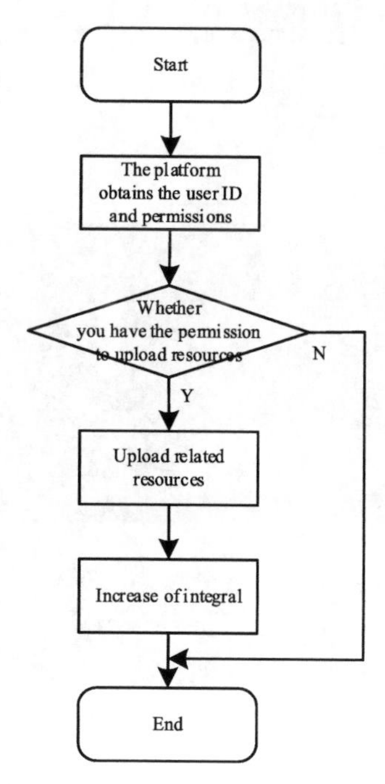

Fig. 3. Business logic diagram of the integral incentive mechanism contract

As shown in Fig. 3, the intelligent contract with the incentive mechanism of points designed in this paper gives corresponding rewards mainly through the user's resource uploading behavior. When a teacher registers an account on the platform, the platform will call the insertScore contract as the initial score under the current account, and then give corresponding rewards according to the initial score. When teachers log in to the platform and choose to upload education resources of employment education courses, the platform will call updateScore interface function to modify user credits and increase corresponding credits according to the number of resources uploaded by teachers [7]. Of course, in the teaching resource sharing function designed in this paper, if the administrator finds infringement or other problems in the employment education course resources uploaded by the teacher, the platform can implement the remove Course interface function to delete the uploaded course resources and return the result to the teacher's account. The above intelligent contract of incentive mechanism is used to realize the business logic of the teaching resource sharing function. The platform will call the corresponding contract to ensure the smooth implementation of resource sharing operations such as uploading and downloading by users.

3.3 Employment Education Data Recommendation Function

In order to improve the utilization rate of teaching resources, the data recommendation function is set up on the employment education network platform designed in this paper. By studying the corresponding teaching data recommendation algorithm, this paper decides to use residual neural network to realize the employment education data recommendation function of the platform [8]. Residual neural network is a special deep learning model. The implementation of employment education data recommendation function is shown in Fig. 4.

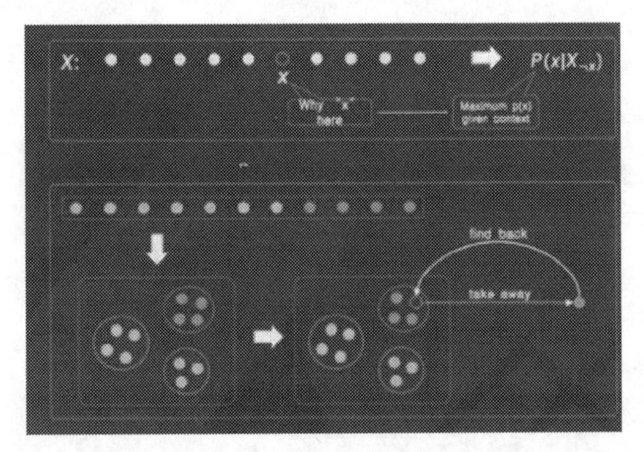

Fig. 4. Recommendation function of employment education data

In the process of training and learning, problems such as model degradation and gradient dissipation often occur in the ordinary deep neural network model, which affects the recommendation effect of employment and education data. Therefore, this paper introduces residual neural network to alleviate problems such as gradient dissipation and model degradation caused by the deepening of network layers. Improve the operation performance of the employment education network platform [9, 10]. When the employment education data recommendation function of the residual neural network design platform is used, the text information features of the education resources of the employment education course are extracted through the network. In this process, two parallel extractors with independent weights are adopted in the design platform, that is, the title and abstract information of the education resources of the employment education course are extracted respectively as the final text features. As follows:

$$T = Q([B; Z]) \tag{1}$$

where, T is the extracted text information characteristics of educational resources of employment education courses; Q is the fully connected layer of the residual neural network model [11, 12]; B is the title features of teaching resources. Z is the abstract characteristics of teaching resources. Then, the potential of students and teaching resources are obtained through the residual neural network. As students' learning interest, the potential factors of educational resources of employment education courses are assumed to be Y, and the calculation formula is as follows:

$$Y_{t \leftarrow T}^{L+1} = f\left(\omega \cdot Y_{t \leftarrow T}^{L} + p\right) \tag{2}$$

where, $Y_{t \leftarrow T}^{L+1}$ is the potential factor of text information feature T of educational resource t of employment education course in layer $L+1$ of residual neural network; $Y_{t \leftarrow T}^{L}$ is the potential factor of text information feature T of educational resource t of employment education course in layer L of residual neural network; f is the aggregate function; ω is the weight matrix; p is biased. At the same time, according to the dynamic preferences of students within a certain period of time, the potential factor of students' learning interests is determined to be X [13]. Finally, according to the students' learning interest, the recommendation of educational resources of employment education courses can be realized. The expression is as follows:

$$H_{i,t} = Loss(X, Y) \tag{3}$$

where, $H_{i,t}$ is student i preference for educational resources t of employment education course [14, 15]; *Loss* is the loss function of residual neural network. As shown in Formula (3), students' preference for target resources can be predicted by embedding the educational resources of students and employment education courses in the residual neural network model.

4 Platform Test

In order to verify whether the employment education network platform based on blockchain technology designed in this paper has practical effects, this paper tests the platform. The final test results will be developed in the form of a comparison between the ideological and political education platform based on network multimedia technology proposed in literature [2], the multi-concurrent high-speed communication online education platform proposed in literature [3], and the employment education network platform based on blockchain technology designed in this paper. The specific platform test process and results are as follows.

4.1 Test Procedure

Before the platform test, this paper first debugs the platform hardware, mainly the mobile scene debugging. In a mobile phone or tablet computer, test the movement of the ARM-7 chip. The ARM-7 chip is a 32-bit processor based on RISC architecture, which is characterized by high performance, low power consumption and low cost, and generally runs at about 100 MHz. In order to debug the Linux controller, the serial cable is used to connect the computer to the serial interface of the device, and ensure that the correct serial port type and correct data cable are selected. Open a terminal tool on your computer and select the correct serial port number and baud rate. Platform interface operation processing is required to enter the platform's write mode by sending specific commands or pressing specific buttons on the device (such as the RESET button).After the hardware debugging is completed, connect it with the platform software and enter the login interface as shown in Fig. 5.

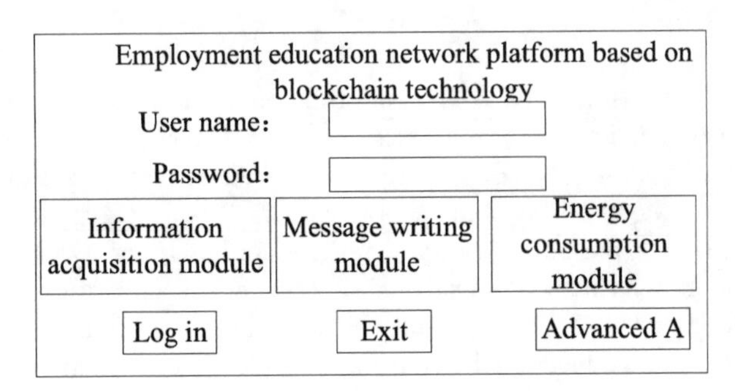

Fig. 5. Platform login interface

As shown in Fig. 5, employment and education data information can be collected in real time in the information collection module; Information writing module can modify, write, Delete employment and education data information at will; The energy consumption module can analyze the energy consumed by the platform. Each module will have a template of corresponding information data. If the template is consistent with the actual information data, it proves that the software debugging is complete; otherwise, the software needs to continue debugging until the actual information data is consistent with the template. After the hardware and software debugging of the platform is completed, the normal operation of the platform can be guaranteed.

4.2 Test Result

In order to test the employment and education network platform based on blockchain technology as a whole, this chapter conducts functions and performance from two aspects. One is to verify whether the platform designed in this paper can meet user expectations, and the other is to verify whether various indicators of the platform are normal. The functional test is mainly based on the earlier use-case analysis. The following is the main functional test content of the employment education network platform, as shown in Table 1.

Table 1 shows that the functions of the employment education network platform designed in this paper can be used normally. On this basis, in order to ensure the smooth and efficient operation of the employment education network platform, the platform performance will be tested in detail below. Taking the response time as an example, the ideological and political education platform based on network multimedia technology proposed in literature [2] and the online education platform based on multiple concurrent

Table 1. Functional test cases of employment education network platform

Function	Test content	Expected output	Test results
User registration	Input necessary information as required by the platform, such as user name and password, to verify the expression information	The authentication is successful, prompting the user	Normal
User login	Log in to the platform using a registered user account	Log in successfully and go to the page	Normal
User management	Modify and delete the personal information of registered users	The deletion or modification succeeds, and a result message is displayed	Normal
Integral management	Users view personal credits and administrators modify user credits	Show credits and details, modify successfully and update the modified credits	Normal
Resource upload	Upload local employment education and teaching resources to the platform	Successful upload	Normal
Resource download	Download platform resources to a local directory	Successful download	Normal
Data recommendation	Browse the resources on the platform and click the "Data recommendation" button	The platform can recommend appropriate course resources according to the user's learning situation	Normal
Resource search	Use keywords to search for the course resources you want to view	Displays the resource information matching the keyword	Normal

high-speed communication proposed in literature [3] are selected as the control group to compare with the platform designed in this paper, and the performance of each platform is recorded in time during the operation process, as shown in Table 2:

It can be seen from the data in the table that the response time of the designed platform in this paper is the fastest 1.24 s, and the response time of the platform in literature [2] and the platform in literature [3] is the fastest 2.32 s and 3.29 s. The performance of the designed employment education network platform in this paper is superior to that of the control platform in all aspects. With the increasing number of user requests, the platform designed in this paper can not only maintain a fast response speed stably, but also run stably, which indicates that the platform designed in this paper has excellent performance and can effectively solve the problems of inaccurate information matching and inaccurate information that often exist in traditional employment information platforms.

Table 2. Performance test results of employment education network platform

Performance index	Platform	Number of requests/times		
		10	50	100
Response time /s	Design platform	1.24	1.31	1.28
	Platform of literature [2]	2.32	2.86	3.17
	Platform of literature [3]	3.29	4.31	5.93

Under the above testing environment, this paper selects different user visits, and compares and tests the ideological and political education platform based on network multimedia technology proposed in literature [2], the online education platform based on multiple concurrent high-speed communication proposed in literature [3], and the maximum amount of data that can be managed simultaneously by the employment education network platform based on blockchain technology designed in this paper. The test results are shown in Table 3.

Table 3. Test results

User access	The maximum amount of data /Byte that can be managed simultaneously by the platform in reference [2]	The maximum amount of data /Byte that can be managed simultaneously by the platform in reference [3]	This paper designs the maximum amount of data /Byte that the platform can manage simultaneously
5,000 times	11000	12000	40000
10,000 times	10000	11000	35000
15,000 times	9000	8000	30000
20,000 times	8000	7000	25000
25,000 times	7000	6000	20000
30,000 times	6000	5000	20000
35,000 times	5000	4000	20000

As shown in Table 3, the maximum amount of data that can be simultaneously managed by the platform in reference [2] is small. When the user visits are 5,000 times, the maximum amount of data that can be simultaneously managed by the platform is 11,000 Byte. When the user visits are 35,000 times, the platform can manage the least amount of data simultaneously, which is 5000 bytes. In reference [3], the maximum amount of data that can be managed by the platform at the same time is small. When the user visits are 5,000 times, the maximum amount of data that can be managed by the platform at the same time is 12,000 bytes. When the user visits are 35,000 times, the platform can manage the least amount of data simultaneously, which is 4000 bytes. This proves that the maximum amount of data that can be simultaneously managed by the platform of literature [2] and the platform of literature [3] will decrease with the increase of user access, and the effect of data management will decline accordingly. However, the platform designed in this paper can manage a large amount of maximum data at the same time. When the user visits are 5,000 times, the platform can manage the maximum amount of data at the same time, which is 40,000 bytes. When the user visits are 25,000 times, 30,000 times and 35,000 times, the amount of data that can be managed by the platform at the same time is maintained at a stable state of 20,000 Byte. This proves that the maximum amount of data that can be managed by the design platform in this paper will not decrease with the increase of user access, and the data management effect is good, which is in line with the purpose of this paper.

In the above test environment, the test interval was set as 2 s, and the consumption of calculated and stored resources was calculated during the test period of 100 requests. The test results are shown in Fig. 6.

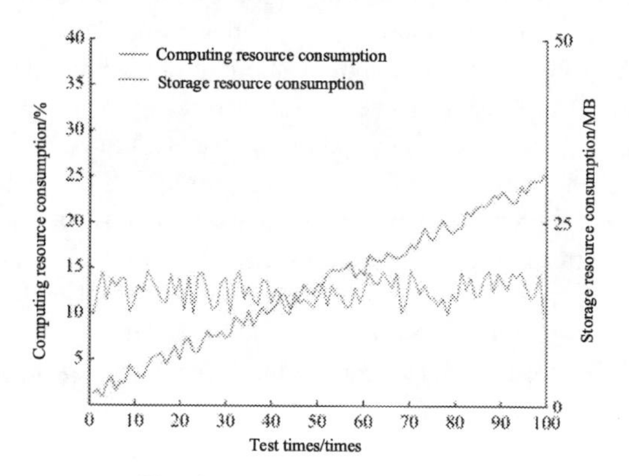

Fig. 6. Performance test results

As shown in Fig. 4, during 100 tests, the computing resource consumption remained within the range of 10% ~ 15%, and the storage resource consumption basically had a linear relationship with the total number of requests initiated. In conclusion, under the above technical framework, the request confirmation time of second time level and low resource consumption can basically meet the application requirements of the employment education network platform on a certain scale. However, in a larger scenario scale, there is still a triangle contradiction of decentralization, scalability and security, and the blockchain technical architecture still needs to be further optimized.

5 Conclusion

At present, the development of Internet technology has brought new opportunities and challenges to employment education. Various online employment education platforms have emerged one after another, but these platforms have problems with security and credibility, such as user privacy disclosure, certificate falsification and other risks. Therefore, this paper designs the employment education network platform architecture based on blockchain technology. Based on blockchain technology, the employment education network platform composed of network layer, infrastructure layer and business application layer is constructed. Through P2P network technology to provide services for communication between nodes in the blockchain, to achieve effective point-to-point communication. The smart contract is compiled via the online editor Remix, the corresponding binary-coded file and application Binary interface (ABI) are obtained, and the deployment contract request is submitted on any blockchain node. Teaching resource sharing includes personal data sharing, teaching resource sharing, teaching results sharing and evaluation system sharing. The potential factors and teaching resources of students are extracted by residual neural network. A three-layer employment education network platform is designed, which can not only manage the background easily and quickly, but also realize the intelligent contract, teaching data sharing, data recommendation and other functions of the employment education network. At the same time, this paper verifies through platform testing that the platform has high reliability, can maintain efficient operation efficiency, and provides strong help for the development of employment education information management in our country. However, this study has some limitations, due to the decentralized nature of the blockchain and a large amount of encrypted computing, its processing power is limited and processing speed is slow. The popularization of blockchain technology requires more advanced hardware equipment and higher technical level, thus increasing the cost of operation and maintenance of the platform and the threshold of use. At present, the penetration rate of blockchain application is low in China, and many users do not understand blockchain technology, which also limits the promotion and application of the employment and education network platform based on blockchain. In the follow-up research, we can consider the integration of blockchain technology, artificial intelligence technology and 5G communication technology from the perspective of development, giving full play to the advantages of 5G transmission bandwidth and the characteristics of artificial intelligence self-learning, guide blockchain technology into more traditional industries and help industrial upgrading.

References

1. Haji, S., Hafidh H. A . Youth employment creation factors in the EA countries data and policy analysis. Asian J. Educ. Soc. Stud. 2021, 18(3) (2021):13–29
2. Liu, X.J.: Design of ideological and political education platform based on network multimedia technology. Tech. Autom. Appl. **41**(3), 177–181 (2022)
3. Peng, X.H., Li, K.L., Zhong, L.H., et al.: Design of online education platform for multi-concurrent High-speed communication. Mod. Electron. Tech. **44**(18), 92–96 (2021)
4. Hasan, M.K., Alkhalifah, A., Islam, S., et al.: Blockchain technology on smart grid, energy trading, and big data: security issues, challenges, and recommendations. Wirel. Commun. Mob. Comput. **2022**(9), 1–26 (2022)
5. Esen, M., Seren, G.Y.: The impact of gender inequality in education and employment on economic performance in Turkey: evidence from a cointegration approach. Equality Diversity Inclusion: Int. J. **41**(4), 592–607 (2022)
6. Scanlon, G., Doyle, A.: Transition stories: Voices of school leavers with intellectual disabilities. Br. J. Learn. Disabil. **49**(4), 456–466 (2021)

7. Shenfu, C., Guofu, L., Changzai, R., et al.: Remote resource sharing simulation of interactive experiment platform based on blockchain. Comput. Simul. **39**(6), 233–237 (2022)
8. Mcway, R.: Cruising through school: general equilibrium effects of cruise ship arrivals on employment and education. Am. Economist **67**(1), 5–23 (2022)
9. Krause, J.S., Dismuke-Greer, C.L., Rumrill, P., et al.: Job retention among individuals with multiple sclerosis: relationship with prediagnostic employment and education; demographic characteristics; and disease course, severity, and complications. Arch. Phys. Med. Rehabil. **103**(12), 2355–2361 (2022)
10. Elsayed, M.S., Le-Khac, N.A., Jurcut, A.D.: Dealing with COVID-19 network traffic spikes [cybercrime and forensics]. IEEE Secur. Priv. Mag. **19**(1), 90–94 (2021)
11. Fan, L., Xia, M., Huang, P., et al.: Research on educational information platform based on cloud computing. Secur. Commun. Networks **2021**(1), 1–11 (2021)
12. Liu, H.: Application and strengthening strategies of network resources in the construction of teaching platform. J. Phys. Conf. Ser. **1915**(4), 042056 (2021)
13. Hoi, V.N., Hang, H.L.: Student engagement in the facebook learning environment: a Person-centred study. J. Educ. Comput. Res. **60**(1), 170–195 (2022)
14. Milton, P.: Intelligent and deep learning collaborative method for E-Learning educational platform using Tensorflow. Turkish J. Comput. Math. Educ. (TURCOMAT), 2021, 12(10):https://turcomat.org/index.php/turkbilmat/article/view/4881
15. Gheorghiu, C. I., Deaconu, I. D., Chirila, A. I., et al. Educational platform for an underground smart parking using programmable logic controllers. In: 2021 International Conference on Applied and Theoretical Electricity (ICATE). 2021
16. Diguiseppi, G., Clomax, A., Rampton Dodge, J., et al.: Social network correlates of education and employment service use among youth experiencing homelessness: a longitudinal study. Child. Youth Serv. Rev. 129 (2021)
17. Mladenov, V., Chobanov, V., Seritan, G. C., et al.: A flexibility market platform for electricity system operators using blockchain technology.Energies, 15 (2022)
18. Mishra, M. K., Balamurugan, M., Roy, R. R., et al.: Blockchain technology empirical studies on the demand of distributed network. J. Phys. Conf. Ser., 1964(4):042012 (2021)

Research on Blockchain Based Data Sharing of Teaching Resources in Higher Vocational Mobile Education

Xiaoli Wang[1](✉) and Mengxing Niu[2]

[1] Sanmenxia College of Social Administration, Sanmenxia 472000, China
16603989677@163.com
[2] Sanmenxia Polytechnic, Sanmenxia 472000, China

Abstract. The secure sharing of mobile education resources in vocational colleges refers to sharing student learning materials or other academic materials who have received vocational education in a secure and authorized manner to meet students' learning resource needs. Against the background of mobile internet and information-based society, vocational education needs to pay more attention to the sharing of educational resources to improve students' academic abilities and comprehensive qualities and enhance the quality of education and teaching. Secure sharing addresses the security issues of shared information. Therefore, a blockchain-based method for sharing data on vocational mobile education has been proposed. A compound chaotic password was constructed using the wavelet compound chaotic password matrix. Combined with blockchain technology, a secure model for sharing vocational mobile education teaching resource data was established. Experimental results showed that the computation time of this method was less than 150 s and can effectively increase operational efficiency. Furthermore, this system has low time and space overheads and high throughput, indicating that the proposed method can ensure secure sharing of educational resources.

Keywords: Blockchain Technology · Higher Vocational Mobile Education Teaching Resources · Data Sharing · Chaos Encryption · Digital Signature

1 Introduction

Vocational mobile education is one of the current development trends and future key directions, and has become a hot field of education development and teaching reform. Faced with the rapid development of vocational mobile education and the rapid updating of teaching resources, educational resource sharing has become an important solution [1, 2]. However, achieving the goal of sharing teaching resources in vocational mobile education still faces various challenges and difficulties. Due to the centralized operation and management method requiring the use of third-party platforms for data exchange, there is also a risk of data being copied or resold by third parties in the data sharing of vocational mobile education teaching resources. Education resources cannot be effectively protected, and the risk of leakage is relatively high. In order to effectively address

L. Yun et al. (Eds.): ADHIP 2023, LNICST 549, pp. 110–122, 2024.
https://doi.org/10.1007/978-3-031-50549-2_8

the above issues, relevant experts have conducted extensive research on the security sharing of teaching resources in vocational mobile education.

For example, Miao et al. [3] proposed a secure data sharing method for the Internet of Things based on joint deep reinforcement learning, using sensitive task decomposition and a layered asynchronous joint learning framework to achieve efficient and secure data sharing. At the same time, using deep reinforcement learning technology, select participants with sufficient computing power and high-quality data sets, and share local data models to achieve reliable data sharing, while protecting data privacy. Zhang et al. [4] proposed a cloud trust driven hierarchical sharing method for IoT information resources, which utilizes complementary judgment matrices and minimum non negative deviation values to obtain the optimal weight vector. At the same time, information security is considered and the information resource acquisition process is designed to ensure data reliability. By preprocessing the heterogeneous data collected by RFID devices and using trust based adaptive detection algorithms, evaluate the credibility of trust driven algorithms in limited resources and heterogeneous network environments, and achieve the sharing of data resources. Manogaran et al. [5] proposed a blockchain assisted secure data sharing model for the intelligent industry based on the Internet of Things, which manages the security of data collection and dissemination, including inbound and outbound. Identify bad data sequences through recursive learning techniques. Outbound security measures use blockchain information based on reputation and sequence differentiation for end-to-end authentication, in order to achieve resource information sharing.

Combining the above research methods, a blockchain based data sharing method for vocational mobile education teaching resources is proposed. In the wavelet transform space, the wavelet coefficients of plaintext information are encrypted by compound chaos, and the security model of educational resource data sharing is established by combining blockchain technology and hash function. The hash function is used to verify the data integrity, and the digital signature and hash value are uploaded to the blockchain for storage, so as to realize the safe sharing of teaching resource data of higher vocational mobile education. The experimental results show that this method completes the calculation in a relatively short time and effectively improves the system's operational efficiency. In addition, the proposed method has small time and space costs, high throughput, and can ensure the safe sharing of educational resources.

2 Safe Sharing Methods for Teaching Resources in Vocational Mobile Education

2.1 Information Resource Encryption

In order to effectively ensure the safe sharing of teaching resources in higher vocational mobile education, two one-dimensional chaotic systems are selected to establish a complex chaotic system. In order to improve the encryption performance of the chaotic system, the correlation coefficient and uniformity index are selected to establish a comprehensive objective function, and the adaptive chaos immune particle swarm optimization algorithm is introduced to optimize the control parameters in the system.

Logistic mapping is a one-dimensional discrete dynamical system, and the analytical equation corresponding to Logistic mapping is shown in formula (1):

$$z_{n+1} = \beta z(1 - z_n) \tag{1}$$

In the equation, z represents a chaotic variable; β represents chaotic parameters; n represents a constant.

After determining the control parameters, select corresponding time series for different initial values z_0. For different control parameters, the system exhibits different characteristics, and after continuous bifurcation operations, the system ultimately reaches a chaotic state.

The analytical equation corresponding to Cubic mapping is shown in formula (2):

$$z_{n+1} = xz_n^3 - yz_n \tag{2}$$

In the equation, x and y represent control parameters.

By establishing a complex chaotic system, information resources can be encrypted. Analyze the independence and convenience of chaotic systems, and effectively ensure the diversity and convenience of the initial particle swarm through chaotic initialization operations. The fitness value ω of different particles is calculated as follows:

$$\omega = \frac{\omega_{\min}}{(\omega_{\max} - \omega_{\min})} \tag{3}$$

where, $\omega_{\max}$ and $\omega_{\min}$ are the maximum and minimum fitness values respectively.

After obtaining the value of fitness, it is necessary to adaptively adjust the inertia weight coefficient in the evolution process to ensure that the convergence speed and convergence accuracy of the algorithm are effectively improved and find a better global optimal solution. Chaotic mutation operation can effectively screen out inert particles and remove them all, avoiding the algorithm from falling into local optima. The control parameter coordination optimization process ψ_1 is given through formula (4):

$$\psi_1 = D_{oef}\,\omega(x_1, y_1) + u_1(x) + u_2(x) \tag{4}$$

In the formula, D_{oef} represents the correlation coefficient; $u_1(x)$ and $u_2(x)$ represent different uniformity indicators; (x_1, y_1) represents the degree of concealment of ciphertext.

After completing the above operations, effectively combine it with wavelet composite chaotic encryption to encrypt teaching resource information [6–8]. The detailed operation steps are as follows:

The basic wavelet function $\varpi(n)$ in Fourier wavelet transform is transformed into a function cluster $\varpi_{e,\tau}(n)$ by scaling and translation:

$$\varpi_{e,\tau}(n) = \frac{\psi_1}{\sqrt{\phi}}\alpha\left(\frac{t - \alpha}{\phi}\right) \tag{5}$$

In the formula, ϕ represents the stretching parameter; α represents the translation parameter; t represents the operating cycle.

The Fourier wavelet transform satisfies the constraint condition I_ψ in formula (6):

$$I_\psi = \int_R \frac{|\varpi(n)|^2}{\omega} d\omega \tag{6}$$

where, $\varpi(n)$ represents the basic wavelet function; R represents a constant; d represents the wavelet coefficient.

Put function $s(t)$ in the two-dimensional real space into the wavelet transform space, and obtain the corresponding basic wavelet function through a series of operations to realize the wavelet transform. Combining the cryptography of wavelet transform and composite chaotic sequences to form a wavelet composite chaotic cryptosystem. Expand and optimize all system parameters to obtain 2 initial keys and 1 composite chaotic block key corresponding to the system. The corresponding calculation formula is:

$$\begin{cases} \partial_1 = \dfrac{\varpi_{e,\tau}(n)(c_1 + c_2) \cdot c_3}{s_{\min}} \\ \partial_2 = \dfrac{\varpi_{e,\tau}(n)(c_1 + c_2 + c_3)}{s_{\max}} \\ \partial_3 = \dfrac{(s_{\max} - s_{\min})}{\varpi_{e,\tau}(n)(c_1 + c_2 - c_3)} \end{cases} \tag{7}$$

In the formula, ∂_1 and ∂_2 represent different initial keys; ∂_3 represents a mixed chaotic block key; c_1. ∂_2 and c_3 represent different ciphertext information, respectively; $s_{\max}$ and $s_{\min}$ represent the maximum and minimum approximate entropy.

Two initial keys of the system are iteratively operated to obtain a brand new discrete chaotic sequence, and related operations such as displacement transformation are carried out to obtain a composite chaotic cryptographic sequence $Z_{(c)}$, as shown in formula (8):

$$Z_{(c)} = \partial_1 p(n - E_{m1}) I_{m1} \otimes \partial_2 p(n - E_{m2}) I_{m2} \tag{8}$$

where, p represents step function; E_{m1} and E_{m2} represent different time delays; I_{m1} and I_{m2} represent different composite parameters.

By using a block key to perform bandpass filtering on the composite chaotic cryptographic sequence, a composite chaotic encryption sequence with the same length as the wavelet transform sequence is obtained [9–11]. At the same time, binary or encryption operations are performed on it to obtain the ciphertext sequence $Z_{(ci)}$:

$$Z_{(ci)} = \partial_3 \left[p(n - E_{m1}) - p(n - E_{m2} - I_{m1}) \right] Z_{(c)} \otimes Z_{(swt)} \tag{9}$$

In the formula, $Z_{(swt)}$ represents a wavelet transform sequence.

After performing wavelet transform on plaintext sequence $Z_{(I)}$, continue to perform proportional transform on it to obtain wavelet transform sequence $Z_{(swt)}$ as shown in formula (10):

$$Z_{(swt)} = B_c \cdot \varpi(n) \tag{10}$$

In the equation, B_c represents the scaling factor.

By calculating the fitness value corresponding to the ciphertext sequence, determine whether the fitness value meets the set constraints, and if so, encrypt it; On the contrary, it is necessary to modify the control parameters of the composite chaotic system, re constrain it, and then encrypt the teaching resource information. Assuming completion, directly output the results to achieve encryption of teaching information resources for vocational mobile education. Otherwise, the fitness value corresponding to the ciphertext sequence needs to be recalculated.

2.2 Construction of a Security Sharing Model for Teaching Information Resources

Due to the random distribution of shared information in the network environment, the process of information sharing is more complex and cumbersome, which affects the security of information sharing. Blockchain has the characteristics of decentralization, tamper resistance, traceability, and high security, enabling efficient and secure sharing of data. Storing encrypted data on the blockchain can ensure the integrity and authenticity of the data, prevent data from being tampered with, and eliminate the intervention of intermediate institutions to achieve point-to-point data sharing. This not only improves the efficiency of data sharing, but also reduces risks in the data sharing process and enhances data security. In order to improve the security of information sharing, blockchain technology is introduced to construct a teaching information resource security sharing model [12–14].

The blockchain architecture is divided into five levels, namely application layer, contract layer, data layer, blockchain layer, and logical layer. The blockchain architecture is shown in Fig. 1, and each level is defined and its specific functions are as follows:

(1) Logic layer

It is mainly responsible for abstraction and analyzing various functions corresponding to intelligent cooperation, and providing users with simple and clear function calls. The work of related modules is mainly communicated by the Restful interface.

(2) Blockchain layer

This level is mainly responsible for completing related operations such as data sharing and permission interaction. Considering the needs of actual business scenarios, it is necessary to choose private chain as the blockchain network $P(x, y, z)$ of the original system, and the corresponding calculation formula is as follows:

$$P(x, y, z) = \{u, t, \beta, \delta\} \tag{11}$$

In the formula, u represents the blockchain unit; t represents the sampling period; β represents the incentive mechanism on the blockchain; δ represents the total number of public chains.

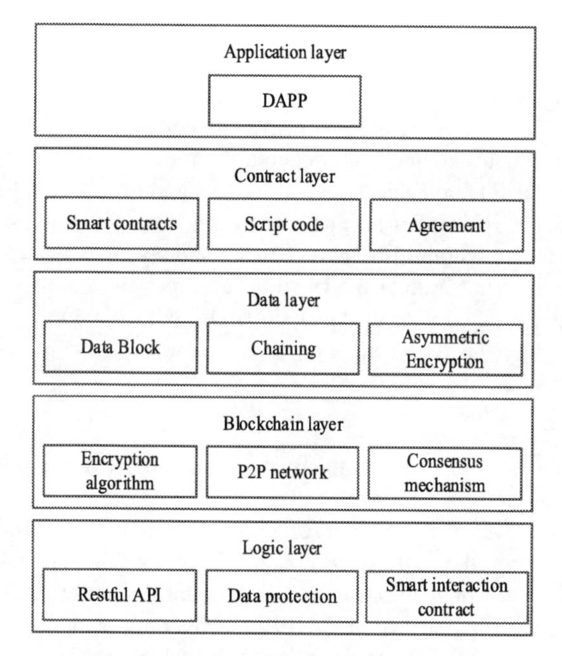

Fig. 1. Blockchain Architecture Diagram

This level also undertakes the task of defining content related to blockchain technology, such as block structure, digital signatures, information transmission, consensus mechanisms, etc. Information sharing involves multiple aspects of information data, and there are significant differences in information and different requirements for security. Through the processing of the blockchain layer, normalization processing of different information can be achieved [15, 16]. The normalization processing formula for shared information is:

$$Y_i = \frac{X_i - X_{\min}}{X_{\max} - X_{\min}} \tag{12}$$

In Eq. (12), X_i and Y_i respectively represent the shared information before and after normalization processing; $X_{\min}$ and $X_{\max}$ represent the minimum and maximum values of shared information, respectively.

(3) Data layer

The main purpose of the data layer is to achieve different types of data transactions, encapsulating the processed data into blocks, and then directly transmitting them to the corresponding blockchain database. The transaction order is composed of timestamp and hash values.

(4) Contract layer

In the model, it plays a logical control role. As different functions are coordinated by different contracts, corresponding contracts need to be constructed based on actual needs;

(5) Application layer

After receiving the requirements from the demand side, this level confirms and analyzes the shared information requirements, extracts corresponding shared information based on the requirements content, and integrates it.

In blockchain, each node needs to synchronize the entire chain information, as the storage capacity of each block data is limited. If the transaction iteration volume is large, it will affect the efficiency of consensus. In order to effectively improve resource utilization, it is necessary to store all critical information resources on the chain, determine the correlation between data on and off the chain, and establish a corresponding teaching information resource security sharing model $I_{(p)}$, as shown in formula (13):

$$I_{(p)} = Y_i\left\{t_1 \times Z_{(ci)} + t_2 \times Z_{(ci)}\right\} \times N_{(r)} \tag{13}$$

In the formula, t_1 and t_2 respectively represent different sampling stages; $N_{(r)}$ represents a random number generator.

The application layer also incorporates a data integrity verification mechanism. In the verification process, the hash function is used to verify the data integrity [17–19], and SHA-256 is used to obtain the corresponding digital signature; Upload all encrypted data to the private network, store it in the IPFS distributed file manager, and generate a blockchain hash value $H_{(a,b,c)}$, corresponding to the calculation formula:

$$YD = \left\{M_D, K_D, Z_{(ci)}, f_P\right\} \tag{14}$$

$$H_{(a,b,c)} = \frac{P(x, y, z) - N_{(r)} \times I_{(p)}}{Z_{(ci)}} \tag{15}$$

where, *YD* represents the meta Information set storing data; Formula (15) represents the process of storing data in IPFS to generate a blockchain contract hash value. K_D represents the keyword of the data; f_P is the unique access policy for encrypted data; M_D indicates the name of the data element.

Upload all digital signatures and generated hash values onto the blockchain. Assuming that the owner of teaching information resources agrees to share securely, the data requester will obtain the address corresponding to the data file, and at the same time expand the digital signature calculation through the hash function to compare the digital signature obtained by the calculation with the digital signature of the data owner, thus realizing the verification of data integrity [20–23]. The specific verification rules are as follows:

$$\begin{cases} \vartheta_i \geq \xi \text{ legal} \\ \vartheta_i < \xi \text{ Illegal} \end{cases} \tag{16}$$

In the formula, ϑ_i represents the digital signature of the i-th party requesting shared information; ξ represents the threshold for setting digital signatures for authentication.

In summary, complete the secure sharing of teaching information resources in vocational mobile education.

3 Experimental Analysis

In order to verify the effectiveness of the proposed blockchain based teaching resource data sharing method for vocational mobile education, the experiment used reference [3] on the blockchain based model for sharing cultural relics information resources (referred to as the "reference [3] method") and reference [4] on the decentralized ciphertext data security sharing method with anti-attribute tampering (referred to as the "reference [4] method") as comparative methods. The experiment was conducted in the Windows 10 system environment using IDEA programming software for simulation testing, and experimental analysis was conducted from the following aspects.

3.1 Result Analysis

The experiment uses system operating efficiency, operating expenses, and throughput as evaluation indicators to verify the security sharing performance of teaching information resources in vocational mobile education.

Operational efficiency refers to the time and computing resources required for a data sharing method to actually run. For the data sharing method of mobile education teaching resources, the operating efficiency directly affects the user experience and the response speed of the system. If the operation efficiency of the data sharing method is low, the delay of data transmission and processing may be increased, which will affect the real-time performance of teaching resources and user experience.

The operating overhead includes resources such as hardware devices, software platforms and network bandwidth required by the data sharing method. The data sharing method of mobile education teaching resources needs to take into account the computing power, storage capacity and network connection stability of mobile devices. The data sharing method with low operating cost can improve the utilization efficiency of resources, reduce the cost and adapt to the environment requirements of various mobile devices.

Throughput refers to the volume of requests or data traffic that a data sharing method can handle. For the data sharing method of mobile education teaching resources, high throughput means that more users can access and share resources at the same time, and improve the scalability and capacity of the system. Therefore, the data sharing method with higher throughput can meet the needs of large-scale users and provide stable and efficient teaching resource sharing service.

Operation efficiency, operation cost and throughput are the key indicators to evaluate the performance of mobile education teaching resource data sharing method, which directly affect the practical application and user experience of the system.

(1) System operational efficiency testing

System operational efficiency is one of the indicators for evaluating system operational performance. Usually refers to the amount of tasks processed by the system under a specific workload. The higher the operational efficiency, the greater the workload and task volume that the system can handle, and the better the system performance. A higher value indicates that the system is processing more tasks. If the operational efficiency of the system decreases, it indicates potential faults or other safety issues in the system. The experiment mainly measures the operational efficiency of different methods by calculating time. The experimental results of the operational efficiency of different methods are shown in Fig. 2.

From Fig. 2, it can be seen that in the initial stage of the experiment, the calculation time of different methods is less than 100 s. However, as the sample continues to increase, the calculation time of the methods in reference [3] and [4] significantly increases, especially when the sample size is large, the maximum calculation time exceeds 150 s. The overall calculation time of the proposed method is less than 150 s and is less affected by the number of samples. Through comparison, it can be seen that the proposed method has the shortest calculation time, fully demonstrating its advantages in improving system operation efficiency.

(2) System running cost testing

Running cost refers to the cost of resources and time consumed by a computer when executing a specific program. This typically includes system hardware resource consumption (such as CPU usage, memory consumption, etc.) and time consumption (such as response time and processing time, etc.). The lower the running cost, the better the system performance. A larger value indicates that the system consumes more resources and time, resulting in higher operating costs. Meanwhile, if the operating cost of the system increases, it indicates that the system faces more security risks, as some security measures can increase the operating cost of the system. The experimental results of comparing overhead in time and space using different methods are shown in Fig. 3.

Further analysis of (a) in Fig. 3 shows that as the amount of shared data during system operation increases, there is a significant difference in the time cost of the system. The overall time cost of the proposed method system is less than 5 s. The system time cost of the methods in reference [3] and [4] is higher than that of the proposed methods. This indicates that the proposed method can process data quickly and efficiently in data sharing scenarios, thereby improving the system's running speed and service quality. Analyzing Fig. 3 (b), it can be seen that under the same amount of data sharing, the spatial overhead of the proposed method system is smaller than the methods in reference [3] and [4]. The smaller the spatial overhead of a data sharing system, the higher the likelihood that the system can support larger scale data sharing in the future. It can be seen from the comparison that the system operation of the proposed method provides less space for data sharing services, which can effectively reduce overhead and improve the system operation speed.

In summary, the proposed method has excellent performance in data sharing and can provide efficient data sharing services for the system.

(3) Throughput testing

Throughput is an indicator that measures the number of tasks a system can complete per unit of time. Usually refers to the amount of data that a computer can process per unit of time. The higher the throughput, the stronger the system's concurrency and efficiency. The larger the value, the more concurrent tasks the system can handle. It also reflects the efficiency of information sharing. The higher the value, the higher the efficiency of information sharing. If shared resources are attacked by hackers or abused by some

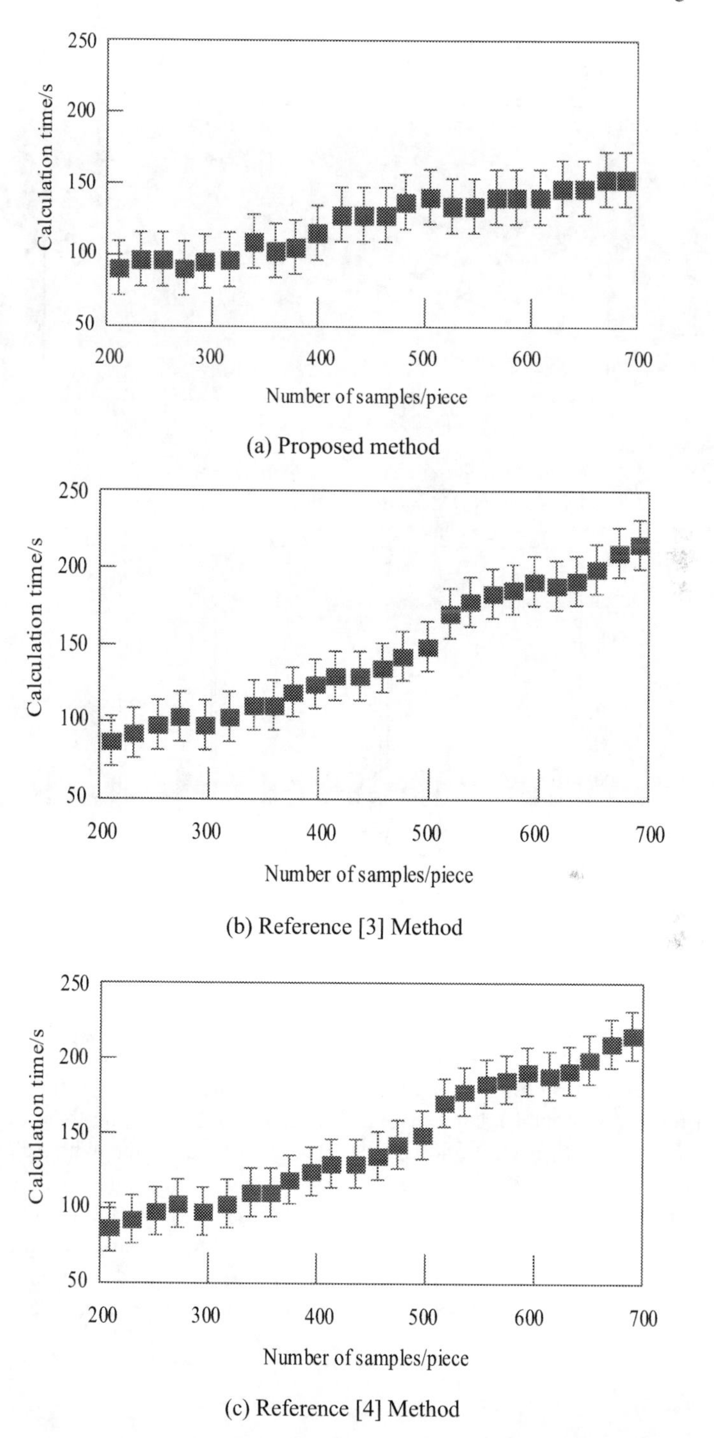

Fig. 2. Comparison of experimental results on operational efficiency of different methods

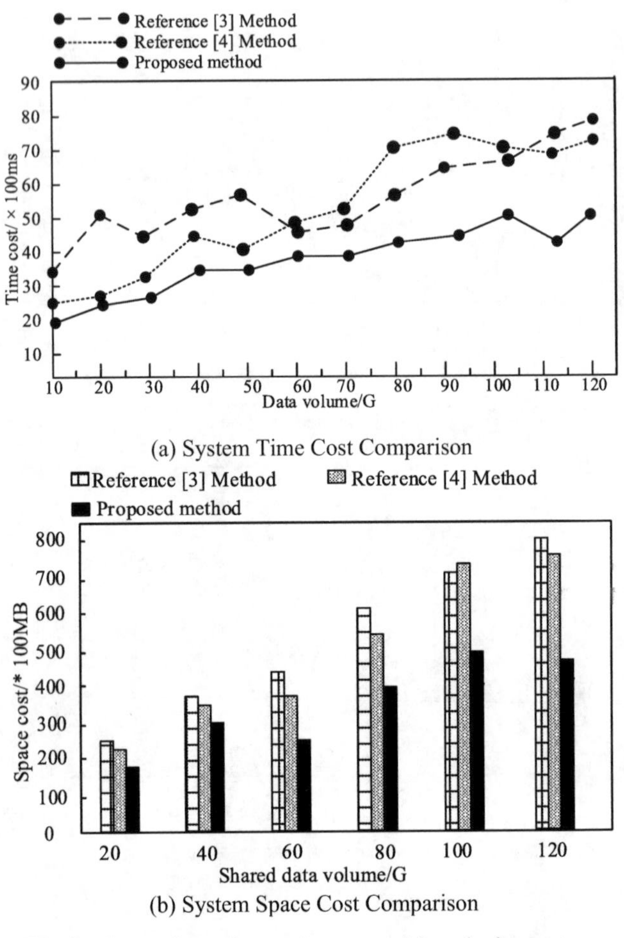

Fig. 3. Comparison of operating costs of data sharing systems

malicious users, it can lead to a decrease in the system's throughput. The experimental results of information sharing throughput using different methods are shown in Fig. 4:

From the experimental results in Fig. 4, it can be seen that the throughput of the proposed method has been continuously increasing in a straight line, with a clear upward trend. However, the throughput of the methods in reference [3] and [4] is significantly lower than that of the proposed method. This shows that the proposed method can more effectively utilize system resources and achieve more efficient data sharing, indicating that the teaching information sharing efficiency of the proposed method is high and can effectively reduce hacker attacks, Ensure the secure sharing of teaching information resources.

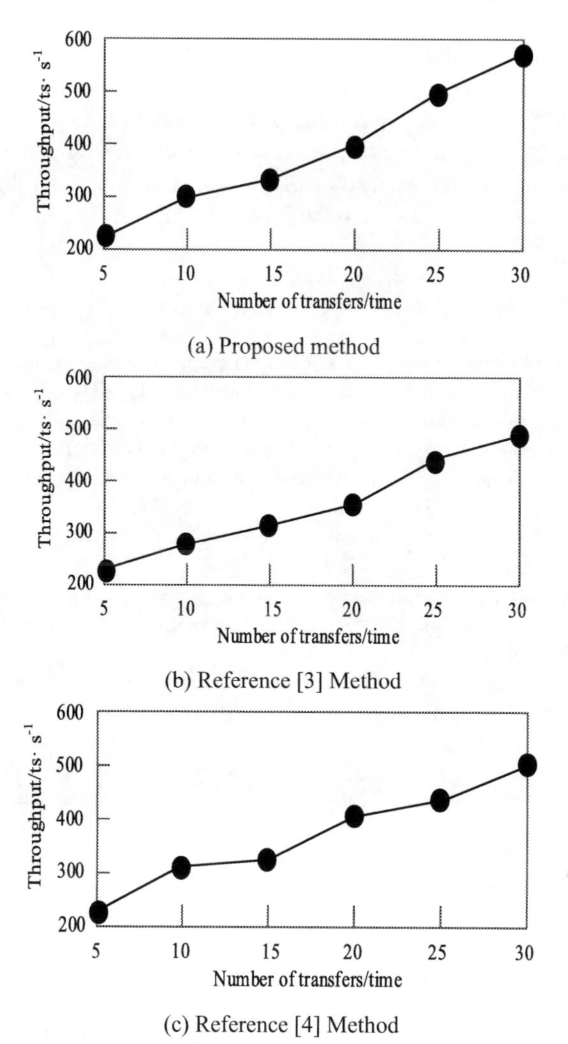

(a) Proposed method

(b) Reference [3] Method

(c) Reference [4] Method

Fig. 4. Comparison of experimental results of information sharing throughput using different methods

4 Conclusion

In order to ensure the safe sharing of teaching resource data of mobile education in higher vocational colleges, a method based on blockchain technology is proposed. In this study, a method of data security sharing of vocational mobile education teaching resources based on blockchain technology is designed. Chaotic sequences are used to initialize the fitness values of different particles, which can increase the randomness and security of cryptosystems. A more complex and secure cryptosystem is constructed by combining wavelet transform with chaotic cryptosystem. Wavelet transform can improve the data concealment and anti-interference ability, while chaotic cryptosystem can increase the randomness and unpredictability of cryptography. By combining the two, stronger data encryption and decryption capabilities can be achieved. Blockchain technology can provide a reliable data verification mechanism, by hashing the data and recording the hash value on the blockchain, you can verify whether the data has been tampered with. At the same time, blockchain can also be used to generate and verify digital signatures, ensuring the authenticity of data and trusted sources.

References

1. Xu, M., Ma, S., Wang, G.: Differential game model of information sharing among supply chain finance based on blockchain technology. Sustainability **14**(12), 7139 (2022)
2. Singh, C.E.J., Sunitha, C.A.: Chaotic and Paillier secure image data sharing based on blockchain and cloud security. Expert Syst. Appl. **198**, 116874 (2022)
3. Miao, Q., Lin, H., Wang, X., et al.: Federated deep reinforcement learning based secure data sharing for internet of things. Comput. Netw. **197**, 108327 (2021)
4. Zhang, J.: Cloud trust-driven hierarchical sharing method of internet of things information resources. Complexity **2021**, 1–11 (2021)
5. Manogaran, G., Alazab, M., Shakeel, P.M., et al.: Blockchain assisted secure data sharing model for internet of things based smart industries. IEEE Trans. Reliab. **71**(1), 348–358 (2021)
6. Liu, Y., Ko, Y.C.: Image processing method based on chaotic encryption and wavelet transform for planar design. Adv. Math. Phys. **2021**, 1–12 (2021)
7. Welba, C., et al.: Josephson junction model: FPGA implementation and chaos-based encryption of sEMG signal through image encryption technique. Complexity **2022**, 1–14 (2022). https://doi.org/10.1155/2022/4510236
8. Lingamallu, N.S., Veeramani, V.: Secure and covert communication using steganography by wavelet transform. Optik **242**, 167167 (2021)
9. Liu, L., Meng, L., Peng, Y., et al.: A data hiding scheme based on U-Net and wavelet transform. Knowl.-Based Syst. **223**, 107022 (2021)
10. Deb, N., Elashiri, M.A., Veeramakali, T., Rahmani, A.W., Degadwala, S.: A metaheuristic approach for encrypting blockchain data attributes using ciphertext policy technique. Math. Probl. Eng. **2022**, 1–10 (2022). https://doi.org/10.1155/2022/7579961
11. Rahman, M.S., Khalil, I., Moustafa, N., et al.: A blockchain-enabled privacy-preserving verifiable query framework for securing cloud-assisted industrial internet of things systems. IEEE Trans. Industr. Inf. **18**(7), 5007–5017 (2021)
12. Sifah, E.B., Xia, Q., Agyekum, K.O.B.O., et al.: A blockchain approach to ensuring provenance to outsourced cloud data in a sharing ecosystem. IEEE Syst. J. **16**(1), 1673–1684 (2021)
13. Wu, C., Ke, L., Du, Y.: Quantum resistant key-exposure free chameleon hash and applications in redactable blockchain. Inf. Sci. **548**, 438–449 (2021)
14. Miao, W.H., Wang, J.X., Zheng, Z.H.: Identity authentication scheme based on blockchain and multi factor combination. Comput. Simul. **39**(5), 402–408 (2022)
15. Tang, G., Zhang, Z.: Two-party signing for ISO/IEC digital signature standards. Comput. J. **66**(5), 1111–1125 (2022). https://doi.org/10.1093/comjnl/bxac001

Blockchain Based Logistics Tracking and Traceability Method for E-Commerce Products

Zimin Bao(✉) and Yanqing Han

Department of Management, Xi'an Jiaotong University City College, Xi'an 710018, China
zimin3537@163.com

Abstract. To improve the traceability effect of e-commerce product logistics and shorten its response time, this study proposes a blockchain based e-commerce product logistics traceability method. This method first conducts theoretical research on the traceability of e-commerce product logistics information, and then obtains data through the collection of e-commerce product logistics information. The ProVOC model is used to complete the design of e-commerce product logistics information storage model. Finally, based on this, the process design of e-commerce product logistics traceability is completed, achieving the function of e-commerce product logistics traceability. The experimental results show that the proposed method has better application effect and shorter response time, which has greater application value.

Keywords: Blockchain Technology · Electronic Commerce · Product Information · Physical Distribution Management · Traceability

1 Introduction

With the development of internet technology, e-commerce has flourished in China, and the efficient and convenient shopping experience it brings has been increasingly loved by consumers. In recent years, with the popularization of mobile devices and the rapid development of mobile internet technology, the proportion of mobile e-commerce in the e-commerce market has continued to increase. By 2021, mobile e-commerce has accounted for 72.9% of all e-commerce transactions [1, 2]. With the booming development of e-commerce, people's consumption patterns have also changed greatly. Online shopping, takeout and other ways have become an important part of People's Daily life, but the quality problems in online shopping are also increasing. Due to the lack of a unified and effective management mechanism, there are many phenomena in the field of e-commerce, such as counterfeiting, shoddy goods, and false advertising, which seriously affect consumers' trust in the quality of e-commerce products. Due to the fact that most of the products sold on e-commerce platforms are standardized, it is difficult to determine the responsible party when there are product quality issues, and it is also difficult to find a solution in the first place when there are quality issues [3, 4]. In recent years,

L. Yun et al. (Eds.): ADHIP 2023, LNICST 549, pp. 123–135, 2024.
https://doi.org/10.1007/978-3-031-50549-2_9

relevant national departments have been committed to exploring the quality and safety assurance system of e-commerce products, and logistics information traceability is one of the most direct and effective means to ensure the quality and safety of e-commerce products. Therefore, it is very important to track the logistics traceability of e-commerce products.

The so-called logistics traceability is a method to track the whole logistics process through technical means, which can obtain logistics information accurately and in real time, and provide consumers with more safe, efficient and reliable logistics services [5]. In the logistics tracking and tracing system, each logistics package is endowed with a unique identification code, which can be used to track the entire transportation process of the package and understand the location, status and other relevant information of the package. The application of logistics tracing technology can optimize logistics services, improve logistics efficiency and reduce liability disputes. Its application in e-commerce products can play an important role in guaranteeing product safety and other aspects. At present, logistics traceability technology has been widely applied in the food industry. By identifying and recording the entire process data of products from production to transportation, it can achieve full supervision of products and greatly prevent the occurrence of various food safety issues [6]. In addition, logistics traceability technology has been widely applied in industries such as pharmaceuticals, chemicals, and textiles, achieving the goals of full traceability, rapid response, and loss reduction. In short, logistics traceability is a logistics management method with broad application prospects. It can provide timely, accurate, and fully recorded logistics information, help improve logistics efficiency and safety, thereby improving consumer satisfaction, enhancing the competitiveness and market reputation of enterprises.

At present, there are mainly two methods for e-commerce product traceability, one is based on product coding system, the other is based on product identification system. Among them, the former is mainly used in food, medicine and other fields; The latter is mainly used in cosmetics, clothing and other fields. However, due to its higher cost and higher technical requirements, it is difficult to popularize comprehensively in our country. Many scholars have carried out research on this issue and have achieved certain research results. Some scholars proposed the design of cluster agricultural product supply chain traceability model based on blockchain relay technology in view of the characteristics of product logistics supply chain, such as long and multiple chains, dispersed production, heterogeneous information sources and complex network chain structure [7]. This method first analyzes the business composition of the clustered agricultural product supply chain, utilizes blockchain relay technology to connect the upstream and downstream blockchains of the supply chain, and constructs the relay multi chain topology structure and traceability model of the clustered agricultural product supply chain; Then, the application chain utilizes relay chains and cross chain routing to achieve cross chain interaction, and verifies the effectiveness of cross chain transactions through the Hyperledger Fabric endorsement strategy. Finally, taking the clustered kiwifruit supply chain as an example, a prototype system of the model was designed and implemented, and functional, performance, and scalability testing and analysis were conducted. Some scholars have proposed a study on the storage method of fruit and vegetable blockchain traceability data based on smart contracts to address issues such as untimely sharing of product data upstream and downstream traceability enterprises in product logistics traceability systems, difficulty in ensuring traceability data security, and inability of regulatory authorities to monitor all traceability data in real-time [8]. Firstly, a fruit and vegetable product traceability framework and traceability data storage model based on blockchain multi-chain architecture are designed. Intelligent contracts are used to realize the encryption storage on the classification chain of traceability data, authorization access between chains, decryption and query based on authorization vouchers, so as to ensure the ciphertext link of traceability privacy data in the whole supply chain of fruit and vegetable. Reduce the block chain storage pressure, realize the upstream and downstream enterprise privacy data traceability without island authorization sharing, based on Hyperledger Fabric channel technology to achieve product multi-chain traceability. Some scholars proposed a blockchain based reliable traceability system for agricultural product quality safety, aiming at the problems of centralized data storage, easy data tampering and data trust existing in the existing product quality traceability system. On the basis of analyzing the business processes and key blockchain technologies of the agricultural product industry chain, this method designs a trustworthy traceability block structure for agricultural products; Then, the "On Chain + Off Chain" collaborative management and storage strategy for agricultural product quality and safety traceability information was proposed to solve the problems of high data storage pressure, low query efficiency, and data explosion at each node in the agricultural product traceability blockchain network. The Kafka consensus mechanism was adopted to achieve consensus operation involving multiple agents, providing real-time data processing capabilities with high throughput and low latency; Finally, a set of intelligent contract rules and contract triggering conditions for agricultural product traceability were developed to ensure the reliability of agricultural product data and the credibility of the traceability platform; A trustworthy traceability system for agricultural product quality and safety has been developed based on the Hyperledger Fabric blockchain platform [9]. In the application process of the above three methods, due to the large amount of overall calculation, the response time is long, and the traceability results are not accurate. In view of the above deficiencies, this study introduces blockchain technology and RFID to complete the design of e-commerce product logistics traceability method based on blockchain.

2 Theoretical Research on Traceability of E-commerce Product Logistics Information

E-commerce product logistics information traceability refers to the phased tracking and recording of logistics information starting from the logistics transportation of e-commerce products, in order to achieve the purpose of consulting and tracing logistics process information. With the rapid development of e-commerce, more and more consumers are starting to shop online, making logistics one of the core links of e-commerce and promoting the research of logistics information traceability theory. The theory of logistics information traceability for e-commerce products mainly includes the following aspects:

1) Logistics information technology. Logistics information can be traced through the Internet of Things, RFID, GPS, bar code and other technical means to collect and record the whole process of logistics information [10].
2) Logistics process management. Standardize the logistics process to ensure that every link is effectively monitored and managed, and respond to abnormal situations in a timely manner.
3) Information system integration. The logistics information system will be integrated and unified management, so as to effectively avoid information island and information lag.
4) Data security protection. Efficient encryption, backup and recovery of logistics information to ensure the security and integrity of logistics information.
5) Data analysis and mining. By analyzing and mining logistics information, data support is provided for optimizing logistics management and improving customer satisfaction.

In short, the research on the traceability theory of e-commerce product logistics information aims to improve logistics efficiency, optimize logistics service quality, and achieve lean management of logistics processes. At the same time, it also provides consumers with a safe and reliable trading experience, promoting the sustained and rapid development of e-commerce.

3 Design of E-commerce Product Logistics Information Collection Module

The logistics information collection of e-commerce products needs to be realized with the help of RFID technology. In this study, the high-frequency long-distance reader produced by Texas Instruments is selected for data collection. Its main performance is shown in Table 1 below.

Table 1. RFID Parameter Table

Operating Frequency	13.56 MHz
Support label types	Tag-it HF/Tag-it HF-1/ISO15693-2
Supply Voltage	24V DC + 5%/−1%
Power consumption	Maximum 60 W
Transmitter power	0.5 W–10 W
Radiation modulation method	AM (10%–30%)
Communication protocol	ISO Host Protocol
Communication parameters	Maximum 115 KBits
Storage	EEPROM 1 kByte

The label selected for the production of a standard label package can be compatible with global open standards to provide user readable storage space. This type of label is more suitable for application and product security, library, supply chain management, property tracking ticket and other fields. Table 2 lists the main features of labels.

Table 2. Label Characteristics Table

Project	Parameter
Supported Standards	ISO/IEC15693-2
Recommended operating frequency	13.56 MHz
UID Digits	64 bit
User available storage capacity	2 Kbit
Accumulated access times	100000 times
Data storage specimen	>10 years
Size	48 × 101 mm
Operation temperature	−25 °C– +75 °C

Each reader is connected to a desktop computer and the reader works as a data acquisition tool. Considering the characteristics of generality and equipment communication between them through the RS232 serial port. Read/write device configuration commands, speaking, reading and writing control command is used to adjust to read and write device parameters to adapt to specific application host command, buffer read mode is used for data exchange between the host and the label.

4 Design of E-commerce Product Logistics Information Storage Model

The decentralized and immutable characteristics of blockchain provide a reliable storage environment for data traceability information. This makes it almost impossible for malicious nodes to tamper with traceable data stored in the blockchain. This paper proposes a blockchain data tracing storage model based on ProVOC model. The data tracing information in the ProVOC model is described as a triplet of "performing entities," "activities," and "data," but the relationships between the components are not well represented in the triplet, and "relationships" need to be introduced to document the relationships between the components. Therefore, this paper defines a data traceability information as a quad consisting of "execution entity", "activity", "data" and "relationship" between components. The container storing the quintuple is called traceability *Document*, and its metadata description method is shown in Formula (1):

$$Document = (Agent, Activity, Data, \mathrm{Re}lations) \tag{1}$$

In the formula, *Agent* represents the "executing entity" component; *Activity* represents the "activity" component; *Data* represents the "data" component; Re*lations* represents the relationship between various components.

The metadata description method for a 'executing entity' component is shown in Formula (2):

$$Agent = (ID, Name, Type, Parameter) \tag{2}$$

where *ID* is used to globally identify an "execution entity" component. *Name* Identifier used to identify a human-readable "executing entity". *Type* is used to identify whether the "executing entity" is human or non-human. *Parameter* Indicates the creation time and place of the execution entity.

The metadata description method of an "active" component is shown in Formula (3):

$$Acivity = (ID, Type, Parameter) \tag{3}$$

Among them, *ID* is used to globally identify an "activity" component. *Type* is used to identify the type of "activity", such as copying, adding, modifying, etc. *Parameter* is used to identify information such as the time, location, and conditions of the occurrence of the 'activity'.

The metadata description method for a "data" component is shown in Formula (4):

$$Data = (ID, Name, Type, HashType, Hash, Parameter) \tag{4}$$

where, *ID* is used to determine the unique data globally. *Name* is used to identify a piece of data in a human-readable manner so that users can understand the actual meaning of a piece of data by its name. The *Type* of data identifies the type of data, for example, docx, mp3, or png. *HashType* identifies the hash value of the data, which can quickly verify the integrity of a piece of data, and *HashType* identifies the hash algorithm.

The hash value obtained. *Parameter* is used to represent information such as the time, location, and condition of data generation. The 'parameter' component is possessed by other first-level components and is used to describe the time, location, and conditions of the existence of other components.

The metadata description method for a "parameter" component is shown in Formula (5):

$$Parameter = (ID, Temporal, Spatial, Condition) \tag{5}$$

where, *ID* is used to globally identify the "parameter" component, *Temporal* is used to represent "time parameter", *Spatial* is used to represent "spatial parameter", and *Condition* is used to represent "condition parameter".

The term 'relationship' is used to describe the 'relationship' records between components in the serialized ProVOC model. The metadata description method for a 'relationship' record is shown in formula (6):

$$\mathrm{Re}lation = (ID, Name, UpperComponent, LowerComponent) \tag{6}$$

ID is used to globally identify a relational record. *Name* a name used to identify Relationship. In addition, each "relationship" record contains *UpperComponent* (the upper component) and *LowerComponent* (the lower component). *UpperComponent* identifies the main component of "relationship" and F identifies the object component of "relationship". ProVOC model relationship names are shown in Table 3.

Table 3. ProVOC Model Relationship Table

Upper component	Relationship	Lower component
Agent	*Associate*	*Activity*
Activity	*Input*	*Data*
Activity	*Output*	*Data*
Activity	*Trigger*	*Activity*
Data	*Derive*	*Data*
HumanAgent	*Control*	*NonHumanAgent*
Agent *Activity* *Data*	*With*	*Parameter*

The structured modeling of the ProVOC model is incorporated into the storage model, as shown in Fig. 1. The user's upload of the ProVOC model's serialization module converts the user's upload of the Provoc model into the JSON format of the traceability, which is then uploaded to the blockchain. At the same time, the system needs to monitor the block information on the blockchain in real time. When it is confirmed that new traceability information is packaged into blocks and written into the blockchain, the system converts the traceability information into structured data through the structured modeling module of the ProVOC model and stores it in the relational database.

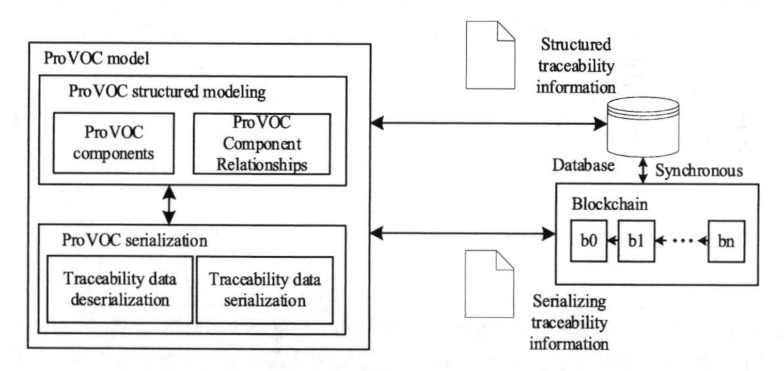

Fig. 1. Improved ProVOC storage model diagram

In the blockchain data traceability storage model, the same data traceability information needs to be written into both the blockchain and the relational database. The main function of data traceability information stored in relational database is to facilitate complex operations of traceability business. Its essence is a copy of traceability information stored in the blockchain, and the process of storing data traceability information in the database is shown in Fig. 2. The data traceability system detects that new data traceability information has been written to the blockchain network, and obtains the newly written data traceability information in the blockchain. The serialized data traceability information in the blockchain is structured through the ProVOC model and stored in the relational database.

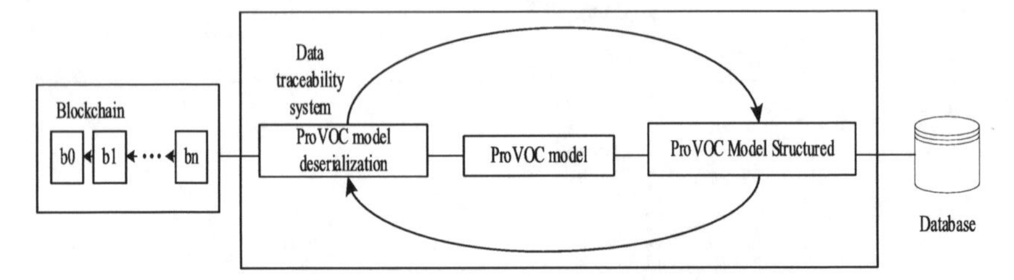

Fig. 2. Flow Chart of Data Traceability Information Entry

5 Design of Logistics Traceability of E-commerce Products

E-commerce product logistics traceability is to return the data traceability information that meets the query conditions, and the query process is shown in Fig. 3.

Blockchain is not suitable for complex query operations, so we put the query operations in a traditional relational database. The local relational database stores the structured data of the traceability information of the upper link, and the system will synchronize the traceability information of the upper link to the local database when it monitors the link of new traceability information. The traceability information in the local database is not up-to-date due to problems such as network delay. Therefore, in order to ensure the timeliness of the queried traceability information, it is necessary to check whether the latest traceability information in the local database is consistent with that in the blockchain during each query operation. If the traceability information in the local database is detected to be not up-to-date, the latest data traceability information in the blockchain should be obtained and stored in the local database to ensure the timeliness of the traceability information in the local database. After ensuring the timeliness of the data in the database, query the eligible data traceability information based on the user's query criteria, serialize the queried data traceability information into JSON format, and return it to the user. The data traceability system supports querying traceability documents through the ID of the traceability document, querying traceability documents through the "executing entity" component ID, and querying traceability documents through the "data component ID." By entering the "executing entity" component ID, information about all data traceability documents that the "executing entity" component is involved in can be queried. Similarly, by entering the "data" component ID, You can query the information of all data traceability documents involved in this' data 'component. After the data traceability system queries the data traceability information that meets the requirements, it will be displayed on the front-end page. Click the "View" button after the corresponding traceability information to view the retrieved JSON format traceability document. After clicking the save button, the viewed data traceability information can be downloaded as a JSON file and saved locally.

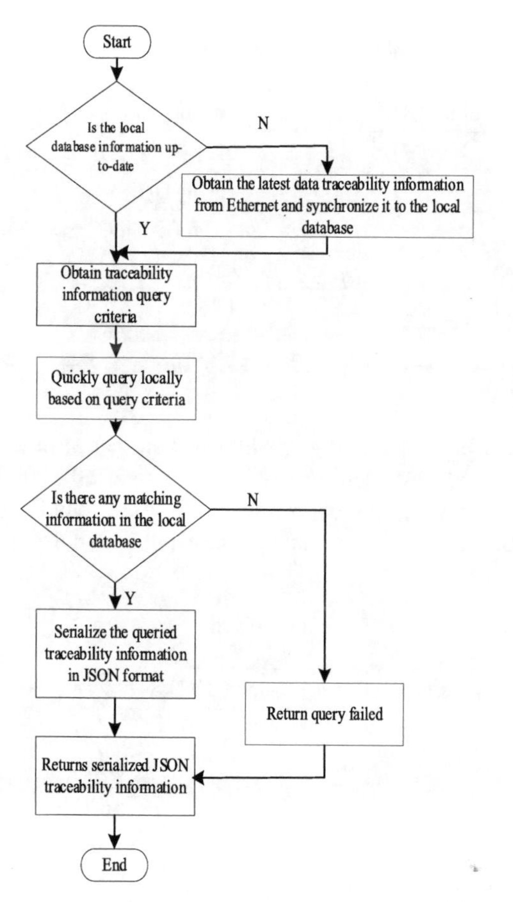

Fig. 3. Flow Chart of Data Traceability Information Query

6 Experimental Analysis

6.1 Experimental Environment Settings

This test chooses to build and start a private Geth Ethereum client locally, simulate the real operating environment of blockchain, and generate blockchain of different heights to complete the test. The specific test environment is shown in Table 4.

6.2 Performance Testing

Hyperledger Fabric was tested using the Caliper testing framework. The Caliper testing Framework is a tool developed specifically for Hyperledger federated chain testing. After setting the configuration parameters, we can test the performance of the traceability system network built in this paper. The tests mainly include write test and query test. Write test is the write performance test of blockchain ledger, and query test is the read performance test of query ledger. According to the actual application, the number of

Table 4. Test Environment Settings

Configuration information	Version parameters
CPU	Inter i7-8750H 2,20 GHz
Memory	16 GB DDR4
Hard disk	512 GB HDD
Operating system	Ubuntu 18.04 LTS
Co language	1.15.7
Hash algorithm	SHA256

traceability query of blockchain ledger is higher than that of product write operation. There are six rounds of test, each round initiates a total of 2,000 transactions on the network, and the number of requests per second of each round is set to 50, 100, 200, 400, 800 and 1600. The write performance test results are shown in Table 5.

Table 5. Write performance test results

Number of tests	Success/times	Transmission rate/TPS	Mean delay/s	Throughput/TPS
1	2000	45	1.67	43
2	2000	100	10.14	75
3	1997	198	16.73	105
4	1998	405	19.01	142
5	1999	430	19.25	145
6	1998	415	21.12	137

Based on the comprehensive data in Table 5, it can be seen that the overall network successful transaction rate is greater than 99%, ensuring the normal operation of the system. When the actual transmission rate approaches around 430TPS, network congestion occurs, and the actual transmission limit under the local configuration is 430TPS. When the sending rate reaches around 400TPS, the transaction throughput reaches a peak of 145TPS. Continue to increase the transmission rate, but the actual transmission rate will not increase, and the throughput will basically remain stable. When the sending rate is below 50 times/s, the average transaction delay is 2 s, which can quickly respond to user operations and meet a large number of write operation requirements. However, when the request sending rate exceeds 200 times/s, the average transaction delay increases to over 16 s, and the highest average delay in the test reaches 21 s.

The query type test was tested in 6 rounds. A total of 2000 transaction requests were launched to the network in each round, and the number of requests per second in each round was set to 50, 100, 200, 400, 800 and 1600, respectively. The test results of query performance were shown in Table 6.

Table 6. Query Performance Test Results

Number of tests	Success/times	Transmission rate/TPS	Mean delay/s	Throughput/TPS
1	2000	48	1.67	49
2	2000	100	0.08	97
3	2000	199	4.12	193
4	2000	397	14.21	321
5	2000	425	18.35	297
6	2000	451	17.62	305

Based on the data in Table 6, it can be seen that the success rate of query requests is 10%, ensuring the normal operation of the system. When the system transmission rate reaches about 440TPS, it is the actual transmission upper limit. The results show that when the number of requests reaches 400TPS, the actual send rate is 397TPS, and the throughput reaches a peak of about 321TPS. When the number of requests is not more than 200 times/s, the maximum transaction delay is about 4 s, while the lower transmission rate will have a transaction delay of less than 1 s, and the response speed is fast, which can meet the requirements of users to query product data when tracing the source. When the transmission rate exceeds 400 times per second, the transaction delay approaches 19 s.

6.3 Comparative Testing

In order to verify the progressiveness of the proposed method, literature [6] method and literature [7] method are selected as the comparison methods to carry out comparative tests. Three methods are used to jointly conduct logistics traceability, and response time is selected as an indicator to compare the performance of the three methods. Using three methods to jointly trace 10 e-commerce products, the comparison of traceability accuracy is shown in Fig. 4.

As shown in Fig. 4, during this test, there are 300 concurrent users in the simulation system, and all functions of the operation and maintenance management system are executed simultaneously. The proposed system starts to enter the transaction processing scene from 30 s, and the lowest transaction processing response time of the system is 0.36 s at the 6th minute. At around 4 min and 10 s, the maximum transaction response time is 1.12 s. When tracing the logistics of e-commerce products in the actual operation scenario, the average response time of transactions is about 0.69 s, while the comparison method enters the transaction processing scenario at 1.25 s and 2.00 s respectively, and the response time of transaction processing is much higher than that of the proposed method. Therefore, it can be proved that the proposed method has shorter response time and better performance. It can meet the requirement of real-time logistics inquiry.

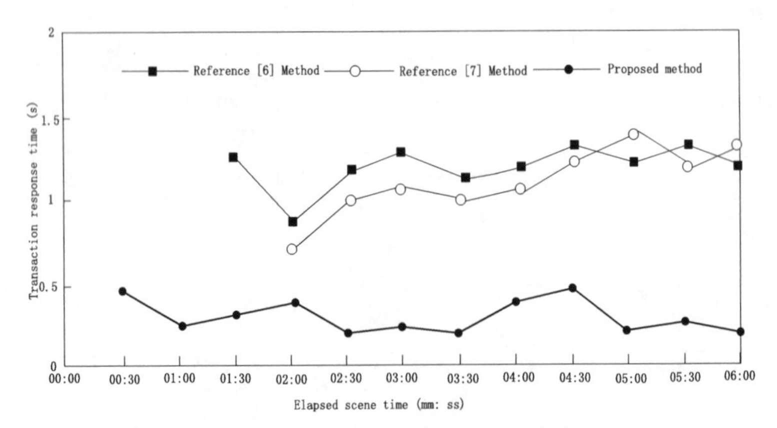

Fig. 4. Comparison of Response Times of Three Methods

7 Conclusion

This study proposes and designs a blockchain based e-commerce product logistics tracking and traceability method to address the issues of long response time and poor traceability effectiveness of traditional logistics tracking and traceability methods. This study is based on RFID technology to obtain commodity logistics information, and then uses the ProVOC model to design an e-commerce product logistics information storage model. Finally, the traceability process is designed to achieve its functionality. Experimental results show that the proposed method has good performance, and its minimum Transaction processing response time is 0.36 s, which is better than the comparison method, and has certain application value. Although this study has achieved certain results, its data confidentiality was not designed in the study, which is also the next research direction.

Acknowledgement. 2022 Annual Project of the "14th Five-Year Plan" Education and Science Planning of Shaanxi Province: Research on the improvement of digital literacy and skills of college students majoring in economics and management in the digital economy era (SGH22Y1715).

References

1. Panprommin, D., Manosri, R.: DNA barcoding as an approach for species traceability and labeling accuracy of fish fillet products in Thailand. Food Control **136**, 1–6 (2022)
2. Guo, T., Chen, Y., Zhu, X.: An optimization model for process traceability in case-based reasoning based on ontology and the genetic algorithm. IEEE Sens. J. **21**(22), 25123–25132 (2021)
3. Zhang, S., Liao, J., Shuangcheng, W., Zhong, J., Xue, X.: A Traceability public service cloud platform incorporating IDcode system and colorful QR code technology for important product. Math. Probl. Eng.Probl. Eng. **2021**, 1–15 (2021). https://doi.org/10.1155/2021/5535535
4. Zhang, Y., Liu, Y., Zhang, J.: Development and assessment of blockchain oT-based traceability system for frozen aquatic product. J. Food Process EngEng **44**(1), 106079 (2021)

5. Pandey, V., Pant, M., Snasel, V.: Blockchain technology in food supply chains: review and bibliometric analysis. Technol. Soc. **69**, 101954 (2022)
6. Liu, Z., Wu, L., Meng, W.: Accurate range query with privacy preservation for outsourced location-based service in IoT. IEEE Internet Things J. **8**(18), 14322–14337 (2021)
7. Ahmad, R.W., Salah, K., Jayaraman, R., et al.: Blockchain in oil and gas industry: applications, challenges, and future trends. Technol. Soc. **68**, 101941 (2022)
8. Bhutta, M., Ahmad, M.: Secure identification, traceability and real-time tracking of agricultural food supply during transportation using internet of things. IEEE Access **9**, 65660–65675 (2021)
9. Wang, S.: Artificial intelligence applications in the new model of logistics development based on wireless communication technology. Sci. Program. **10**(9), 211189 (2021)
10. Azzouz, K., Arif, J., Benboubker, M.B.: Exploratory study of the role of logistics service providers in terms of traceability in the process of outsourcing of logistics' activities: case of Moroccan LSP. Int. J. Eng. Res. Afr. **54**, 187–208 (2021)

Mobile Monitoring, Civilian Audio and Acoustic Signal Processing

Personalized Recommendation Method of Rural Tourism Routes Based on Mobile Social Network

Yi Liu and Qingqing Geng(✉)

Chongqing College of Architectural and Technology, Chongqing 401331, China
gqq123cq@163.com

Abstract. The existing personalized recommendation methods for tourism routes have the problem of low tourist satisfaction, so a personalized recommendation method for rural tourism routes based on mobile social networks is proposed. According to the mobile social network model, calculate the number of mobile message hops and define a set of social information paths to complete the processing of travel route data based on mobile social networks. On this basis, implement denoising of tourism route data, determine personalized route recommendation schemes by deriving route sequences, and complete the design of personalized rural tourism route recommendation methods based on mobile social networks. The experimental results show that under the influence of the above methods, the number of tourists choosing fixed tourism routes significantly increases, and the satisfaction level of tourists with the recommended routes also increases, which meets the practical application needs of personalized recommendation of rural tourism routes.

Keywords: Mobile Social Network · Rural Tourism Routes · Personalized Recommendation · Message Hops · Geographical Factor Characteristics · Vectorization · Data Denoising · Route Sequence

1 Introduction

As the main carrier of mobile intelligent devices, human's social attributes have also been given to mobile intelligent devices. The social attributes of nodes are one of the factors that must be considered. Therefore, the concept of mobile social network has also been proposed on the basis of opportunity network. Mobile social network is an opportunity network that considers the social attributes of nodes. It also transmits data among nodes in the mode of "store carry forward". This network fully considers the social attributes of nodes, and uses this as a basis to help select appropriate intermediate nodes to help forward messages. Mobile social networks have broad research prospects in the future. With the rapid change of intelligent devices and the concept of smart city and smart home, many intelligent devices have entered thousands of households, and the popularity of intelligent devices has also increased rapidly; The development of communication technology also makes mobile social network have a broader development prospect under 5G network. At the same time, with the improvement of the performance of mobile

L. Yun et al. (Eds.): ADHIP 2023, LNICST 549, pp. 139–152, 2024.
https://doi.org/10.1007/978-3-031-50549-2_10

intelligent devices, their computing power and storage capacity are bound to continue to increase. Communication technology, node storage capacity and node computing capacity are the key technologies of mobile social network. The development of these three will also provide a broader development platform for mobile social network. "Store carry forward" is the most basic mode of mobile social network data transmission [1]. The main research content of mobile social network is also around these three points. In the "storage" direction, it mainly focuses on the cache space management of nodes, and uses appropriate cache management strategies to improve the network performance; In the "carrying" direction, whether the node's mobility model is practical is also an important research direction. On this basis, the research on the social attributes of nodes is also one of the research directions.

In recent years, more and more intelligent mobile terminals are equipped with positioning function. With intelligent mobile devices becoming a necessity in public life, people's demand for location-based services has shown explosive growth, and location-based social networks have attracted a large number of users. Users share their travel photos or "check-in" data through social networks to record access history and share life experience, thus accumulating a large number of access footprints or "check-in" record data with geographical markers. These users' historical access data provide an opportunity to understand people's behavior, which can be effectively applied to personalized travel recommendations based on social networks, including that with the continuous growth of self-help tourism groups, more and more users can organize their travel materials from the network [2]. However, with the rapid increase of social network users, the rapid growth of network information has resulted. When users face massive network data, they cannot quickly select information.Users prefer to automatically obtain personalized travel recommendations that meet their specific needs, help users quickly filter useless information from a large number of travel information, and improve the efficiency and comfort of users in integrating information. The hybrid tourism route recommendation algorithm based on DGKDK determines the value range of forgetting coefficient according to the prediction results of user preferences, and then recommends tourist attractions with the help of preference model. Tourism route recommendation method based on potential interest spots of users plans the starting point and end point of the tourism route by analyzing the historical tourism footprint, and on this basis, realizes the on-demand planning of the route track. However, in rural tourism, the number of tourists attracted by the tourist routes recommended by the above two methods is relatively limited, which cannot effectively improve the tourist satisfaction and does not meet the personalized recommendation needs.

To address the above issues, design a new personalized recommendation method for rural tourism routes based on mobile social networks. Based on the mobile social network model, the data processing of tourism routes has improved the quality of the data. Based on the processing results of tourism route data, denoise it and recommend personalized routes by deriving route sequences.

2 Travel Route Data Processing Based on Mobile Social Network

Rural tourism route data processing is based on the mobile social network model to solve the vectorization characteristics of geographical factors. This chapter will carry out in-depth research on the above contents.

2.1 Mobile Social Network Model

The mobile social network model is a mathematical model used to describe and analyze people's social interactions in a mobile environment. This model is usually based on graph theory and Complex network theory, which can help us understand people's behaviors and relationships in mobile social networks.

In the mobile social network model, people are viewed as nodes, and their social connections are represented as edges connecting these nodes. These edges can represent actual social connections between people, such as friends or family relationships, or virtual social connections, such as relationships of mutual interest or mutual interest.

The mobile social network model can also consider people's location information during the movement process. By integrating location information and social relationships, one can understand people's social behavior at different times and spaces, such as the frequency of social activities at different locations, and the social distance between people.

(1) Mobile message hop count

The number of hops of a message refers to the number of intermediate nodes through which the message is delivered from the source node to the destination node. The smaller the hops of a successfully delivered message, the higher the accuracy of each node selected for forwarding the message.The formula for calculating the average hops of messages is as follows:

$$\overline{p} = \frac{1}{\beta \hat{O}} \sum_{\substack{\alpha=1 \\ \chi=1 \\ \delta=1}} \frac{p_\alpha + p_\chi + p_\delta}{3} \tag{1}$$

Among them, $\hat{O}$ represents the delivery characteristics of data messages at the intermediate node, β represents the message delivery parameters between the source node and the intermediate node, α, χ, and δ represent three randomly selected data message labeling coefficients, and the inequality value condition of $\alpha \neq \chi \neq \delta$ remains true. p_α represents the data message jump vector based on parameter α, p_χ represents the data message jump vector based on parameter χ, and p_δ represents the data message jump vector based on parameter δ.

In mobile social networks, unnecessary or inefficient message forwarding results in a waste of network resources. However, setting a small limit on the number of message hops can lead to a limited number of preferred paths and other shortcomings, such as a lack of references. Therefore, it is crucial to impose an appropriate limit on the number of message hops.

The theory of "six degrees of separation", also known as "small world phenomenon", stipulates that if any two strange individuals want to establish a connection, they need at most five intermediate individuals to achieve the goal. It is reflected in the mobile social network. If the message spreads to the destination node with all efforts, it can reach the destination node [3] through up to five intermediate nodes.

Therefore, when the average hops are known, the mobile messages at the five intermediate nodes should be sampled separately to achieve accurate calculation of mobile message hops.

The calculation formula of mobile message hops of mobile social network is as follows:

$$O = \tilde{I}\frac{\hat{o}}{\overline{p}} \times \left(1 - \sqrt{\frac{P_1}{i_1} \cdot \frac{P_2}{i_2} \cdot \frac{P_3}{i_3} \cdot \frac{P_4}{i_4} \cdot \frac{P_5}{i_5}}\right) \tag{2}$$

In the equation, P_1, P_2, P_3, P_4, and P_5 represent the sampling results of mobile messages at the five intermediate nodes, i_1 represents the sampling parameters of mobile messages at the P_1 node, i_2 represents the sampling parameters of mobile messages at the P_2 node, i_3 represents the sampling parameters of mobile messages at the P_3 node, i_4 represents the sampling parameters of mobile messages at the P_4 node, i_5 represents the sampling parameters of mobile messages at the P_5 node, and the inequality sampling condition of $i_1 \neq i_2 \neq i_3 \neq i_4 \neq i_5 \neq 0$ remains true, $\hat{o}$ represents the real-time forwarding characteristics of data messages in mobile social networks, and $\tilde{I}$ represents the node diffusion vector in mobile social networks.

For the processing of travel route data by mobile social network, the value of message hop index must belong to (0, 1) Within the range of "0" It means that the operation stability of mobile social network is weaker; On the contrary, the closer the jump value is "1", which means that the operation stability of mobile social networks is stronger.

(2) Collection of social information paths

The mobile social network determines the preferred path for a message based on the path with a high average encounter intensity. The construction of the preferred path set is the core aspect of the mobile social network model. Traversal is used to ensure that the optimal path in the network is filtered, but it is only suitable for networks with a small number of nodes. In networks with a large number of nodes, filtering through traversal becomes unrealistic. The extensive calculation of optimal paths would consume significant network resources, thereby impacting the overall network performance. To address this, the construction process of the preferred path in the mobile social network is divided into the following steps:

(1) Define the encounter strength $\tilde{u}$ for the number of mobile message hops.
(2) Calculate the processed encounter strength u' based on the hop count O and encounter strength $\tilde{u}$ of the joint mobile message.

$$u' = |O \cdot \tilde{u}|^{-1} \Big/ \left(\frac{\gamma^2 - \gamma}{2}\right) \cdot |\Delta U|^2 \tag{3}$$

where, ΔU represents the unit Cumulant of data message samples in mobile social network, and γ represents the data message recognition coefficient based on mobile social network model.

(3) The construction process of the optimal path in the mobile social network model is essentially the process of adding intermediate nodes in the two nodes. For example, if there is a 1-hop path ① → ⑦, the construction process of the 2-hop path is essentially to select an appropriate intermediate node to serve as a bridge from node ① to node ⑦; If there is a 2-hop path ① → ⑤ → ⑦, the construction process of the 3-hop path is to select a suitable intermediate node as the bridge from node ① to node ⑤.
(4) Construction of hop preferred path. There is only one path from the message to the destination node after only one hop, namely ① → ⑦. And node ⑦ is the destination node of the message. Regardless of the average strength of each hop of the path, it should be included in the preferred path set, so ① → ⑦ is the 1-hop preferred path of the network.
(5) The construction of a 2-hop optimal path. Traverse the encounter intensity of node ⑦ and select nodes with an encounter intensity of no less than u' as the preset intermediate nodes, including nodes ④ and ⑧.

Based on formula (3), the solution result of the social information path set Δ can be expressed as:

$$\Delta = \left\{ y|y = \left(\frac{u'}{\varphi \overline{Y}} + 1\right) \cdot \frac{\sum_{-\infty}^{+\infty} \vec{\varepsilon}}{\sqrt{Y_1 \cdot Y_2 \cdot Y_3}} \right\} \tag{4}$$

Among them, y represents a random variable in the set of social information paths, $\overline{Y}$ represents the preset mean of mobile social information in the preferred path, φ represents the data sample optimization parameter in the traversal path, Y_1 represents the preset parameter value of mobile social information in the 1-hop preferred path, Y_2 represents the preset parameter value of mobile social information in the 2-hop preferred path, and Y_3 represents the preset parameter value of mobile social information in the 3-hop preferred path, $\vec{\varepsilon}$ represents the traversal filtering vector of social information within the mobile network path.

It is worth mentioning that due to the randomness of the movement of nodes in mobile social networks and the unpredictability of the future, the path in the optimal path set cannot be guaranteed to be the most correct path. There are many paths for messages from the source node to the destination node, and other paths with lower average strength per hop may deliver messages to the destination node faster [4]. In short, as the reference basis for message forwarding, the preferred path can only select the relatively excellent path, not necessarily the most correct path. As a means to reduce this impact, in the construction of the preferred path set, the more preferred paths, the smaller the impact, and the most correct path is easier to appear in the preferred path set.

2.2 Geo Factor Feature Vectorization Solution

Mobile social networks stipulate that geographical distance affects the choice of human access. The closer to the center of the frequent activity area, the greater the possibility of user selection. Therefore, geographical factor feature is an effective feature to be extracted. It can be seen from the data that each geographic factor feature vector has an independent longitude and latitude [5]. Longitude and latitude can not only calculate the distance between factor parameters, but also reflect the geographical preference of users for tourism. However, the longitude and latitude of geographical factor characteristics belong to singular data in mobile social networks, that is, compared with other characteristic data, the longitude and latitude value is far greater than other data, which not only makes the recommendation model unable to effectively learn the geographical factor characteristics of users. It also affects the model's effective learning of other features, leading to the decline of the model's learning ability.

The following table records the topic vector distribution structure of tourism users.

Table 1. Theme vector distribution structure of tourism users

Tourism users	Food	Park	College	History	Shopping
User-01	0.43	0.05	0.32	0.03	0.21
User-02	0.06	0.51	0.08	0.07	0.17
User-03	0.03	0.05	0.77	0.06	0.13
User-04	0.22	0.25	0.07	0.12	0.30
User-05	0.55	0.07	0.17	0.07	0.04
User-06	0.41	0.03	0.15	0.14	0.28

According to Table 1, different tourism users focus on different themes in the process of rural tourism. The recommended directions of common rural tourism routes generally include food, parks, schools, history and shopping. In mobile social networks, these recommended items will affect the quantitative solution results of geographical factor characteristics.

On the basis of formula (4), calculate the longitude and latitude vectors of rural tourism geographical factor characteristics respectively. The specific calculation formula is as follows:

$$\begin{cases} R_1 = \iota_1\left(\sqrt{\left|r_1 \cdot r_2 \cdot \cdots \cdot r_n / t_{\max} - t_{\min}\right|}\right) \\ R_2 = \sqrt{\iota_2 \cdot \frac{(t_{\max} - t_{\min})}{r_1^2 + r_2^2 + \cdots + r_n^2}} \end{cases} \quad (5)$$

Among them, R_1 represents the longitude vector of rural tourism geographical factor features, R_2 represents the dimension vector of rural tourism geographical factor features, $r_1, r_2, ..., r_n$ is the sampling parameter of n rural tourism routes that meet the conditions of the mobile social network model, and the sampling condition of $r_1, r_2, \cdots, r_n \in \Omega$ is constant. ι_1 represents the longitude value sampling parameter, ι_2 represents the dimension value sampling parameter, and $t_{\max}$ represents the maximum value result of the user's tourism geographical preference weight, $t_{\min}$ represents the minimum value result of the user's travel geographical preference weight.

Assuming e represents the initial value of the vectorization parameter, λ represents the personalized arrangement parameter of geographical factor features, T represents the average value of the personal theme vector of tourism users, κ represents the geographical distance sampling conditions of rural tourism attractions, and the simultaneous formula (5) can be used to express the vectorization solution results of geographical factor features of rural tourism routes based on mobile social networks as follows:

$$T = \left| \frac{1}{\kappa} \times \overline{E} \right| \cdot \int_{e=1}^{+\infty} \lambda \cdot (R_1 \times R_2)^2 \tag{6}$$

The vectorization solution condition is used to vectorize the travel user's sign in data. After matrix decomposition, the user's sign in record matrix is transformed from high-dimensional sparse data to low-dimensional user potential vector. For users with similar check-in records, the user potential vector obtained through matrix decomposition is closer in the vector space. That is, the closer the user vector is, the higher the access similarity between users.

3 Design of Personalized Recommendation Method for Rural Tourism Routes

According to the application requirements of the mobile social network model, the tourism route data is denoised, and then the route sequence expression is combined to achieve personalized recommendation of rural tourism routes.

3.1 Data Denoising of Tourism Routes

Location information (longitude and latitude) in mobile social networks is an important feature to determine the specific location of tourism users and calculate the distance between geographical factors.

First of all, the data items with missing location information in the data set are removed. The loss of location information makes it impossible to determine the location of the user's clock in record, and thus the travel route visited by the user cannot be determined. In the field of personalized recommendation, location information can effectively mine users' access preferences, and then recommend travel routes that meet users' preferences when recommending to users.

Secondly, the travel route visited by the user is determined according to the location information of the check-in record. When the user visits a place, multiple check-in records will be generated. For example, when the user is at a point of interest or a mall, multiple record data will be generated, but multiple check-in records belong to the same place. Match the route of user access according to the location information of the check-in record, as shown in Fig. 1. Using the location information of the travel user, set the location of the route as the center, and determine the coverage of the complete travel route with a radius of 200 m. By calculating the European distance between the longitude and latitude of the two, determine whether the user sign in place belongs to the route track [6]. If the sign in record is within the coverage range of the route track, confirm that the user's sign in place belongs to the tourism route, otherwise the user's sign in place does not belong to the tourism route.

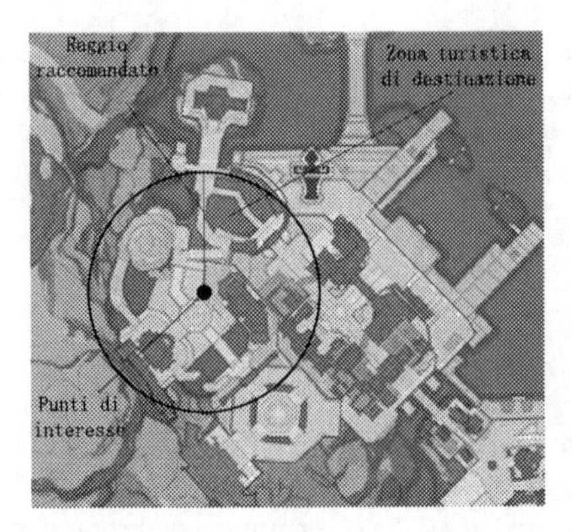

Fig. 1. Matching Diagram of Rural Tourism Routes

Rural tourism should have the following characteristics: 1. It has obvious regional characteristics; 2. "Rural" means rural resources, rural environment and rural industries; 3. Urban residents are the main tourist source market of rural tourism; 4. Rural tourism integrates economic benefits, leisure and entertainment functions, and can bring economic and other benefits to the local area; 5. Rural tourism has a multi-level product system and is a changing space-time concept. Rural landscape artistic conception tourism resources include corridor rural landscape artistic conception tourism resources and regional rural overall landscape artistic conception tourism resources. The rural landscape artistic conception tourism resources are the most easily damaged resources in the rural tourism resource system and the most difficult to maintain and develop in the process of rural tourism planning [7]. The rural landscape is characterized by large dispersion and small concentration. Each scenic spot is connected by corridors, such as rural traffic channels, rivers, rural landscape ecological corridors, etc. Corridor is an important component of landscape and an indispensable part of landscape planning and design.

The recommended rural tourism routes mine individual preference for tourism scenes according to the frequency of users' historical visits to different topics. The denoising expression for rural tourism route data is:

$$Q = (a + 1) \times \frac{\log \tilde{w} + \log \tilde{A}}{-\kappa \cdot T} \tag{7}$$

In the formula, $\tilde{w}$ represents the connection characteristics between rural tourism attractions, $\tilde{A}$ represents the historical visit frequency characteristics of rural tourism route data, and a represents the denoising weight.

The popularity of tourist routes and the activity of tourists are different and distributed.

Therefore, it is of great objective significance to consider the impact of both on the quality of recommendation results in the recommendation algorithm of tourism routes to improve the quality of tourism route recommendation.

3.2 Route Sequence

In the data construction layer, the mobile social network retrieves geotagged photos directly and utilizes the Haversine formula to calculate the distance between the user's shared photos and each point of interest along the rural tourism route. If the distance is less than 200 m, it is determined that the user's photo location belongs to the point of interest. This method allows for the calculation of places visited by users, thus creating a collection of users' historical access records.

Rural tourism planning is a kind of tourism planning. It is based on the environment of villages, countryside, and pastoral areas. It sets goals according to the development law of rural tourism and market characteristics, and makes overall planning to achieve this goal, forming a distinctive development direction [8]. The development of rural tourism first needs to investigate the types, existing conditions and characteristics of resources in order to protect and use resources efficiently and reasonably. Only on the premise of resource protection, the development and utilization of resources can realize the sustainable development of rural tourism. When the development of rural tourism destination is at a standstill stage, it shows that the resources have been seriously damaged and the situation is not optimistic. At this time, the development and utilization of resources should be limited, and more effective measures should be taken to increase the protection of resources. When rural tourism destination goes into decline, it shows that tourism resources are about to be exhausted, and the remaining rural tourism resources are not enough to support the development of rural tourism. New tourism resources must be developed to enhance the attraction of tourism destination in order to revive rural tourism.

The calculation results of the personalized characteristics of rural tourism routes are as follows:

$$\dot{D} = \frac{\sum\limits_{-\infty}^{+\infty} \left(S_{\mu}^{2} - S_{\nu}^{2}\right)}{\sqrt{Q \cdot \left[1 - \left(d_{\mu} - d_{\nu}\right)^{2}\right]}} \tag{8}$$

Among them, μ and ν represent two randomly selected interest point labeling parameters for rural tourism routes, and the inequality value condition of $\mu \neq \nu$ remains true. S_{μ} represents the tourism route access record feature based on parameter μ, d_{μ} represents the related tourism route data occupation parameter, S_{ν} represents the tourism route access record feature based on parameter ν, and d_{ν} represents the related tourism route data occupation parameter.

Compared with other tourism projects, rural tourism projects are unique and pay more attention to the participation of tourists. It promotes tourism projects from exhibition viewing to participatory experience. Tourists can not only enjoy the beautiful pastoral scenery, but also participate in labor and rural folk culture activities. Experiential rural tourism helps tourists to experience their hometown, integrate into a healthy, simple and simple rural lifestyle, and improve their understanding of rural life and agricultural production [9]. The development of rural tourism has a clear tourist source orientation. Rural tourism uses rural unique resources as tourism attractions to attract urban residents living in a fast-paced, stressful life. Tourists mainly come from the surrounding urban markets.

On the basis of formula (8), the definition formula of rural tourism route sequence is derived as follows:

$$G = \frac{1}{2\varpi} \exp \sqrt{\frac{\left|f \times \dot{D}\right|^{-1}}{g}} \tag{9}$$

In the formula, ϖ represents the personalized scheduling parameter of rural tourism route data, f represents the real-time sharing coefficient of tourism route data, and g represents the attractiveness evaluation parameter of rural tourism resources to tourists.

There are different types of villages, with local culture, rural environment, clothing, eating habits and other rural characteristics caused by different history, culture, economic development and climate. Rural characteristics are the most essential attraction of rural tourism. It is necessary to select unique tourism resources for rural tourism planning on the basis of analyzing the tourist market.

3.3 Personalized Route Recommendation

$\vec{Z}$ represents the real-time recommendation vector of rural tourism route data in mobile social networks, whose value directly affects the adaptation relationship between personalized tourism route recommendation results and user personal behavior.

For the solution of real-time recommendation vector $\vec{Z}$, the following expression is satisfied:

$$\vec{Z} = \dot{X} \frac{1}{\sum_{-\infty}^{+\infty} b \times \tilde{M}} \tag{10}$$

In the formula, $\dot{X}$ represents the rural tourism search features selected based on personalized principles, $\tilde{M}$ represents the user's personal search behavior vector, and b represents the search parameters of tourism route data in mobile social networks.

The personalized tourism route recommendation problem is optimized according to the route sequence orientation problem. The personalized tourism route recommendation is expressed as the following integer planning problem. By maximizing the final score of the tourism route, the final recommended tourism route is determined, as shown in the following formula.

$$h = \left(l\vec{Z}\right) - \sqrt{\sum_{c=1}^{+\infty} \frac{1}{\vec{j}} \left(\dot{k} \times G\right)} \tag{11}$$

Among them, l represents the personalized corresponding parameters of travel users in mobile social networks, c represents the minimum value of the exported parameters of travel route data, $\vec{j}$ represents the execution vector of personalized recommendation behavior, and $\dot{k}$ represents the planning features of personalized recommendation behavior.

Villagers are the main body of the countryside. The main purpose of developing rural tourism is to excavate, protect and inherit rural culture, and make full use of rural tourism resources, adjust and optimize the rural industrial structure, expand agricultural functions, promote the employment rate of farmers, and increase farmers' income [10]. Tourists participating in rural tourism can visit and experience the villagers' lifestyle, but it is necessary to ensure the orderly development of the villagers' daily life. Therefore, when planning rural tourism, the design of tourism activities should conform to the villagers' lifestyle and habits, and fully respect the wishes of farmers.

In addition, due to the different search behaviors of user objects, before recommending rural tourism routes for them, mobile social networks will also summarize the user's search habits in the previous period of time, so as to achieve on-demand recommendation of tourism route data on the basis of ensuring personalized needs.

4 Example Analysis

To highlight the differences in the use of personalized rural tourism route recommendation methods based on mobile social networks, DGKDK based hybrid tourism route recommendation algorithms, and DGKDK based hybrid tourism route recommendation algorithms, the following comparative experiments are designed.

4.1 Experimental Environment

Define 6 scenic spots in a rural tourist attraction, as shown in Fig. 2.

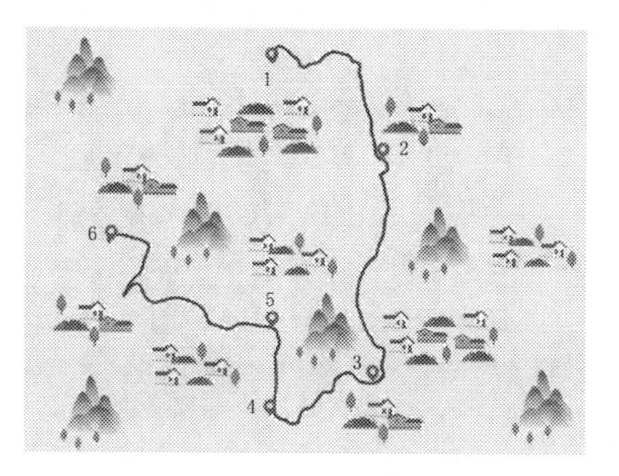

Fig. 2. Planning of rural tourist attractions

Establish tourism routes with these six scenic spots as the core, and input these routes into the Internet platform for users to choose. For Internet hosts, their preference for each scenic spot is exactly the same when they make travel routes.

When a user logs into the Internet platform, he or she will retrieve the travel routes according to his or her personal preferences. At this time, the host component will recommend personalized routes for the user according to his or her search habits. In the later stage of the summary work, according to the user's choice of the recommended route, the tourist satisfaction can be determined (the more people choose the personalized route recommended by the Internet platform, the higher tourist satisfaction will be).

First, use the personalized recommendation method of rural tourism routes based on mobile social network to control the Internet platform, record the user's choice of recommended routes under the effect of this method, and the results are the experimental group data.

Secondly, the mixed tourism route recommendation algorithm based on DGKDK is used to control the Internet platform, record the user's choice of the recommended route under the effect of this method, and the result is the first control group data.

Then, use the tourism route recommendation method based on the user's potential interest scenic spots to control the Internet platform, record the user's choice of the recommended route under the effect of this method, and the result is the data of the second control group.

Finally, the experimental data are collected and the experimental rules are summarized.

4.2 Data Processing

The figure below reflects the specific number of rural tourism routes recommended by the Internet hosts selected by the experimental group and the control group (Fig. 3).

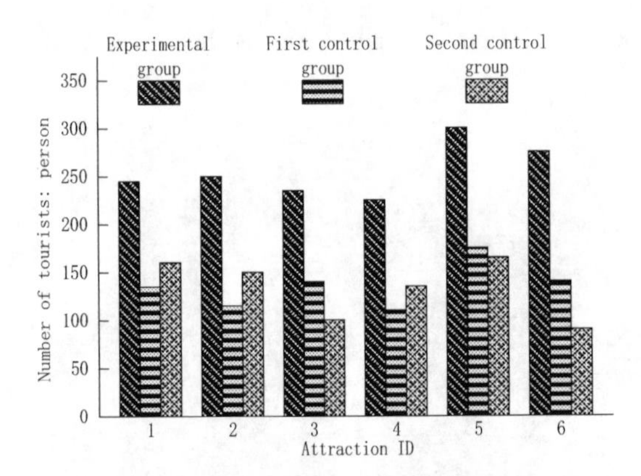

Fig. 3. Statistics of tourists

Experimental group: under the effect of the experimental group method, the number of people who chose the Internet host to recommend the tourist routes for No. 5 scenic spot reached 300; The number of people who chose the travel route recommended by the Internet host for No. 4 scenic spot was the smallest, but it also reached 225.

The first control group: under the effect of the first control group method, the number of people who choose the tourist routes recommended by the Internet host for No. 5 scenic spot is the largest, 175 people; The number of people who chose the travel route recommended by the Internet host for No. 4 scenic spot was the smallest, only 112.

The second control group: under the effect of the second control group method, the number of people who choose the tourist routes recommended by the Internet host for No. 5 scenic spot is the largest, 161 people; The number of people who chose the travel route recommended by the Internet host for No. 6 scenic spot was the smallest, only 88 people.

To sum up, during the whole experiment, the average number of tourists in the experimental group was the highest, the average number of tourists in the first control group was in the middle, and the average number of tourists in the second control group was the lowest.

4.3 Experimental Conclusion

The conclusion of this experiment is:

(1) The application ability of the hybrid tourism route recommendation algorithm based on DGKDK is weak. Under its recommendation, the number of tourists who choose fixed tourism routes is relatively small, which is not enough to improve tourists' tourism satisfaction.
(2) Compared with the hybrid tourism route recommendation algorithm based on DGKDK, the personalized rural tourism route recommendation method based on mobile social network has a slightly stronger application ability, but it still cannot maintain a high level of tourist satisfaction.
(3) The application of personalized recommendation methods for rural tourism routes based on mobile social networks can effectively solve the problem of fewer tourists choosing fixed tourism routes, improve tourists' tourism satisfaction, and thus set more personalized rural tourism routes in line with practical application needs.

5 Conclusion

With the increasing number of users of location-based social networks, the "check-in" record data and travel information shared by users have generated a large number of social network data containing geographical location information. These social network data contain a wealth of user travel time and space and context information, providing an excellent opportunity to understand user behavior and realize personalized travel recommendations. Although the traditional methods of personalized travel recommendation based on social networks have achieved some results, the personalized travel recommendation method based on mobile social networks proposed by the above research has great advantages in feature mining in social networks and extraction of users' personalized travel preferences.

The feature analysis and vectorization research of social network data. The theme features, geographical factor features and user access features are identified as the effective features that affect user preferences. In addition, in order to alleviate the problem of data sparsity, and enable discrete data in social networks to be sent into the deep learning model for effective training, the topic features, geographical factor features and user access features are vectorized.

Personalized travel route recommendation algorithm. On the basis of personalized recommendation of mobile social networks, users prefer to recommend a travel route composed of multiple points of interest that users are interested in. The similarity between the user interest vector and the context vector of interest points is taken as the feature of the tourism target area, and the dynamic interest preference of the user in the tourism process is determined by weighting these two aspects.

References

1. Ghaderian, S., Wan, M.: The factors affecting personal information disclosure and usage continuance intention on mobile social networking services. Int. J. Adv. Res. **9**(5), 235–244 (2021)
2. Kurikala, G., Gupta, G.: Mobile social networking below side-channel attacks: sensible security challenges. Int. J. Sci. Res. Comput. Sci. Eng. Inf. Technol. **2**(2), 1076–1084 (2021)
3. Sleptsov, Y.A., Nikiforova, S.V., Meshcheryakov, K.Y., et al.: Features of tourist routes in the republic of Sakha: extreme tours, unique natural sites, archaeological and ritual attractions. Int. J. Agric. Ext. **9**(4), 13–20 (2021)
4. Li, X., Li, J.W., Yu, N.: Tourist route recommendation method based on user needs. Comput. Eng. Des. **42**(05), 1339–1345 (2021)
5. Sun, Z.Q., Luo, Y.L., Zheng, X.Y., et al.: Intelligent travel route recommendation method integrating user emotion and similarity. Comput. Sci. **48**(S1), 226–230 (2021)
6. Nitu, P., Coelho, J., Madiraju, P.: Improvising personalized travel recommendation system with recency effects. Big Data Min. Anal. **4**(3), 139–154 (2021)
7. Ilic, J., Lukic, T., Besermenji, S., et al.: Creating a literary route through the city core: tourism product testing. J. Geograph. Inst. Jovan Cvijic SASA **71**(1), 91–105 (2021)
8. Hu, B.B., Lu, J.L., Zheng, C.Y.: Application of improved PrefixSpan algorithm on popular travel routes. J. Yunnan Minzu Univ. Nat. Sci. Ed. **31**(01), 94–102 (2022)
9. Guo, H., Jordan, E.J.: Social exclusion and conflict in a rural tourism community: a case study from Likeng Village, China. Tour. Stud. **22**(1), 42–60 (2022)
10. Liu, Y., Cao, Y., Liu, J., et al.: Research on heterogeneous information network recommendation algorithm based on dynamic iterative sampling. Comput. Simul. **39**(05), 324–328 (2022)

A Personalized Recommendation Method for English Online Teaching Video Resources Based on Machine Learning

Hua Sui[1](✉) and Yan Liu[2]

[1] Shenyang Aerospace University, Shenyang 110136, China
yidajiang11112@163.com
[2] Dalian University of Science and Technology, Dalian 116038, China

Abstract. In order to optimize the utilization of learning resources, provide learners with English online teaching video resources that meet their learning needs and interests, and achieve intelligent recommendation of English online teaching video resources, a machine learning based personalized recommendation method for English online teaching video resources is proposed to address the problem of poor accuracy in personalized recommendation of traditional methods for English online teaching video resources. Using clustering algorithms to classify English online teaching video resources and extract the features of English online teaching video resources. Using graph convolutional neural networks in deep learning to predict users' preference for English online teaching video resources, obtain recommendation results, and achieve personalized recommendation of English online teaching video resources based on deep learning. The experimental results show that the method has a recommendation accuracy of 92% for 300 users, a video resource recommendation recall rate of 91%, a recommendation path completion rate of 95%, and a recommendation time of 5 s, resulting in better recommendation performance.

Keywords: Machine Learning · Online English Teaching · Teaching Videos · Personalized Recommendation of Resources

1 Introduction

With the rapid development of internet technology, in the field of English learning, as more and more people begin to accept the new mode of online education, English online education has become the first choice for many people to learn English. On online learning platforms, the massive amount of English teaching video resources provides students with rich learning content and resources. However, how to select suitable content for students from the massive video resources has become an important issue faced by the education industry. Traditional online English education platforms mostly rely on manual or conventional recommendation algorithms to classify videos. Students need to spend a lot of time and energy to find the content they are interested in or suitable for, and they

L. Yun et al. (Eds.): ADHIP 2023, LNICST 549, pp. 153–165, 2024.
https://doi.org/10.1007/978-3-031-50549-2_11

are likely to miss other high-quality learning resources [1]. At the same time, traditional recommendation algorithms fail to fully consider students' learning needs and interests, and the recommended content often fails to meet students' expectations. Therefore, in order to solve the above problems, more and more researchers are exploring the use of personalized recommendation technology to improve the user experience and learning effectiveness of online English education platforms. By analyzing users' learning history and behavior, personalized recommendation systems can provide customized teaching video resources for each user, better meeting their learning needs and improving English learning effectiveness [2].

Reference [3] proposes a personalized English resource recommendation method based on knowledge graphs. By analyzing some problems that arise in traditional classroom teaching and online English teaching, various knowledge points in English teaching are classified and corresponding knowledge graphs are formed. Secondly, annotate relevant knowledge points in online teaching. Then, the daily learning behavior of students was analyzed to obtain a characteristic graph of their learning behavior. Finally, collaborative filtering technology and content-based recommendation technology are combined to provide more English teaching resources for learners to better meet their personal needs. Reference [4] proposes a massive online learning resource recommendation method based on multi-objective optimization. Model the learning resource recommendation task as a multi-objective optimization problem. Based on multi-objective evolutionary algorithm, the massive online learning resource recommendation is achieved through four steps: recommendation, clustering, optimization goal setting, individual representation, and calculation of genetic operators. Reference [5] proposes an immersive contextual teaching method for college English based on machine learning in artificial intelligence and virtual reality technology. Through a comparative teaching experiment on two classes of freshmen in a certain university, the experimental class conducted immersive virtual situational teaching based on VR technology from a constructivist perspective, while the control class used common multimedia devices and traditional teaching methods. The experimental results indicate that the combination of constructivist theory and virtual reality technology in immersive contextual teaching of college English can indeed improve students' English proficiency. The above methods have certain effectiveness, but the accuracy of personalized recommendation for English online teaching video resources still needs to be improved.

In recent years, with the continuous maturity of artificial intelligence technologies such as deep learning and machine learning, personalized recommendation algorithms have also been continuously innovated and optimized. Based on this, this article proposes a personalized recommendation method for English online teaching video resources based on machine learning, aiming to provide English learners with a more personalized, efficient, and comprehensive learning experience. This method combines the clustering algorithm and graph Convolutional neural network in deep learning, and combines the two to achieve the classification and feature extraction of English online teaching video resources, and predicts users' preferences for resources based on the deep learning model. Deep learning models can better capture users' interests and behavior patterns, thereby achieving more accurate personalized recommendations. By integrating different technical methods together, more comprehensive and accurate recommendation results are provided.

2 Preprocessing of English Online Teaching Video Resources Based on Machine Learning

This article uses clustering algorithms to classify English online teaching video resources, which can classify similar video resources into one category and extract the features of English online teaching video resources, better reflecting the content and characteristics of video resources. The graph Convolutional neural network in deep learning is introduced to predict users' preferences for English online teaching video resources. By modeling the relationship between users and resources, the graph convolution neural network can better capture users' interests and behavior patterns, so as to achieve more accurate and personalized recommendations. This method can effectively extract common features between video resources and provide a foundation for subsequent recommendation processes.

2.1 Analysis of the Categories of English Online Teaching Video Resources

The clustering algorithm is used to analyze the collected English online teaching video resources and divide different objects into different clusters. Because the clustering algorithm belongs to unsupervised learning, the classification analysis is completed under the premise of unknown target conditions. The most famous clustering algorithm is the K-Means algorithm. The K in the K-Means algorithm represents the number of sample classifications, and the specific values depend on the actual situation. The implementation process of the algorithm is as follows:

Assuming that given the training sample $\{x^{(1)}, x^{(2)}, \cdots, x^{(n)}\}$ of English online teaching video resources, each $x^{(i)} \in R^n$, does not include label information y.

Randomly select K cluster centroid point and represent it as $u_1, u_2, \cdots, u_n \in R^n$.

Loop through the following steps until convergence:

Calculate the class to which each video resource sample i belongs:

$$c^{(i)} = \arg\min_j \left\| x^{(i)} - u_j \right\|^2 \tag{1}$$

Calculate the similarity between each sample and the centroid point [6], and use Euclidean distance as the evaluation criterion for similarity. Select centroid points with a small distance from the sample, and divide the samples with the same centroid distance into the same cluster to complete the initial classification of the samples.

Recalculate the centroid points for each cluster j:

$$u_j = \frac{\sum_{i=1}^{m} 1\{c^{(i)} = j\} x^{(i)}}{\sum_{i=1}^{m} 1\{c^{(i)} = j\}} \tag{2}$$

For video resource sample data of the same category, recalculate the centroid points of that category to complete the data update. This article uses the average value of sample data from the same category [7–9] as the basis for updating centroids. Repeat the above operation until convergence occurs.

2.2 Extracting Features of English Online Teaching Video Resources

In order to improve the processing speed of matching user search terms with English online teaching video resources, the features of English online teaching video resources are extracted and output in the form of feature vectors. The word frequency feature refers to the number of times a given word appears in English online teaching video resources, and its expression is:

$$TF_c = \frac{T_c}{T} \tag{3}$$

Among them, T and T_c are the total number of words in the English online teaching video resources and the number of times words c appear in the English online teaching video resources, respectively. Due to the large amount of data in English online teaching video resources, there may be extraction bias in the process of word frequency feature extraction. Therefore, the concept of inverse document word frequency is introduced [10–12]. In the process of extracting inverse document word frequency features, it is believed that the higher the frequency of a word appearing in an English online teaching video resource, the lower the frequency of that word appearing in all resources, Indicates the thematic prominence of the word in the designated English online teaching video resources [13–15]. In addition to keywords, the authority and citation of English online teaching video resources can also reflect the characteristics of the resources to a certain extent, and their feature vector expression is:

$$\begin{cases} Authority = \frac{1}{2}Level + \frac{1}{2}Cite \\ Cite = \frac{Cites}{\max Cite} \end{cases} \tag{4}$$

Among them, *Level* and *Cite* are the quantitative results of the publication level and citation volume of English online teaching video resources, respectively, while *Cites* and max *Cite* correspond to the citation volume of the resources and the largest citation volume in the resource source database [16]. Use the same method to extract and fuse feature vectors, and ultimately obtain the comprehensive feature extraction results of English online teaching video resources [17].

2.3 Calculate the Similarity Between User Interests and English Online Teaching Video Resources

In the process of calculating the similarity between user interests and English teaching video resources, the keywords searched by learners are used as user interest features to obtain the similarity between user interests and English online teaching video resources. The formula for calculating the shortest path between search keywords and user interests in English online teaching video resources is as follows:

$$\text{ShortestPath}(\gamma_2, \gamma_1) = \begin{cases} \text{ShortestPath}(\gamma_2, \gamma_1) \\ 1 \quad (\mathit{if}\ \gamma_2 \in \lambda\phi_i \text{ and } \gamma_1 \in \lambda\phi_i) \\ 0 \quad (\mathit{if}\ \gamma_2 = \gamma_1 \text{ or } \gamma_2 \text{ and } \gamma_1 \text{ no path exists}) \end{cases} \tag{5}$$

According to the collaborative filtering principle, the definition formula of the interest correlation degree between user interest point γ_2 and interest point γ_1 is:

$$KCD(\gamma_2, \gamma_1) = \begin{cases} \frac{in|\gamma_2|}{\text{ShortestPath}(\gamma_2, \gamma_1)} \\ 0 \quad (\textit{if} \ \text{ShortestPath} \ (\gamma_2, \gamma_1) = 0) \end{cases} \tag{6}$$

From the above equation, it can be seen that when the shortest distance between user interest points γ_2 to γ_1 is not equal to zero, the interest correlation degree between user interest points γ_2 to γ_1 is the shortest distance ShortestPath(γ_2, γ_1) between interest points γ_2 to γ_1. When the shortest distance between interest points γ_2 to γ_1 is zero, it indicates that there are no identical interest points for English online teaching video resources between user interest points γ_2 to γ_1. When there is no relationship between interest points, it can be considered that the interest point has no value to the user and its similarity can be set to 0.

In the collaborative filtering English online teaching video resource database $\Omega(\vartheta, \chi)$, ϑ represents the combination of user interest points, and χ represents the set of similarity between user interest points. For user ξ with historical search behavior, their point of interest $\lambda(\xi)$ is a collection of interest points for all English online teaching video resources that the user has searched for historically. The English online teaching video resource database $\lambda(\xi)$ includes a combination of interest points for all English video teaching ς. The connection between Resource ς and the English online teaching video resource database represents the resource connection between User Interest ς and English online teaching video resource ξ. The calculation formula is as follows:

$$KCD(\varsigma, \xi) = \sum_{\varepsilon_i \in \lambda(\varsigma)} \frac{in|\varepsilon_i|}{\text{ShortestPath} \ (\varepsilon_i, \lambda(\xi))} \tag{7}$$

Among them, ε_i represents any user interest point in the English online teaching video resource database, the depth of the user interest point $|\varepsilon_i|$ represents the importance of the user interest point, and ShortestPath $(\varepsilon_i, \lambda(\xi))$ represents the shortest path of all user interest points [18–20].

After calculating the connectivity between user interest points and English online teaching video resources, the similarity between user interest points and English online teaching video resources is calculated using the connectivity of English online teaching video resources. The calculation formula is as follows:

$$\text{Interest}(\xi, \varpi) = \frac{1}{|\Re(\xi)|} * \sum_{j \in \Re(\xi)} \frac{\overrightarrow{\omega_i} \cdot \overrightarrow{\omega_j}}{|\overrightarrow{\omega_i}| \cdot |\overrightarrow{\omega_j}|} \tag{8}$$

Among them, ξ represents the set of user interest points, ϖ represents the set of keywords that users search for in English online teaching video resources, $\Re(\xi)$ represents the set of user historical interest points, $|\Re(\xi)|$ represents the number of historical English teaching video resources, and $\overrightarrow{\omega_i}$ and $\overrightarrow{\omega_j}$ represent the similarity between English online teaching video resources i and j, respectively. In the calculation process of similarity, the higher the similarity between user interests and English online teaching video resources, the closer the preferences of English online teaching video resources and user interest points are, indicating that the English online teaching video resources are more worthy of recommendation.

3 Personalized Recommendation of English Online Teaching Video Resources Based on Deep Learning

Using graph convolutional neural networks in deep learning to predict users' preference for English online teaching video resources. A heterogeneous graph constructed using $P = \langle W, U \rangle$ representation, where set W represents all nodes, nodes in W are divided into two categories: X and Y, X represents the set of English online teaching video resource nodes x, and Y represents the set of user nodes y; Set U represents all edges, and the edges in set U are also divided into two categories. The interaction behavior a_{xy} generated by user y and English online teaching video x constitutes one type of edge set A, and the switching behavior generated by user b_{ij} from English online teaching video i to English online teaching video j constitutes another type of edge set B. The weight of this edge θ is determined by the number of times the user switches in the dataset. If $\theta = 0$, no switching behavior occurs, that is, the edge does not exist.

Design an embedding layer based on unique hot coding based on user identity identification. Construct an embedding layer matrix C for m user and an embedding layer matrix E for n English online teaching video resources. If the dimension of C is $m \times D$, the dimension of E is $n \times E$, where D represents an adjustable low dimensional spatial dimension. Set unique hot coding ϕ_y and ϕ_x for the user and English online teaching video resources respectively, and combine matrices C and E, The low dimensional representation vectors c_y and e_x for users and English online teaching video resources can be obtained as follows:

$$\begin{cases} c_y = C^T \phi_y \\ e_x = E^T \phi_x \end{cases} \tag{9}$$

In the personalized recommendation algorithm for English online teaching video resources, users' preferences need to be divided into two parts: long-term preferences and short-term preferences, and modeled separately. Among them, long-term preferences are fixed preferences that users do not easily change over time, and vice versa, short-term preferences.

Modeling the user's long-term preferences through the representation vector of all English online teaching video resources nodes that the user has watched. This involves propagating and aggregating all English online teaching video resources that the user has watched to the user nodes. B_y represents the set of all English online teaching video resources that the user y has interacted with, and Y_x represents the set of all users who have watched English online teaching video resources x, $\alpha_l^{(k+1)}$ and $\beta_l^{(k+1)}$ respectively represent the network parameters and offset parameters when propagating from layer 1 to layer $k + 1$, ϑ represents the nonlinear activation function, $aggregate(\cdot)$ represents the vector aggregation operation, $g_{y-l}^{(k)}$ and $g_{y-l}^{(k+1)}$ represent the user y layer k and layer $k + 1$ representation vectors, $h_{x-l}^{(k)}$ and $h_{x-l}^{(k+1)}$ represent the video x layer k and layer

$k + 1$ representation vectors, and the propagation formulas of $g_{y-l}^{(k+1)}$ and $h_{x-l}^{(k+1)}$ are as follows:

$$\begin{cases} g_{y-l}^{(k+1)} = \vartheta\left(\alpha_l^{(k+1)}\left(g_{y-l}^{(k)} + aggregate\left(h_{x-l}^{(k)}|x \in Y_x\right)\right) \right. \\ \left. \qquad +\beta_l^{(k+1)}\right) \\ h_{x-l}^{(k+1)} = \vartheta\left(\alpha_l^{(k+1)}\left(h_{y-l}^{(k)} + aggregate\left(g_{x-l}^{(k)}|y \in B_y\right)\right) \right. \\ \left. \qquad +\beta_l^{(k+1)}\right) \end{cases} \tag{10}$$

Modeling the short-term preferences of users through the representation vector of their previous interaction behavior, that is, the directed edge from the previous English online teaching video resource to the next English online teaching video resource serves as the propagation path of the vector. This can decouple the modeling of long-term preferences and short-term preferences, avoiding mutual influence between the two.

Use set R_x to represent all videos with x directed edges, $\alpha_s^{(k+1)}$ and $\beta_s^{(k+1)}$ to represent network parameters and bias parameters when propagating from layer 1 to layer $k + 1$, $aggregate(\cdot)$ represents vector aggregation operation combined with weights, $h_{x-s}^{(k)}$ and $h_{x-s}^{(k+1)}$ represent video x's k and $k + 1$ layer representation vectors, $h_{i-s}^{(k)}$ represents the next English online teaching video resource i's k layer representation vector, and the propagation formula for $h_{x-s}^{(k+1)}$ is as follows:

$$h_{x-l}^{(k+1)} = \vartheta\left(\alpha_s^{(k+1)}\left(h_{y-s}^{(k)} + aggregate\left(h_{i-s}^{(k)}|i \in R_x\right)\right) + \beta_s^{(k+1)}\right) \tag{11}$$

Based on the long and short term preference vector propagation, two components of English online teaching video resources were obtained, and the two components were merged to obtain a user preference comprehensive vector $h^{(k+1)}$ as follows:

$$h^{(k+1)} = h_{x-l}^{(k+1)} + h_{x-s}^{(k+1)} \tag{12}$$

After obtaining user preferences, calculate the cosine similarity between user preference feature vector $h^{(k+1)}$ and target English online teaching video resource cosine similarity ω of feature vector η_{s_i}:

$$\omega = \frac{\sum_{k=0,i=1}^{N} \left(h^{(k+1)} \times \eta_{s_i}\right)}{\sqrt{\sum_{k=0}^{N} \left(h^{(k+1)}\right)^2} \sqrt{\sum_{i=1}^{N} \left(\eta_{s_i}\right)^2}} \tag{13}$$

Among them, N is the vector dimension.

Combining user preferences for English online teaching video resources with ω calculating user preferences for target English online teaching video resources, f_{pq} setting the similarity between English online teaching video resources i and j as ω_{ij}, user y preference for v types of English online teaching video resources as $T(y, v)$, and user y

preference for good English online teaching video resources as B_y, The set of L English online teaching video resources with the highest similarity to English online teaching video resources i is $K(i, L)$, and the recommended result f_{pq} is as follows:

$$f_{pq} = \underset{j \in v}{T}(y, v) \sum_{i \in B_y} \omega_{ij} \tag{14}$$

Obtain the top f_{pq} user preference based on actual needs, generate a recommendation list, and complete personalized recommendation of English online teaching video resources based on deep learning.

In summary, the flowchart of the personalized recommendation method for English online teaching video resources based on machine learning is shown in Fig. 1.

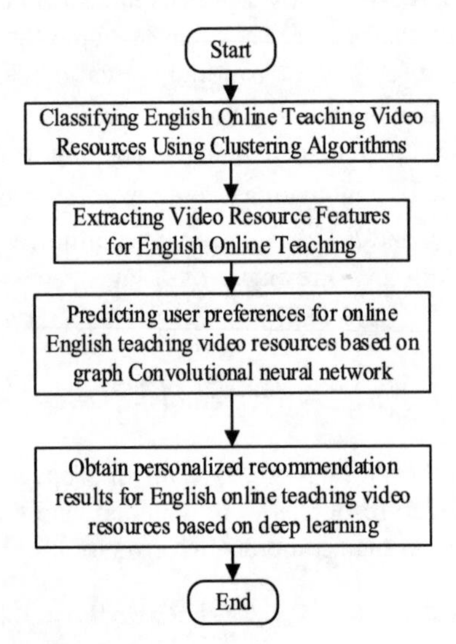

Fig. 1. Flow chart of personalized suggestion method for online teaching video resources based on machine learning

4 Experimental Analysis

4.1 Experimental Preparation

In order to test the performance of the personalized recommendation method for English online teaching video resources based on machine learning, 300 users were selected for testing, and their data sources covered multiple majors in a certain university. Among them, students majoring in computer science, Electronic engineering and information technology accounted for 30% of the total, students majoring in business management, marketing and international trade accounted for 30% of the total, students majoring in education, psychology and sociology accounted for 20% of the total, and students majoring in other fields of literature, art, medicine and engineering accounted for 20% of the total. The data sources of these 300 users can include multiple channels such as user registration information, user behavior data, user interest surveys, user reviews and feedback, and social media data. By comprehensively analyzing these data, we can better understand users' needs and interests, and provide data support for the testing and optimization of personalized recommendation methods based on machine learning.

And the following testing environments were deployed:

Hardware environment: a processor with a frequency of 2.4 GHz and two computers with a Windows 7 operating system;

Software environment: 32-bit server, Windows 7 operating system, and MySQL database.

The specific testing process is as follows: users first browse all English video teaching resources according to the operating steps, select the teaching video resources they are interested in, and record the recommended results based on the learning process.

4.2 Setting Evaluation Indicators

In order to verify the recommendation effectiveness of the recommendation system in the article, accuracy, recall, and recommendation path completion are selected as evaluation indicators, and the calculation formula is:

$$A = \frac{TP + TN}{TP + FP + FN + TN} \tag{15}$$

$$D = \frac{TP}{TP + FN} \tag{16}$$

Among them, A represents accuracy, D represents recall, TP represents the amount of correctly recommended teaching video resources, TN represents the amount of correctly recommended other data, FP represents the amount of incorrectly recommended teaching video resources, and FN represents the amount of unrecommended teaching video resources.

The higher the accuracy and recall rate of the system, the better the accuracy and effectiveness of the system in the application process. The higher the completion of the recommended path, the higher the success rate of teaching video resource recommendation, and the better the recommendation effect can be achieved.

4.3 Analysis of Test Results

In order to highlight the effectiveness of the recommendation system in the article, the methods of reference [3] and reference [4] were compared to test the accuracy, recall, and recommendation path completion of three methods in recommending English video teaching resources. The results are as follows.

The accuracy test results of English video teaching resource recommendation using three methods are shown in Fig. 2.

The results of Fig. 2 show that the accuracy of recommending English video teaching resources using the methods of reference [3] and reference [4] is below 80%, while the

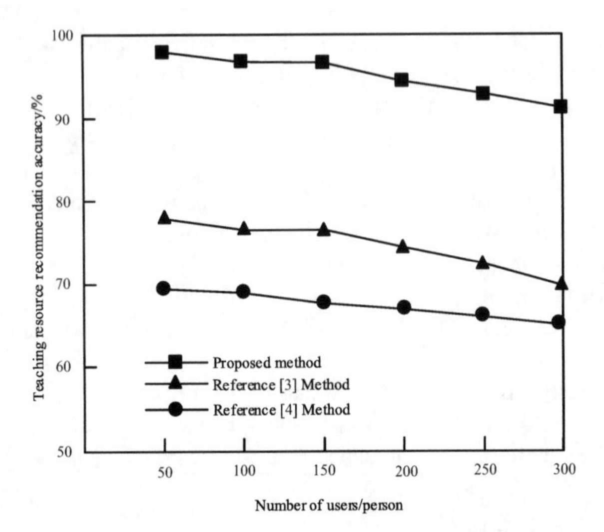

Fig. 2. Test results of accuracy in recommending English online teaching video resources

accuracy of recommending English video teaching resources using the proposed method is 92% for 300 users. This indicates that the proposed method can accurately classify English online teaching video resources using machine learning algorithms and extract English online teaching video resources that users are interested in, Effectively improving the accuracy of recommendations.

The recall rate test results of English video teaching resource recommendations using three methods are shown in Fig. 3.

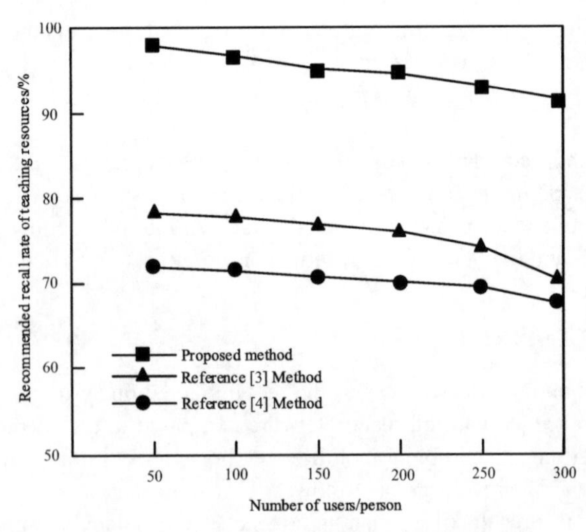

Fig. 3. Results of the recall rate test for English online teaching video resources recommendation

From the results in Fig. 3, it can be seen that in the recall test, the video resource recommendation recall rates of the methods in reference [3] and [4] are both below 80%, while the video resource recommendation recall rate of the proposed method is 91%, indicating that the recommendation effect of the proposed method is better.

The recommended path completion test results of the three methods are shown in Fig. 4.

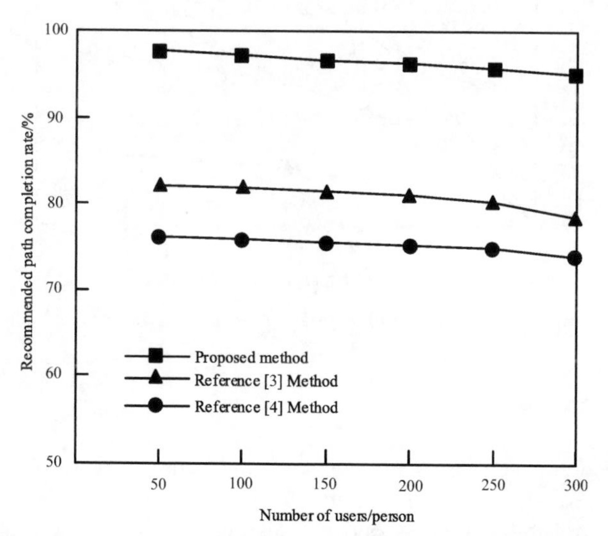

Fig. 4. Completion Test Results of Recommended Path for English Online Teaching Video Resources

According to the results in Fig. 4, it can be seen that the recommended path completion test results of the methods in reference [3] and reference [4] are relatively close, both between 70% and 85%. However, when using the proposed method to recommend English video teaching resources, the recommended path completion rate for 300 users is 95%, which can accelerate the success rate of English video teaching resources recommendation and have better recommendation effects.

On this basis, the time consumption of personalized recommendation of English online teaching video resources using three methods was tested, and the experimental comparison results are shown in Fig. 5.

Analyzing Fig. 5, it can be seen that the recommended English online teaching video resources for the method in reference [3] take 20 s, the recommended English online teaching video resources for the method in reference [4] take 14 s, and the recommended English online teaching video resources for the method in this article take 5 s. The English online teaching video resource recommendation method proposed in this article takes less time and the efficiency of the recommendation algorithm is good.

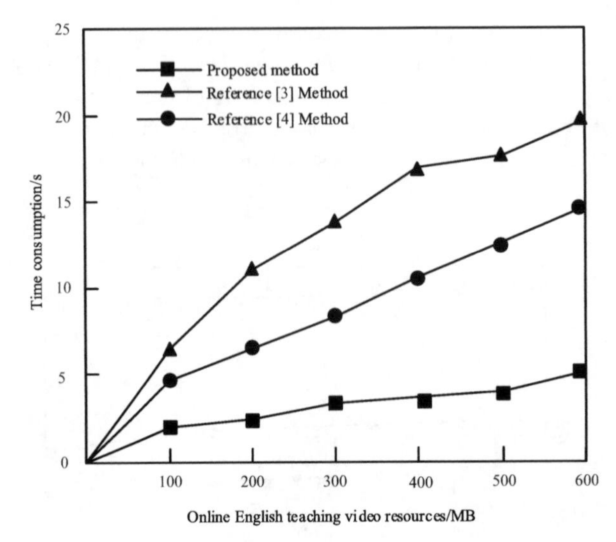

Fig. 5. The time-consuming test results of recommending English online teaching video resources

5 Conclusion

This study proposes a personalized recommendation method for English online teaching video resources based on machine learning technology. By preprocessing and analyzing multidimensional data such as students' learning behavior, a personalized recommendation model is constructed to achieve personalized recommendation of English online teaching video resources. Compared with traditional English online teaching video resource methods, this method has better recommendation accuracy, recall rate, and recommendation path completion.

In addition, this study explores the application of machine learning technology in the recommendation of English online teaching video resources by optimizing and validating the model, providing new ideas and methods for related research. In the future, we will continue to improve this method, improve recommendation accuracy and coverage, and strive to achieve greater practical results in students' English learning and online English education.

References

1. Zou, F., Chen, D., Xu, Q., et al.: A two-stage personalized recommendation based on multi-objective teaching–learning-based optimization with decomposition. Neurocomputing **452**, 716–727 (2021)
2. Zhou, L., Wang, C.: Research on recommendation of personalized exercises in English learning based on data mining. Sci. Program. **2021**, 1–9 (2021)
3. Huang, Y., Zhu, J.: A personalized English learning material recommendation system based on knowledge graph. Int. J. Emerg. Technol. Learn. (Online) **16**(11), 160 (2021)
4. Li, H., Zhong, Z., Shi, J., et al.: Multi-objective Optimization-based recommendation for massive online learning resources. IEEE Sens. J. **21**(22), 25274–25281 (2021)

5. Ma, L.: An immersive context teaching method for college English based on artificial intelligence and machine learning in virtual reality technology. Mob. Inf. Syst. **2021**, 1–7 (2021)
6. Honrubia-Escribano, A., Villena-Ruiz, R., Artigao, E., et al.: Advanced teaching method for learning power system operation based on load flow simulations. Comput. Appl. Eng. Educ. **29**(6), 1743–1756 (2021)
7. Paul, A., Wu, Z., Liu, K., et al.: Robust multi-objective visual Bayesian personalized ranking for multimedia recommendation. Appl. Intell. 1–12 (2022)
8. Naserian, E., Wang, X., Dahal, K.P., et al.: A partition-based partial personalized model for points-of-interest recommendations. IEEE Trans. Comput. Soc. Syst. **8**(5), 1223–1237 (2021)
9. Wang, X., Zhang, D., Asthana, A., et al.: Design of English hierarchical online test system based on machine learning. J. Intell. Syst. **30**(1), 793–807 (2021)
10. Liu, Z., Wang, L., Li, X., et al.: A multi-attribute personalized recommendation method for manufacturing service composition with combining collaborative filtering and genetic algorithm. J. Manuf. Syst. **58**, 348–364 (2021)
11. Prakash, R., Anoosh, R.K., Sharon, S.: Comparative study of machine learning algorithms for product recommendation based on user experience. ECS Trans. **107**(1), 19551–19562 (2022)
12. Pilani, A., Mathur, K., Agrawal, H., et al.: Contextual Bandit approach-based recommendation system for personalized web-based services. Appl. Artif. Intell. **35**(7), 1–16 (2021)
13. Li, X.: A new evaluation method for English MOOC teaching quality based on AHP. Int. J. Continuing Eng. Educ. Life-Long Learn. **32**(2), 201–215 (2022)
14. Zhang, Z.: A method of recommending physical education network course resources based on collaborative filtering technology. Sci. Program. **2021**(10), 1–9 (2021)
15. Chiu, M.C., Huang, J.H., Gupta, S., et al.: Developing a personalized recommendation system in a smart product service system based on unsupervised learning model. Comput. Ind. **128**(10), 103421 (2021). https://doi.org/10.1016/j.compind.2021,128:103421-103440
16. Li, M., Bao, X., Chang, L., et al.: Modeling personalized representation for within-basket recommendation based on deep learning. Expert Syst. Appl. **192**(8), 1–14 (2021)
17. Yang, X., Zhou, Z., Xiao, Y.: Research on students' adaptive learning system based on deep learning model. Sci. Program. **2021**(13), 1–13 (2021)
18. Liang, X., Yin, J.: Recommendation algorithm for equilibrium of teaching resources in physical education network based on trust relationship. J. Internet Technol. **3**(1), 133–141 (2022)
19. Wang, M., Wang, Y.: Research on English teaching information pushing method based on intelligent adaptive learning platform. Int. J. Continuing Eng. Educ. Life-Long Learn. **31**(2), 133–151 (2021)
20. Bian, L.: Integration of 'Offline + Online' teaching method of college English based on web search technology. J. Web Eng. **20**(4), 1207–1217 (2021)

Personalized Recommendation Method for Tourist Attractions Based on User Information Mixed Filtering

Hongshen Liu(✉) and Honghong Chen

Heilongjiang Polytechnic, Haerbin 150080, China
okjhn12300@126.com

Abstract. In order to improve the effectiveness of tourist attraction recommendation, this article proposes a personalized recommendation method for tourist attractions based on mixed filtering of tourist information. This method includes two parts: the construction of a tourist attraction information database and the personalized recommendation method for tourist attractions. Among them, the construction of the tourist attraction information database includes three steps: mining tourist attraction information based on association rules, updating tourist attraction data, and constructing a tourist attraction information feature vocabulary based on topic similarity clustering. The personalized recommendation methods for tourist attractions mainly include two aspects: describing the semantic association of tourist attraction information and selecting the optimal personalized recommendation path for tourist attraction information. The experimental results show that the proposed method improves the accuracy and efficiency of personalized recommendation for tourist attractions.

Keywords: User Information · Mixed Filtration · Scenic Spot · Theme Similarity Clustering · Personalized Recommendations

1 Introduction

Nowadays, tourism has become a common way of leisure and entertainment for people. However, tourists usually need to spend a lot of time and effort planning their itinerary before traveling, and due to a lack of sufficient travel information, they often find it difficult to make the best choice [1]. Meanwhile, with the continuous growth of personalized user needs, traditional tourism recommendation systems are no longer able to meet the needs of tourists. In response to these issues, tourist attraction recommendation systems have emerged [2]. It can improve users' travel experience and satisfaction by analyzing their travel preferences and behavioral data, recommending scenic spots that match their interests and needs [3].

The reference [4] method utilizes convolutional neural networks (CNN) to extract text comment features for sentiment classification, and calculates similar user groups using Pearson similarity formula to achieve tourism recommendation. Reference [5]

L. Yun et al. (Eds.): ADHIP 2023, LNICST 549, pp. 166–178, 2024.
https://doi.org/10.1007/978-3-031-50549-2_12

proposed a multi-objective tourism route recommendation method that integrates user features and group intelligence. Firstly, obtain scenic spot information and corresponding group intelligence data through websites such as Ctrip, Wanglu Travel, Baidu Index, etc.; Secondly, combining user characteristics and group intelligence data, construct the comprehensive attractiveness of scenic spots to users with different characteristics and calculate the attractiveness index of tourism routes; Finally, define a multi-objective optimization function for tourism route recommendation and generate a route recommendation list using the multi-objective genetic algorithm NSGA2. Reference [6] Method By summing up the impact of 11 situational elements on scenic spot recommendation and discussing the difference in their impact degree, a tourism scenic spot recommendation model integrating situational awareness and random forest algorithm is proposed, and the situational elements are modeled as the characteristic attributes to be considered when the decision tree splits in random forest to achieve tourism scenic spot recommendation.

The personalized recommendation system for tourist attractions places more emphasis on user satisfaction with the recommendation results, and can flexibly and accurately recommend based on user behavior and interest information. Personalized recommendation of tourist attractions can not only provide better tourism choices for tourists, but also promote the development of the tourism industry. By accurately recommending scenic spots, tourists can improve their stay time and consumption level, and promote local economic development. Therefore, this paper proposes a personalized recommendation method for tourist attractions based on mixed filtering of user information. On the basis of constructing a tourist attraction information database, personalized recommendation of tourist attractions based on user information mixed filtering is achieved by describing the semantic association of tourist attraction information and selecting the optimal personalized recommendation path for tourist attraction information. In order to enhance the experience of tourists, promote the development of the tourism industry, and promote the management and development of tourist attraction information through this study.

2 Construction of Tourist Attraction Information Database

2.1 Mining Tourist Attraction Information Based on Association Rules

Using association rule mining algorithm [7–9] to mine tourist attraction information, based on the principle of association rule mining algorithm, strong association rules with the minimum support and minimum confidence are found in numerous tourist attraction information databases to achieve tourist attraction information mining.

The definition of HGD refers to the percentage of the thing containing the Information set of tourist attractions in the whole tourist attraction information database, recorded as $h(i)$, and DKL refers to the union of two Information set of tourist attractions in the whole transaction data. Suppose that in the Information set $I = (i_1, i_2, \cdots, i_n)$ of tourist attractions, there is a weight value for any tourist attraction information i. This weight value is used to measure the importance of tourist attraction information in the whole set. The greater the weight value, the more important the tourist attraction information is. On this basis, the tourist attraction information in the set is sorted according to the weight value, and a combination of ranking from large to small is obtained to form a linear ordered set.

Use z, x to represent the element in the Information set I of the tourist attraction. If $z < x$, it means that z is in front of x. If the weighted support of z is defined as H(z), then the minimum weighted support of the tourist attraction information is:

$$\mathrm{H}(z) = \frac{s - f(z)}{D} \tag{1}$$

In formula (1), D represents the number of tourist attraction information database data, s represents the weighted frequent tourist attraction Information set, and $f(\cdot)$ represents the calculation factor of weighted support.

According to the above formula, calculate the minimum weighted support of the tourist attraction information database. Based on this, calculate the data confidence using the association rule mining algorithm [10] as follows:

$$Z = \frac{k(z \cup x)}{n(x)\mathrm{H}(z)} \tag{2}$$

In formula (2), $k(z \cup x)$ represents the number of times two tourist attraction information simultaneously appears in the tourist attraction information database, and $n(x)$ represents the degree of data correlation.

According to the above calculation, find out the problem of frequent Information set of tourist attractions in the tourist attraction information database, determine the reliable relationship rules in the tourist attraction information database, so as to complete the information mining of tourist attractions, and determine the process of information mining of tourist attractions as shown in Fig. 1:

2.2 Update of Tourist Attraction Data

2.2.1 Differences in Tourist Attractions

Firstly, the evaluation of different tourist attractions by the same user is calculated by subtracting the scores to determine the degree of dissimilarity of tourist attractions, which is a measure of the dissimilarity of tourist attractions.

Set x as the evaluation set after recommending tourist attractions, i and j as any two tourist attractions, with the same user U rating them U_1 and U_2 respectively. For the currently selected tourist attractions i and j, set $S(x)$ as the user who overrated, N as the number of users, and the similarity between tourist attractions i and j is:

$$d = \frac{Z(U_1 - U_2)}{N \times S(x)} \tag{3}$$

According to formula (3), if the user's ratings for i and j are very close, the degree of difference between tourist attractions is very small, while the opposite is true. The dissimilarity calculation of tourist attractions is the result of analyzing from the user rating dimension. If analyzed from the tourist attraction rating dimension, the dissimilarity calculation of users will be obtained.

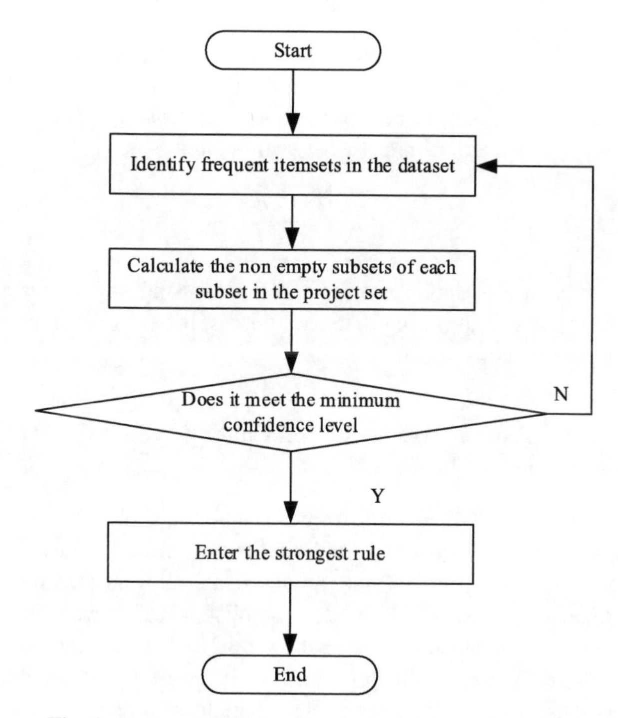

Fig. 1. Process of Mining Tourist Attraction Information

2.2.2 User Dissimilarity

Extract absolute values by subtracting the ratings of different users for the same tourist attraction, and obtain the degree of user dissimilarity. This calculation can be used to measure the degree of difference in user preferences.

Set X as the training set, A and B as any two users, and the same tourist attraction i is rated by users A and B respectively. User A has a rating of A_1, user B has a rating of B_1, and the tourist attraction that is overrated by A and B is $S'(x)$, V is the number of tourist attractions. The difference between user A and user B is:

$$d' = \frac{A_1 - B_1}{M \times S'(X)} \tag{4}$$

According to formula (4), if the ratings of users A and B in the same tourist attraction are very close, then the dissimilarity of users is small, while the dissimilarity is large.

2.2.3 Tourist Attraction Update Process

From the above calculation, it can be concluded that the dissimilarity matrix can perform local online updates on newly added data, and only one rating vector is used for each update, without the need for mutual operations with other vectors, thus saving computational time. The specific update method is as follows:

After calculating the tourist attraction i, several new user ratings are added. If n_1 is the new user of the tourist attraction, n_2 is the scoring set of old users, A_1 and B_1 are selected as any two users, and users A_1 and B_1 score tourist attraction i as a_1 and b_1, then the update result of the dissimilarity between A_1 and B_1 is:

$$d'' = d' + \frac{d}{|b_1 - a_1|} \tag{5}$$

In summary, because updating only requires processing one vector at a time and does not require the use of all rating data for updates, it has the characteristics of less workload and fast processing speed.

2.3 Constructing a Tourism Attraction Information Feature Thesaurus Based on Theme Similarity Clustering

Separate tourist attraction information from conventional information and design an algorithm that can filter such information. Using topic similarity clustering as the core algorithm [11, 12], design a model that can aggregate relevant information and filter irrelevant information. Based on this model, establish a feature vocabulary for tourist attraction information, and cluster based on the similarity between these vocabulary and tourist attraction information. In this process, it is necessary to first establish a suitable parameter interval using the existing vocabulary, which can directly affect the clustering effect of tourist attraction information and the accuracy of the feature vocabulary. Among them, the selection of clustering centers is the most cautious step, and the feature lexicon obtained through different clustering centers has completely different characteristics. At this point, vectors can be used as the basis for discriminating information similarity and as clustering tools for tourist attraction information, and the frequency of tourist attraction information occurrence can be treated as independent semantic units. Among them, the K value of topic similarity clustering is a very important calculation node, and the topics of different clustering centers are cross connected to calculate the information similarity in pre training through clustering algorithms. This article uses the method of extracting feature categories to establish a feature text space, and appropriately introduces the feature factor vocabulary of tourist attraction information. By increasing the weight of low-frequency words, a cross channel of word frequency is established, enabling the algorithm to quickly collect vocabulary of tourist attraction information with different weights. In the dataset of the feature vocabulary, the frequency calculation expression of a vocabulary can be obtained as follows:

$$\lambda = \frac{d'' \sum_{x=1}^{n} b_x}{\sum_{y=1}^{a} \sum_{x=1}^{b} b_{yx}} \tag{6}$$

In the formula, λ represents the characteristic frequency coefficient of a certain vocabulary; y represents the number of irrelevant information related vocabulary; b_x represents the number of times feature words appear in the category of tourist attraction information. The element feature similarity of each word frequency can be obtained through formula (6) as follows.

$$w = \frac{n(p+q)}{(p+v)(q+v)(u+v)} \times \lambda \tag{7}$$

In the formula, p represents the number of documents containing information about tourist attractions and belonging to irrelevant information; q represents the number of documents containing information about tourist attractions that are not irrelevant; u represents the number of documents that do not contain tourist attraction information and belong to irrelevant information; v represents the number of documents that do not contain tourist attraction information and do not belong to irrelevant information.

The larger the word frequency coefficient obtained, the higher the correlation between the word and tourist attraction information, and the more suitable it is to be included in this feature vocabulary; The smaller the word frequency coefficient obtained, the smaller the correlation between the word and tourist attraction information, and the less suitable it is to be included in the feature vocabulary. By traversing all the information vocabulary in the tourist attraction information database, the latest feature words in the tourist attraction information can be automatically obtained and stored in the feature vocabulary database.

3 Personalized Recommendation Methods for Scenic Spots

3.1 Describe the Semantic Association of Tourist Attraction Information

Using the related technologies of Ontology Semantic Web, realize the digitization of tourist attraction information stored in the tourist attraction information database, describe terms, descriptors, titles and other normative documents, obtain descriptive metadata, and send it to users as associated data. First, according to the Cool URIs naming specification formulated by the Semantic Web, URI naming is carried out for the tourist attraction information of the tourist attraction museum. With the help of various description methods provided by FRBR, an associated data vocabulary set is created to describe the semantic ontology of tourist attraction information. The specific types are shown in Table 1:

Based on Table 1, construct a progressive transformation mechanism for the semantic association of tourist attraction information, and select the corresponding subnet type involved. Then describe the associated semantics of tourist attraction information, use entity extraction mechanism, and use D2RQ transformation tool to transform tourist attraction information into RDF metadata form. Based on this, create new semantic descriptive metadata. Finally, based on the characteristics of user dissimilarity, the publishing mode of associated data is selected to expand the information of tourist attractions, gather the Open APIs provided by local tourism, and build a network environment with stronger correlation and scheduling of tourist attraction information. By combining services and associated data, the integrated system links internal and external tourist attraction information. This completes the description of the semantic association of tourist attraction information.

Table 1. Semantic Classification Standards for the Association of Tourist Attraction Information

Semantic name	Subnet type involved	Segmentation criteria
Hierarchical relationship	P-P; K-K; M-M	Genus
Citation relationship	P-P; M-M	Entity
Related relationships	P-P; K-K; M-M	Overall part
Equivalence relationship	P-P; K-K; M-M	Synonymous
Attribute relationship	P-P; P-K; K-K; M-M	Synonymous
Discussing Relationships	K-K; K-M; M-M	Synonym
See relationship	P-K; K-K; K-M; M-M	Antonym

3.2 Selecting the Personalized Recommendation Path for the Best Tourist Attraction Information

After publishing the associated data of tourist attraction information, it is necessary to optimize the associated data network based on user access and retrieval information, select the optimal personalized recommendation path for tourist attraction information, that is, the link path of the associated data, and perform operations such as adding and modifying tourist attraction information data. Firstly, standardize the association semantics [13], count the frequency of different semantic query words, determine the core query words, determine the four attributes of the core query words, classify them, and calculate the similarity between different tourist attraction information and the word. Using the similarity distance formula, the similarity Q calculation formula is:

$$Q = w\sum_{j=1}^{4}\left(a_k - a_j\right) \tag{8}$$

In the formula, j is the four attributes of the core query word, a is the associated data of tourist attraction information, and k is the visual spatial dimension of this data [14, 15]. When $k = 1$ is used, it represents the true distance between the core query word and the spatial dimension. When $k \neq 1$ is used, it is the definite distance, representing the total absolute wheelbase on the spatial dimension.

Transform formula (8) to obtain the optimal path S for the transformation boundary distance in the visualization spatial dimension as follows:

$$S = \frac{1}{cQ} \tag{9}$$

In the formula, c is the frequency at which the associated data of tourist attractions and core query words appear together. When $0 < c < 1$, the optimal path S value is between (0, 1), the smaller c, the closer S value approaches 0, and when $c > 1$, the smaller c, and the closer S value approaches 1.

Based on the frequency of the occurrence of data related to different tourist attraction information, determine the values of the k, j two link parameters, so that the S value reaches the limit value, and obtain the final path of the transformation data boundary distance. Use this path to transform the related data of tourist attraction information, and obtain personalized recommendation results of tourist attraction information.

4 Experiments

4.1 Experimental Setup

The experimental environment settings required for the experiment are shown in Table 2.

Table 2. Experimental Environment Settings

Classification	Facilities	Parameter
Computer hardware	CPU	Lenovo i5-2340@3.40 GHz
	Running memory	4 GB
	Hard drive	2T
	Graphics card	GTX 1080
Computer software	Background development language	Python
	Information processing framework	Keras
	Tourist attraction information database	SQL
	Operating system	Windows 10
	Result calculation	MATLAB

The methods used in this article, reference [4], and reference [5] were used for testing.

4.2 Result Analysis

(1) Comparing the time required to recommend tourist attraction information using different methods under the same tourist attraction information, the test results are shown in Fig. 2.

Analyzing Fig. 2, it can be seen that under the same amount of data information, the testing time of this method is shorter than that of the two comparative methods, and the highest recommended time for tourist attraction information is 4 s. In summary, it can be seen that the method in this article accurately excavates the information of tourist attractions by constructing a tourist attraction information database, reduces the time for recommending tourist attraction information, and improves recommendation efficiency.

(2) Under the same amount of information data, three methods were tested for the accuracy of recommending tourist attraction information, and the test results are shown in Fig. 3.

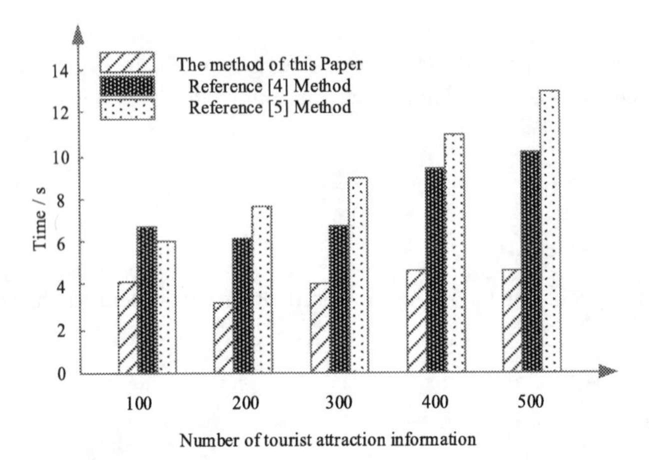

Fig. 2. Test Results of Recommended Time for Tourist Attraction Information

By analyzing Fig. 3, it can be seen that in the same amount of data information, compared with the two comparison methods, the accuracy of the proposed method in recommending tourist attraction information is higher and the detection accuracy tends to be stable. Among them, the accuracy of the proposed method for recommending tourist attraction information is higher than 88%. This is because the proposed method constructs a tourist attraction information feature vocabulary based on topic similarity clustering, which describes the semantic association of tourist attraction information and improves the accuracy of tourist attraction information recommendation.

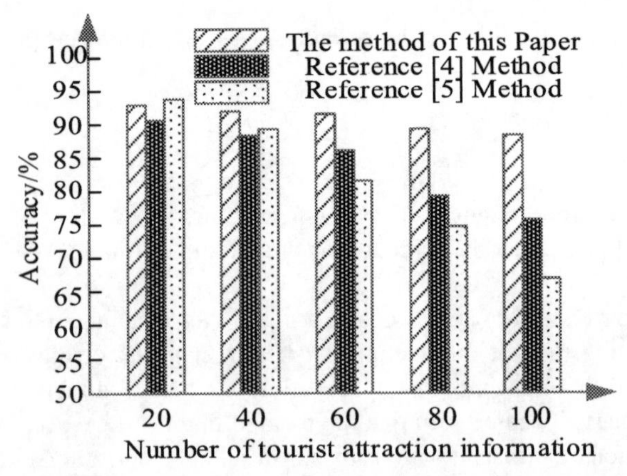

Fig. 3. Test Results of the Accuracy of Tourist Attraction Information Recommendation

(3) Three sets of experiments were conducted on 9 different tourism node elements, and their recommended tourist attraction information and comparison results are shown in Fig. 4:

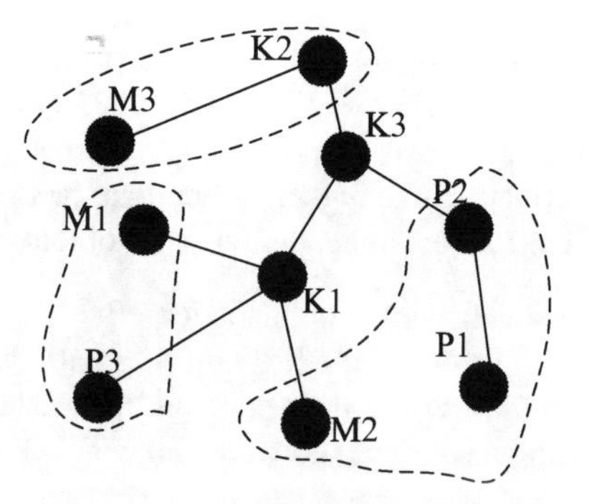

(a) Results of the method construction in this article

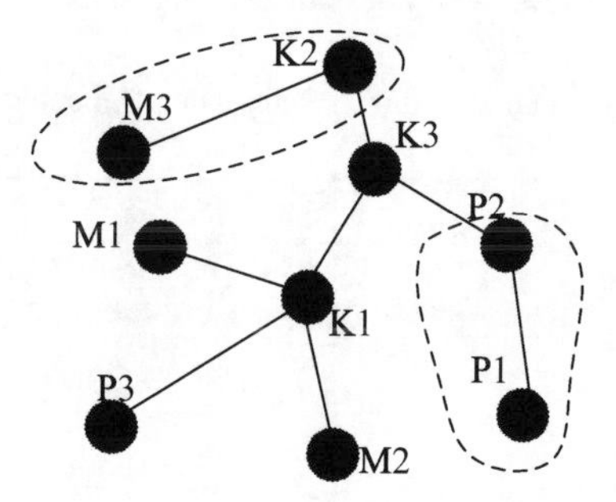

(b) Reference [4] Method Construction Results

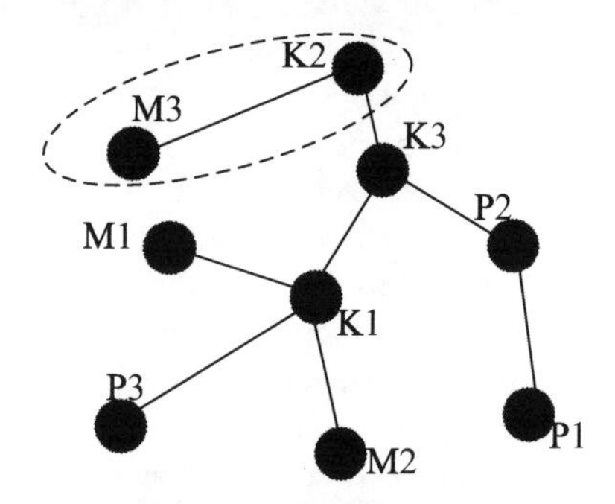

(c) Reference [5] Method Construction Results

Fig. 4. Experimental Comparison Results

It can be seen from Fig. 4 that the node set of the Information set of tourist attractions integrated by the reference [4] method and the reference [5] method is relatively loose. The reference [4] method constructs the association semantics of P2, M2 and K2, M3, and the reference [5] method value only constructs the association semantics of K2, M3, and the nodes P3, M2 do not establish a clear relationship with other nodes, while the association relationship between the nodes of this method,

There are more methods than those in reference [4] and reference [5], and three associated semantics of P1P2M2, K2M3, and M1P3 have been constructed. The results indicate that the method proposed in this paper reveals the relevance of tourist attraction information to a higher extent than the two comparative methods, improving the personalized recommendation effect of tourist attraction information.

To further test the personalized recommendation effect of the proposed method for tourist attractions, the satisfaction level of 100 tourists with different indicators of the recommended route results using the three methods was calculated using a 10 scale scoring system. The scoring results are shown in Table 3.

Table 3. User satisfaction results

Evaluating indicator	The method of this paper	Reference [4] method	Reference [5] method
Overall satisfaction	8.9	6.8	7.4
Overall coordination	8.6	6.7	6.8
Reflectance of Scenic Area Characteristics	9.2	7.8	6.7
Personalization level	9.3	7.4	6.9
Average	9.0	7.2	7.0

The above results indicate that the overall satisfaction, overall coordination, reflection of scenic spot characteristics, and degree of personalization of the personalized recommendation results of tourist attractions using this method are all rated at least 8.5 points by the respondents, with an average score of 9.0 points for each indicator; The average scores for various indicators in the methods of reference [4] and reference [5] are 7.2 and 7.0 respectively. The above results show that the recommended tourist attractions using this method can satisfy more users and have better design effects.

Carrying capacity is an important indicator for measuring the operational performance of corresponding methods. Complete the bearing capacity test from two aspects: different number of scenic spots and different number of tourists. Three methods are used to test the carrying capacity of different tourist attractions and numbers of tourists. The comparison results are shown in Fig. 5.

Analyzing Fig. 5, it can be seen that under the same number of scenic spots, the number of bytes transmitted per second by our method is higher than that by the methods in reference [4] and [5]. As the number of scenic spots increases, the number of bytes transmitted shows a steady growth trend, and the number of bytes transmitted changes steadily; The methods in reference [4] and reference [5] show significant fluctuations in the number of byte transfers as the number of tourist attractions increases. As the number of tourists continues to increase, the byte transmission quantity of the method proposed in this article has remained stable. In addition, the byte transmission quantity of the methods in reference [4] and reference [5] has shown a downward trend with the increase of tourists and shows significant fluctuations. The experimental results indicate

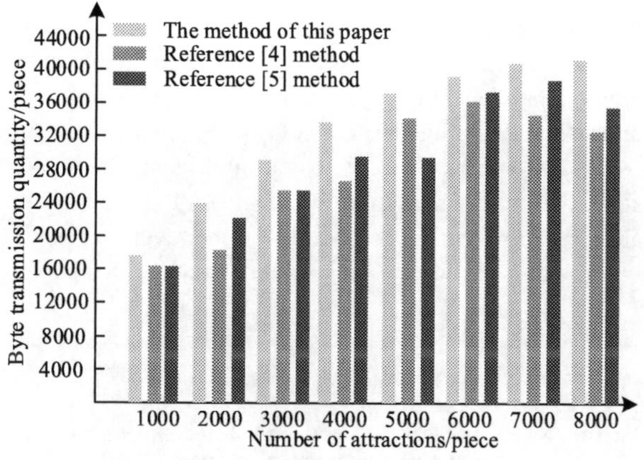

(a) Results of the method construction in this article

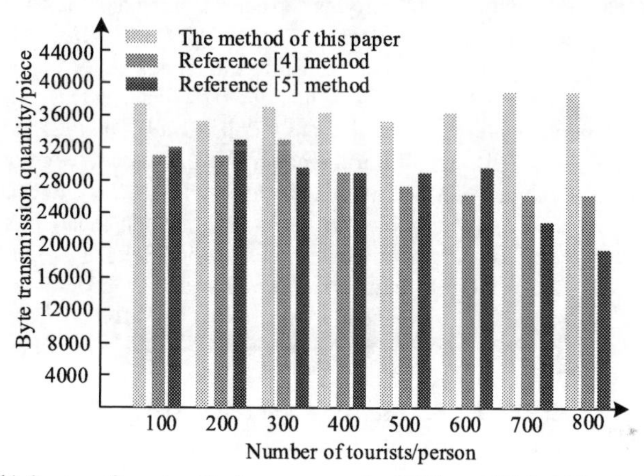

(b) System Carrying Performance under Different Visitor Numbers

Fig. 5. Load bearing performance test results of three methods

that the method proposed in this paper has a good carrying capacity, and the carrying capacity is not affected by the increase in the number of scenic spots and tourists.

5 Conclusion

This paper proposes a personalized recommendation method for tourist attractions based on mixed filtering of user information, and draws the following conclusion: this method can effectively improve users' selection experience and satisfaction with tourist attractions. This method combines user behavior data with user personal information and utilizes a hybrid filtering algorithm for user information to achieve personalized recommendation of tourist attractions. At the same time, this method optimizes recommendation results and filters user personal information to improve the accuracy of recommendations. The experimental results show that this method has higher recommendation accuracy and better user satisfaction. In summary, this method will be one of the important development directions for future tourist attraction recommendation systems and is expected to be widely applied and promoted. In the future, with the continuous development of big data and artificial intelligence technology, personalized recommendations for tourist attractions will become more accurate and intelligent. Recommendations based on location. Personalized tourist attraction recommendations can be further optimized based on information such as the user's current location, travel time, and real-time traffic conditions, combined with user history and social network information.

References

1. Qin, P., Jia, H., Huo, X., et al.: User personalized POI recommendation method integrating big data mining. Comput. Simul.**39**(6), 355–358,385 (2022)
2. Wu, X.: An ecological environmental carrying capacity estimation of tourist attractions based on structural equation. Int. J. Environ. Technol. Manag. **25**(4), 310–323 (2022)
3. Syahputri, J., Dharmowijoyo, D.B., et al.: Effect of travel satisfaction and heterogeneity of activity-travel patterns of other persons in the household on social and mental health: the case of Bandung Metropolitan area. Case Stud. Transp. Policy **10**(4), 2111–2124 (2022)
4. Zhang, J., Bai, S., Liu, S.: Design of travel recommendation model based on convolutional neural network. J. Terahertz Sci. Electron. Inf. Technol. **18**(6), 1128–1132 (2020)
5. Ma, Z., Chen, C., Huang, Z.: Multi-objective travel-route recommendation method based on integration of user features and group-intelligence. J. Geo-Inf. Sci. **24**(10), 2033–2044 (2022)
6. Wu, X., Yang, X., Zhu, F.: Tourist attractions recommendation model integrating context-awareness and random forest. Mod. Electron. Tech. **46**(6), 154–160 (2023)
7. Zhao, Z., Jian, Z., Gaba, G.S., et al.: An improved association rule mining algorithm for large data. J. Intell. Syst. **30**(1), 750–762 (2021)
8. Chen, J., Becken, S., Stantic, B.: Assessing destination satisfaction by social media: an innovative approach using importance-performance analysis. Ann. Tour. Res. **93**(Mar.), 103371.1–103371.19 (2022)
9. Lakshmi, N., Krishnamurthy, M.: Association rule mining based fuzzy manta ray foraging optimization algorithm for frequent itemset generation from socialmedia. Concurr. Comput. Pract. Exp. **34**(10), e6790.1–e6790.9 (2022)
10. Kaur, G., Datta, R.K.: Predicting the formation of tornadoes using association rule mining by studying a real life tornado event: Georgia, USA January, 2013. Mausam: J. Meteorol. Dept. India **72**(4), 813–820 (2021)
11. Guo, H.X., Wang, J.R., Peng, G.C., et al.: A data mining-based study on medication rules of Chinese herbs to treat heart failure with preserved ejection fraction. Chin. J. Integr. Med. **28**(9), 847–854 (2022)
12. Binding, C., Gnoli, C., Tudhope, D.: Migrating a complex classification scheme to the semantic web: expressing the integrative levels classification using SKOS RDF. J. Doc. **77**(4), 926–945 (2021)
13. Wan, S., Kamis, N.H., Ahmad, S., et al.: Similarity–trust network for clustering-based consensus group decision-making model. Int. J. Intell. Syst. **37**(4), 2758–2773 (2021)
14. Law, R., Pylkkanen, L.: Lists with and without syntax: a new approach to measuring the neural processing of syntax. J. Neurosci. Off. J. Soc. Neurosci. **41**(10), 2186–2196 (2021)
15. Martin, M.B.D.J., et al.: Novel methodology to visualize biomass processing sustainability & cellulose nanofiber product quality. ACS Sustain. Chem. Eng. **10**(11), 3623–3632 (2022)

Monitoring Method of Permanent Magnet Synchronous Motor Temperature Variation Signal Based on Model Prediction

Li Liu[1,2], Jintian Yin[1,2(✉)], Dabing Sun[3], Hui Li[1,2], and Qunfeng Zhu[1,2]

[1] Hunan Provincial Key Laboratory of Grids Operation and Control on Multi-Power Sources Area, Shaoyang University, Shaoyang 422000, China
yinjintian112@yeah.net

[2] School of Electrical Engineering, Shaoyang University, Shaoyang 422000, China

[3] Hunan Jianwang Shengke New Energy Technology Co., Ltd., Shaoyang 422000, China

Abstract. In order to improve the monitoring accuracy and quality of permanent magnet synchronous motor (PMSM) temperature variation signal, and achieve the ideal effect of high-precision monitoring of PMSM temperature variation signal, model prediction is introduced, and the monitoring method of PMSM temperature variation signal based on model prediction is studied. The wireless sensor technology is used to collect the temperature signals of various parts of the motor, integrate, clean, replace and protocol the original data, establish a deep learning network model to predict the characteristics of the motor temperature variation, identify the motor temperature variation signal, combine the variation pruning and variation interval, and use the delayed reporting strategy to monitor the early warning of the motor temperature variation signal, complete the monitoring of temperature variation signal of permanent magnet synchronous motor based on model prediction. The experimental analysis results show that the recall rate and accuracy rate of the design method are above 90%, and maintain detection efficiency above 97%, the monitoring accuracy of the temperature variation signal of the permanent magnet synchronous motor is high.

Keywords: Model prediction · Permanent magnet synchronous motor · Abnormal temperature signal · Monitor · Wireless sensor technology

1 Introduction

Permanent magnet synchronous motors (PMSM) are widely used as driving motors for electric vehicles, ships, etc. due to their excellent power density, efficiency and initial torque. But in actual use, its high power density will bring serious temperature rise problems, which will affect the working efficiency of the motor, and even damage the core components of the motor. For example, when the temperature exceeds a certain limit, it will cause aging or damage of the stator winding insulation layer, causing damage to the motor; It will lead to irreversible demagnetization of the permanent magnet,

L. Yun et al. (Eds.): ADHIP 2023, LNICST 549, pp. 179–194, 2024.
https://doi.org/10.1007/978-3-031-50549-2_13

which will affect the performance of the entire motor, resulting in high maintenance costs [1]. Therefore, the monitoring of temperature variation signal has always been the focus of the research of permanent magnet synchronous motor. Installing sensors inside the permanent magnet synchronous motor is the most direct method [2] to obtain the abnormal temperature signal of components. This method has high requirements for the installation position, quantity and accuracy of sensors. The internal wiring of the motor will be more complex, and the cost will also increase. Therefore, the monitoring method of indirect temperature variation signal of permanent magnet synchronous motor has always been a research hotspot in this field. Therefore, in order to ensure the safe, stable and efficient operation of the permanent magnet synchronous motor in the work, an effective and reliable method is needed to monitor the temperature variation signal of the permanent magnet synchronous motor. The temperature variation signal monitoring and thermal protection of the permanent magnet synchronous motor provide certain technical support.

Based on the above background, in recent years, more research has been done on the monitoring methods of temperature variation signal of permanent magnet synchronous motor. The existing traditional monitoring methods of temperature variation signal of permanent magnet synchronous motor can be summarized into the following two categories: first, finite element method. Finite element analysis [3] is a common method for electromagnetic design and calculation in 2D and 3D models. The temperature of permanent magnet synchronous motor [4] is obtained by discretizing the finite element of the motor model, setting the physical field boundary, conducting nonlinear transformation on the thermal boundary, establishing the heat balance equation of each element, and solving the domain total equation to simulate the heat dissipation calculation. The finite element analysis method can accurately calculate the uneven distribution of power loss inside the motor, and obtain the specific distribution of temperature field inside the motor. However, the modeling of the finite element analysis method is more complex and requires a lot of computing resources, which is difficult to meet the actual engineering needs.

Second, magnetic flux observation method. With the increase of temperature, the residual magnetic flux density of permanent magnetic materials will decrease significantly. The flux observation rule is to obtain the change of the flux through the flux observer, establish an accurate parameterized motor temperature model, and obtain the temperature of the permanent magnet synchronous motor rotor [5]. Takashi et al. proposed a magnetic temperature estimation method for variable flux leakage built-in permanent magnet synchronous motor based on flux linkage observer, which consists of Gopinath flux observer, flux linkage observer and magnet temperature estimator based on lookup table. Min et al. used stator current, permanent magnet flux and stator resistance as state variables to construct extended Kalman filter, estimate permanent magnet flux chain and stator resistance and indirectly obtain stator temperature rise [6]. The flux observation method is applicable to the whole torque range of the motor, and can track the aging effect of the permanent magnet. However, the magnetic flux observation method has high requirements for the accuracy of observation instruments in practical applications, and the magnetic flux deviation due to measurement errors has a greater impact on the prediction results of temperature, so attention must be paid to reducing the magnetic flux deviation in use.

Among the existing traditional methods, the accuracy of finite element analysis method is the highest compared with other methods, but due to the limitations of large amount of calculation, long time consumption and so on, it is not much in practical use. The flux observation method has high requirements for the accuracy of the signal observer. These classic traditional methods require researchers to have a lot of motor thermal knowledge and experience in parameter selection, and the traditional methods can no longer meet the actual needs. Therefore, a monitoring method based on model prediction for temperature variation signal of permanent magnet synchronous motor is proposed. The use of mathematical modeling to predict the temperature variation of a permanent magnet synchronous motor allows for temperature monitoring without the need for sensors. Model predictive algorithms can accurately forecast the motor's temperature based on historical data and environmental parameters. This enables appropriate control actions based on the predicted results. By continuously updating the model parameters and input variables, dynamic monitoring and adjustment of the motor's temperature variation can be achieved, ensuring its normal operation within a safe temperature range.

2 Design of Monitoring Method for Abnormal Temperature Signal of Permanent Magnet Synchronous Motor Based on Model Prediction

2.1 Temperature Data Acquisition of Permanent Magnet Synchronous Motor

In order to monitor the abnormal temperature signal of permanent magnet synchronous motor (PMSM), it is necessary to obtain the temperature monitoring data of PMSM during operation. This time, the wireless sensor technology is used to collect the temperature signal of the permanent magnet synchronous motor [7]. Suppose that the temperature wireless sensor sequence of a permanent magnet synchronous motor [8] is S, which is expressed by the formula:

$$S = \begin{bmatrix} S_1^1 & S_1^2 & \cdots & S_1^t \\ S_2^1 & S_2^2 & \cdots & S_2^t \\ \cdots & \cdots & & \cdots \\ S_n^1 & S_n^2 & \cdots & S_n^t \end{bmatrix} \tag{1}$$

where, S_n^t Represents the No n Temperature sensors at monitoring points t The status value of the time step. Given the multi-dimensional sensor sequence of the permanent magnet synchronous motor [9] in the whole day in combination with the actual scene, the permanent magnet synchronous motor [10] temperature variation signal is measured and recognized in real time using the data driven method, and the temperature variation signal record [11] is formed. In particular, the text is mainly based on model prediction for research, so the abnormal signal of motor temperature needs to be monitored online and in real time, rather than a global comparison check after obtaining all data. The following figure is the schematic diagram of motor temperature data acquisition.

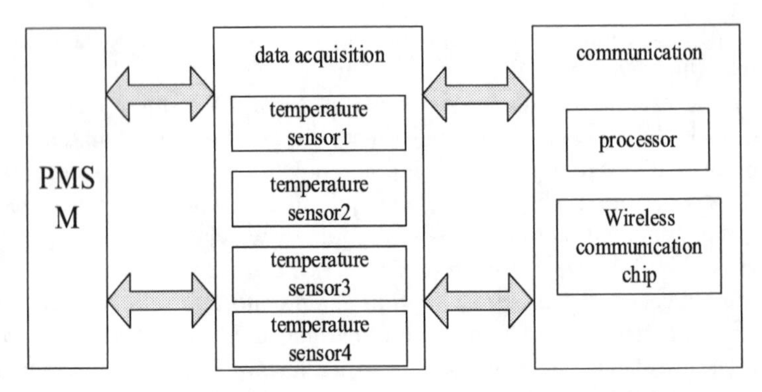

Fig. 1. Schematic diagram of motor temperature data acquisition

As shown in Fig. 1, during the operation of the motor, wireless sensors are introduced to collect the temperature data of the motor. The temperature monitoring area includes the generator winding, generator bearing, rectifier, tower base control cabinet of the permanent magnet synchronous motor [12] and the environment of the motor. The communication standard of temperature wireless sensor is based on IEEE802.15.4 standard, also known as Zigbee standard, which is promoted by international Zigbee alliance. Compared with common wireless communication standards, IEEE802.15.4 standard protocol stack is compact and simple, with low specific implementation requirements. As long as the 8-bit processor is equipped with 5 kb ROM and 64 kb RAM, it can meet its minimum needs, thus greatly reducing the cost of the chip [13]. In wireless sensor networks, the channel contention of nodes is realized by CS-MA/CA (Carrier Sense Multiple Access/Collision Avoidance) mechanism. CSMA/CA has two modes to choose from: slotted CSMA/CA and slotless CSMA/CA, which are respectively applied to beacon enabled communication and non enabled communication [14]. In this design method, the beacon enabled time slot CSMA/CA mode is used to compete for the channel access. CSMA/CA with timeslots involves the timeslot guarantee mechanism - GTS oGTS, which is similar to the time division multiple access (TDMA) mechanism, but it can dynamically allocate timeslots for devices with receiving and sending requests. The use of time slot guarantee mechanism requires time synchronization between devices. The time synchronization in IEEE 802.15.4 is achieved through the "super frame" mechanism.

The superframe divides the communication time into active and inactive parts. During inactivity, the devices in the PAN network will not communicate with each other, so they can enter the sleep state to save energy. The active period of the superframe is divided into three stages: beacon frame transmission period, competitive access period (CAP) and non competitive access period (CEP).

In the contention access period of superframe, IEEE 802.15.4 network equipment has a time slot CSMA/CA access mechanism, and any communication must be completed before the end of the contention access period. In the non competitive period, the coordinator divides the non competitive period into several CTS according to the CTS application of the device in the last superframe PAN network. Each CTS consists of several time slots, and the number of time slots is specified when the device applies for CTS. If the application is successful, the application device will have the specified number of time slots. The timeslot in each CTS is assigned to the timeslot application device, so there is no need to compete for channels. The IEEE 802.15.4 standard requires that any communication must be completed within the CTS assigned by the user.

It is specified in the superframe that the non contention period must follow the contention period. The functions of the competition period include that network devices can send and receive data freely, devices in the domain apply to the coordinator for the CTS period, and new devices join the current PAN network. In the non competitive phase, the equipment designated by the coordinator sends or receives data packets, and sends the received motor temperature signal to the computer for subsequent prediction and analysis of temperature variation signals.

2.2 Motor Temperature Signal Conversion

After obtaining the required monitoring data, process the data. The data extracted from the PMSM temperature database is often missing or the value of the data is too low. Therefore, in order to ensure that there are sufficient data support conditions for subsequent temperature variation signal prediction and monitoring, the original data shall be integrated, cleaned, replaced and protocol processed. The first step is to summarize the heterogeneous data in the monitoring data and refine the original data at the bottom. The second step is to clean the data [15] by filling missing data and processing noise. Step 3, perform decimal scaling transformation on the data, and the expression is:

$$V(n) = v^S(n) \Big/ 10^k \tag{2}$$

where, $V(n)$ Indicates the monitoring point after transformation n Temperature signal sample; $v^S(n)$ Indicates the original temperature signal; k Indicates the minimum ratio [16] to ensure that the maximum value of the transformed temperature signal is less than 1. Use the above formula to complete the data transformation. Step 4: Since the data set obtained from the wireless sensor monitoring database is large, it will consume a lot of time in the later modeling, so the data needs to be processed by specification. Reduce the amount of data in the camphor form by extracting data at 10 min intervals.

2.3 Temperature Anomaly Signal Recognition Based on Model Prediction

The recurrent neural network in deep learning is a mainstream learning model used to serialize data modeling. This time, the deep learning technology is used to establish a prediction model, expand the input sequence data in the direction of time or event evolution, and establish the structure of the deep learning network model as shown in the following Fig. 2.

As shown in Fig. 2, x Represents the input matrix; e Represents hidden layer vector; a Represents the output matrix; V Represents the weight matrix of the output layer; U Represents the weight matrix of the input layer; W Represents the weight matrix [17] of the hidden layer. In the deep learning network model, the hidden layer of the current time of the cyclic network depends not only on the input of the current time, but also on

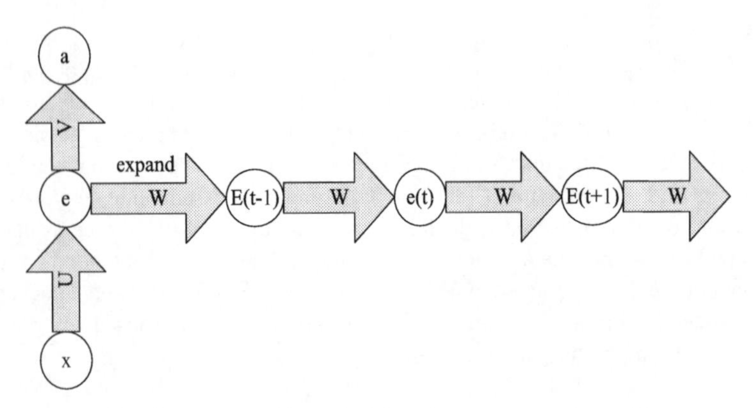

Fig. 2. Structure diagram of deep learning network model

the hidden layer of the previous time, and the weight matrix of the hidden layer determines how much information will be retained to the current time at the previous time. This structure has a natural advantage for processing serialized data. It can mine the relationship between the front and back sequences and give weight, form memory and participate in the following operations [18]. Input the transformed motor temperature signal into the deep learning network model, use the network model to predict the temperature variation of each motor component, and determine the memory and forgetting of information through the forgetting gate of the memory block, which is expressed by the formula:

$$C(t) = f \odot C(t-1) + i^{V(n)} \odot \widetilde{C} \tag{3}$$

where, $C(t)$ Indicates the current cell state; f Indicates the forgotten door; $C(t-1)$ Indicates the cell state at the previous moment; $i^{V(n)}$ Indicates the input door; $\widetilde{C}$ Indicates the internal cell status [19]. The forgetting door needs to be activated by the activation function, which is expressed by the formula:

$$f = \varepsilon\big(W_i\big[g_{t-1}, V(n)\big] + eab\big) \tag{4}$$

where, ε Represents the activation function; g_{t-1} Indicates the hidden state at the previous time; b Represents the corresponding offset matrix. The value of the forgetting gate is limited to (0,1), which can directly control how much information of the cell state at the previous time can participate in the operation at the current time, that is, 0 is discarded and 1 is retained, playing the role of selection and forgetting. According to the current cell state, the output gate is calculated to calculate the characteristics of the temperature signal variation of the permanent magnet synchronous motor. The calculation formula is:

$$y = a \odot \tanh[C(t)] \tag{5}$$

where, y Represents the output vector of the deep learning network model, that is, the characteristics of the motor temperature variation signal; tanh Indicates the activation operation [20]. According to the characteristics of the temperature abnormal signal calculated above, the mean square error of the motor temperature signal is predicted, which is expressed by the formula:

$$h = \frac{1}{N}\sum (y - y^*)^2 \tag{6}$$

where, h Mean square error of motor temperature signal; N Represents the number of training samples of deep learning network model; y^* Indicates the maximum allowable limit of motor temperature signal. If h If it is greater than zero, it means that the current temperature signal is abnormal. If h If it is less than zero, it means that the current temperature signal is a normal signal, so the model can be used to predict the characteristics of motor temperature variation, and identify the abnormal signal.

2.4 Alarm Monitoring of Abnormal Motor Temperature Signal

Judging based on the prediction results of the deep learning network model alone will produce many false positives and false alarms. This is because the essence of the deep learning network model prediction is to judge abnormal changes based on the density between samples. In this way, the temperature value of a motor measurement point is too different from that of most other motors, and the motor temperature signal will be considered as abnormal changes. Although this is correct, the actual situation is that the temperature values of different motors will change slightly when the motor is running, but it is still fluctuating up and down in the normal range as a whole. In addition, it is only considered that the abnormal change is valid when the temperature of one measuring point is higher than that of other measuring points. The "abnormal change" caused by temperature lower than other measuring points is not considered in this paper. Therefore, we will eliminate these false positives through a series of rules, that is, pruning.

The initialization matrix is used to evaluate the abnormal degree of motor temperature signal from different feature dimensions. The calculation formula is as follows:

$$flag \begin{cases} 1 & h \geq \max\left(95\_percentile(h), y^*\right) \\ 0 & else \end{cases} \tag{7}$$

where, $flag$ Indicates the abnormal index value of the motor based on different characteristic dimension variables; If the temperature of the first measuring point of the motor is greater than 95% of the quantile and the maximum temperature limit of all motor temperature measuring points at the same time point, then the abnormal temperature signal indicator of this measuring point of the motor is set to 1, indicating that the current temperature value of this measuring point of the motor is at a high level, that is, the temperature signal of this measuring point is a abnormal signal; otherwise, the signal indicator is set to 0, indicating that the temperature signal of this measuring point is a normal signal.

After the aforementioned pruning operation, the state of the motor monitoring results at each time may be discontinuous. Therefore, if the starting point of an abnormal change is determined directly based on the original results, a large number of records will be generated. How to determine the start of a mutation record and balance the detection accuracy and false alarm rate requires a reasonable design of the monitoring process. In this paper, in order to accurately determine the starting point, duration and other key information of the temperature anomaly signal, the strategy of delayed reporting is designed for the real-time monitoring of the flow type. The specific way is as follows.

If there is no abnormal change reported in the front interval, when the front 1 bit is merged into the front interval to form a abnormal change record, the starting point of the abnormal change record is the first non-zero bit in the motor lead time window. If there is an exception in the leading interval, the leading 1 bit is merged into the abnormal change record in the leading interval. If an abnormal change is reported continuously, and then all the changes in the motor lead time window are set to normal, the record of the abnormal change ends. At this time, the starting point of the motor lead time window is the end point of the abnormal change record. Since then, the newly reported changes have been regarded as another record of changes. According to the above strategies, the warning and monitoring of the motor temperature change signal is carried out to realize the monitoring of the permanent magnet synchronous motor temperature change signal based on the model prediction.

3 Experimental Analysis

3.1 Experiment Preparation

In order to verify the reliability and feasibility of the monitoring method for temperature variation signal of permanent magnet synchronous motor based on model prediction designed in this paper, a permanent magnet synchronous motor is taken as an experimental object, and the following table shows the parameters of permanent magnet synchronous motor (Table 1).

Table 1. Parameters of Permanent Magnet Synchronous Motor

S/N	Parameter	Numerical value
1	Rated current	4 A
2	Rated speed	400 rpm
3	DC Bus Voltage	36 V
4	Measured stator winding resistance	0.33 Ω
5	External wiring resistance	0.04
6	Self perception	2.91 mH
7	Mutual inductance	−0.33 mH
8	Flux linkage	77.6 mWb
9	Polar logarithm	5
10	D-axis inductance	3.24 mH
11	Q axis inductance	3.24 mH

The experiment takes the temperature signal of permanent magnet synchronous motor in one day as the unit, and the sampling interval is 1 min. It contains the temperature sensor signal sequence of five permanent magnet synchronous motors in 43 days, including stator temperature, driving end temperature, non driving end temperature, and the temperature difference between driving end and non driving end. Therefore, the one-day sequence includes a total of 32 * 4 dimensional features. According to the specific manifestation of signal abnormal change, the temperature signal of permanent magnet synchronous motor presents abnormal mainly in two types, point abnormal and context abnormal. The value of one time step in the sequence is significantly higher than that of other points in the adjacent time step, which is considered as point anomaly, generally referred to as isolated point, and may be caused by sensor failure. Context exception means that within the length of a window, the sequence value obviously deviates from other sequence values of the same type. It is considered as a context exception and is usually related to the abnormal working state of the motor. According to the above rules, the real abnormal record labels that can be captured from the perspective of post inspection are shown in the table below (Table 2).

Table 2. Abnormal Change Record of Temperature Signal

Date	ID	Starting point	End point
0109	21	643	639
0110	22	1084	1167
0111	23	350	400
...	...	...	...
0116	27	1000	1100

As shown in the above table, there are 9 records of temperature signal variation, and each record includes the date of occurrence and the start and end time points. If the sensor sequences under different dates are marked according to the true abnormal records, the data description is shown in the table below (Table 3).

Table 3. Abnormal sequence count (by date)

Category	Number	Average length	Characteristic dimension
Normal sequence	36	1010	128
Aberrant sequence	7	1213	128
Total	43	1043	128

As shown in the above table, the normal sequence has 36 days and the abnormal sequence has 7 days. The abnormal sequence here refers to the abnormal records that have occurred within the time range of that day. If there is no abnormal change in the whole day, the sequence is considered normal. The monitoring results are shown in the following Fig. 3.

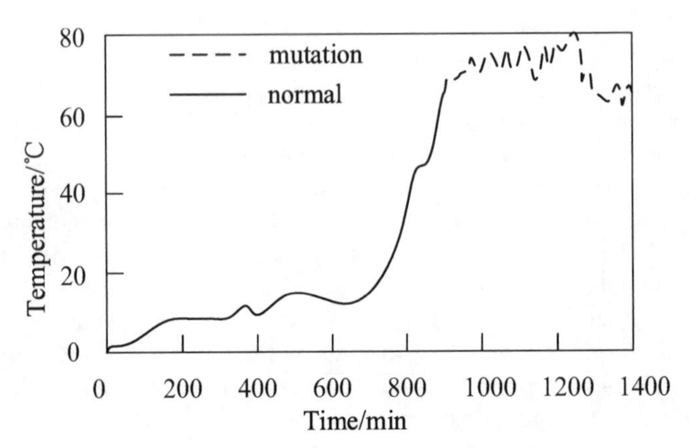

Fig. 3. Signal Diagram of Abnormal Temperature Change of Motor

The experimental data set contains a number of sensor data collected from the PMSM on the test bench, with a total of 13 temperature related characteristics. The specific characteristics are shown in the following Table 4.

Table 4. Characteristics of Experimental Data Set

S/N	Characteristic tag name	Describe
1	u_q	Voltage q component in d-q coordinate
2	Coolant	Coolant temperature
3	Stator_winding	Stator winding temperature
4	u_d	Voltage d component in d-q coordinate
5	Stator_tooth	Stator tooth temperature
6	Motor_speed	Motor speed
7	i_d	Q component of current in d-q coordinate
8	i_q	D component of current in d-q coordinate
9	Pm	Permanent magnet temperature
10	Ststor_yoke	Stator yoke temperature
11	Ambient	Ambient temperature
12	Troque	Motor torque
13	Profile_id	Measurement Session ID

The results of the entire data set are obtained with a measurement frequency of 2 Hz, including 69 mutually independent measurement sessions, with a total of more than 1.3 million pieces of data. Due to the huge data set, there are inevitably some distorted data due to measurement errors. In order to avoid the impact of these data on the prediction accuracy of the deep learning model and reduce the calculation cost. The whole benchmark data set is de sampled, and the training set and test set are divided. About 32000 pieces were selected as the training set and 544 pieces were selected as the test set. All deep learning models use five fold cross validation in training to avoid model over fitting as far as possible.

According to the actual situation of the permanent magnet synchronous motor in the experiment, the parameters of the deep learning network model are set. The time step of the deep learning model is 5. The specific super parameter settings are shown in the table below (Table 5).

Table 5. Parameters of deep learning network model

S/N	Parameter	Numerical value
1	Hide Layer	3
2	Neurons in each layer	(64, 16)
3	Weight	Random-normal
4	Optimizer	Nadam
5	Learning rate	0.01
6	Number of training rounds	100
7	Noise	1 × 10–3

In addition, the number of neurons in the hidden layer of the deep learning network model is set to 32, and the deep learning network model has a layer of 1D-CNN convolution before entering the gated cycle unit, and the number of convolution cores is 16. The input features of the deep learning network model are filtered, and the filtered features are put back, and the weight initialization adopts rand_Normal mode. In the experiment, the random variance of all Gaussian noises is uniformly set to 1×10^{-3}. In the actual motor operating environment, there is a lot of noise. Although part of the hardware circuit settings have been partially filtered, the noise will still have a certain impact on the signal acquisition. In traditional methods, basically a series of filtering operations are carried out on the collected signals to filter out the noise. In this experiment, the noise is filtered by the nonlinear transformation of deep learning, which is also a test of the robustness of the deep learning network model. The deep learning network model predicts the temperature of the four core components of the permanent magnet synchronous motor, namely, the stator teeth, the stator yoke, the stator winding and the permanent magnet. Due to certain differences in the electrical structure and temperature rise properties of the four core components, the temperature prediction models of each target component are trained separately, that is, each component corresponds to a model, but the super parameters remain the same.

3.2 Result Analysis

After completing the above experimental preparation, the experimental evaluation indicators are selected. In order to make the experimental results and experimental data have a certain explanation and reliability, two traditional methods are selected for comparison. The two traditional methods are the monitoring method based on the finite element method and the monitoring method based on the magnetic flux observation method. The following are represented by the traditional method 1 and the traditional method 2 respectively. In the monitoring carried out on a daily basis, the sequence that actually contains abnormal changes is defined as a negative case under the positive case period. Then the accuracy of abnormal signal monitoring can be measured by classification related indicators. This selection is marked with confusion matrix, whose definition is shown in the following Table 6.

Table 6. Confusion matrix

Project	Negative example of monitoring value	Positive example of monitoring value
Negative example of actual value	TN	FP
Positive example of actual value	FN	TP

As shown in Table 6, TN (True Negative) refers to the correct negative case monitored, that is, the normal sequence correctly monitored; FPC False Positive) refers to the positive example of monitoring error, that is, the error monitoring is a normal sequence of abnormal sequence; FN (False Negative) refers to the negative case of monitoring error, that is, the abnormal sequence with error detection as normal sequence; TPC True Positive) means to monitor the correct positive example, that is, the correctly monitored abnormal sequence. Based on the above description, the monitoring accuracy is defined as follows:

$$\Pr ecision = TP \big/ (TP + FP) \tag{8}$$

where, $\Pr ecision$ It indicates the monitoring accuracy of motor temperature abnormal signal. The recall rate monitored is defined as follows:

$$\mathrm{Re}call = TP \big/ (TP + FN) \tag{9}$$

where, $\mathrm{Re}call$ Indicates the recall rate of motor temperature abnormal signal monitoring. In this experiment, the accuracy rate and recall rate are used as the evaluation indicators of the monitoring accuracy of the three methods of temperature variation signal. According to the way of identifying the abnormal records in the experimental scheme setting, the number of correctly hit anomalies and the number of falsely reported anomalies under different methods are respectively counted, and the accuracy rate of the three methods of temperature variation signal monitoring is calculated using formula (8), with time as the variable, according to the statistical data, the accuracy comparison chart of the three methods is shown below.

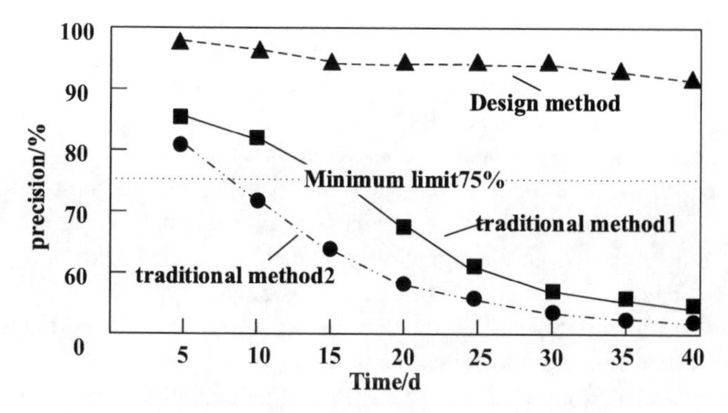

Fig. 4. Comparison Diagram of Accuracy of Three Methods

It can be seen from the above figure that the design method has relatively high accuracy in monitoring the temperature variation signal of permanent magnet synchronous motor, with an average accuracy rate of 96.48%. The average accuracy rates of traditional method 1 and traditional method 2 are 75.46% and 64.18% respectively. Moreover, the accuracy rate of the design method is less affected by the monitoring time, and does not decrease significantly with the increase of the monitoring time. When the monitoring time is 40, the accuracy rate is 94.46%, which is far higher than the two traditional methods, and is greater than the minimum limit value, indicating that the design accuracy monitoring accuracy is relatively high. In order to further verify the monitoring accuracy of the design method, take the number of permanent magnet synchronous motor temperature signal samples as the variable, compare the recall rate of the three methods of motor abnormal signal monitoring, use formula (9) to calculate the accuracy rate of the three methods of temperature variation signal monitoring, and draw the comparison chart of the recall rate of the three methods according to the statistical data as shown below.

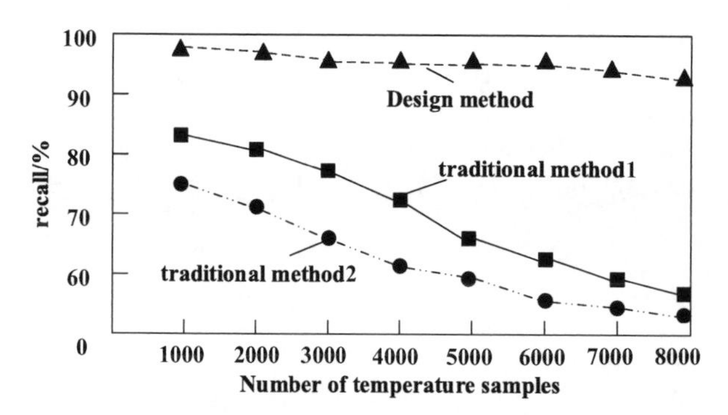

Fig. 5. Comparison of recall rate of three methods

It can be seen from the above figure that the recall rate of the design method is relatively high in this experiment. Although the recall rates of the three methods are constantly decreasing with the increase of the number of samples of permanent magnet synchronous motor temperature signals, the reduction of the recall rate of the design method is relatively small. When the number of samples of motor temperature signals reaches 8000, the recall rate of the design method is 92.45%, nearly 31% higher than that of the traditional method 1, it is nearly 35% higher than traditional method 2. Therefore, this experiment proves that the design method in this paper has a high monitoring accuracy of temperature variation signal of permanent magnet synchronous motor, and can monitor the temperature variation signal of permanent magnet synchronous motor with high accuracy. The design method in this paper has good feasibility and reliability, and is more suitable for monitoring the temperature variation signal of permanent magnet synchronous motor than the traditional method.

Compare the detection efficiency of the three methods based on the accuracy and recall results obtained in Figs. 4 and 5. The comparison diagram is shown below (Fig. 6).

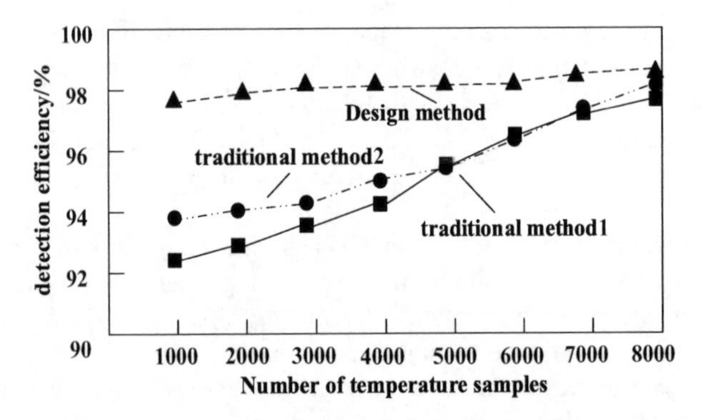

Fig. 6. Comparison of detection efficiency among three methods

From the above figure, it can be seen that the detection efficiency of the design method in this experiment remains above 97% as the number of temperature signal samples of the permanent magnet synchronous motor increases; The detection efficiency of traditional method 1 in this experiment remained above 92%; The detection efficiency of traditional method 2 in this experiment remained above 94%. Therefore, the design method in this article can efficiently detect the temperature variation signal of permanent magnet synchronous motors, and has high reliability.

4 Conclusion

As the power source, the safe and reliable operation of permanent magnet synchronous motor is very important. Motor temperature is the key parameter to reflect the running state of the motor. It is of great and positive significance to realize the monitoring of temperature variation signal to ensure the operation safety of permanent magnet synchronous motor. In this paper, a monitoring method of temperature variation signal of permanent magnet synchronous motor based on model prediction is proposed, aiming to address the key issue of ensuring the safe and reliable operation of these motors. The speed, current, torque and other state parameters of the permanent magnet synchronous motor are taken as the input of the prediction model, and the motor temperature is taken as the output of the prediction model. The high-precision training of the prediction model is carried out to achieve the accurate prediction of the temperature variation signal of the motor, on this basis, the residual error of temperature prediction is used to alarm the abnormal change of motor temperature. The experimental data show the correctness and effectiveness of the proposed method. Compared with the traditional method, it can accurately monitor the abnormal temperature signal of the permanent magnet synchronous motor, ensure its healthy and safe operation, and has positive significance for improving the ability of the permanent magnet synchronous motor to operate safely. Future research can further explore the following directions:

Firstly, it can consider introducing more state parameters and environmental information to improve the accuracy and stability of model predictions.

Secondly, it can combine methods such as machine learning and deep learning to further optimize the training and predictive capabilities of the model.

Additionally, research can be conducted on how to achieve joint monitoring and control of multiple motors to address temperature variation issues in complex operating conditions.

Acknowledgement. Hunan Provincial Natural Science Foundation of China (2023JJ50270, 2023JJ50267); Hunan Provincial Department of Education Youth Fund Project (21B0690); Shaoyang City Science and Technology Plan Project (2022GZ3034); Hunan Provincial Department of Science and Technology Science and Technology Plan Project (2016TP1023).

References

1. Pasqualotto, D., Zigliotto, M.: A comprehensive approach to convolutional neural networks-based condition monitoring of permanent magnet synchronous motor drives. IET Electr. Power Appl. **15**(7), 947–962 (2021)
2. Tsai, M.F., Tseng, C.S., Li, N.S., et al.: Implementation of a DSP-based speed-sensorless adaptive control for permanent-magnet synchronous motor drives with uncertain parameters using linear matrix inequality approach. IET Electr. Power Appl. **16**(7), 789–804 (2022)
3. Sahebjam, M., Sharifian, M.B.B., Feyzi, M.R., et al.: Novel methodology for direct speed control of a permanent magnet synchronous motor with sensorless operation. IET Electr. Power Appl. **15**(6), 728–741 (2021)
4. Parvathy, M.L., Eshwar, K., Thippiripati, V.K.: A modified duty-modulated predictive current control for permanent magnet synchronous motor drive. IET Electr. Power Appl. **15**(1), 25–38 (2021)
5. Taherzadeh, M., Hamida, M.A., Ghanes, M., et al.: Torque estimation of permanent magnet synchronous machine using improved voltage model flux estimator. IET Electr. Power Appl. **15**(6), 742–753 (2021)
6. Vahid, Z.F., Akbar, R.: 2-D analytical on-load electromagnetic model for double-layer slotted interior permanent magnet synchronous machines **16**(3), 394–406 (2022)
7. Fu, R., Cao, Y.: Hybrid flux predictor-based predictive flux control of permanent magnet synchronous motor drives. IET Electr. Power Appl. **16**(4), 472–482 (2022)
8. Qu, J., Jatskevich, J., Zhang, C., et al.: Improved multiple vector model predictive torque control of permanent magnet synchronous motor for reducing torque ripple. IET Electr. Power Appl. **15**(5), 681–695 (2021)

9. Ahn, J.M., Lim, D.K., Jeong, G.W.: Performance analysis of the outer-rotor surface-mounted permanent magnet synchronous motor for high altitude long endurance unmanned aerial vehicle applied grain-oriented electrical steel. J. Korean Magn. Soc. **32**(2), 93–99 (2022)
10. Lv, K., Dong, X., Zhu, C.: Research on fault-tolerant operation strategy of permanent magnet synchronous motor with common dc bus open winding phase-breaking fault. Energies **15**(8), 1–12 (2022)
11. Yue, H., Wang, H., Wang, Y.: Adaptive fuzzy fixed-time tracking control for permanent magnet synchronous motor. Int. J. Robust Nonlinear Control **32**(5), 3078–3095 (2022)
12. Li, T., Ma, R., Bai, H., et al.: A compare research on third harmonic current control for five-phase permanent magnet synchronous motor. J. Northwestern Polytech. Univ. **39**(4), 865–875 (2021)
13. Cheng, Y., Wang, Y.H., Li, C., et al.: Sliding mode control of PMSM based on Luenberger observer. Comput. Simul. **40**(4), 231–235 (2023)
14. Larbaoui, A., Chaouch, D.E., Belabbes, B., et al.: Application of passivity-based and sliding mode control of permanent magnet synchronous motor under controlled voltage. J. Vibr. Control **28**(11/12), 1267–1278 (2022)
15. Cui, P., Zheng, F., Zhou, X., et al.: Current harmonic suppression for permanent magnet synchronous motor based on phase compensation resonant controller. J. Vibr. Control **28**(7–8), 735–744 (2022)
16. Bai, C., Yin, Z., Zhang, Y., et al.: Multiple-models adaptive disturbance observer-based predictive control for linear permanent-magnet synchronous motor vector drive. IEEE Trans. Power Electron. **37**(8), 9596–9611 (2022)
17. Si, J., Feng, C., Nie, R., et al.: A novel high torque density six-phase axial-flux permanent magnet synchronous motor with 60 degrees phase-belt toroidal winding configuration. IET Electr. Power Appl. **16**(1), 41–54 (2022)
18. Hou, X., Wang, M., You, G., et al.: Study on speed sensorless system of permanent magnet synchronous motor based on improved direct torque control. Trans. Inst. Meas. Control **44**(9), 1744–1754 (2022)
19. Tola, O.J., Obe, E.S., Obe, C.T., et al.: Finite element analysis of dual stator winding line start permanent magnet synchronous motor. Przeglad Elektrotechniczny **98**(4), 47–52 (2022)
20. Babaei, M., Feyzi, M., Marashi, A.N.: Extended Poincare' model and non-linear analysis of permanent-magnet synchronous motor scalar drive system. IET Power Electron. **15**(9), 855–864 (2022)

Research on Personalized Recommendation of Mobile Social Network Products Based on User Characteristics

Min Zhou[1](✉), Wei Xu[2], and Xinwei Li[3]

[1] Department of Employment and Entrepreneurship, Xiamen Ocean Vocational College, Xiamen 361000, China
zhoumin693@163.com
[2] Department of International Trade, Xiamen Ocean Vocational College, Xiamen 361000, China
[3] Human Resources Department, Changchun University of Architecture and Civil Engineering, Changchun 130000, China

Abstract. In the process of conducting online product recommendations, the lack of comprehensive user profiling in constructing user personas has led to low Top-10 hit rate, average reciprocal rank, and normalized discounted cumulative gain of recommended products. To effectively address this issue, a user feature-based personalized recommendation method for mobile social networks (MSN) is proposed. By analyzing the basic attributes, interaction attributes, feedback attributes, and interest attributes of MSN users, user attribute features are extracted to build user personas. Based on these user personas, personalized recommendations for mobile social network products are achieved using MetaEE. This involves updating the recommended products based on the collection of user interactions with historical items until there is overlap between the support set and the query set of the personalized recommendation meta-learning samples. The corresponding products are then considered as the final recommended results. Experimental results demonstrate that the designed recommendation method outperforms the comparison methods in terms of Top-10 hit rate, average reciprocal rank, and normalized discounted cumulative gain across multiple experimental scenarios, indicating a promising recommendation performance.

Keywords: User characteristics · Mobile social network goods · Personalized recommendation · Attribute characteristics · MetaEE · Losses

1 Introduction

With the rapid development of Internet technology and mobile device applications, online services have become an indispensable part of people's life [1]. However, the e-commerce platform will release a large number of commodity content and user shopping

Research Project: Xiamen Ocean Vocational College's Institute-level Research Project (High-level Talent Project) KYG202004.

L. Yun et al. (Eds.): ADHIP 2023, LNICST 549, pp. 195–209, 2024.
https://doi.org/10.1007/978-3-031-50549-2_14

browsing records and other data every day [2]. Faced with a large number of commodities, how to quickly select the commodities that users are interested in and how to quickly recommend commodities to potential users on e-commerce platforms have become the difficulties that platform providers need to urgently solve [3]. Therefore, personalized recommendation technology is becoming a research hotpot in the field of product recommendation. The recommendation system can accurately locate users' interests and commodity characteristics [4] by analyzing users' behavior data, and realize the matching between commodity providers and users. It has been applied in many network services in the current society. Some shopping platforms predict users' preferences based on users' browsing records, purchase records and other information, launch various services such as "Guess what you like" to recommend products [5]. The performance of the recommendation system is determined by the algorithm. A good recommendation algorithm can directly and quickly help users locate interesting goals, optimize users' browsing experience, and bring higher user loyalty and economic benefits to the e-commerce platform [6]. Content based recommendation, collaborative filtering based recommendation and hybrid recommendation, as classic methods, have been widely used in the industry by virtue of their respective advantages [7]. However, these recommendation algorithms all have problems such as cold start caused by the historical interaction between users and commodities and data sparsity caused by low effective data. At the same time, due to many factors such as the increasing personalized needs of users, the recommendation services provided by these traditional algorithms have been severely tested [8]. According to different recommendation methods, recommendation algorithms can be divided into three types: the first is collaborative filtering recommendation algorithm, which uses the same interests of users or similar characteristics of goods to recommend. Collaborative filtering recommendation algorithm has the advantages of high accuracy, finding new items and not requiring item feature information. However, it also has drawbacks such as cold start issues, data sparsity, algorithm scalability, and recommendation preference issues; The second is content based recommendation, which uses users' interest preferences and content information to label users and goods and then uses supervised learning or deep learning methods to recommend. Content based recommendation algorithms have advantages such as independence, interpretability, and adaptability to cold start problems. However, it also has drawbacks such as feature representation issues, lack of diversity, and inability to capture dynamic changes in user interests; The third is hybrid recommendation. This method combines one or more of the previous two recommendation algorithms and combines the advantages of each algorithm to recommend [9]. Hybrid recommendation algorithms have the advantages of improving recommendation accuracy, diversity, novelty, robustness, and scalability. However, it also has drawbacks such as complexity and computational overhead, parameter selection and adjustment, as well as interpretability and comprehensibility [10]. The hybrid recommendation algorithm can effectively remedy some defects of the single use of the above two methods by retaining the advantages of the above two recommendation algorithms, so it has been studied by many scholars at present.

Therefore, this paper proposes a user feature-based personalized recommendation method for mobile social networks. This method analyzes the basic attributes, interaction attributes, feedback attributes, and interest attributes of mobile social network users to extract user attribute features, thereby achieving user persona construction. Based on user personas, personalized recommendations for mobile social network products are achieved using MetaEE. Research on personalized recommendations for mobile social network products helps improve recommendation accuracy, enhance user experience, facilitate commercial conversion, and drive innovation and development in the field of recommendation systems. These studies provide important support and guidance for the development of mobile social networks and e-commerce.

2 Design of Personalized Recommendation Method for Mobile Social Network Products

2.1 Construction of User Profile Model Based on Multi-attribute Features

The existing user portrait research rarely considers multiple types of data at the same time in practical applications to build a portrait model with more attribute characteristics, making the constructed portrait model relatively incomplete, which will lead to certain deviations in the process of mining user characteristics, distinguishing user groups, and searching for similar users. In order to better improve the user model, this paper proposes a user portrait model with multiple attribute features, aiming to build a more complete portrait by taking into account the four attributes of users: basic attributes, interactive attributes, feedback attributes, and interest attributes. The construction of user profile first depends on the integrity of data collection. The more user related data, the more favorable it is for mining user preferences, and thus the user profile will be more effective; Secondly, a good tag system can take all aspects of users into consideration, which can make the data fully mined and make the user profile more complete and relevant to users; Finally, the model design of user profile is essential. The collected data is analyzed and processed through various data mining technologies, and the final user profile is obtained based on the constructed indicator system. Among them, the integrity of data collection directly affects the accuracy of user profile construction. Most existing studies obtain data by writing programs or directly using data collectors to crawl data from the API portal, which is effective compared with traditional methods such as questionnaires and in-depth interviews.

Improve the difficulty of data acquisition, and reduce the phenomenon of small amount of data or data errors caused by users' failure to answer or random answer due to boredom. The data required in this paper is mainly divided into three parts: user basic information, domain data and external environmental factors.

(1) Basic user information: refers to the static data of the user, which generally does not change and is relatively stable. It mainly includes user ID, gender, age, education level, home address, occupation, etc.
(2) Domain data: user dynamic data, which can reflect users' personal preferences and display users' characteristics. It mainly includes clicking, browsing, collecting, sharing, adding, comments, ratings, likes, etc.
(3) External environment data: This data mainly refers to the situational data of users when the above behaviors occur. Such data can reflect some regularity to a certain extent, which is helpful for predicting user behaviors and judging the direction of

user interests. It generally includes the occurrence time, region, month, month, day, temperature, humidity, weather conditions, equipment used, network connection, etc.

On this basis, only a complete and effective user tag system can show the full picture of users, three-dimensional users, and ensure their integrity and accuracy. In order to better display our user attributes, better distinguish the differences between their attribute characteristics, and better use our model. Therefore, based on the user profile label system designed above, each feature is quantified to show the degree of differentiation between feature levels, and the online shopping user profile model is expressed in the form of a vector, that is

$$MUP = \{B, A, F, P\} \tag{1}$$

Among them, *MUP* represent online shopping user portrait model; *B* represents the user's basic attributes; *A* indicates user interaction sex; *F* indicates user feedback attribute; *P* indicates the user interest attribute.

For the basic attribute characteristics of users, it mainly refers to the demographic information of users, usually referring to the user's ID, gender, age and occupation. In order to better standardize its presentation results, its indicators need to be quantified by grades to achieve the normalization of the results. For example, gender can be divided into two categories, male and female, represented by 0 and 1 respectively; Similarly, age can be divided into four sections according to the classification standards of the United Nations World Health Organization. Children are 18 years old and below, young people are 19–35 years old, middle-aged people are 36–59 years old, and old people are 60 years old and above. They are represented by 1, 2, 3, and 4 respectively. Marriage can be divided into two categories: 0 means married and 1 means unmarried. See Table 1 for details.

Table 1. Classification of user basic attribute characteristics

index	classification	Quantitative representation
Gender	male	0
	female	1
Age	18 years and below	1
	19–35 years old	2
	36–59 years old	3
	60 and above	4
marriage	married	0
	unmarried	1

The interaction attribute characteristics of users mainly refer to the interaction behavior data between users, mainly including the number of clicks, additional purchases,

purchases, and collections of users. This attribute is used to determine the activity of online shopping users, so that online shopping user groups can be divided and services can be provided for different types of users. In this paper, the entropy weight method is used to calculate the overall interaction value according to the collected index values. The specific calculation formula is

$$A = \sum w_j A_{ij} \tag{2}$$

Among them, w_j represents the weight value corresponding to each feature, A_{ij} that is, each feature under the interaction attribute of a single user a_{ij} the formula obtained from data standardization is

$$A_{ij} = \frac{a_{ij} - \min a_j}{\max a_j - \min a_j} \tag{3}$$

Among them, $\min a_j$ for j the minimum value of indicators, similarly, $\max a_j$ for j the maximum value of indicators.

As for the feedback attribute characteristics of users, this paper conducts emotional analysis on all comments of users on the product, so as to obtain an emotional value that can better display the user's attitude towards the product compared with the score, which can effectively improve the impact caused by the huge difference between the score and the text in the comments. Thus, the rating and the emotional value of comments are fused to get the user's attitude towards the purchased products in the past. This article also considers user ratings sc and user comment emotional value se as an indicator of user feedback attributes. The formula for calculating the feedback value is

$$F = \alpha * sc + \beta * se \tag{4}$$

Among them, when the user only has ratings but no comments, $\alpha = 1$; Otherwise, $\alpha = \beta = 0.5$.

User rating sc it refers to the average value of all scores of a single user, that is

$$sc = \frac{\sum sc_i}{n_{sc}} \tag{5}$$

Among them, sc_i represents all scoring results of a single user, n_{sc} indicates the total number of all user ratings.

User rating sc it refers to the average value of all scores of a single user, that is

$$se = \frac{\sum se_i}{n_{se}} \tag{6}$$

Among them, se_i represents all scoring results of a single user, n_{se} indicates the total number of all user ratings.

The last is the analysis of user interest attribute characteristics, which mainly refers to the deep mining and extraction based on the user's historical viewing data and comment text data.

Refined features. In order to better distinguish the impact of different features on user interest, how to quantify the user refined tags is the key to quantitative analysis.

The category of user interest attribute is mainly a feature of text form, so it needs to be transformed. UGC, as the user's comment data on the product, well expresses the user's preferences for different attributes. The TF-IDF (Term Frequency – Reverse Document Frequency) method is used here to quantify the weight of features. This is because the more users pay attention to a feature, the more they will repeatedly mention the feature, which is just similar to the principle of this method. Therefore, here the characteristic attributes are quantified by using this method. The method is used to distinguish the important features that users pay attention to, and get the weight values of each indicator feature, thus showing the differences between users. The calculation method is

$$k_{ij} = tf_{i,j} * IDF \tag{7}$$

Among them, k_{ij} indicates that the user is interested in k preferences of attributes, $tf_{i,j}$ refers to a word k in comments d number of occurrences in $count(w, d)$ and comments d total words in $size(d)$ ratio of i.e.

$$tf_{i,j} = \frac{count(w, d)}{size(d)} \tag{8}$$

IDF the value can be determined by the total number of comments N divide by word k number of comments (k, d), and then take the logarithm of the quotient obtained i.e.

$$IDF = \log[\frac{N}{docs(k, d)}] \tag{9}$$

With the help of Eq. (9), the calculated weight value of the keyword is displayed according to its size, and each keyword corresponds to a weight value. This value reflects the user's concern for each product, and the degree of differentiation is shown.

In the way shown above, build a user profile model with multi attribute features, and provide an execution basis for subsequent product recommendation.

2.2 Personalized Recommendation

Combined with the multi-attribute user profile model built in Sect. 2.1, this paper uses the general learning paradigm MetaEE to achieve personalized recommendation for mobile social network products. The overall recommendation process is shown in Fig. 1.

It can be seen from Fig. 1 that the embedding module of users and items generates the embedding vector of users and items as the input of the model, and then combines MAML to train and update the recommended parameters.

In the process of training the recommendation process, first, the user commodity data set is divided into independent support sets and query sets. The parameter combination of the recommended model is 0, which is also the goal of meta learning training. Based on the inner and outer layer circulation of MAML algorithm, the task parameters are updated using two levels of local and global updates. Through local updates, the corresponding preference network is trained for specific users, and global updates generalize its specific network to the preference network adapted by all users. On the basis of the user preference network, this paper focuses on users and objects. The embedded representation of products is enhanced to generate an initial embedding suitable for

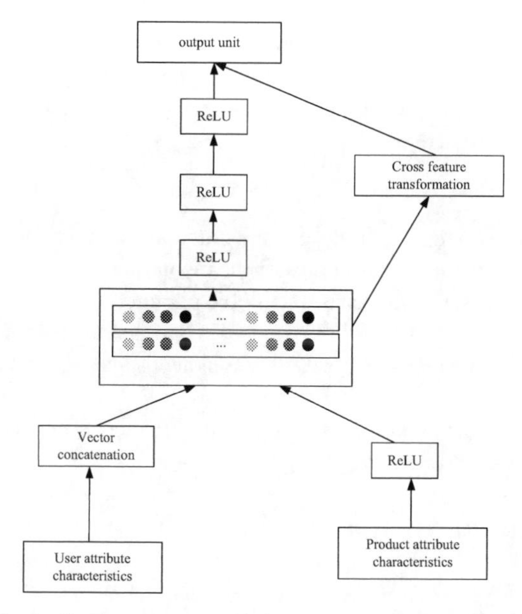

Fig. 1. Product recommendation process based on MetaEE

the recommendation model. The local update and global update of feature parameters correspond to the inner loop and outer loop of the MAML framework respectively.

Learn the meta knowledge learned in the tasks of meta training and meta testing to guide the rapid adaptation of target tasks. First, based on the meta learning sampling support set and query set, a pair of data in the support set query set is composed for each user and historical interactive items, as well as the corresponding tags under the items. The specific calculation method can be expressed as

$$D_1 \cup D_2 = \{X_{ij}, Y_{ij}\}_{j=1}^{n} \tag{10}$$

Among them, X_{ij} and Y_{ij} a collection composed of users and historical interactive items, representing D_1 represents a set of meta learning sampling support for personalized recommendation of mobile social network products, D_2 represents the meta learning sampling query set for personalized recommendation of mobile social network products, n it is a manually set size, representing the number of items sampled by a specific user, so one user samples a total of n as the support set and query set θ random initialization to obtain an initial parameter user bias evaluation function f_θ, which can be expressed as

$$p = f_\theta(X) \tag{11}$$

According to the way shown in Eq. (11), and according to the evaluation function p training is conducted on the support set. Each user's preference evaluation represents a meta training task. The loss of the corresponding task is obtained by evaluating the preference on each task. The loss calculation method can be expressed as

$$L(f_\theta) = -Y \log p - (1 - Y) * \log(1 - 9) \tag{12}$$

Among them, $L(f_\theta)$ represents the loss function. According to the loss calculation gradient, the random initial model parameters θ update to integrate specific tasks, that is, parameters that adapt to specific user preferences. The specific update steps in the support set can be expressed as

$$\theta' \leftarrow \theta - \lambda \nabla L(f_\theta) \tag{13}$$

Among them, $\nabla L(f_\theta)$ represents the change gradient parameter of the loss function, λ indicates the set super parameter, θ' indicates the recommended parameters that adapt to specific user preferences. On this basis, the global parameters are updated progressively by calculating the loss of each user in Eq. (11). Ensure that the recommendation results found in the final formula (10) can meet the actual needs and preferences of users to the maximum extent.

Complete personalized recommendation of mobile social network products in the way shown above.

3 Test Experiment Analysis

3.1 Test Environment

Experimental equipment and programming language: The proposed model is written by the Keras depth framework and Python language, and Pycharm is used as the development tool to conduct experimental tests on the Windows 10 operating system and the server platform with the GPU model of GeForce 1660Ti 6G and above.

3.2 Test Data Set Preparation

In the experiment, MovieLens - 1M and Tafeng datasets were used. Both datasets contain user statistics (MovieLens-1M includes gender, age, occupation, etc., Tafeng includes consumer ID, age, region, etc.) and commodity attribute information (MovieLens-1M includes movie type, Tafeng includes original ID, subcategory, quantity, price, etc.). Table 2 and Table 3 describe the characteristics of the two data sets, as follows.

Table 2. Data set statistics

data set	user	commodity	interactive
MovieLens-1M	6060	3620	1003410
Tafeng	32690	23695	952010

Tables 4 and 5 describe the data set format, demographic and commodity statistics. The details are as follows.

Based on the above data set, we carried out test analysis on the effect of personalized recommendation of goods. Through analyzing the test results under different recommendation conditions, we evaluated the practical application value of designing personalized recommendation methods for goods on mobile social networks based on user characteristics.

Table 3. Format of MovieLens-1M Dataset

User ID	Item ID	score
1	1193	5
1	662	4
1	915	3
1	3412	5
2	1369	3

Table 4. Tafeng Dataset Format

parameter	information			
Consumer ID	1	1	1	1
Age	K	K	K	K
region	E	E	E	E
Subclass ID	5059	6230	4789	7266
Commodity classification	100312	1102649	12036	100349
number	1	2	2	1
cost	194	142	79	46
selling price	20	131	50	100

Table 5. Demographic information format of MovieLens-1M data set

User ID	Gender	Age	occupation
1	F	1	10
2	M	56	16
3	M	25	15
4	M	46	7
5	F	52	9

3.3 Test Scheme Settings

While comparing the overall model of EACoupledCF in this paper, this paper further divides into three models: DCCF (only considering the implicit feedback part), i-EACoupledCF (local EACoupledCF, considering the local coupling of spatial CNN and implicit feedback part), and g-EACoupledCF (global EACoupledCF, considering the global coupling of spatial CNN and implicit feedback part). The classic comparison models of commodity recommendation: NeuMF, DeepCF, Wide&Deep and CoupeCF, and two new research models DeepICF and UCC-OCCF will be used to conduct comparative experiments with the recommendation methods proposed in this paper.

In the stage of setting evaluation indicators for recommendation results, this paper uses the widely used leave one out performance verification to evaluate all comparison methods. Randomly select one user's interaction with the product as the test item for each user, and the rest of the interactions as the training data. In addition, 99 commodities not in the user's commodity interaction set are randomly selected to form the user's test data together with the above selected test commodities. This model evaluates the performance by ranking 100 products of each user. Hit rate using Top-K ($HR@K$), average reciprocal ranking (MRR@K) and normalized discounted cumulative income $NDCG@K$) as an evaluation indicator. $HR@K$ it is a recall based metric, which is used to measure whether the test item is before all items in the test set K bit. MRR@K and $NDCG@K$ it is a weighted index, which is used to assign higher scores to the highest ranked goods in the given recommendation list. Among them, hit rate is an indicator to measure recall rate in Top-K recommendation, and its specific calculation method can be expressed as

$$HR@K = \frac{\#hits@K}{|GT|} \tag{14}$$

Among them, $|GT|$ is the total number of test sets, $\#hits@K$ indicates the total number of test sets in the Top-K recommendation list of each user.

Normalized Discounted Cumulative Gain is an indicator to evaluate the sorting results, which is used to evaluate the accuracy of sorting. It is usually used to indicate the gap between the recommended sorting list and the user's real interaction list.

$$NDCG@K = Z_k \sum\nolimits_{i=1}^{k} \frac{2^{rel_i} - 1}{\log_2(i+1)} \tag{15}$$

Among them, rel_i represents the i the "grade relevance" score of commodities in three locations, generally speaking $rel_i \in \{0, 1\}$. That is, if the goods at this location are in the test set, then rel_i is 1, otherwise it is 0. Z_k is the normalization coefficient, which is used to represent the reciprocal of the sum of the subsequent summation formula in the best case.

4 Results and Analysis

Figure 2 and Fig. 3 show that in the scenario where only implicit feedback data is considered, this paper designs a recommended method to HR@10, NDCG@10 and MRR@10 Compared with the baseline methods DeepCF, NeuMF and DeepICF, the effect of.

It can be seen from the comparative experimental data in Fig. 2 and Fig. 3 that the personalized recommendation method of mobile social network products designed in this paper based on user characteristics only considers the implicit feedback dataHR@10, NDCG@10 and MRR@10. The performance under the indicators is better than the baseline DeepCF, NeuMF and DeepICF models:

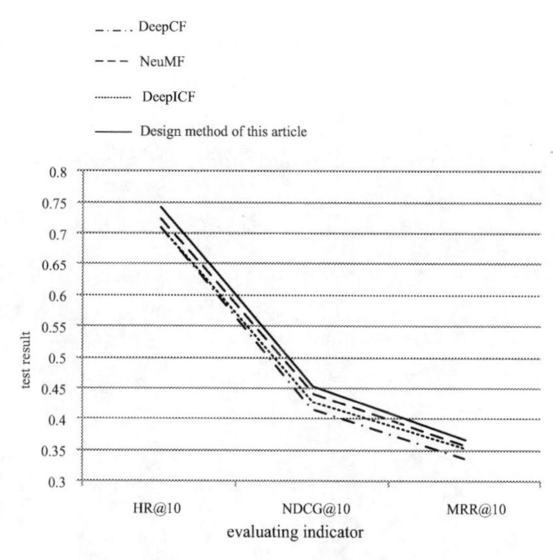

Fig. 2. Effect improvement on MovieLens-1M (only implicit feedback data is considered)

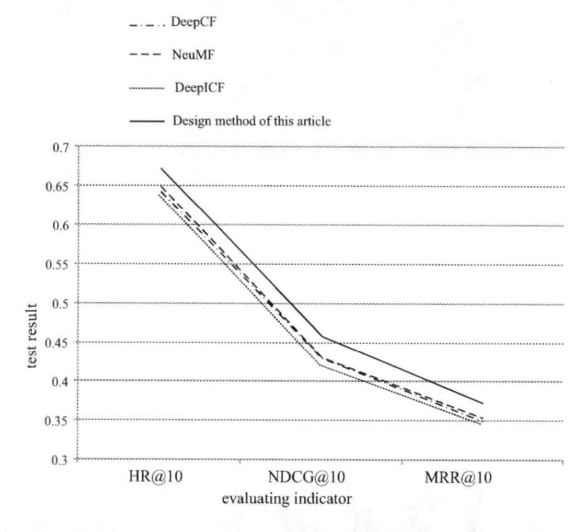

Fig. 3. Effect improvement on Tafeng (only implicit feedback data is considered)

(1) Compared with DeepCF on the MovieLens-1M dataset, the three indicators increased by 4.5% respectively 54%, 7.21%, 6.86%; Compared with NeuMF, it increased by 2.60%, 2.61% and 2.75% respectively; Compared with DeepICF, it has increased by 4.07%, 6.47% and 3.81% respectively.
(2) On Tafeng dataset, compared with the baseline DeepCF, the method in this paper achieves 4.65%, 6.74% and 5.56% on three indicators respectively; Compared with NeuMF, it has increased by 3.60%, 6.37% and 5.44% respectively; Compared with DeepICF, it increased by 4.83%, 8.10% and 6.47% respectively. These results show that the personalized recommendation method based on user characteristics can

achieve better performance improvement on the two datasets in comparison with the recommendation methods that directly use implicit feedback information.

Figure 4 and Fig. 5 show the effect improvement under the scenario of considering both explicit attributes of users and goods and implicit feedback data.

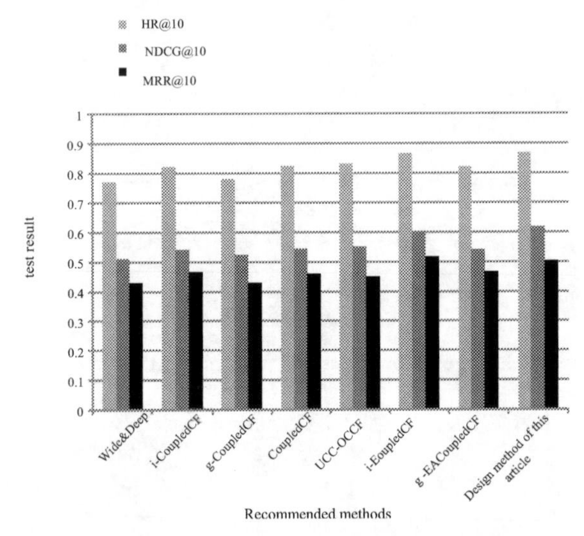

Fig. 4. Effect improvement on MovieLens-1M (considering both explicit attributes and implicit feedback data)

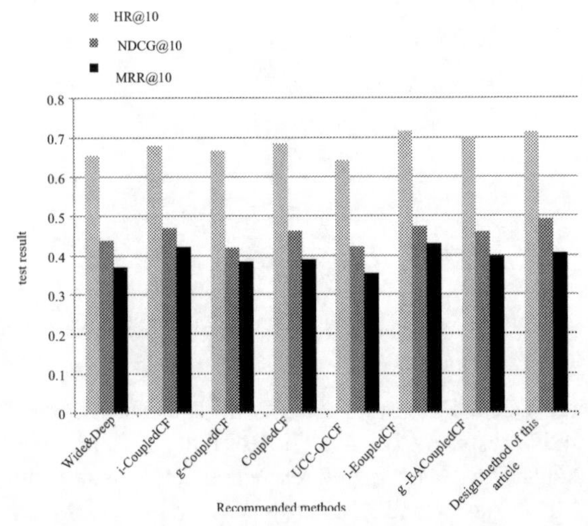

Fig. 5. Effect improvement on Tafeng (considering both explicit attributes and implicit feedback data)

It can be seen from the comparative experimental data in Fig. 4 and Fig. 5 that the personalized recommendation method for mobile social network products designed in this paper based on user characteristics has also improved under the three indicators compared with CoupledCF, Wide&Deep, and UCC-OCCF in the scenario of considering both the explicit attributes and implicit feedback data of users and products:

(1) On the MovieLens-1M dataset, it can be seen from the data that whether it is a local i-EACoupledCF model (CNN considering local explicit coupling relationships combined with DCCF, local EACoupledCF) or a global g-EACoupledCF model (CNN considering global explicit coupling relationships combined with DCCF, global EACoupledCF). As well as the overall model effect of local+global combination, it is significantly better than the three baseline methods (where the g-EACoupledCFHR@10. Its performance is basically the same as that of UCC-OCCF). For HR@10 indicators, i-EACoupledCF increased by 5.27%, 12.17% and 3.97% respectively; G-EACoupledCF increased by 6.64%, 8.29% and 6.04% respectively; The evaluation index of the recommended results of the recommended method designed in this paper increased by 5.19%, 12.62% and 4.39% respectively. ForNDCG@10 indicators, i-EACoupledCF increased by 11.19%, 17.60% and 9.39% respectively; G-EACoupledCF increased by 9.82%, 12.81% and 4.93% respectively; The overall model EACoupledCF increased by 13.71%, 20.99% and 12.53%. For MRR@10 indicators, i-EACoupledCF increased by 10.65%, 19.82% and 14.38% respectively; G-EACoupledCF increased by 14.32%, 13.90% and 8.73% respectively; The evaluation index of the recommended results of the recommended method designed in this paper increased by 9.80%, 17.29% and 11.96% respectively.
(2) It can be seen from the data that the performance of UCC-OCCF on the MovieLens-1M dataset is basically close to the evaluation indicators of the recommended results of the recommended methods designed in this paper, but on the Tafeng data, the three indicators of the UCC-OCCF model perform poorly, even not up to the level of the NeuMF model that only applies to implicit feedback data. Compared with the baseline, the evaluation index of the recommendation results of the recommended method designed in this paper has basically improved by more than 5%, 10%, and 10% in three aspects, which fully shows the adaptability of the recommended method designed in this paper to Tafeng dataset, and also shows that the personalized recommendation method of mobile social network products based on user characteristics proposed in this paper has better generalization ability.

Based on the above test results, it can be concluded that the personalized recommendation method of mobile social network products designed in this paper based on user characteristics can effectively ensure that the final recommendation effect can meet the objective needs of users, and has good practical application value for actual online product marketing.

5 Conclusion

According to the statistics of the report of China Internet Information Center, the current domestic network service penetration rate has reached more than 70%, and the Internet users on the mobile application end use short videos, shopping and other applications more than a quarter of the time every day. The market share of mobile e-commerce has increased significantly. It can be seen that the e-commerce platform has become more and more popular with the development of the Internet by virtue of its convenience, online comparison, wide range of choices, price concessions and many other advantages. Electronic commerce has greatly changed people's lives and integrated into all aspects of people's lives. How to better improve the service according to the feedback of users, and recommend products according to the interests of users, thus improving the revenue of businesses and user experience, has become a very important issue for businesses and users. In order to improve the user experience, e-commerce platforms generally provide the function of commodity reviews. The user reviews that follow are not only the most intuitive expressions of users' views, but also make communication between different users more convenient. Accurately extracting users' interests and preferences from users' purchase behavior and comments has gradually become the key to product recommendation. However, the traditional methods have the problems of low Top-10 hit rate of recommended products, low Mean reciprocal rank and low cumulative benefits of standardized discount. Therefore, this paper proposes personalized recommendation research on mobile social network products based on user characteristics, in order to greatly improve the satisfaction of users with recommended products. The experimental results show that, in various scenarios, the Top-10 hit rate, Mean reciprocal rank and normalized discount cumulative income of the products recommended by the design recommendation method are better than the comparison method, which fully solves the problems existing in the traditional methods, indicating that the design recommendation method has higher accuracy, personalization and commercial value, and has a positive impact on the performance of the recommendation algorithm and user satisfaction. The future research directions for personalized recommendation of mobile social network products include social relationship modeling, multi-source data fusion, user privacy protection, real-time recommendation and dynamic adaptation, user interaction and participation, and cross platform recommendation. These research directions will further improve the accuracy, degree of personalization, and user satisfaction of personalized recommendation of mobile social network products.

References

1. Liu, Y.C., Cao, Y.Y., Liu, J.X., et al.: Heterogeneous information network recommendation algorithm based on dynamic iterative sampling. Comput. Simul. **005**, 039 (2022)
2. Guo, Z., Yu, K., Li, Y., et al.: Deep learning-embedded social Internet of Things for ambiguity-aware social recommendations. IEEE Trans. Netw. Sci. Eng. **PP**(99), 1 (2021)
3. Su, C., Zhou, H., Gong, L., et al.: Viewing personalized video clips recommended by TikTok activates default mode network and ventral tegmental area. Neuroimage **237**(6), 118136 (2021)
4. Fan, J., Jiang, Y., Liu, Y., et al.: Interpretable MOOC recommendation: a multi-attention network for person6alized learning behavior analysis. Internet Res. Electron. Netw. Appl. Policy **2**, 32 (2022)
5. Jacobi, A., Hunter, G., Whelan, K., et al.: P.092 successful implementation of a supported conversation program on an acute stroke unit. Can. J. Neurol. Sci./Journal Canadien des Sciences Neurologiques **48**(s3), S45–S46 (2021)
6. Wu, H.: Application of collaborative filtering personalized recommendation algorithms to website navigation. J. Phys. Conf. Ser. **1813**(1), 012048 (2021)

7. Zhou, Q., Su, L., Wu, L., et al.: Deep personalized medical recommendations based on the integration of rating features and review sentiment analysis. Wirel. Commun. Mob. Comput. **2021**(7), 1–9 (2021)
8. Xiao, X., Sun, R., Yao, Z., et al.: A novel framework with weighted heterogeneous educational network embedding for personalized freshmen recommendation under the impact of COVID-19 storm. IEEE Access **9**, 67129–67142 (2021)
9. Pokushalov, E., Losik, D., Kozlova, S., et al.: Association between personalized evidence-based anticoagulation therapy and outcomes at 1-year follow-up in patients with atrial fibrillation: an analysis from the Atrial Fibrillation registry. Eur. Heart J. (Suppl. 1) (2021)
10. Szamreta, E.A., Wayser, G.R., Prabhu, V.S., et al.: Drivers and barriers to information seeking: qualitative research with advanced cervical cancer patients in the United States. J. Clin. Oncol. **39**(15_suppl.), e18705 (2021)

A Personalized Recommendation Method for Online Painting Education Courseware Based on Hyperheuristic Algorithm

Rong Yu(✉) and Bomei Tan

Nanning University, Nanning 530200, China
yyrr22@yeah.net

Abstract. Aiming at the problem of high similarity in online education courseware for painting, which leads to poor recommendation effectiveness. In order to optimize the recommendation effect of online education courseware for painting and improve the effectiveness of online education for painting, a personalized recommendation method for online education courseware for painting based on hyper heuristic algorithm is proposed. This paper introduces the calculation of keyword weight of personalized recommendation, the calculation of similarity of online painting education courseware, the calculation of user similarity, and the Committed step of the calculation of similarity between users and online painting education courseware. The Ant colony optimization algorithms in the super heuristic algorithm is used to design the personalized recommendation process and complete the theoretical research on personalized recommendation of online painting education courseware. The experimental results indicate that this method the average accuracy of this method is 91% with interest and 86% with no interest. When the number of people in the same period increases to 1000, the growth rate of system throughput slows down and the growth rate is relatively small. This method can recommend high-quality online education courseware for painting, which helps improve the learning experience and effectiveness of users for online education courseware for painting.

Keywords: Hyperheuristic Algorithm · Online Education · Painting Courses · Courseware Recommendation · Personalized Recommendation · Ant Colony

1 Introduction

Currently, online education has become one of the important ways for people to learn, and online education for painting is also showing a thriving trend. However, online education is mostly based on universal curriculum settings, lacking personalized recommendations tailored to individual differences among learners [2], resulting in low learning outcomes and inability to fully explore learning potential. Therefore, how to achieve personalized recommendation for online painting education courseware has important research value.

Reference [3] designed a personalized learning resource recommendation system that includes user profiles. The system consists of a data layer, a data analysis layer,

L. Yun et al. (Eds.): ADHIP 2023, LNICST 549, pp. 210–223, 2024.
https://doi.org/10.1007/978-3-031-50549-2_15

and a recommendation computing layer. Among them, the recommendation computing layer discovers users' learning behavior patterns through similarity analysis and clustering algorithms, uses TF-IDF method to mine users' resource preferences, and provides personalized learning suggestions based on this. Reference [4] proposed a personalized recommendation algorithm for network information that integrates LDA and attention. Using the LDA model to induce the distribution of topics and words in documents, introducing HowNet to process word semantics and calculate semantic similarity between them. Adopting attention mechanism to achieve personalized recommendation of online information. Reference [5] extracts biological gene fragments, social gene fragments, and behavioral gene fragments from user online learning information, calculates the proportion of each gene fragment using the CRITIC weight method, and constructs a user's social media gene map to help users quickly obtain personalized online education services.

The learners of online painting education have a diverse group, with differences in learning interests, painting abilities, and learning habits. Therefore, it is not possible to simply recommend universal painting textbooks, but rather to have a deep understanding of students' personalized needs in order to better recommend textbooks that are suitable for them. At the same time, personalized recommendations can improve students' online learning enthusiasm and learning effectiveness. By analyzing students' learning data, establishing personalized learning models, and recommending targeted painting textbooks, it can meet students' learning needs, stimulate their interest in learning, and improve their learning enthusiasm. At the same time, personalized learning recommendations can better improve learning outcomes, maximize students' learning abilities, and truly achieve personalized customization in online painting education. In summary, based on the personalized needs of students, this article proposes a personalized recommendation method for online painting education courseware based on hyper heuristic algorithm.

2 Personalized Recommendations

Extract courseware keywords based on online education courseware for painting, sort them according to keyword frequency, and obtain different keyword weights for online education courseware for painting. Use the obtained weights to calculate the similarity of online education courseware for painting. Based on this, a personalized recommendation process for online painting education courseware is designed using hyper heuristic algorithms.

2.1 Keyword Weight Calculation

Obtain keyword weights through the TF-IDF algorithm [6]. TF and IDF represent word frequency and reverse word frequency, respectively, reflecting the importance of keywords in online painting education courseware and the general importance of words. Assuming F and K respectively represent the number of all extracted keywords and the number of times keywords appear in the designated online education courseware for painting, the calculation result of word frequency TF is shown in formula (1):

$$\mathrm{TF} = \frac{K/F}{N} \tag{1}$$

Assuming K and N respectively represent the number of online education courseware containing a certain keyword and the total number of all online education courseware containing a certain keyword, the IDF calculation result can be obtained as shown in formula (2):

$$\text{IDF} = \lg \frac{N}{|K+1|} \tag{2}$$

Obtain the final TF-IDF calculation result, that is, the keyword weight of the online education courseware for painting, as shown in formula (3):

$$\text{TF-IDF} = \text{TF} \times \text{IDF} \tag{3}$$

2.2 Calculation of Similarity in Online Education Courseware for Painting

If the number of common keywords in the online education courseware i and j for painting is t, the similarity calculation results of the two online education courseware for painting can be obtained as shown in formula (4):

$$\text{sim}(i,j) = \text{TF-IDF}_{\max} \sum_{i,j=1}^{t} W_i \times W_j \tag{4}$$

Among them, W represents the keyword weight.

2.3 User Similarity Calculation

Calculate the total similarity of users by calculating their attribute similarity and user activity similarity.

(1) User attribute similarity

Let C_i represent the numerical attribute similarity between users obtained based on the values of different user attribute intervals [7]. The calculation result of user attribute similarity is shown in formula (5):

$$\text{sim}_{num} = \text{sim}(i,j) \sum_{i=1}^{n} C_i \tag{5}$$

The text type attributes of users are taken as 1 and 0 respectively when they are the same or different. By accumulating the similarity sim_t of each user's text type attribute, the calculation result of user attribute similarity is shown in formula (6):

$$\text{sim}_{att} = \text{sim}_{num} + \text{sim}_t \tag{6}$$

From the above formula, it can be seen that user similarity increases with the increase of sim_{att} value, otherwise it is the opposite.

(2) User activity similarity

By obtaining the similarity sim_{act} of user activity through dynamic information of users, assuming that the number of common keywords between user U_i and user U_j is t, sim_{act} can be obtained as shown in formula (7):

$$\text{sim}_{act} = \frac{1}{\sum_{i,j=1}^{t} W_i \times W_j} \tag{7}$$

The similarity of user activity is higher when the sim_{act} value is higher, otherwise the opposite is true.

(3) User similarity

Linear weighted user attribute similarity and active similarity are used to obtain the final user similarity, as shown in formula (8):

$$\text{sim}_{A,B} = \text{sim}_{att} + \text{sim}_{act} \tag{8}$$

2.4 Similarity Between Users and Online Education Courseware for Painting

Assuming that the number of common keywords between user U_i and online education courseware j for painting is t, the similarity between user U_i and online education courseware for painting can be obtained as shown in formula (9):

$$\text{sim}_{U_i} = \text{sim}_{A,B} \sum_{i=1}^{t} W_i \times W_j \tag{9}$$

The similarity between users and online education courseware for painting is higher when the sim_{U_i} value is higher, otherwise the opposite is true.

2.5 Personalized Recommendation Based on Ant Colony Algorithm

The hyper heuristic algorithm based on meta heuristic algorithm adopts the existing meta heuristic algorithm (as a high-level strategy) to select LLH. This kind of super heuristic algorithm can be divided into Tabu search [8], genetic algorithm [9], genetic programming, ant colony algorithm, etc. Among them, the ant colony algorithm [10] is used for personalized recommendation. The pheromone mechanism of the ant colony when searching for food is used to express the preference of users for online painting education courseware, which can better consider the diversity and personality of users, so as to improve the coverage and personalization of recommendations. Therefore, this article utilizes ant colony algorithm to achieve personalized recommendation of online painting education courseware.

Based on the specific values of the similarity between users and online education courseware for painting obtained above, the picking probability and abandonment probability for influencing factors are obtained as formula (10):

$$\begin{cases} p_1 = \left[\lambda_2/\mathrm{sim}_{U_i}(\lambda_1 + \lambda_2)\right] \\ p_2 = 1 - p_1 \end{cases} \tag{10}$$

In the formula: p_1 represents the probability of picking up; p_2 represents the probability of abandonment; λ_1 represents the degree of correlation between shallow impact data; λ_2 represents the degree of correlation between deep impact data. In order to ensure recommendation efficiency, based on this, a personalized recommendation ant colony optimization control model for painting online education courseware is constructed as formula (11):

$$A = P + \frac{p_2(w + y)}{G} \tag{11}$$

Among them, P represents the optimization extremum for personalized recommendation of online education courseware for painting, G represents the optimal solution for personalized recommendation of online education courseware for painting, and w and y represent the gradient global extremum and individual extremum obtained for personalized recommendation of online education courseware for painting in each iteration. The ant colony optimization result for personalized recommendation of painting online education courseware is formula (12):

$$Z = A\left[v(\tau + 1) + p(\tau + 1)\right] \tag{12}$$

Among them, $v(\tau + 1)$ and $p(\tau + 1)$ are the efficiency parameters for personalized recommendation of online education courseware for painting recommended by the ant colony in the next moment.

Based on the probability of picking up, factors that can improve the efficiency of the enterprise are obtained. Let the ant colony be ant_M, and the number of ants in this algorithm is M. The location of each ant is a m-dimensional vector that can be represented as $ant_\rho(t)$, where $1 \leq \rho \leq M$ and ρ are natural numbers. Make the ant colony search from any position, i.e. assign an initial value to $ant_{\rho 1}(t), ant_{\rho 2}(t), \ldots, ant_{\rho n}(t)$; Then allow ants to freely search within the set range, allowing the $ant_\rho(t)$ value to continuously change during n iteration until clustering is completed. The clustering process of this factor can be divided into four stages as shown in Fig. 1.

Based on the above clustering process, a clustering model for personalized recommendation of painting online education courseware is constructed, and the output is formula (13):

$$B = f(z) + Z[\varphi + h(u)] \tag{13}$$

Among them, $f(z)$ represents the rating value of personalized recommendation for online education courseware, φ is the hierarchical scheduling model parameter for personalized recommendation output of online education courseware, and $h(u)$ is the feature information that is beneficial for detection.

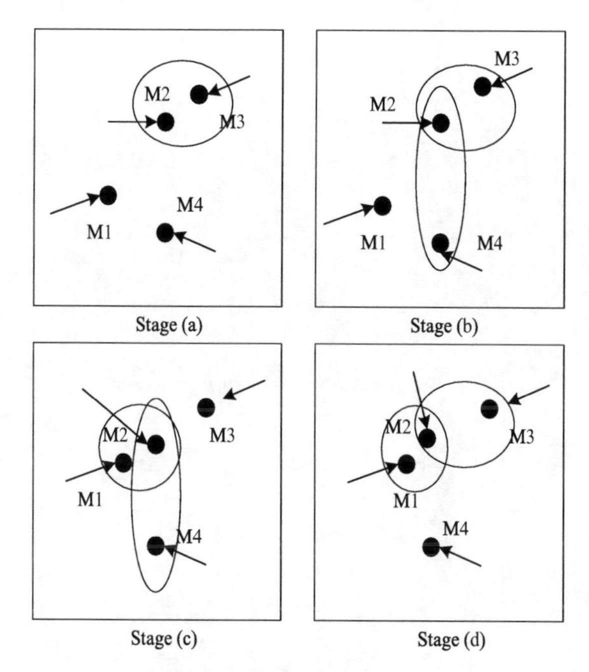

Fig. 1. Clustering process

By using the method of weighted sum per pixel, an adaptive adjustment formula for personalized recommendation of painting online education courseware is obtained as shown in formula (14):

$$R = B|W_{\min} + W_{\max}| \tag{14}$$

Among them, $[W_{\min}, W_{\max}]$ is the distribution range of top-down paths in ant colony optimization, generally taken as [0.5, 0.6]. Based on the above analysis, the algorithm implementation process for personalized recommendation of online education courseware for painting is shown in Fig. 2.

Step 1: Initialize ant colony information;
Step 2: $i = 1$, and select the i-th ant;
Step 3: Solve the ant route;
Step 4: Bring the results of the solution into the global ant colony;
Step 5: Solution optimization. After optimization, proceed to step 6. If not optimized, return to step 4;
Step 6: Update global ant colony information, $i = i + 1$;
Step 7: Whether to traverse all ants, if not, return to Step 3; If yes, output personalized recommendation results for online education courseware on painting.

3 Experimental Testing

3.1 Experimental Setup

The experimental platform parameters are shown in Table 1.

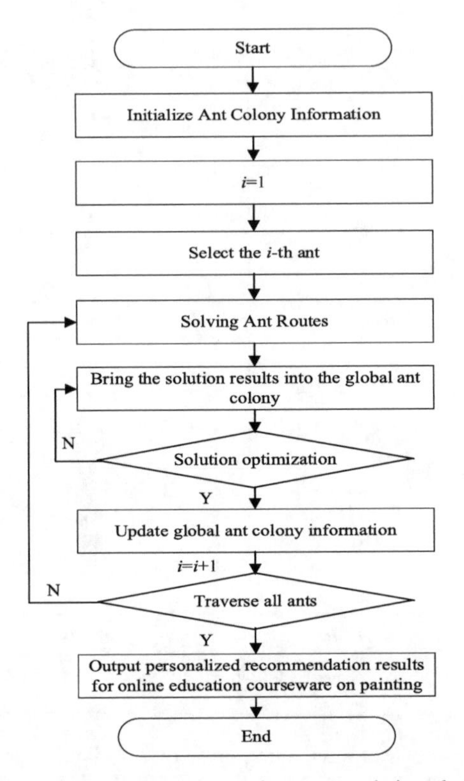

Fig. 2. Implementation process of recommendation algorithm

Table 1. Experimental Platform Parameters

Category	Parameter
Number of courses	100 painting courses
Course Type	Sketching, colors, characters, landscapes, still life, animals, etc.
Number of students	1000 students using the platform
Number of teachers	20 teachers provide painting courses and assessments
Evaluation method	After students submit their works, the teacher conducts a comprehensive evaluation, including grading and targeted suggestions

The dataset parameters are shown in Table 2.

Table 2. Dataset parameters

Category	Parameter
Dataset size	100000 images of painting works
Classification quantity	10 different painting categories, such as characters, landscapes, and imitations
Dataset Source	Obtained from 5 famous art museums and galleries
Annotation method	Include artist, work name, era, creative location, and painting style labels
Image feature dimension	RGB color feature dimension, edge feature dimension, texture feature dimension, etc.

Under the above preparation, the reference [3] method, the reference [5] method and the algorithm in this paper are respectively used for personalized recommendation of online painting education courseware. By comparing the indicator data of each algorithm, the effectiveness and feasibility of the algorithm in this paper are verified.

3.2 Result Analysis

(1) Qualitative experiments

The experimental indicators include accuracy, recall, comprehensive average and average absolute error. The details are as follows:

Precision refers to the ratio of the recommended number of correct online education courseware for painting recommended by the algorithm to the total number of online education courseware for painting that actual users are interested in. It is expressed as follows: Precision = recommended number of correct online education courseware for painting/recommended total number of online education courseware for painting. The higher the accuracy value, the higher the proportion of users who are interested in the painting online education courseware recommended by the recommendation algorithm.

Recall rate refers to the ratio between the recommended number of correct painting online education courseware by the algorithm and the total number of painting online education courseware of interest to the actual user. It is expressed as: Recall = recommended number of correct painting online education courseware/total number of painting online education courseware of interest to the actual user. The higher the recall rate, the more likely the recommendation algorithm can find the online education courseware of painting that users are interested in the F1 score, which combines accuracy and recall, is a harmonic mean expressed as F1 score = 2 * (Precision * Recall)/(Precision + Recall). The comprehensive average takes into account both accuracy and recall, and is a commonly used evaluation indicator.

Average absolute error (MAE) refers to the average of the absolute value of the difference between the recommended value and the true value, which is expressed by the formula: MAE = (|recommended value - true value|)/number of samples. The smaller the average absolute error is, the smaller the error between the recommended value and the true value of the recommended algorithm is, and the more accurate the recommendation result is the comparison results of experimental data are shown in Fig. 3.

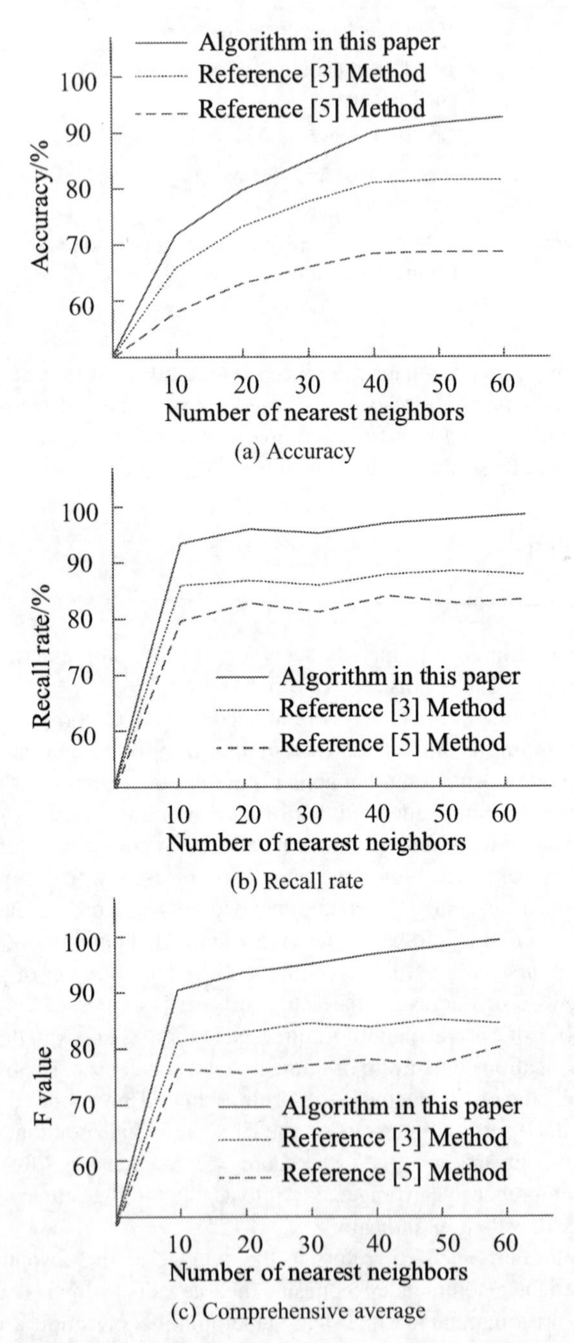

Fig. 3. Comparison of experimental data for performance indicators of various algorithms

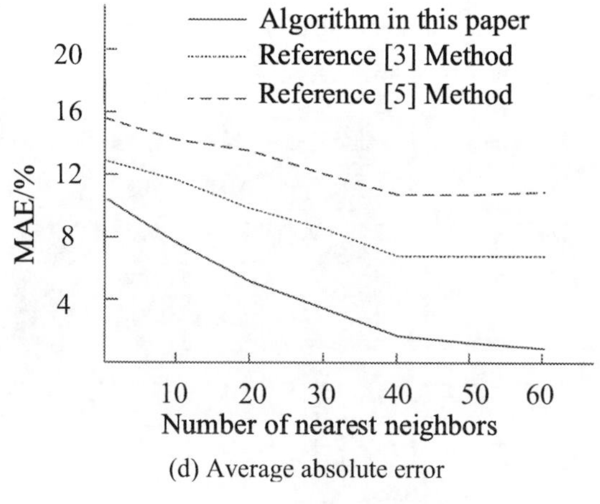

(d) Average absolute error

Fig. 3. (*continued*)

From the curve trend in Fig. 3, it can be seen that compared to the methods in reference [3] and [5], the algorithm proposed in this paper has significant advantages. In Fig. 3 (a), the accuracy of each algorithm is directly proportional to the number of nearest neighbors. When the number of nearest neighbors reaches a certain value, the accuracy of the algorithm in this paper gradually stabilizes, but there is still a small increase in accuracy; From the comparison of recall rates shown in Fig. 3 (b), it can be seen that although the fluctuation of each algorithm is relatively small, the curve of the algorithm in this paper has always been at a high level, indicating that the algorithm has a high probability of recommending online education courseware for painting that users are interested in; According to the weighted harmonic average trend in Fig. 3 (c), it is found that the algorithm proposed in this paper has good comprehensive performance and relatively ideal recommendation performance; The average absolute error value of each algorithm in Fig. 3 (d) shows that the error of the algorithm recommendation results in this paper is small and has been declining, which has certain feasibility.

(2) Quantitative experiments

A personalized recommendation experiment was conducted on online education courseware for painting based on the browsing records of each user, and the experimental results are shown in Table 3.

From Table 3, it can be seen that the average accuracy rate for personalized recommendation of online education courseware in painting is 91%, while the average accuracy rate for non interest is 86%. This result fully demonstrates that the method proposed in this article can meet the demand for personalized recommendation of online education courseware in painting.

In addition to the above indicator tests, the LoadRunner load testing tool is used to predict the recommended performance of this method through the throughput test. The basic steps for testing are as follows:

(1) Setting up a testing environment: In order to conduct throughput testing, it is necessary to first establish a testing environment in which various actual operational

Table 3. Personalized Recommendation Results of Online Education Courseware for Painting

User	Items of interest			No interest items		
	Total	Correct number	Accuracy/%	Total	Correct number	Accuracy/%
A	345	304	88.1	248	205	82.6
B	321	301	93.7	565	515	91.1
C	357	312	87.3	350	289	82.5
D	435	398	91.4	198	158	79.8
E	305	268	87.8	352	308	87.5
F	387	362	93.5	201	169	84.1
G	369	329	89.1	358	323	90.2
H	325	303	93.2	265	231	87.2

situations can be simulated. This may require the use of special testing tools and software.

(2) Define test cases: Test cases are a sequence of actions that need to be executed during the testing process, including steps such as request, response, and validation. When defining test cases, it is necessary to consider parameters such as load, number of concurrent users, and data size.
(3) Testing: When conducting testing, it is necessary to execute defined test cases and record indicator data during the testing process, such as processing time, throughput, etc.
(4) Analysis of test results: After the test is completed, it is necessary to analyze the test results to evaluate the performance of the equipment or system. During the analysis process, various factors need to be considered, such as hardware, software, network environment, etc.

According to the above process, the number of simultaneous operation users in the simulation setting gradually increased from 100 to 1000. The performance test of the method in this paper, the method in Reference [3] and the method in Reference [5] was conducted through the throughput, and the results are shown in Table 4.

According to the statistical results in Table 4, it can be seen that the recommended users of the system, reference [3] method, and reference [5] method in this article increased from 100 to 1000 at the same time, and the throughput increased with the increase of the number of users. Among them, the methods in reference [3] and reference [5] have lower throughput. Due to the slow response of the methods in reference [3] and reference [5], the number of successful recommendations is relatively low. When the number of people in this method increases to 1000 during the same period, the growth rate of system throughput slows down, and the growth rate is relatively small. The system operation status remains stable and all users successfully complete the recommendation, without any system access failure or lag phenomenon. Therefore, it can be concluded that the recommendation performance of this method is good.

Table 4. Throughput testing

Number of simulated users/person	Algorithm in this paper/Mbps	Reference [3] Method/Mbps	Reference [5] Method/Mbps
100	17.72	14.22	15.62
200	21.26	17.76	19.16
300	26.50	23.12	24.51
400	29.56	26.06	27.46
500	33.88	30.38	31.78
600	38.02	34.52	35.92
700	41.38	37.88	39.28
800	45.92	42.42	43.82
900	49.54	46.04	47.44
	54.62	51.12	52.52

The similarity calculation ablation experiment is an experiment conducted on the influence of similarity calculation methods in the personalized recommendation system for online painting education courseware based on hyperheuristic algorithms. Set the similarity calculation method as an independent variable and set different experimental groups, each using a similarity calculation method. Use the correct number of courseware to evaluate the quality and effectiveness of recommendation results. The results are shown in Table 5.

Table 5. Experimental result

	Algorithm in this paper	Reference [3] Method	Reference [5] Method
Experimental courseware	368	368	368
Correct courseware	367	352	346
Accuracy	99.73%	95.65%	94.02%

From Table 5, it can be seen that the recommendation results analyzed using this method are significantly higher than those of traditional methods. The recommendation of the method in this article will not completely filter out bottlenecks, and this reason is due to the similarity calculation of online education courseware for painting.

4 Conclusion

Through the study of personalized recommendation methods for painting online education courseware based on hyper heuristic algorithms, the following conclusion has been drawn: personalized recommendation systems can greatly improve users' learning experience and effectiveness for painting online education courseware. Especially in the era of big data, recommendation algorithms can track user learning records and interest preferences, accurately recommend courseware and content suitable for users, avoid the shortcomings of traditional education methods, and improve the effectiveness of courseware and user satisfaction. Moreover, with the continuous optimization and update of recommendation algorithms, it can also help users discover interesting content that they have not discovered, enhance the fun of painting learning, and stimulate users' interest in learning. In summary, personalized recommendation systems will become one of the important methods for future education, which can help users comprehensively improve the learning effect and fun of painting education. With the rapid development of online education in painting, personalized recommendation systems have gradually become an important trend. In the coming years, personalized recommendation of online education courseware for painting will continue to maintain rapid development. Here are several possible research prospects:

(1) Multimodal recommendation system: Integrate multimedia resources in online education courseware for painting, and use multiple modal information such as voice, image, and text to design multimodal recommendation algorithms to improve recommendation effectiveness and user satisfaction.
(2) Recommendation system based on deep learning technology: Deep learning (AL) is currently one of the most promising research fields. Combining the characteristics of the painting field, through deep learning technology, more accurate and intelligent recommendation algorithms are established. For example, deep learning technology can be used to construct an image recognition system that accurately recognizes and analyzes the user's painting style to support recommendation algorithms.

References

1. Merritt, E.G., Powell, R.B., Stern, M.J., Frensley, B.T.: A systematic literature review to identify evidence-based principles to improve online environmental education. Environ. Educ. Res. **28**(5), 674–694 (2022)
2. Katz, D., Huggins-Manley, A.C., Leite, W.: Personalized online learning, test fairness, and educational measurement: considering differential content exposure prior to a high stakes end of course exam. Appl. Meas. Educ. **35**(1), 1–16 (2022)
3. Li, X.R., Liang, H.W., Feng, J.Y., Xiao, J.P., Peng, W.F.: Design of personalized learning resource recommendation system for online education platform. Comput. Technol. Dev. **31**(2), 143–149 (2021)
4. Zhang, Y.B., Zhao, J.L.: Personalized recommendation method of network information integrating LDA and attention. Comput. Simul. **39**(12), 528–532 (2022)
5. Zhang, J.D., Jiang, L.P.: Research on recommendation of online education service based on social media gene mapping. Inf. Stud. Theory Appl. **44**(8), 131–138 (2021)

6. Yu, H., Ji, Y., Li, Q.: Student sentiment classification model based on GRU neural network and TF-IDF algorithm. J. Intell. Fuzzy Syst. Appl. Eng. Technol. **40**(2), 2301–2311 (2021)
7. Shen, X., Junwei, D.U., Gong, D., Yao, X.: Developer cooperation relationship and attribute similarity based community detection in software ecosystem. Chin. J. Electron. **32**(1), 39–50 (2023)
8. Rao, K.S., Sridhar, M.: A tabu search algorithm for general threshold visual cryptography schemes. Ingénierie des Systèmes D Information **26**(3), 329–335 (2021)
9. Alves, D., Oliveira, D., Andrade, E., Nogueira, B.: GPU-brkga: a GPU accelerated library for optimization using the biased random-key genetic algorithm. Lat. Am. Trans. **20**(1), 14–21 (2022)
10. Pramanik, S., Goswami, A.: Discovery of closed high utility itemsets using a fast nature-inspired ant colony algorithm. Appl. Intell. Int. J. Artif. Intell. Neur. Netw. Complex Probl. Solving Technol. **52**(8), 8839–8835 (2022)

Reliability Analysis of Aeroengine Teaching System Based on Virtual Reality Technology

Mingfei Qu(✉) and Xin Zhang

College of Aeronautical Engineering, Beijing Polytechnic, Beijing 100176, China
qmf4528@163.com

Abstract. The emergence of virtual reality technology provides new technological support for the development of teaching, enabling students to immerse themselves more in the teaching environment. However, there is relatively little research on the application of virtual reality technology in aviation engine teaching, which cannot guarantee its reliability. Therefore, a reliability analysis study of aviation engine teaching system based on virtual reality technology is proposed. Select the professional virtual reality modeling and simulation software LabVIEW to construct a three-dimensional model of an aviation engine, design a human-computer interaction teaching process, build a three-dimensional vision model based on the principle of human eye stereo vision, analyze the generation of three-dimensional graphics in the computer, and thus achieve the operation of the aviation engine teaching system. The experimental data shows that under different experimental conditions, the maximum display integrity of the teaching scene obtained after the application of the designed system is 98%, and the maximum success rate of human-machine interaction in the teaching process is 96%, fully confirming the stronger reliability of the designed system.

Keywords: Aeroengine · Reliability · Virtual Reality Technology · 3D Model · Human Machine Interaction

1 Introduction

With the rapid development of information technology, the existing aviation engine teaching mode cannot meet the needs of practical applications. New teaching media are constantly emerging, and after multimedia, a new type of teaching media has emerged in the field of teaching technology, which is virtual reality technology. Traditional aviation engine teaching experiments generally include viewing real engine prototypes in the showroom and conducting engine test runs. However, the internal structure of aviation engines is very complex and difficult to observe, and it is not possible to observe the internal structure solely by observing the actual engine prototype. Moreover, it is difficult to have a comprehensive understanding of the connection relationship between the entire engine and various components. Conducting an aviation engine test drive experiment is not only expensive, but also incurs significant engine losses, making it too expensive

L. Yun et al. (Eds.): ADHIP 2023, LNICST 549, pp. 224–238, 2024.
https://doi.org/10.1007/978-3-031-50549-2_16

to use as a teaching experiment. For students majoring in science and engineering, the practical aspect has become an important aspect for them to deeply understand theoretical knowledge. However, currently, there are problems with outdated equipment and high experimental expenses in aviation engine teaching experiments, making it difficult to meet the practical needs of students in the learning process and keep up with the rapid development of science and technology. Given the above reasons, it is urgent to adopt a new teaching experimental method to ensure a certain level of teaching quality.

Virtual reality technology transforms multimedia teaching platforms from two-dimensional planes to three-dimensional spaces, creating a virtual, realistic, and interactive teaching environment. It can build a three-dimensional virtual space, and students will become a member of this virtual space, and can interact with this space to effectively simulate human behavior in the natural environment, such as seeing, listening, and moving. It is an advanced human-computer interaction technology, which is the intersection and integration of many disciplines, integrating multimedia technology, artificial intelligence, computer graphics, multimedia technology, sensor technology High speed parallel Real-time computing technology and human behavior research and other technologies.

From the existing research results, virtual reality technology is mainly applied in the process of physics teaching, such as fluid real-time simulation systems based on virtual reality technology and physics laboratory teaching systems based on virtual reality technology. However, there are few research achievements related to the application of virtual reality technology in the field of aviation engine teaching.

Therefore, this article applies virtual reality technology to aviation engine teaching and experimentation, proposing a new type of aviation engine teaching and experimentation method, which has important theoretical significance and application value for the digitization and modernization construction of aviation engine teaching and experimentation.

2 Aeroengine Teaching System Design Research

2.1 Aeroengine 3D Modeling Module

According to the performance requirements of the aeroengine teaching system, the professional virtual reality modeling and simulation software LabVIEW is selected to conduct three-dimensional modeling of aeroengines.

The general steps for modeling the solid model aeroengine are:

Step 1: Get modeling data

The internal structure of an aeroengine is very complex, and there are many complicated accessories besides a few main components. The model used in teaching should be as simple as possible to facilitate observers to understand the internal structure of the engine; Moreover, if the amount of model data is very large, it will affect the real-time performance of the system to a certain extent [3]. Therefore, it is necessary to simplify the model. The geometric shape data of the engine model mainly comes from the engineering drawings and photos of the real prototype. Combining with the components we

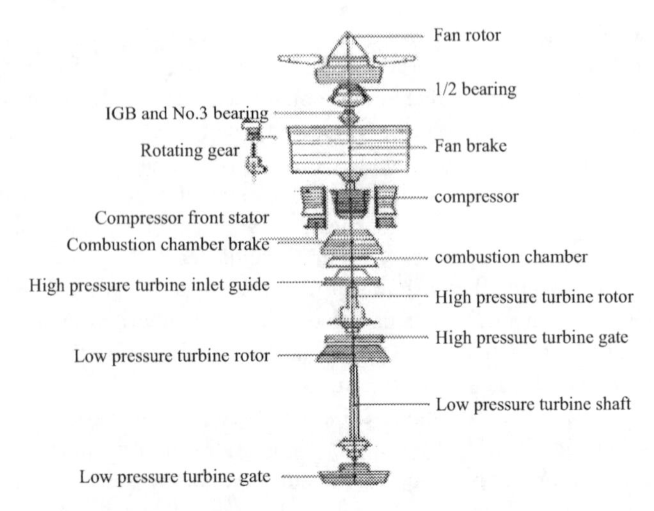

Fig. 1. Schematic Diagram of Aeroengine Structure

care about, we further simplify the engineering drawings to obtain a three-dimensional model of the engine suitable for teaching experiments, as shown in Fig. 1.

There are five main parts of aeroengine: inlet, compressor, combustion chamber, turbine and tail nozzle. We only care about the core engine parts, namely compressor, turbine and combustion chamber, so we omit the inlet port in modeling. Since the tail nozzle will produce tail flame when the engine is working, we need to simulate the effect of tail flame in modular experiments, so we do not omit it, but simplify the original prototype and build a contraction nozzle. The original prototype also has an afterburner, but because the afterburner has a long length and occupies a large space, if the afterburner is not removed, the overall observation effect of the engine model will be affected, so the afterburner is partially removed, and the tail nozzle is directly connected behind the low-pressure turbine support plate. The structure of a real engine is very complex, consisting of thousands of parts. In addition to some important parts, there are many accessories and accessories. The guiding principle of our simplified model is to establish key components that have an important impact on the engine performance, have a great relationship with the shape, and play an important role in rotor rotation, such as blades, disks, casings, flame tubes, shafts, bearings, engine housings, etc. Components such as accessory drive, lubricating oil system, fuel system and control system will be omitted during modeling. When the engine is working, the low-pressure turbine rotor drives the low-pressure compressor to rotate, and the high-pressure turbine rotor drives the high-pressure compressor to rotate. In order to realize the rotation of the engine rotor blade, 7 bearings are built, including 5 roller support bearings and 2 ball bearings.

Step 2: Determine the database hierarchy of the model

3D models are managed and operated in the way of model database. The model database reflects the geometric space position of each entity in the real environment, as well as the structural relationship between models and within models, and determines the hierarchy of all entity models in the virtual scene. Hierarchical division of models can facilitate the division of modeling and the organization and management of entity models. Hierarchical division of entity models can decompose complex models into several basic units from top to bottom, clarify the objectives of model construction, and greatly reduce the workload of modeling [4]. LabVIEW modeling provides a tree hierarchy structure to organize and manage the model. When modeling, you should first decompose the model according to the hierarchy structure, and arrange the parts at the appropriate node locations for management.

The hierarchy diagram of aeroengine model is shown in Table 1.

Table 1. Aero engine structure hierarchy

Level 1	Level II	Level 3
compressor	High pressure compressor	Rotor blade
		stator vane
		Wheel disc
	Low pressure compressor	–
turbine	Turbine guide vane	–
	High pressure turbine	–
	Low pressure turbine	–
combustion chamber	Cross flame tube	–
	Flame tube	Flame tube wall
	Combustion chamber shell	–
	Lower wall of expander gate	–
Machine brake	External brake	Low pressure compressor brake
		Outer channel housing
	Internal machine brake	High pressure compressor brake
		Turbine gate
axis	axis	Inner shaft
		Outer shaft
	Bearing	Low pressure bearing
		High pressure bearing
	Head fairing	–
	Nozzle tailstock	–

Step 3: Establish the model

According to the hierarchical structure and drawings planned in advance, establish a 3D model [5] in LabVIEW. When modeling, the complexity of internal parts of the engine is taken into consideration, and the modeling is carried out according to the size scale of 10:1 (model size: real size). The full model of aeroengine is shown in Fig. 2

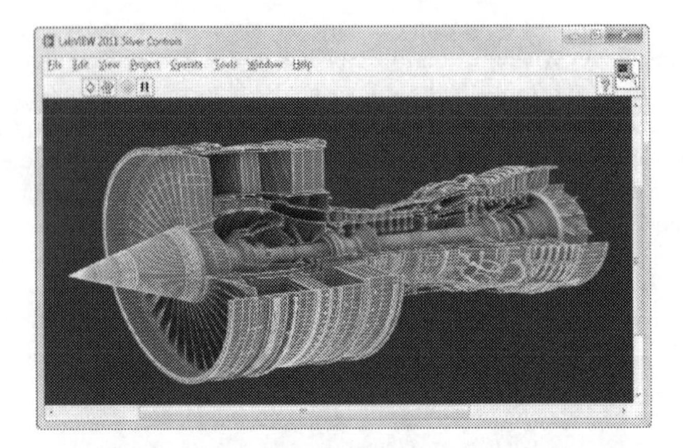

Fig. 2. Schematic Diagram of Aeroengine Full Model

As shown in Fig. 2, the components of 3D visual simulation model are as follows:

Low pressure compressor: 5-stage axial flow type. The rotor adopts a drum disk structure, and the rotor adopts two roller bearing;

High pressure compressor: 12 stage axial flow type. The rotor adopts a drum disk structure, the inlet is provided with an inlet guide vane, and the rotor adopts a roller bearing and a ball bearing;

Combustion chamber: annular tube type reflux combustion chamber. There are 8 flame tubes, 8 fuel nozzles, 2 igniters and 8 cross flame tubes;

High pressure turbine: two-stage axial flow type. There is a roller bearing at the front end;

Low pressure turbine: two-stage axial flow type. There is a roller bearing at the rear end;

Shaft: double shaft rotor;

Outer casing: LP compressor casing, HP compressor casing, diffuser casing, combustion chamber casing, turbine guide casing.

The above process completed the construction of the three-dimensional model of the aeroengine, which laid a solid foundation for the follow-up research.

2.2 Human Computer Interaction Module in Teaching Process

Human computer interaction is the key to the realization of aeroengine teaching process. Its main application technology is virtual hand technology, which mainly uses computers to create a virtual environment, and makes the virtual hand model as the replacement of human hands in the virtual environment. The operator transmits the operation information of human hands to the virtual hand model in the virtual scene through virtual peripherals (data gloves). Realize the action mapping from human hand to virtual hand, and get information feedback (force/tactile feedback) from the virtual environment, so as to achieve interaction with the virtual environment. Virtual hand technology is a new technology with multi - discipline. It involves gesture input technology, position tracking technology, virtual hand modeling technology, stable grasping technology, etc.

Virtual hand models can be presented to users in a visual manner, enabling them to intuitively understand the position, posture, and movements of the hand. This helps to improve users' understanding of system operations and interaction processes, and promotes effective communication with computer systems. Through the virtual hand model, users can operate and interact with their hands in a natural way in the virtual environment. This simulation experience can enhance users' perception and control of operations, providing a more realistic interaction experience. Virtual hand technology is the key to realize the engine virtual assembly in this teaching simulation system [6].

For human-computer interaction based on virtual hand, it is to use data gloves to measure the angle of each joint on the human hand, use position trackers to measure the spatial position of the human hand, control the motion and state of the virtual hand model in the computer according to the measured motion data, and use virtual hands to grasp, release, and translate virtual objects. And calculate the position and attitude of the moved virtual object. Finally, the operator judges whether to complete the operation according to various sensory feedback, and then starts a new task. Human computer interaction based on virtual hand consists of operator, virtual peripheral and virtual environment, as shown in Fig. 3.

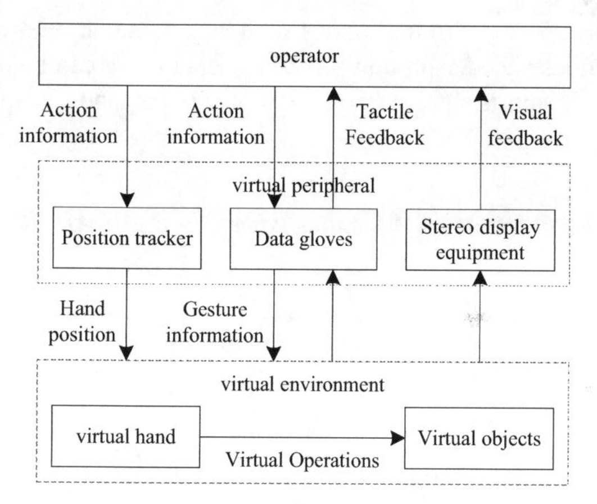

Fig. 3. Schematic diagram of human-computer interaction framework in the teaching process based on virtual hands

Among them, gesture input technology is one of the key technologies of virtual hand technology. At present, there are many methods of gesture recognition, mainly including gesture input based on data gloves and gesture input based on computer vision. This paper mainly applies gesture input technology based on data glove. This technology uses data gloves as an interactive tool for gesture input. The data glove used in this system is 5DT Data Glove 5 produced by 5DT Company. The 5DT Data Glove 5 data glove is a black elastic double-layer textile. The interlayer is a leather sensor made of optical fibers. There is a sensor on each finger to measure the average flexion and extension of the finger (that is, the flexion and extension of the joint in the middle of the finger). There are also two sensors at the wrist to measure the pitch angle and tilt angle of the palm. The position distribution of sensors is shown in Fig. 4.

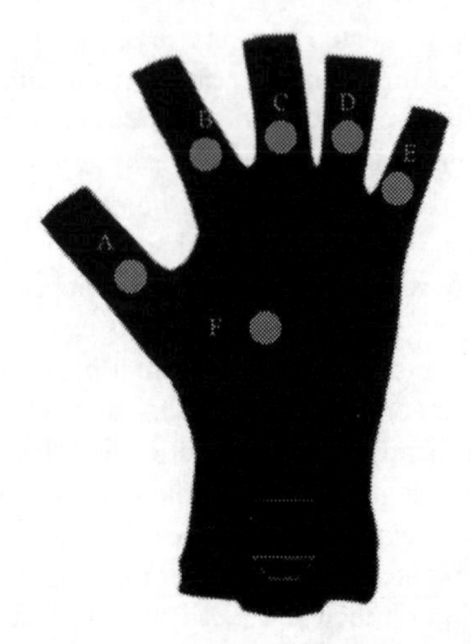

Fig. 4. 5DT Data Glove Sensor Distribution Diagram

5DT Data Glove 5 has working modes such as command, report data, continuous data, and analog mouse. The sampling rate of the data glove can be up to 200 Hz. The data string sampled from the glove has a total of 9 bytes, and the specific meaning is shown in Table 2.

Table 2. Meaning of Data String Bytes Collected by Data Gloves

Byte sort	byte	meaning
1	Header	The leading byte indicates the beginning of a new data string
2	F1	Flexion and extension of thumb
3	F2	Flexion and extension of index finger
4	F3	Flexion and extension of middle finger
5	F4	Flexion and extension of ring finger
6	F5	Flexion and extension of little finger
7	pitch	Tilt angle of hand swing up and down
8	roll	Tilt angle of palm rotation
9	checksum	Checksum

Since the data glove has no position sensor, the position information of the hand in the three-dimensional space cannot be obtained from the input data of the data glove. Therefore, in this application system, the position change of the hand in the three-dimensional space is not obtained through the data glove, but through the position tracker.

In order to better control the state of the virtual hand grasping the virtual object, we add the grasping state flag bit to the virtual object. True indicates that the virtual hand has grasped the object, and false indicates that the virtual hand has not grasped the object; We set the initial value of the grab status flag bits of all virtual objects in the scene to false. The position, orientation and posture of the virtual hand are obtained by acquiring the spatial position and orientation of the human hand and the angle values of each joint from the SDT data glove and FOB electromagnetic tracker; Then, collision detection is carried out to determine the grabbing status of virtual objects according to the grabbing rules, set the grabbing status flag bit of virtual objects in combination with the current grabbing status, and correspondingly realize the grabbing or releasing of virtual objects [7].

The algorithm flow of the whole virtual hand grabbing virtual objects is shown in Fig. 5.

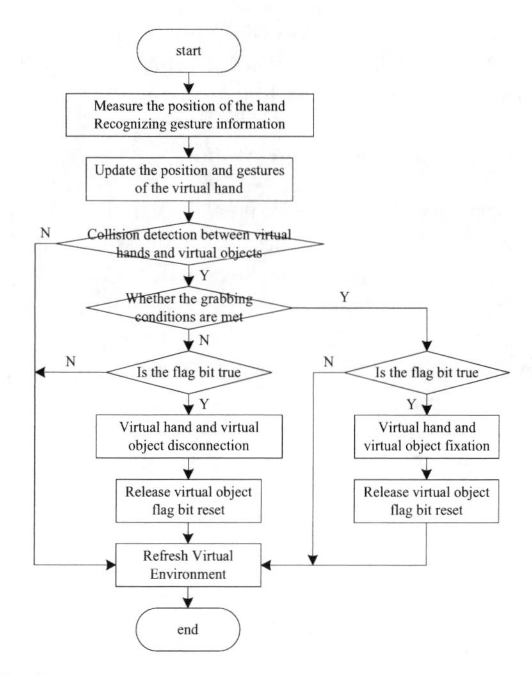

Fig. 5. Flow Chart of Virtual Hand Grasping Virtual Objects

As shown in Fig. 5, the detailed steps of the virtual hand grasping algorithm are as follows:

Step 1: execute actions according to assembly requirements; Obtain the value of the electromagnetic tracker, that is, the absolute spatial position and orientation value of the hand, obtain the value of each sensor of the data glove, and convert it into the corner value of each joint of the finger.

Step 2: The computer calculates the parameters required for the motion of the virtual hand according to the obtained spatial position and orientation values of the human hand, as well as the rotation angles of the finger joints, and updates the position and gesture of the virtual hand in the virtual operation space according to these parameters.

Step 3: Collision detection between virtual hand and virtual object. If there is no collision between the two, go to step 6; If there is a collision between the two, judge whether the virtual hand and the virtual object meet the snatch conditions [8].
Step 4: if the snatch conditions are met, judge the marking position. If it is true, the coordinate system of the virtual hand is fixed with the coordinate system of the virtual object, the virtual object is grabbed by the virtual hand, and the virtual object pair is translated and rotated with the virtual hand. Mark position 1, and turn to step 6; Otherwise, go to step 6 directly.
Step 5: if the conditions for snatch are not met, judge the mark position. If it is true, remove the connection between the virtual hand and the virtual coordinate system, release the virtual object with the virtual hand, mark position 0, and turn to step 6; Otherwise, go to step 6 directly.
Step 6: Refresh (redraw) the virtual environment and turn to step 1.
The above process details the whole process of teaching human-computer interaction, which provides support for aeroengine teaching experiments.

2.3 Teaching Scene Stereoscopic Display Module

In virtual reality, stereoscopic display technology is one of the key technologies. It is a necessary condition for a virtual reality based system. Without in-depth stereoscopic visual effects, it is impossible to feel immersive, and it is also impossible to achieve the goal of virtual reality. In order to generate stereoscopic images and enable users to see high-quality stereoscopic effects, we need to conduct in-depth research on the hardware implementation and software algorithm of stereoscopic display. This section focuses on introducing the principle of human eye stereoscopic vision, so as to propose a stereoscopic vision model, then analyze the generation of three-dimensional graphics in the computer, and finally give an example of stereoscopic effects.

Because people's eyes are a certain distance apart, the image of the same object in the left and right eyes will be slightly different, and this difference will form the main clue to judge the depth information of the static scene. Binocular parallax is actually the angle difference caused by the binocular imaging of two objects with a certain distance in the depth direction in space, and the calculation formula is

$$\delta = 2\left(\arctan\frac{L}{2D} - \arctan\frac{L}{2(D+\Delta D)}\right) \tag{1}$$

In formula (1), δ it represents an object O_1 and O_2 poor perspective; L is the distance between pupils; D it indicates the distance between the object and the eye.

When D much greater than L Eq. (1) can be written as:

$$\delta \approx \frac{L}{D} - \frac{L}{D+\Delta D} = \frac{L\Delta D}{D(D+\Delta D)} \tag{2}$$

If the object O_1 object O_2 be relative to O_1 the depth position of δ the size of the corner. In fact, the human eye cannot directly feel δ. However, the vision system can get the information of depth direction by comparing the position difference of the object imaging on the retina. Binocular parallax is only effective in a small range in depth perception, which is generally considered to be 0–100. When the visual distance of an object is less than 380mm, that is, when the angle of the object to both eyes is less than a certain value, diplopia will occur, and the stereoscopic feeling cannot be formed. This effect must be taken into account in the design of stereoscopic display to avoid the stereoscopic image pair being difficult to fuse and thus unable to produce stereoscopic sense [9].

Three algorithms are usually used to generate three-dimensional images in vision systems: left and right eye view generation based on projection transformation principle; Stereogram generation algorithm based on correlation theory; Fast holographic imaging algorithm. The algorithm based on correlation theory is an improvement of the algorithm based on projection transformation, and their basic principles are the same; Fast hologram generation algorithm is the only image generation technology that can provide all depth clues. It has high computing efficiency, but requires special hologram imaging equipment. The principle of left and right eye view generation algorithm based on projection transformation principle is briefly described below.

a. Single view perspective projection

Single viewpoint perspective is a single vanishing point perspective, which takes the viewpoint as the projection center, and the line of sight starts from the viewpoint. Unlike parallel projection, the line of sight is not parallel. The line of sight crosses the projection plane and intersects the object. The intersection point with the projection plane is the corresponding image point of the intersection point of the line of sight and the object. The viewpoint of perspective projection is $P_c(x_c, y_c, z_c)$, the projection plane is XOY plane, a point on the body $P(x, y, z)$ the projection of is $P_s(x_s, y_s, z_s)$, the expression is

$$\begin{cases} x_s = x_c + (x - x_c) \times t \\ y_s = y_c + (y - y_c) \times t \\ z_s = z_c + (z - z_c) \times t \end{cases} \tag{3}$$

In Eq. (3), t it refers to auxiliary calculation parameters, which are calculated by the following formula:

$$t = -\frac{z_c}{z - z_c} \tag{4}$$

Substitute Formula (4) into Formula (3) to get the projection line equation, which is expressed as

$$\begin{cases} x_s = \frac{x_c - x z_c}{z - z_c} \\ y_s = \frac{y_c - y z_c}{z - z_c} \end{cases} \tag{5}$$

To facilitate the research, set the perspective projection viewpoint at Z negative half axis, the distance from the origin is d, there are: $x_c = y_c = 0$, $z_c = -d$, and substitute it into formula (5) to obtain the final projection line equation, which is expressed as

$$\begin{cases} x_s = \frac{xd}{z+d} \\ y_s = \frac{yd}{z+d} \end{cases} \tag{6}$$

b. Double viewpoint projection of left and right eyes

For stereoscopic images, it is required to generate different views relative to the left and right eyes. Each eye observes the scene from different positions, so that two eyes can see different pictures. These views are generated according to different projection centers [10]. Set the coordinate system OXYZ as the world coordinate system, the projection plane is located at Z = 0, and the distance between two eyes is e, the projection center corresponding to the left view is $L\left(-\frac{e}{2}, 0, -d\right)$, the projection center of the right view is $R\left(\frac{e}{2}, 0, -d\right)$, d Is the distance from the viewpoint to the projection plane.

Spot $P(x, y, z)$ taking the left viewpoint as the projection center, the expression of the projection point generated by projection onto the XOY plane is

$$\begin{cases} x_{sL} = \frac{xd - {}^{ze}/_{2}}{z+d} \\ y_{sL} = \frac{yd}{z+d} \end{cases} \tag{7}$$

In Eq. (7), (x_{sL}, y_{sL}) it represents a point $P(x, y, z)$ the projection point generated by taking the left viewpoint as the projection center and projecting it onto the XOY plane [11].

Same $P(x, y, z)$ the point takes the right viewpoint as the projection center, and the expression of the projection point on the XOY plane is

$$\begin{cases} x_{sR} = \frac{xd + {}^{ze}/_{2}}{z+d} \\ y_{sR} = \frac{yd}{z+d} \end{cases} \tag{8}$$

In Eq. (8), (x_{sR}, y_{sR}) it represents a point $P(x, y, z)$ the projection point generated by taking the right viewpoint as the projection center and projecting it onto the XOY plane.

When the projection map is displayed on the CRT screen, the coordinates of the projection point on the XOY plane need to be converted to the local viewport coordinate system [12]. The coordinate axes of the left and right viewport coordinate systems are parallel to the world coordinate system respectively, and the left viewport coordinate system moves left along the X axis from the world coordinate system $\frac{e}{2}$ the right viewport coordinate system is moved right along the X axis from the world coordinate system $\frac{e}{2}$ get. Thus, it is obtained that $P(x, y, z)$ the coordinates of the left image point in the left viewport coordinate system are $\left(x'_{sL}, y'_{sL}\right)$, $P(x, y, z)$ the coordinates of the right image point in the right viewport coordinate system are $\left(x'_{sR}, y'_{sR}\right)$. Only the projection points in the left and right viewports are displayed on the screen, and the contents in the whole left and right viewports will be displayed in the same area of the screen. At this time, the formula for calculating the relative distance between the two image points is

$$S = x'_{sL} - x'_{sR} = e\left(1 - \frac{z}{z + d}\right) \tag{9}$$

So, point $P(x, y, z)$ the image in the left and right eyes has a certain sense of distance S, form parallax, and form stereoscopic sense through brain fusion. S the size of, which affects the quality of image fusion, depends on the distance between two viewpoints and the relative position of the object, projection center, and projection plane [13]. In practical applications, the size of the final screen display area should also be considered. It is one of the key tasks of the stereoscopic display system to select the projection center spacing according to the different sizes of the screen display area, so as to form a better image fusion effect.

Through the design and development of the above three modules, the operation of the aeroengine teaching system is realized, which provides effective support for the aeroengine experiment teaching.

3 Design System Reliability Analysis

3.1 Selection of Virtual Instrument

In the virtual instrument system, hardware is only for signal acquisition, and software is the key of the whole instrument. When the user's test requirements change, or the test items need to be added or reduced, the user only needs to change the software program appropriately to get the test instrument system that meets the test requirements. This is the meaning of the slogan "The Software Is The Instruments" put forward by National Instruments, the initiator of virtual instruments. The software provides three main functions: an integrated development environment, an advanced interface with the instrument hardware, and a graphical user interface. The main software modules include data acquisition, data analysis, data display, file management and report output. The main development environments of virtual instruments include C, C++ Builder, VB, Delphi, LabVIEW, LabWindows/CVI, etc. LabVIEW, which is a fully graphical special development tool for virtual instruments, integrates a large number of instrument function modules internally. The development environment based on graphical programming language G enables developers to focus only on instrument functions and liberate themselves from complex, tedious and time-consuming language programming. Compared with the traditional programming method, using LabVIEW to design virtual instruments can improve the efficiency by 4–10 times.

The design steps of LabVIEW are shown in Fig. 6.

The virtual instruments selected above are used as experimental equipment to facilitate the subsequent experiments.

3.2 Analysis of Experimental Results

In order to intuitively display the reliability of the design system, the fluid real-time simulation system based on virtual reality technology and the physical laboratory teaching system based on virtual reality technology are selected as comparison systems 1 and 2, and the integrity of teaching scene display and the success rate of human-computer interaction in the teaching process are set as the evaluation indicators of system reliability. The calculation formula is

$$\begin{cases} Q = \frac{q_1}{q_{total}} \times 100\% \\ K = \frac{k_1}{k_{total}} \times 100\% \end{cases} \tag{10}$$

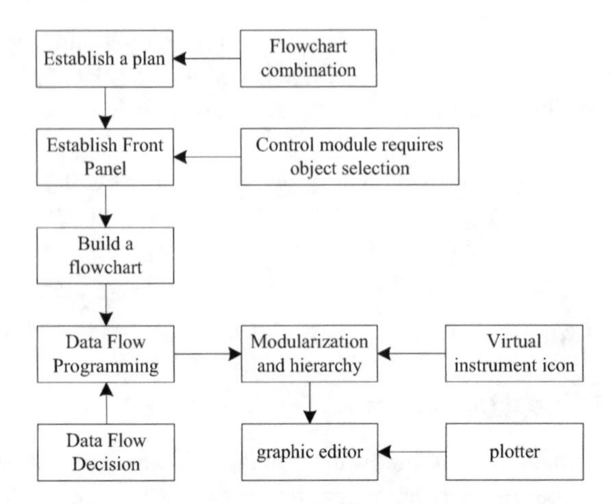

Fig. 6. Schematic diagram of LabVIEW design steps

In Eq. (10), Q it represents the display integrity of teaching scenes; K it represents the success rate of human-computer interaction in the teaching process; q_1 it represents the display area of the teaching scene; q_{total} it represents the total area displayed in the teaching scene; k_1 it indicates the number of successful human-computer interaction in the teaching process; k_{total} it represents the total number of human-computer interactions in the teaching process.

The display integrity of teaching scenes obtained through experiments is shown in Fig. 7.

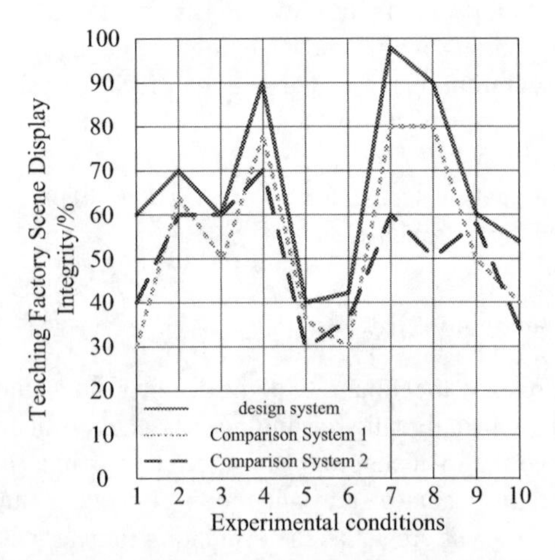

Fig. 7. Schematic Diagram of Teaching Scene Display Integrity

As shown in the data in Fig. 7, under different experimental conditions, the completeness of the teaching scene display was obtained by comparing System 1 with 30%–80%, and by comparing System 2 with 30%–70%. However, after the design and application of the system, the completeness of the teaching scene display was obtained by 40%–98%. The completeness of the teaching scene display was much higher than that of Comparing System 1 and System 2, providing more complete teaching scene support for aviation engine teaching.

The success rate of human-computer interaction in the teaching process obtained through experiments is shown in Table 3.

Table 3. Data sheet of human-computer interaction success rate in teaching process/%

Test conditions	design system	Comparison system 1	Comparison system 2
1	89	56	45
2	94	45	44
3	95	64	41
4	96	52	56
5	91	58	54
6	80	63	57
7	84	62	50
8	95	64	41
9	91	51	42
10	90	49	38

As shown in Table 3, under different experimental conditions, the success rate of human-machine interaction in the teaching process obtained by comparing System 1 is 45%–64%, and the success rate of human-machine interaction in the teaching process obtained by Comparing System 2 is 38%–57%. However, the success rate of human-machine interaction in the teaching process obtained by designing the system application is 80%–96%. The success rate of human-machine interaction in the teaching process is much higher than that of Comparing System 1 and Comparing System 2, indicating that the success rate of the method in this paper is relatively high, It can provide more stable performance support for aviation engine teaching.

4 Conclusion

Virtual reality technology has become the fastest developing multidisciplinary comprehensive technology in the field of computer science. This technology can provide new technical support for teaching development. However, the application research of this technology in aviation engine teaching is relatively limited, and its reliability cannot be guaranteed. Therefore, a reliability analysis study of aviation engine teaching systems based on virtual reality technology is proposed. By combining aviation engine teaching and experimentation with virtual reality technology, a visual simulation system for aviation engine teaching and experimentation with high real-time, high interactivity, and deep immersion is designed, which intuitively displays the entire process of the engine during operation. And through experiments, it has been verified that the design system can greatly improve the completeness of teaching scene display and the success rate of human-machine interaction in the teaching process, providing more effective system support for aviation generator teaching.

Acknowledgement. School level project of Beijing Polytechnic, Project Name: Application of virtual reality technology in aeroengine teaching (2022X007-SXZ).

References

1. Qin, T., Cook, M., Courtney, M.: Exploring chemistry with wireless, PC-less portable virtual reality laboratories. J. Chem. Educ. **98**(2), 521–529 (2021)
2. Bu, X., Ng, P., Chen, Q., et al.: Effectiveness of virtual reality-based interventions in rehabilitation management of breast cancer survivors: protocol of a systematic review and meta-analysis. BMJ Open **12**(2), e053745–e053745 (2022)
3. Wang, Y., Chang, F., Wu, Y., et al.: Multi-Kinects fusion for full-body tracking in virtual reality-aided assembly simulation. Int. J. Distrib. Sens. Netw. **18**(5), 625–636 (2022)
4. Schleuinger, M.: Information retrieval interfaces in virtual reality-a scoping review focused on current generation technology. PLoS ONE **16**(2), e0246398–e0246398 (2021)
5. Chen, S.C.: Multimedia in virtual reality and augmented reality. IEEE Multimedia **28**(2), 5–7 (2021)
6. Karsenty, K., Tartakovsky, L., Sher, E.: A diesel engine with a catalytic piston surface to propel small aircraft at high altitudes—a theoretical study. Energies **14**(7), 1905 (2021)
7. Martin, C., Quinn, D., Murphy, A., et al.: Understanding influence of powerplant component connection strategies on aircraft engine structural deformations. J. Aircr. **58**(5), 1–11 (2021)
8. Pang, S., Jafari, S., Nikolaidis, T., et al.: A novel model-based multivariable framework for aircraft gas turbine engine limit protection control. Chin. J. Aeronaut. **34**(12), 57–72 (2021)
9. Zhang, Y.: Interactive intelligent teaching and automatic composition scoring system based on linear regression machine learning algorithm. J. Intell. Fuzzy Syst. **40**(2), 2069–2081 (2021)
10. Robust, An, Z., Li, T.: Simulation of dynamic response of solid rocket engine under standard transportation conditions. Comput. Simul. **39**(10), 66–70 (2022)
11. Xia, Q.: Application of 3Ds max and virtual reality technology in 3D submarine scene modeling. Microprocess. Microsyst. **80**, 103562.1–103562.6 (2020)
12. Song, J., Jiang, L., Wang, L.: P-3.10: current status and prospects of simulation training equipment based on virtual reality and augmented reality technology. In: SID Symposium Digest of Technical Papers, vol. 52, no. 8, pp. 742–745 (2021)
13. Brown, C., Hicks, J., Rinaudo, C.H., et al.: The use of augmented reality and virtual reality in ergonomic applications for education, aviation, and maintenance. Ergon. Des. Q. Hum. Factors Appl. (6), 106480462110034 (2021)

Research on Software Test Data Optimization Using Adaptive Differential Evolution Algorithm

Zheheng Liang[1,2(✉)], Wuqiang Shen[1,2], and Chaosheng Yao[1,2]

[1] Joint Laboratory on Cyberspace Security of China Southern Power Grid, Guangzhou 510000, China
liangzhaoheng23213@163.com, shenwuqiang@gdxx.csg.cn
[2] Guangdong Power Grid, Guangzhou 510000, China

Abstract. In order to improve the coverage of the target path corresponding to the generated software test data after optimization, and make the data better adapt to the software test process, the adaptive differential evolution algorithm is introduced to design the optimization method for software test data. Using the PSO algorithm to simulate the biological evolution mechanism in nature, and with the help of computer programming, the generated software test data are preliminarily trained; Draw on the path correlation based regression test data evolution of adaptive differential evolution algorithm to generate the relevant path representation method, mark the program to be tested, and construct the test data fitness function based on this; A hybrid model MPSO is proposed to select the best individual data in software test data; The selection criteria of scaling individuals are introduced into the adaptive scaling factor to cluster the optimal data individuals, so as to realize the design of optimization methods. The comparison experiment results show that the designed method has a good effect in practical application. This method can improve the target path coverage corresponding to the generated software test data on the basis of controlling the time length and evolution times required for software test data optimization.

Keywords: Adaptive differential evolution algorithm · Data selection · Individual clustering · Fitness function · Software test data · Optimization

1 Introduction

In order to ensure the quality of software, not only need advanced technical means, perfect R & D process, but also need continuous testing, no matter how advanced the current technology, how perfect the R & D process, can not 100% guarantee the software developed zero defects. Software testing can help scientific research and technology developers find out the errors or potential defects in the process of software development, and reduce the damage caused by software errors and potential defects as much as possible. In order to ensure the high reliability and integrity of software products, it is necessary to design enough test cases. However, the scale of software products is increasing day by day, the structure is becoming more and more complex, and the update

L. Yun et al. (Eds.): ADHIP 2023, LNICST 549, pp. 239–250, 2024.
https://doi.org/10.1007/978-3-031-50549-2_17

cycle is becoming shorter and shorter. It is a huge challenge for software testing to carry out comprehensive testing to find the problems and potential defects in the increasingly tight software development cycle. Therefore, it is necessary to use effective test cases within a limited test time to find out the errors and potential defects of the product as far as possible to ensure the accuracy and reliability of the software product, and the basis of this test is a high degree of target path coverage.

In recent years, the use of artificial intelligence algorithms such as ant colony algorithm, particle swarm optimization algorithm and genetic algorithm to automatically generate test cases has attracted more and more researchers' attention, and has achieved good research results. For example, Li Lu et al. [1] studied the defect optimization method of high-performance traffic analysis software based on particle swarm optimization algorithm, which introduced particle swarm optimization algorithm to analyze the current defect status of high-performance traffic analysis software, reset defect parameters, and build the defect optimization model of high-performance traffic analysis software. However, this method has the problem of long running time. Yang Bo et al. [2] studied a software fault location assisted test case generation method using improved genetic algorithm. Based on genetic algorithm, this method assisted the generation of test cases in the process of software fault location through the ranking of suspected faults in software fault location, but the marking path coverage of this method was low. The research shows that the adaptive differential evolution algorithm has the advantages of fast yield with fewer parameters and fast yield in high dimensions [3, 4]. Therefore, this paper takes the program code of unit test in software testing as the research object, introduces adaptive differential evolution algorithm to calculate all the tested paths, and uses the fitness function of the tested paths to improve the process of output optimal solution by optimizing the mutation operator, and improves the adaptive differential evolution algorithm by improving the mutation operator. Test cases covering each path are generated automatically to optimize software test data. This paper makes use of the convenient and efficient characteristics of artificial intelligence algorithm, which not only reduces the test cost, but also obtains a high efficiency of automatic case generation, moreover, the target path coverage of the generated software test data is more than 90%, which is more than 10% higher than that of the comparison method, which is of great significance to the improvement of product quality.

2 Software Test Data Generation

A lot of work has proved that PSO algorithm can solve the problem of software test data generation better than GA algorithm. Moreover, the PSO algorithm is simple, with fewer parameters to configure and easy to implement [5]. Therefore, this chapter introduces PSO algorithm to design the generation of software test data.

In this process, with the help of computer programming, the problem to be solved is expressed into strings (or chromosomes), that is, binary codes or digital strings, by using the PSO algorithm to simulate the biological evolution mechanism in nature, so as to form a group of strings, and they are placed in the problem solving environment. According to the principle of survival of the fittest, the strings that adapt to the environment are selected for replication, and the two gene operations are crossover and mutation, Create a new generation of clusters that are more adaptable to the environment. After such continuous changes from generation to generation, finally converge to a string that is most suitable for the environment, and obtain the optimal solution of the problem [6]. That is, the training of test software requirement data is preliminarily realized.

On this basis, it should be clear that the essence of PSO algorithm is an optimization algorithm based on swarm intelligence, and its idea comes from artificial life and evolutionary computing theory. The algorithm simulates the behavior of birds flying to find food, and makes the group achieve the optimal goal [7] through the collective cooperation between birds. When using the PSO algorithm to generate software test data, it is necessary to input the software test data into the test environment first, and express the test environment as D, on D in the target retrieval space of dimension, each software test data is regarded as a particle, and the particle is represented as m, then m particles form a population, where the i particles in the d the position of the dimension is x_{id}, x_{id} the corresponding flight speed in space is v_{id}, the optimal location searched by this particle is p_{id}, the current maximum value of the entire particle swarm p_{gd} in this way, the particle position in space is updated, and this process is taken as the data update process, as shown in the following calculation formula.

$$v_{id}^{t+1} = wv_{id}^{t} + c_1 r_1 \left(p_{id} - x_{id}^{t}\right) + c_2 r_2 \left(p_{gd} - x_{id}^{t}\right) \tag{1}$$

In formula (1): v_{id}^{t+1} means on $t + 1$ the corresponding flight speed of particles in space at time; w is the inertial factor, w the larger the value of, the more suitable for large-scale exploration of the solution space, w the smaller the value of is, the more suitable it is to explore the solution space in a small range; v_{id}^{t} means on t the corresponding flight speed of particles in space at time; x_{id}^{t} means on t at the moment i particles in the d dimension position; c_1, c_2 represents normal number, called acceleration factor; r_1, r_2 represents a random number with a value between 0 and 1 [8]. On this basis x_{id}^{t+1} the calculation formula is as follows:

$$x_{id}^{t+1} = x_{id}^{t} + v_{id}^{t+1} \tag{2}$$

On the basis of the above contents, it should be clear that, d the value range of is $1 \leq d \leq D$, for the d position of dimension x constrain the range of change and make it clear x the value range of is $[x_{d\,\min}, x_{d\,\max}]$, for the d dimensional velocity v constrain the range of change and make it clear v the value range of is $[v_{d\,\min}, v_{d\,\max}]$.

In the iteration, if the position and speed exceed the boundary range, the boundary value shall be taken. The particle position in the space can be updated according to Fig. 1 below. In this way, the spatial position of the software test data can be updated to generate more complete and global adaptive test data [9].

The termination condition of the PSO algorithm takes the maximum number of iterations or the predetermined minimum fitness threshold that the optimal position searched by the particle swarm optimization meets according to the specific problem.

Because p_{gd} it is the optimal position of the whole particle swarm. Therefore, the software test data can be generated according to the above steps, or it can be used as the retrieval process of global particles. The global PSO algorithm has a fast convergence speed, but sometimes it falls into a local optimum; The local PSO algorithm

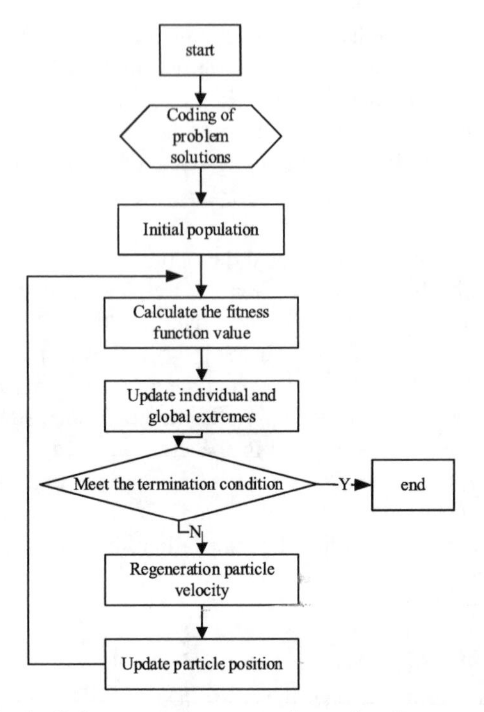

Fig. 1. Software test data generation and update process

converges slowly, but it is relatively difficult to fall into the local optimal value. Therefore, corresponding measures can be taken for data processing according to the specific requirements of software test data generation in practical applications.

3 Construct Test Data Fitness Function Based on Adaptive Differential Evolution Algorithm

On the basis of the above contents, in order to achieve the extraction of the optimal data in the generated software test data and master the fitness of different data in the generated data and the software test process, the adaptive differential evolution algorithm is introduced to construct and design the fitness function of the test data [10].

In this process, it should be clear that the establishment of the mathematical model of the multi-path coverage test data generation problem is closely related to the path representation method, and the tested program generally has multiple paths. In structured programming, the program usually consists of three structures: sequence, selection, and cycle. Among them, selection and cycle structure determine the trend of different branches of the program; For cyclic structure, by introducing *Z* path coverage can decompose it into multiple selection structures. The existing path representation methods include: using the statement number sequence of the program to represent the path, using the branch node sequence to represent the path, and using Huffman code to represent the target path. Using the path correlation based regression test data evolution generation correlation path representation method in adaptive differential evolution algorithm for reference, the program to be tested is marked.

Assume that the program to be tested is Q, insert piles into each branch node in the program, and set the value of "true" to be 1 when crossing the corresponding branch node. If it is false, it is -1, and if it does not pass through the branch node, it is 0. Therefore, the code sequence of 1, -1, and 0 can be used to represent the path of program execution. Run the test program after inserting piles with a set of input data, and the branch node sequence is $D_1, D_2, \cdots, D_n$ the corresponding execution path is expressed as $q(T_i)$, where n is the path $q(T_i)$ length of. In addition, the program to be tested Q the target path of is $q_1, q_2 \cdots, q_N$, N is the number of target paths.

Hypothetical procedure Q the input field of is α, input data is $T_1, \cdots, T_N$, then $T_N \in \alpha$, the path will be executed $q(T_i)$ Compare the coding sequence of and the target path from left to right, find the first different coding position, record the number of the same coding before, and record it as $\left|q(T_i) \cap q_j\right|$, take the ratio between it and the target path code, that is, the total length of the target path, to obtain the path similarity, and express it as $f_i(T)$, then $f_i(T)$ it can be calculated by the following formula.

$$f_i(T) = \frac{\left|q(T_i) \cap q_j\right|}{\max\left\{q(T_i), q_j\right\}} \tag{3}$$

In formula (3): $f_i(T)$ indicates the path similarity. After the above calculation is completed, the problem of solving test data generation can be transformed into solving $f_1(T), f_2(T), f_N(T)$ the fitness function of the maximum value problem is as follows.

$$H = \max(f_1(T), f_2(T), \cdots, f_N(T)) \tag{4}$$

In formula (4): H represents the fitness function of test data. According to the above method, the research on the construction of test data fitness function based on adaptive differential evolution algorithm is completed.

4 Selection of Optimal Individual Data in Software Test Data

After completing the above design, in order to select the best individual data in the software test data and seek the best test data, it should be clear that in the GPSO, particles evolve in the direction of the global optimal particles, and more obviously converge to the current optimal solution. Each time the particle position is updated, the characteristics of all particles in the population are integrated, and the information transmission speed is fast, but the population diversity is easy to lose, and the probability of falling into the local extreme value is large. In the local model LPSO, each particle only shares information with its neighbor nodes, which slows down the speed of information transmission, but the diversity of the population is guaranteed and it is not easy to fall into local extremum.

Therefore, based on the analysis of the impact of different fixed forms of neighbor patterns on the algorithm performance, it is found that the higher the average connectivity of social interactions between particle individuals, the faster the information transmission speed in the population, and the premature convergence phenomenon is more prone to occur. In view of this, a hybrid model MPSO is proposed, that is, by observing the diversity index of particle swarm, each generation of particles selects GPSO or LPSO for evolutionary optimization, so as to realize the selection of the best individual data in software test data. In this process, describe the diversity of particles, that is, analyze the diversity of software test data in space. This process is shown in the following calculation formula.

$$I = \frac{1}{|k| \times |L|} \sum_{i>1}^{N} \sqrt{\sum_{d>1}^{D} \left(x_{id}^{t} - x_{d}^{t}\right)^{2}} \tag{5}$$

In formula (5): I represent the diversity of software test data in the space; k represents the size of particles in the search space; L indicates the maximum diagonal length in the search space. After calculation, output I the calculation results of I it represents the dispersion level of each particle in the community, I the larger the value, the more dispersed the population; I the smaller the value, the more concentrated the population, and the lower the diversity of the population.

Therefore, by observing the group I the feedback information automatically adjusts the topology of particle swarm to maintain the diversity of the community, maintain a certain population density, and avoid the algorithm falling into the local optimal value. On the basis of the above I the population with higher values will be globally and locally optimized, and the process is shown in the following calculation formula.

$$W(I) = \frac{I_i - l_{\min}}{e^{l_{\max} - l_{\min}}} \tag{6}$$

In formula (6): $W(I)$ represents the best individual data in software test data; $l_{\min}$ represents the minimum branch nesting depth; $l_{\max}$ represents the maximum nesting depth; e indicates the depth of the current branch. Complete the selection of the best individual data in the software test data according to the above method.

5 Optimal Individual Clustering and Optimization of Software Test Data Set

On the basis of the above contents, extract the optimal individual data from the software test data selected in the above way, introduce the selection criteria of scaling individuals into the adaptive scaling factor, and cluster the optimal data individuals. The process is shown in the following calculation formula:

$$r = \sum_{i>1}^{n} Mr_i / \sum_{i>1}^{n} r_n \tag{7}$$

In formula (7): r represents the optimal individual clustering; M represents the cluster center. After completing the above calculation, the adaptive differential evolution algorithm is applied to automatically generate the model of test data, as shown in Fig. 2 below.

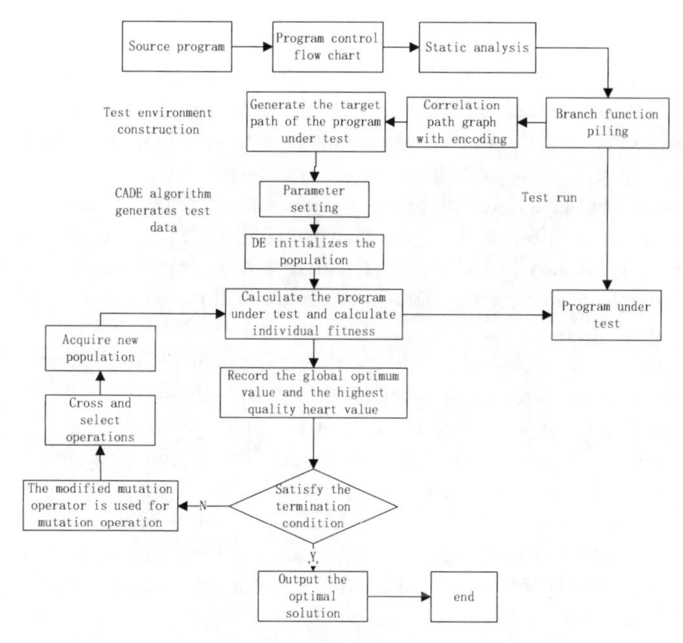

Fig. 2. Optimal solution of output software test data

The construction of the test environment is the basis for the generation of test data. In this stage, the source program is statically analyzed, and the coded test program flowchart generated by branch function instrumentation is used to generate the target path. This method is more concise than the traditional data flowchart.

However, as far as the current research is concerned, the existing methods assume that the mass of each particle in the population is the same, and do not take into account the differences between the particles. Therefore, in the optimization process, the process of outputting the optimal solution is improved. On the basis of the existing algorithms, a mutation operator based on the center of mass is proposed. The improved mutation operator can be calculated by the following formula.

$$A(t) = x_i(t) + F(r + c_1 r_1 + c_2 r_2) \tag{8}$$

In formula (8): $A(t)$ represents the improved mutation operator; F represents the mass of individual particles. After the selection of the improved mutation operator is completed, in order to break the stagnation of evolution, jump out of the local best, let the operator fluctuate in size, and by limiting its value range, make the difference dynamic scaling, and search for optimization in a large range, theoretically it can improve the ability of global optimization. In the early stage of evolution, the operator tries to maintain the diversity of the population, and takes the small value first; in the later stage, the drill takes the large value to speed up the convergence speed. In this paper, the corresponding generation operator of normal distribution random number is used as follows:

$$\beta = normrnd(U \times F) \cdot A(t) \tag{9}$$

In formula (9): β represents the generating operator process; *normrnd* represents the deviation compensation value; U represents a normally distributed random number. By compensating the deviation of the operator generation process, the software test data can be better optimized within the optimal range, which is helpful to jump out of the region where the local extreme value is located quickly.

The cross factor plays a fine tuning role, and the appropriate value can maintain the diversity of the population. If the crossover probability decreases, the candidate will contain more target individuals. If the crossover probability increases, the weight of variant individuals in the newly generated candidate will increase.

At the early stage of the iteration, the diversity of the population is relatively high, so cross operation with a small probability can maintain the diversity of the population. As the number of iterations increases, the diversity of the population is gradually losing and approaching the extreme point. Therefore, local search should be increased to accelerate the convergence of the algorithm, so as to approach the extreme point at a faster speed.

In this process, it should be clear that generating test data is the core part of the whole model based on the adaptive differential evolution algorithm. In this stage, the population is initialized according to the target path and test data range constructed by the test environment, combined with the setting of parameters required by the algorithm, and the population is updated by running the improved mutation operator to guide the population to evolve towards the optimal value; And use the test data to run the program to be tested after pile insertion to calculate the fitness value. When the fitness value meets the needs of software testing, complete the data optimization design.

According to the above method, complete the optimal individual clustering and software test data set optimization, and realize the design and research of software test data optimization method based on adaptive differential evolution algorithm.

6 Comparison Experiment

The research on optimizing software test data using adaptive differential evolution algorithm has been completed from four aspects above. In order to test the application effect of software test data optimization methods, the following will take the tested software program provided by a teaching and research institute as an example to design a comparison experiment and carry out the research as shown below.

In order to ensure that the designed method can play its expected role in the test, the environment in which the software program under test runs is described before the experiment, and the relevant contents are shown in Table 1 below.

According to the above contents, after completing the design of the technical parameters of the comparative experimental environment, the method designed in this paper is used to optimize the data of the tested software. In the optimization process, it assists modern intelligent algorithms such as GA to generate software test data. On this basis, the adaptive differential evolution algorithm is introduced to construct the fitness function of the test data. In order to ensure that the constructed fitness function can be used as the basis for evaluating the software test data, the parameters of the algorithm in the computer can be set according to Table 2 below during the iteration of the adaptive differential evolution algorithm.

Table 1. Technical Parameters of Comparative Experimental Environment

S/N	project	parameter
1	Experimental programming language	Java Language
2	Experimental program running environment	Eclipse environment
3	Microcomputer environment	Windows operating system Intel (R) Core (TM) i3 CPU 2.0 GHz
4	Computer running memory	2 GB RAM

Table 2. Training parameter settings of adaptive differential evolution algorithm

S/N	project	parameter
1	Population iterations (times)	1000
2	Operator crossover probability (%)	0.6
3	Operator compilation probability (%)	0.01
4	Number of independent runs for each case of different programs (times)	50
5	Data cycle branch (s)	6
6	Select nesting quantity (pcs)	3
7	search space	3
8	Learning factor	1.5
9	Inertia weight	0.9
10	Diversity threshold in algorithm	0.1

After setting the experimental parameters, select the best individual data in the software test data, and optimize the software test data set by clustering the best individual data.

On the basis of the above content, software test data optimization method based on particle swarm optimization (reference [1] method) and software test data optimization method based on genetic algorithm (reference [2] method) are introduced, and the introduced method is regarded as traditional method 1 and traditional method 2. The method in this paper and the two traditional methods are used to optimize software test data.

In the optimization process, three software test data populations of different scales are selected, and three methods are used to evolve the software test data respectively. The average length of time required by the three methods to optimize the software test data is compared, which is the key basis for testing the application effect of the method in this paper. The statistical experimental results are shown in Table 3 below.

From the experimental results shown in Table 3 above, it can be seen that the average time required for optimization of software test data using this method is less than 1 s,

Table 3. Average time required for three different methods to optimize software test data

S/N	Population size	Average duration (s)		
		Methods in this paper	Traditional method 1	Traditional method 2
1	50	0.125	0.569	0.639
2	100	0.156	0.693	0.678
3	150	0.269	0.854	1.069
4	200	0.489	0.910	1.598
5	300	0.526	1.025	1.756
6	400	0.963	1.156	1.963
7	500	0.956	1.569	3.051

while the average time required for optimization of software test data using traditional methods is significantly higher than the average time required for optimization of data using this method.

After completing the above experiment, set the input range of software test data to [0, 64], [0512], [1024], use three methods to optimize the software test data, and compare the average evolution times of the three methods on the input data during the optimization process. The results are shown in the following figure:

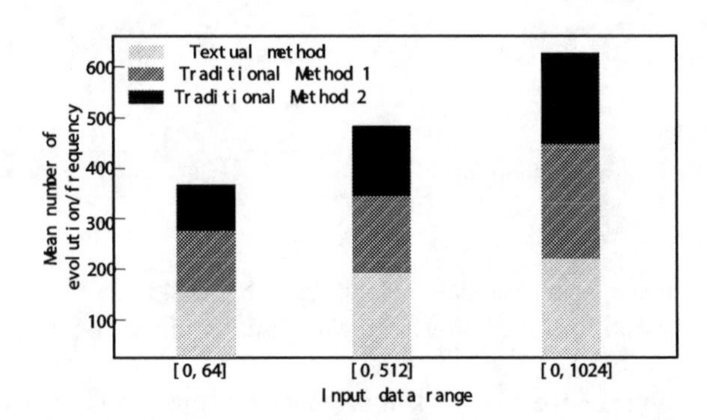

Fig. 3. Average evolution times of three methods for input data in different ranges during optimization

It can be seen from Fig. 3 above that the three methods optimize the training of input data in different ranges. With the increase of input data range, the average evolution times of the three methods show an increasing trend. Use this method to optimize [1024] software test data, and the average number of evolutions in the optimization process is less than 300; The traditional method 1 is used to optimize software test data, and the average evolution times are less than 500; The traditional method 2 is used to optimize the software test data, and the average evolution times are less than 700.Based on the above results, it can be seen that among the three methods, only using this method to optimize software test data can ensure that the average evolution times are at a relatively low level, while using traditional method 1 and traditional method 2 to optimize software test data, the corresponding method has more average evolution times, that is, the data optimization process is more complex.

After the above design is completed, several software test paths are set according to the software test requirements. It is known that different software test paths require different software test data sizes and categories. After comparing the three methods to optimize the software test data, the generated test data can cover the extent of the target test path, that is, compare the availability of the optimized corresponding data in the software testing process. According to the above method, carry out a comparison experiment and make statistics of the experimental results, as shown in Fig. 4 below.

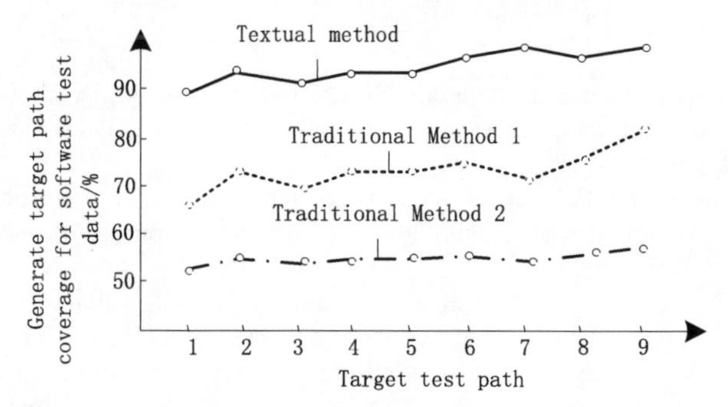

Fig. 4. Target path coverage corresponding to software test data generated by the three optimized methods (%)

From the experimental results shown in Fig. 4 above, it can be seen that the method in this paper is used to optimize the software test data, and the target path coverage corresponding to the generated software test data after optimization is >90%, while the target path coverage corresponding to the generated software test data after optimization in the traditional method 1 is between 60% and 80%; The target path coverage corresponding to the software test data generated after optimization of traditional method 2 is between 50% and 60%.

Based on the above experimental results, the following experimental conclusions are obtained: compared with the traditional methods, the software test data optimization method designed in this paper using adaptive differential evolution algorithm has a good effect in practical application. This method can improve the target path coverage corresponding to the generated software test data on the basis of controlling the time required for software test data optimization and the number of evolutions. In this way, it provides further support and guidance for the standardized implementation of software program testing and other related scientific research work.

7 Conclusion

In order to find potential software defects as much as possible and improve the coverage of target paths corresponding to software test data generation, this paper conducts testing research on the optimization method of software test data through software test data generation, construction of test data fitness function based on adaptive differential evolution algorithm, selection of optimal individual data in software test data, optimal individual clustering and optimization of software test data set. After completing the design of this method, taking the software program under test provided by a teaching and research unit as an example, a comparative experiment was designed to prove that the method designed in this paper has a good effect in practical application. This method can improve the coverage of the target path corresponding to the generated software test data on the basis of controlling the length of time required for software test data optimization and the number of improvements.

References

1. Li, L.: Defect optimization method of high-performance traffic analysis software based on particle swarm optimization. J. Anhui Techn. Coll. Water Resour. Hydroelectr. Power **21**(4), 56–59 (2021)
2. Yang, B., He, Y., Xu, F., et al.: IGA: software fault location assisted test case generation method using improved genetic algorithm. J. Beijing Univ. Aeronaut. Astronaut. **48**(3), 1–13 (2022)
3. Zhang, X., Liu, Q., Qu, Y.: An adaptive differential evolution algorithm with population size reduction strategy for unconstrained optimization problem. Appl. Soft Comput. J. **138** (2023)
4. Niu, D., Liu, X., Tong, Y.: Operation optimization of circulating cooling water system based on adaptive differential evolution algorithm. Int. J. Comput. Intell. Syst. **16**(1) (2023)
5. Li, W., Sun, Y., Huang, Y., et al.: An adaptive differential evolution algorithm using fitness distance correlation and neighbourhood-based mutation strategy. Connect. Sci. **34**(1) (2022)
6. Gouda, S.K., Mehta, A.K.: A self-adaptive differential evolution using a new adaption based operator for software cost estimation. J. Inst. Eng. (India) Ser. B **104**(1) (2022)
7. Chen, J., Chen, X., Zan, T., et al.: An automatic generation of software test data based on improved Markov model. Web Intell. **20**(4) (2022)
8. Ferreira Vilela, R.,Choma Neto, J., Santiago Costa Pinto Victor, H., Lopes de Souza, P.S., do Rocio Senger de Souza, S.: Bio-inspired optimization to support the test data generation of concurrent software. Concurr. Comput. Pract. Exper. **35**(2) (2022)
9. Mohaideen Abdul Kadhar, K., Narayanan, N., Vasudevan, M., et al.: Parameter evaluation of a nonlinear Muskingum model using a constrained self-adaptive differential evolution algorithm. Water Pract. Technol. **17**(11) (2022)
10. Esnaashari, M., Damia, A.H.: Automation of software test data generation using genetic algorithm and reinforcement learning. Expert Syst. Appl. **183** (2021)

Intelligent Monitoring Method of Aircraft Swashplate Plunger Pump Fluidity Based on Different Working Conditions

Chao Ma(✉) and Jinshou Shi

College of Aeronautical Engineering, Beijing Polytechnic, Beijing 100176, China
mczn05082023@163.com

Abstract. The swashplate plunger pump is one of the key components of the aircraft hydraulic transmission system, and its reliable operation is directly related to the flight safety of the aircraft. The key factor affecting the safety of swashplate plunger pumps is flowability. Therefore, an intelligent monitoring method for the flowability of aircraft swashplate plunger pumps based on different operating conditions is proposed. By analyzing the structure and working principle of the swashplate plunger pump, the types of abnormal flow phenomena are identified. Based on this, a multi-scale permutation entropy algorithm is used to extract the features of monitoring data. Due to the large amount of monitoring data features, fusion rules have been developed based on the feature fusion results of its traffic monitoring data. Based on the characteristics of abnormal flow data, an extreme learning machine is used to determine whether the flow of the swashplate plunger pump is abnormal, thereby achieving intelligent monitoring. The experimental data shows that under different experimental conditions, the flow monitoring results of the swash plate plunger pump obtained after the proposed method application are completely consistent with the actual results, with an accuracy of 100%; The monitoring time ranges from 0.20 s to 0.48 s, consistently not exceeding 0.5 s, and the monitoring efficiency is higher, fully confirming the better application performance of the proposed method.

Keywords: Aircraft Swashplate Plunger Pump · Intelligent Monitoring · Condition Monitoring · Mobility · Different Working Conditions

1 Introduction

The plunger pump is the key part of the whole hydraulic transmission system, and its main function is to provide enough hydraulic oil for the hydraulic system when the whole hydraulic system is working. At present, among the positive displacement pumps, only the plunger pump can achieve high pressure and high speed. With the development of hydraulic technology, the development trend of positive displacement hydraulic pump is gradually high-pressure. By comparing the characteristics of the axial pump and the radial pump, it can be concluded that the axial piston pump has the

L. Yun et al. (Eds.): ADHIP 2023, LNICST 549, pp. 251–267, 2024.
https://doi.org/10.1007/978-3-031-50549-2_18

advantages of compact structure, small unit power volume, long service life, high rated working pressure, light weight, convenient arrangement of variable mechanism, etc. [1]. Because of these advantages of plunger pump, the application of axial plunger pump is increasingly extensive. The disadvantages of axial piston pump are high manufacturing cost, more sensitive to oil pollution, and high requirements for oil filtering accuracy.

The axial piston pump can also be divided into two categories, the swashplate type and the inclined shaft type, which have their own characteristics in structure and advantages in practical application. The drive plate is used in the structure of the inclined shaft axial piston pump, which makes the plunger cylinder not bear lateral force during rotation. Therefore, the possibility of the cylinder block overturning the oil distribution plate is small, and it is also conducive to improving the working efficiency and stability of the plunger pair and the oil distribution part. In addition, the allowable inclination angle on the structure is large. However, the structure of the inclined shaft pump is relatively complex, which requires the use of large capacity thrust bearings. The poor workmanship limits its ability to work continuously under high-pressure environment, and the production cost is relatively high. The swashplate axial piston pump, compared with the swashplate axial piston pump, uses a hydrostatic support structure for two pairs of high-speed motion pairs - oil distribution plate and cylinder block, slipper and plunger, which eliminates the large capacity thrust bearing. The swashplate axial piston pump has the advantages of compact structure, good processability, low production cost, small volume and light weight. At present, swashplate axial piston pump has been widely used. The rated working pressure of swashplate axial piston pump is generally within the range of 21–35 MPa, the peak pressure is between 28–40 MPa, the rotating speed of the main shaft is generally below 3000 rpm, and the volume constant is mostly below 300–500 cm/rpm. In recent years, it has reached 2336 cm/rpm or even higher through development. Due to its own structure and performance requirements, the aircraft uses a swashplate axial piston pump.

For swashplate plunger pump, the main Failure cause are wear caused by long-term high load operation, fatigue cracks of key components, failure modes such as reduced efficiency, vibration noise caused by shoe loosening, and even direct stagnation. However, the aircraft Servomechanism adopts closed hydraulic oil circuit, with high motor speed, short single operation time and frequent start stop reversing [2]. Under this working condition, the plunger pump starts, stops, and reverses multiple times, causing sudden changes in internal pressure and causing impacts. The fault mode will mainly be cavitation caused by sudden changes in load and wear and leakage caused by friction pairs after multiple load impacts. The flowability of the swashplate plunger pump is the key factor affecting its normal operation, making it one of the urgent problems to design an effective intelligent monitoring method for the flowability of the aircraft swashplate plunger pump in the current aircraft field. In order to ensure the reliable operation of the swash plate plunger pump, researchers have proposed a method of monitoring the status of the plunger pump by extracting the oil temperature signal of the plunger pump to monitor power loss. But it will fail due to short single operation time, slow heat response, and fast heat loss. Researchers have also analyzed the iron filings content in hydraulic oil using ferrography to determine the wear of plunger pumps, but in practical applications, it is not suitable for plunger pumps installed in closed hydraulic oil circuits.

In addition, the use scenarios and equipment limitations of the aircraft Servomechanism also make the status monitoring of the plunger pump subject to many restrictions. For example, the acquisition of the vibration signal analysis spectrum will be difficult to extract features because of the excessive environmental noise of the Servomechanism, and the acquisition of the oil pressure signal to extract the pulse frequency analysis will be unable to extract the feature frequency band when the speed of the plunger pump is high due to the low sampling rate of the oil pressure sensor.

In response to the problems with the above methods, it is evident that there are significant differences in the effectiveness of state monitoring methods for general plunger pumps under different operating conditions. Therefore, in order to solve the above problems, this article proposes a research on intelligent monitoring method for the flowability of aircraft swashplate plunger pumps.

2 Research on the Intelligent Monitoring Method for the Fluidity of Swashplate Plunger Pump

2.1 Flow Analysis of Swashplate Plunger Pump

By analyzing the structure and working principle of swashplate plunger pump, the types of abnormal flow phenomena of swashplate plunger pump are clarified, which lays a solid foundation for the subsequent intelligent monitoring of swashplate plunger pump flow.

The piston at the hydraulic end of the swashplate plunger pump is driven by the crank connecting rod mechanism at the power end and reciprocates, thereby changing the pressure in the cylinder liner at the hydraulic end, opening and closing the suction valve and the discharge valve, thus realizing the basic function of medium transportation [3]. The structure of swashplate plunger pump is shown in Fig. 1.

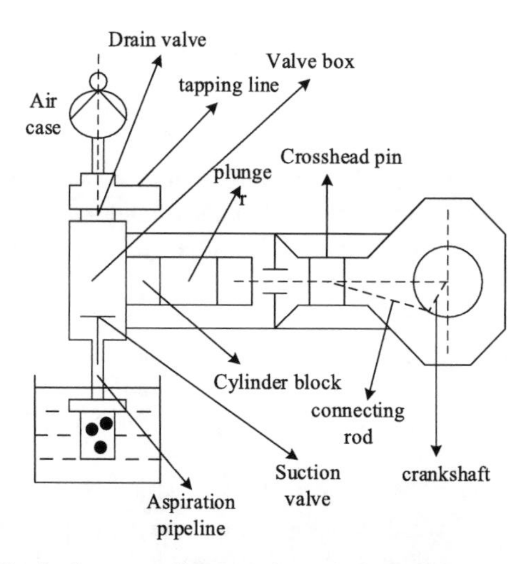

Fig. 1. Structure diagram of swashplate plunger pump

Taking the cylinder body of the pump head as an example, the working principle of its hydraulic end is explained. The crankshaft rotates to drive the plunger to reciprocate once, and the plunger pump executes the working cycle once, as shown in Fig. 2.

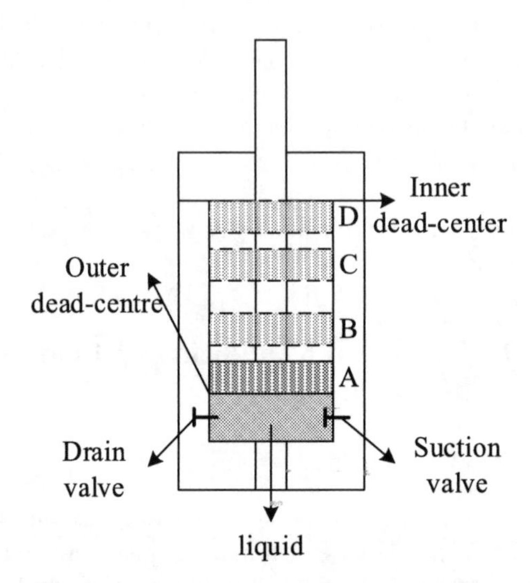

Fig. 2. Schematic diagram of working principle of swashplate plunger pump

As shown in Fig. 2, the working principle of the swashplate plunger pump is the four processes through which the closed volume liquid in the cylinder body passes, namely, expansion (A → B), liquid suction (B → C), compression (C → D), and liquid discharge (D → A), as shown below:

(1) Expansion (A → B)
When the piston is at A, the discharge valve of the pump head is in a critical closed state, but there is still some liquid left in the cylinder. When the plunger moves from the outer dead center to the inner dead center, the spring force makes the drain valve close, the sealed volume in the cylinder body gradually increases, and the vacuum degree in the space also gradually increases. When the force on one side of the suction valve pipeline is greater than the force in the pump, the suction valve opens. At this moment, the plunger moves to position B, and the liquid pressure in the cylinder body is also minimum.
(2) Sucking liquid (B → C)
After passing through point B, as the piston continues to move toward the inner dead center, the liquid flows into the cylinder block through the suction pipeline until it reaches the inner dead center C, and the volume of the cylinder block no longer increases.
(3) Compression (C → D)
When the crankshaft drives the piston to move outward from the inner dead center C, the spring acts on the liquid suction valve to close it, the liquid in the cylinder is compressed, the cylinder volume becomes smaller, and the liquid pressure increases;

When the plunger moves to D, the pressure in the cylinder is enough to push the drain valve open. At this moment, the liquid pressure is the maximum [4].

(4) Drain (D $\rightarrow$ A)

After passing through point D, as the piston continues to move outward to dead center A, the liquid continuously gushes out of the cylinder block through the discharge pipeline until it reaches the dead center, and the volume of the cylinder block no longer decreases. Then enter the next suction and discharge stroke.

It can be seen that the flow at the hydraulic end of the plunger pump is pulsating, and the liquid pressure in the pump changes periodically, and the liquid in the pipeline also belongs to the unstable flow state. The abnormal flow of swashplate plunger pump can be divided into several types, as shown in Table 1.

Table 1. Types of Abnormal Flow Phenomenon of Swashplate Plunger Pump

number	Trigger part	Principle causing abnormal liquidity
1	Pump valve assembly failure	Leakage of valve body, valve seat or valve rubber causes reduction of liquid discharge; The spring of the pump valve breaks and fails, causing serious delay in the closing of the pump valve, uneven discharge flow, and severe vibration
2	Wear of plunger and cylinder liner	The plunger or piston seal is damaged, the working medium leaks, and the liquid discharge is reduced; Plunger, cylinder liner or packing, resulting in increased clearance with cylinder liner, leakage of working medium, and reduced liquid discharge
3	Eccentric wear of plunger and cylinder liner	Due to the misalignment of the plunger or piston rod and crosshead pull rod, the eccentric wear between the plunger or piston and the cylinder liner is serious, and the surface of the cylinder liner rises significantly
4	Suction line/discharge line fault	The pipeline is not sealed tightly, and the circulating liquid leaks; The pipeline is blocked, and the pressure of pump controlled pumping or discharge pipeline increases significantly; Inadequate suction and reduced discharge due to improper installation of suction pipeline

It can be seen from Table 1 that the failure of multiple components in the swashplate plunger pump will lead to abnormal flow, which will affect the normal operation of the swashplate plunger pump. Therefore, intelligent monitoring of the swashplate plunger pump's flow has a vital role and significance.

2.2 Feature Extraction of Flow Monitoring Data of Swashplate Plunger Pump

In order to simplify the flow monitoring process of swashplate plunger pump, multi-scale permutation entropy algorithm is used to extract the characteristics of monitoring data, which provides a basis for subsequent monitoring information fusion.

Permutation entropy is a nonlinear processing method for signal analysis. Permutation entropy is an index to measure the complexity of time series, which is used to study the irregularity and chaos of nonlinear signals. Compared with Lyapunov exponent, sample entropy and other indicators, permutation entropy has the advantages of strong robustness, simple calculation and easy understanding [5]. In the fields of EEG signal, vibration signal and image processing, ideal research results have been achieved. Set the time series of swashplate plunger pump liquidity monitoring data as $\{x(i), i = 1, 2, \cdots, n\}$, and reconstruct the space to get a new time series, the form is as follows:

$$\begin{cases} X(1) = \{x(1), x(1+\alpha), \cdots, x(1+(m-1)\alpha)\} \\ \vdots \\ X(i) = \{x(i), x(i+\alpha), \cdots, x(i+(m-1)\alpha)\} \\ \vdots \\ X(n-(m-1)\alpha) = \{x(n-(m-1)\alpha), x(n-(m-2)\alpha), \cdots, x(n)\} \end{cases} \tag{1}$$

In formula (1), $\{X(i), i = 1, 2, \cdots, n-(m-1)\alpha\}$ represents a new time series; α represents the delay time; m it represents the embedded dimension.

Yes $\{X(i), i = 1, 2, \cdots, n-(m-1)\alpha\}$ arrange them in a gradually increasing order to obtain the corresponding symbol sequence. The expression is

$$\beta(g) = \{j_1, j_2, \cdots, j_m\} \tag{2}$$

In formula (2), $\beta(g)$ represents the symbol sequence corresponding to the new time series; j_m it refers to the no m symbols.

Then calculate the probability of occurrence of any one in all numbering sequences P_g, then the time series can be converted into $\{x(i), i = 1, 2, \cdots, n\}$ The permutation entropy of is defined as:

$$H(m) = -\sum_{g=1}^{k} P_g \ln P_g \tag{3}$$

In Eq. (3), $H(m)$ it represents the new definition of time series permutation entropy; k indicates the maximum number.

In order to facilitate the research, formula (3) is further processed and controlled between 0 and 1, which is expressed as

$$Y(i) = \frac{H(m)}{\ln(m!)} = \frac{-\sum_{g=1}^{k} P_g \ln P_g}{\ln(m!)} \tag{4}$$

In Eq. (4), $Y(i)$ it represents the final time series results of swashplate plunger pump liquidity monitoring data.

$Y(i)$ the size of shows the complexity and randomness of the time series. If $Y(i)$ a smaller value reflects a smaller change in the time series, while a larger value indicates a more complex time series. $Y(i)$ the change of can reflect the small change of a specific position in the time series and amplify it. Therefore, it is regarded as the characteristics of the flow monitoring data of swashplate plunger pump and recorded as $\{Y(i), i = 1, 2, \cdots, n\}$.

The selection of permutation entropy parameter is the key to the performance of feature extraction of swashplate plunger pump flow monitoring data. Therefore, it is necessary to determine the time delay and embedding dimension. Wherein, the delay time of phase space reconstruction α it should be large enough to ensure $x(i)$ and $x(i + \alpha)$ They are basically independent of each other, but they cannot be completely independent, because this will make them have no connection, so that they cannot reflect the dynamic characteristics of the system. Now, for the selection of delay time, there are two most common methods: autocorrelation function method and mutual information method [6]. The autocorrelation function only represents the linear correlation of two variables, while the cross-correlation function represents the overall correlation of two variables. A lot of research shows that the delay time determined by mutual information method α to be more suitable than the time delay determined by the autocorrelation function method, this paper also uses the mutual information method to determine the delay time. The mutual information of the two time series is expressed as:

$$I(\alpha) = \sum_{i=1}^{n} P[x(i), x(i+\alpha)] \log_2 \left[\frac{P[x(i), x(i+\alpha)]}{P[x(i)]P[x(i+\alpha)]} \right] \tag{5}$$

In formula (5), $I(\alpha)$ it means $x(i)$ and $x(i + \alpha)$ mutual information; $P[x(i)]$ it represents the probability of a single event in the first time series; $P[x(i + \alpha)]$ it represents the probability of a single event in the second time series; $P[x(i), x(i + \alpha)]$ it represents the joint probability of two events.

When $I(\alpha)$ equal to 0, the correlation between the two time series is also 0.In the phase space reconstruction process, select $I(\alpha)$ the time delay corresponding to the first minimum is taken as the time delay of phase space reconstruction.

For embedding dimension in phase space reconstruction m selection of, if m too small, the less time series will be obtained after phase space reconstruction, and the characteristics of the system cannot be well represented; If m the value of is too large, the analysis time will be too long, and the influence of interference signal will not be amplified. So embedding dimension m the selection of is equally important. In this study, the method of false adjacent points is selected to estimate the embedding dimension m

determine. When the embedding dimension is low, it may appear that the projection points of two points that were originally far away in the one-dimensional space are very close. These two points are called false proximity points. With the increase of embedding dimension, the distance between false neighboring points gradually increases. When the distance between false neighboring points is fully expanded, the embedding dimension is the minimum embedding dimension.

When the embedding dimension is m, the space points after phase space reconstruction of one-dimensional time series can be expressed as:

$$X(i) = \{x(i), x(i+\alpha), \cdots, x(i+(m-1)\alpha)\} \tag{6}$$

Set to space point $X(i)$, the adjacent space points are $X(j)$, then the Euclidean distance between the two adjacent points is:

$$d_m = \|X(i) - X(j)\|_2 = \sqrt{\sum_{k=1}^{m-1} \left[x(i+k\alpha) - x(j+k\alpha)\right]^2} \tag{7}$$

In Eq. (7), d_m means $X(i)$ and $X(j)$ the Euclidean distance between two adjacent points.

When the embedding dimension is $m+1$, the Euclidean distance between two space points is:

$$d_{m+1} = \|X(i) - X(j)\|_2 = \sqrt{\sum_{k=1}^{m} \left[x(i+k\alpha) - x(j+k\alpha)\right]^2} \tag{8}$$

When d_{m+1} than d_m if it is much larger, it indicates that these two space points are false adjacent points, that is, determine the minimum embedding dimension as d_m corresponding m value.

The determined time delay and embedding dimension are substituted into formula (1), and the final flow monitoring data characteristics of swashplate plunger pump can be obtained through formula (2), (3) and (4) $\{Y(i), i = 1, 2, \cdots, n\}$ to provide a basis for subsequent research.

2.3 Multi Sensor Monitoring Information Fusion

Multi sensor is used to monitor the flow of swashplate plunger pump. Due to the large amount of monitoring information, if it is directly applied, it will waste a lot of manpower and material resources, and reduce the calculation efficiency of the proposed method [7]. Therefore, this section carries out fusion processing for multi-sensor monitoring information.

Multi sensor monitoring information fusion refers to multi-level processing of multiple information. Each level of processing is the reprocessing and abstraction of the upper level information. According to its level in the fusion system, information fusion can be divided into three categories: data level fusion, feature level fusion and decision level fusion, as shown below:

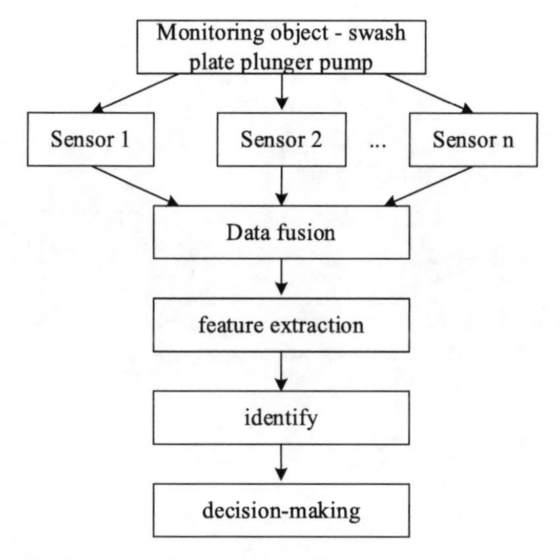

Fig. 3. Schematic diagram of data level fusion principle

First, data level fusion

The principle of data level fusion is shown in Fig. 3.

Data level fusion is the lowest level fusion, which is to directly fuse the data collected by sensors, that is, the data collected by sensors are comprehensively analyzed and processed without preprocessing, and then the subsequent feature extraction is performed from the fused information. Data set fusion requires that the sensors are of the same category, that is, the collected data are the same physical quantity or phenomenon. Data level fusion has less information loss and can provide subtle information, but it has large amount of information, time-consuming processing, large amount of communication and poor anti-interference ability.

Second, feature level fusion

The principle of feature level fusion is shown in Fig. 4.

Feature level fusion belongs to the middle layer. This method first extracts features from the original data collected by sensors, then fuses the feature data obtained by each sensor, and finally carries out the recognition and decision-making process. The method of feature level fusion is used to extract the features of the original information, which to some extent realizes the information compression and facilitates communication and real-time processing. At present, most of the practical application systems adopt the feature level fusion method [8].

Third, decision level integration

Decision level fusion is a high-level fusion. First, feature extraction, recognition and decision-making are carried out independently according to the original information of each sensor, and then the decision results are fused to get the final decision. Decision level fusion is the fusion of decision results. In theory, its effect is more accurate than that of a single sensor, but in practice, it needs to ensure that each sensor signal is independent,

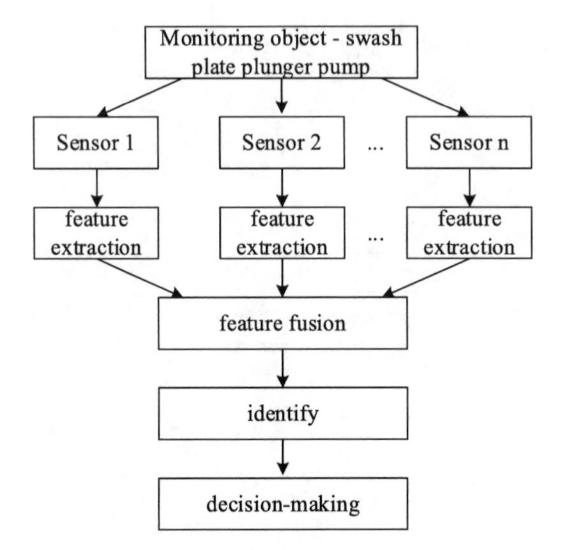

Fig. 4. Schematic diagram of feature level fusion principle

otherwise its performance may be lower than that of feature level fusion. Decision level fusion has small traffic, strong anti-interference ability and little dependence on sensors.

The principle of decision level fusion is shown in Fig. 5.

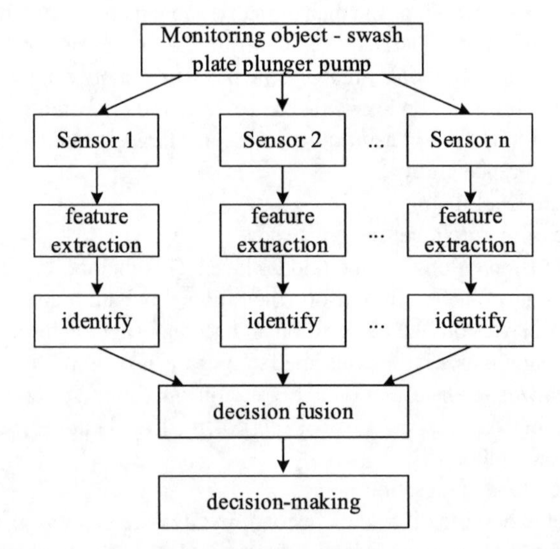

Fig. 5. Schematic Diagram of Decision Level Fusion Principle

Multi sensor information fusion technology can fuse different sensor data in different ways. When feature level fusion is adopted, recognition and decision-making are required after feature fusion extracted from different sensors. This process requires mining potential data information and making decisions. Big data often hides more complex feature information, but the algorithm may not be able to fully learn these information, which will affect the performance of the model. Therefore, the feature information obtained from different algorithms or the feature information at different stages of the algorithm can be fused to form new fusion features, which will greatly improve the feature learning ability. This research proposes a feature fusion method, and the specific principle is shown in Fig. 6.

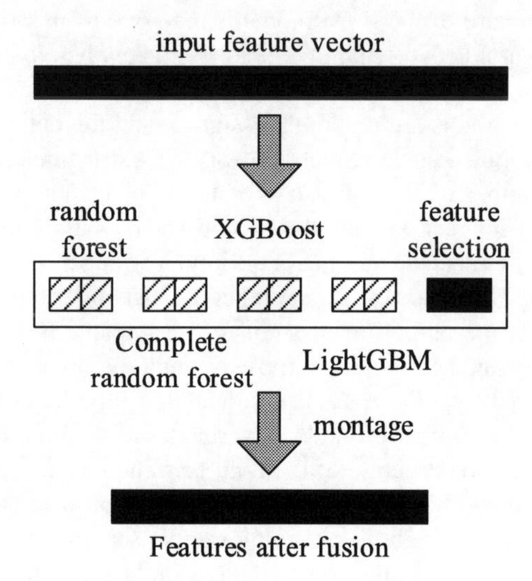

Fig. 6. Schematic diagram of feature fusion principle

As shown in Fig. 6, first train the integrated classifiers of random forest, completely random forest, XGBoost, LightGBM and so on using the original input features, and then train the class probability vector [9]. At the same time, the original input features are screened using random forests, or feature dimensionality is reduced through PCA and other methods to obtain the features after screening or feature dimensionality reduction; Finally, the category probability vector and the processed feature vector are spliced to obtain the final fusion feature. This feature fusion method can not only retain important feature information, but also take into account the global information, while removing redundant features and reducing feature dimensions. Then the fused features are trained using the deep forest model, which is the deep forest of multi feature fusion.

The above process completed the fusion of multi-sensor monitoring information, and obtained the swashplate plunger pump liquidity monitoring data feature fusion results as follows: $\{Z(i), i = 1, 2, \cdots, L\}$, L it represents the total number of fusion features, which facilitates the realization of intelligent monitoring of swashplate plunger pump fluidity.

2.4 Realization of Intelligent Monitoring of Swashplate Plunger Pump Fluidity

Feature fusion results of swashplate plunger pump flow monitoring data obtained above $\{Z(i), i = 1, 2, \cdots, L\}$ on the basis of, obtain the characteristics of liquidity anomaly data from existing literature $\{R(j), j = 1, 2, \cdots, W\}$ (W it indicates the total number of abnormal flow data characteristics), and the extreme learning machine (ELM) is used to determine whether the swashplate plunger pump has abnormal flow, so as to realize the intelligent monitoring of swashplate plunger pump flow and provide assistance for the stable operation of the plunger pump.

The network structure of the ELM model is the same as that of the single hidden layer feedforward neural network, but it is not the gradient based algorithm (backward propagation) in the traditional neural network in the training phase. Instead, the input layer weights and deviations are randomly assigned, and the output layer weights are calculated by the generalized inverse matrix theory. This structure can quickly complete the classification training of the flow state of the plunger pump while ensuring the accuracy. In practical application, only the weights and deviations on all network nodes need to be solved to complete the training of the limit learning machine model. When the characteristic parameters of the test set are input, the network output can be calculated through the weight of the output layer obtained to complete the determination of the piston pump's fluidity and achieve the purpose of liquidity monitoring [10].

Given training set $\{Z(i), T(i)\}$, $T(i)$ represents the i the flow state of the plunger pump at time units. The number of hidden layer nodes of the limit learning machine is N in this way, the network structure of the limit learning machine is established. Like the structure of the single hidden layer feedforward neural network, the input of the neural network from left to right is a training sample set, and there is a hidden layer in the middle [11]. From the input layer to the hidden layer, there is a full connection. Remember that the output of the hidden layer is $G(Z)$, then hide layer output $G(Z)$ the expression of is

$$G(Z) = \left[g_1(Z), g_2(Z), \cdots, g_N(Z)\right] \tag{9}$$

The output of the hidden layer is obtained by multiplying the input by the corresponding weight plus the deviation, and then summing all the node results of a nonlinear function. $G(Z) = \left[g_1(Z), g_2(Z), \cdots, g_N(Z)\right]$ is ELM nonlinear mapping (hidden layer output matrix), $g_k(Z)$ is the first k output of hidden layer nodes. The output functions of hidden layer nodes are not unique, and different output functions can be used for different hidden layer neurons [12]. In general, in practical applications, $g_k(Z)$ it is expressed as follows:

$$g_k(Z) = \zeta(\omega_k, \upsilon_k, Z) = \zeta(\omega_k Z + \upsilon_k) \tag{10}$$

In Eq. (10), $\zeta(\omega_k, \upsilon_k, Z)$ it represents the activation function and is a nonlinear piecewise continuous function satisfying the ELM general approximation ability theorem. Common activation functions include Sigmaid function, sine function, Hardlim function, etc. The activation function used in this modeling is Sigmaid function; ω_k and υ_k represents hidden layer node parameters.

Next, we deduce the transfer relationship from the hidden layer to the output layer. The output of the single hidden layer feedforward neural network ELM for "generalized" is:

$$f_N(Z) = \frac{\sum \psi_k g_k(Z)}{\xi(Z(i), R(j))} \tag{11}$$

In Eq. (11), $f_N(Z)$ it represents the judgment result of the flow state of swashplate plunger pump; ψ_k represents the output weight between the hidden layer and the output layer; $\xi(Z(i), R(j))$ it represents the correlation coefficient between the characteristics fusion result of swashplate plunger pump liquidity monitoring data and the characteristics of abnormal liquidity data [13].

According to the calculation results of formula (11), the piston pump liquidity monitoring rules are formulated as follows:

$$\begin{cases} f_N \leq \Psi^o \text{ Liquidity anomaly} \\ f_N > \Psi^o \text{ Normal liquidity} \end{cases} \tag{12}$$

In Eq. (12), Ψ^o it indicates the determination threshold value of the plunger pump liquidity state.

It should be noted that in the application process of the extreme learning machine, the establishment of the number of hidden layer nodes actually affects the accuracy of the model. The number of nodes is too much. Although the model can express more nonlinear characteristics, at the same time, the compatibility of model prediction is worse. When the data correlation between the test set and the training set is not strong enough, the prediction accuracy is very low; When the number of nodes is too small, the expression ability of nonlinear relationship is limited, that is, the model itself cannot accurately express the relationship of the training set, which will also reduce the prediction accuracy of the model. In practice, the number of nodes selected by default is roughly the same as the sample set size of the input layer, but in practice, the optimal number of nodes cannot be obtained from the algorithm optimization, and it is necessary to test the accuracy optimization of the model several times to finally obtain the optimal number of nodes.

Through the above process, the intelligent monitoring of the swashplate plunger pump's fluidity is completed, which helps the swashplate plunger pump to operate stably.

3 Experiment and Result Analysis

3.1 Establishment of Experimental Environment

In order to verify the application performance of the proposed method, an experimental environment is built, as shown in Fig. 7.

As shown in Fig. 7, the monitoring platform is equipped with a variety of sensors to collect flow related data during the operation of swashplate plunger pump, which meets the requirements of the proposed method application performance test.

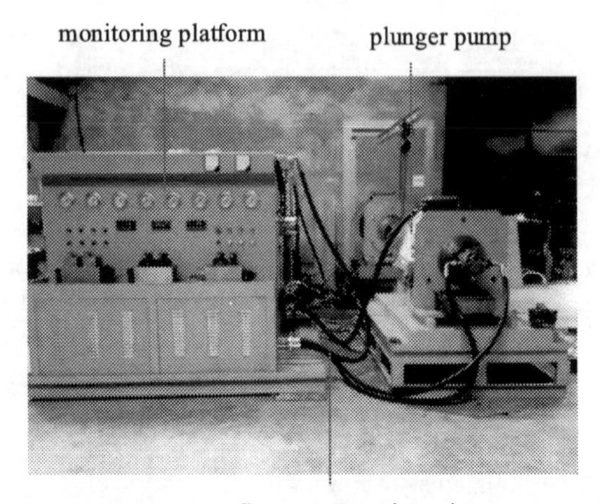

Fig. 7. Schematic Diagram of Experimental Environment

3.2 Setting of Test Conditions

The number of experimental conditions is also the key to affect the reliability of experimental conclusions. In order to improve the credibility of the experimental conclusions, this study has set up 10 different experimental conditions, as shown in Table 2.

Table 2. Setting table of test conditions

Test conditions	Monitoring data volume/MB	Liquidity status
1	4562	abnormal
2	3658	normal
3	4012	abnormal
4	4589	abnormal
5	3569	normal
6	5021	abnormal
7	4578	normal
8	3521	abnormal
9	2659	abnormal
10	3550	normal

As shown in Table 2, the monitoring data volume and liquidity status of the 10 experimental conditions are inconsistent, which indicates that the background conditions of each experimental condition are quite different and meet the requirements of the proposed method application performance test.

3.3 Analysis of Experimental Results

Based on the above built experimental environment and the set experimental conditions, and taking the condition monitoring method of the plunger pump based on the oil temperature signal and the condition monitoring method of the plunger pump based on the ferrography analysis method as the comparison methods 1 and 2, the flow monitoring comparison experiment of the swashplate plunger pump was carried out. In order to clearly show the application performance of the proposed method, the flow monitoring results and monitoring time of swashplate plunger pump are selected as evaluation indicators. The specific analysis process of experimental results is shown below.

The flow monitoring results of swashplate plunger pump obtained through experiments are shown in Table 3.

Table 3. Flow Monitoring Results of Swashplate Plunger Pump

Test conditions	Liquidity status	Propose method	Comparison method 1	Comparison method 2
1	abnormal	abnormal	abnormal	normal
2	normal	normal	abnormal	normal
3	abnormal	abnormal	normal	abnormal
4	abnormal	abnormal	abnormal	abnormal
5	normal	normal	normal	normal
6	abnormal	abnormal	abnormal	normal
7	normal	normal	normal	abnormal
8	abnormal	abnormal	abnormal	abnormal
9	abnormal	abnormal	normal	abnormal
10	normal	normal	normal	abnormal

As shown in Table 3, there are certain discrepancies between the flow monitoring results of the swash plate plunger pump obtained by comparison method 1 and the actual results. There are monitoring errors in three working conditions, with an accuracy rate of 70%; There are certain discrepancies between the flow monitoring results of the swash plate plunger pump obtained by comparison method 2 and the actual results, with monitoring errors occurring in 4 working conditions, with an accuracy rate of 60%; The flow monitoring results of the swashplate plunger pump obtained after the application of the proposed method are completely consistent with the actual results, with an accuracy of 100%; This indicates that the proposed method provides more accurate flow monitoring results for the swashplate plunger pump.

The flow monitoring time of swashplate plunger pump obtained through experiments is shown in Fig. 8.

As shown in the data in Fig. 8, the flow monitoring time of the swash plate plunger pump obtained by method 1 is 0.35 s–0.65 s, while the flow monitoring time of the swash

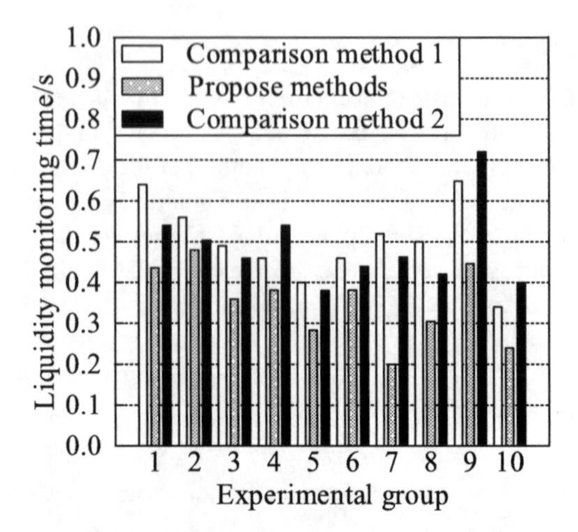

Fig. 8. Schematic diagram of flow monitoring time of swashplate plunger pump

plate plunger pump obtained by method 2 is 0.38 s–0.72 s. However, the flow monitoring time of the swash plate plunger pump obtained by method proposed after application is 0.20 s–0.48 s. The flow monitoring time of the swash plate plunger pump obtained by method proposed after application is lower than that of comparison methods 1 and 2, and always does not exceed 0.5s, this indicates that the proposed method has a higher efficiency in monitoring the flowability of the swashplate plunger pump.

4 Conclusion

The swashplate plunger pump is the key to the smooth flight of aircraft, and whether the swashplate plunger pump can operate stably depends on its own fluidity. Therefore, an intelligent monitoring method for the fluidity of aircraft swashplate plunger pumps based on different operating conditions is proposed. And through experiments, it was verified that the monitoring results of the proposed method are completely consistent with the actual results, with an accuracy of 100%; The monitoring time is 0.20 s to 0.48 s, always not exceeding 0.5 s, with higher monitoring efficiency, providing effective method support for the stable operation of the swashplate plunger pump and the aircraft. In our research, we will continue to improve our design methods, improve the real-time monitoring level of the swashplate plunger pump, and provide certain technical support for the smooth flight of the aircraft.

Acknowledgement. A school-level project of Beijing Polytechnic, Project Name: Research and application of the flowability of the aircraft swashplate plunger pump under different working conditions (2022X050-KXZ).

References

1. Liu, S., Chen, P., Woźniak, M.: Image enhancement-based detection with small infrared targets. Remote Sens. **14**, 3232 (2022)

2. Shao, N., Chen, Y.: Abnormal data detection and identification method of distribution internet of things monitoring terminal based on spatiotemporal correlation. Energies **15**(6), 2151 (2022)
3. Zhang, H., Ge, D., Yang, N., et al.: Study on internet of things architecture of substation online monitoring equipment. MATEC Web Conf. **336**(5), 05024 (2021)
4. Zhang, Y., Zou, X., Zhang, B., et al.: A flexible turning and sensing system for pressure ulcers prevention. Electronics **10**(23), 2971 (2021)
5. Liu, S., Lu, M., Liu, G.: A novel distance metric: generalized relative entropy. Entropy **19**(6), 269 (2017)
6. Santos, R., Leal-Junior, A.G., Ribeiro, M., et al.: Datacenter thermal monitoring without blind spots: FBG-based quasi-distributed sensing. IEEE Sens. J. **21**(8), 9869–9876 (2021)
7. Lim, Y.Y., Smith, S.T., Padilla, R.V., et al.: Monitoring of concrete curing using the electromechanical impedance technique: review and path forward. Struct. Health Monit. **20**(2), 604–636 (2021)
8. Liu, S., Wang, S., Liu, X., et al.: Human inertial thinking strategy: a novel fuzzy reasoning mechanism for IoT-assisted visual monitoring. IEEE Internet Things J. (2022). https://doi.org/10.1109/JIOT.2022.3142115
9. Li, H., Spencer, B.F., Liu, W., et al.: Multi-feature integration and machine learning for guided wave structural health monitoring: application to switch rail foot. Struct. Health Monit. **20**(4), 2013–2034 (2021)
10. Drouaz, M., Colicchio, B., Moukadem, A., et al.: New time-frequency transient features for nonintrusive load monitoring. Energies **14**(1437), 1437–1437 (2021)
11. Sun, Z., Wang, Y., Zhang, H., et al.: Parameter optimization and performance simulation evaluation of new swash plate-plunger energy recovery device. Desalin. Int. J. Sci. Technol. Desalt. Water Purif. **528**, 115598 (2022)
12. Pronyakin, V.I., Skrypka, V.L,, Abykanova, B.T.: Surface microgeometry monitoring of large-sized aircraft elements. IOP Conf. Ser. Mater. Sci. Eng. **1027**, 012024 (2021)
13. Saracyakupoglu, T.: A comprehensive fracture research of an aircraft swash-plate pump. Eng. Fail. Anal. **140** (2022)

Design of Mobile Terminal Music Education Platform Based on Django Framework

Chao Long[1](✉) and Chunhui Liu[2]

[1] College of Art, Hebei University of Economics and Business, Shijiazhuang 050061, China
Longchaochao67@163.com

[2] School of Economics and Management, Hunan Software Vocational and Technical University, Xiangtan 411000, China

Abstract. Music education usually requires experienced professional teachers, but educational resources are limited and cannot meet educational needs. Therefore, this article designs a mobile terminal music education platform based on the Django framework. Using Apache server to interact with users, design the mobile terminal education platform architecture based on Django framework. Based on the platform architecture and the hardware structure of the education platform, select suitable Web servers and hubs. With the support of hardware devices, develop a mobile terminal music education resource library to clarify the functions of the mobile terminal music education platform. Build a music teaching resource sharing service model to meet the interaction needs between resources, and thus complete the design of a mobile terminal music education platform based on the Django framework. The experimental results show that the designed mobile terminal music education platform has a good application effect. The platform can be compared through stress testing in the experiment. Under high concurrent user conditions, the platform's response time still does not exceed 3 s, and the average error rate is only 0. 5%, indicating that the platform in this paper has good stability.

Keywords: Django Framework · Hardware Equipment · Education Platform · Music · Mobile Terminal

1 Introduction

In recent years, the development of computer software technology has changed people's working and living patterns. In terms of teaching, people's way of acquiring knowledge is also changing greatly. The situation of the single mode of traditional education in the classroom is changing, especially with the breakthrough development of Internet technology in recent years. It makes the original online education mode based on B/S mode more abundant and diversified, and people gradually accept the mode of completing various learning tasks through the network. Although mobile Internet access has become more and more powerful over the years, there are not many applications of modern educational technology transferred from PC to mobile terminals such as mobile phones,

L. Yun et al. (Eds.): ADHIP 2023, LNICST 549, pp. 268–282, 2024.
https://doi.org/10.1007/978-3-031-50549-2_19

and they are mainly entertainment websites. As for learning websites [1], especially the course websites for professional learning in colleges and universities, there are few and few, and occasionally there are only theoretical discussions. In this situation, music education still requires professional teachers and learning venues most of the time, but these music education resources are limited and inconvenient. To this end, design a mobile music education platform that provides a wider range of music education resources, enabling more people to easily access and learn music.

Reference [2] suggests that the development of information technology has led to the adoption of more advanced learning technologies. Electronic learning and machine learning are two new teaching methods widely used in the Internet of Things (IoT) and digital education platforms (DEP). Combining machine learning and SEM (structural equation modeling) to achieve the design of an education platform. Reference [3] proposed the construction of Digital art education platform under the "Internet plus" environment. The principle and process of data mining in the The Internet Age are clarified, and an interactive data fusion algorithm and model for the Internet and crowdsourcing are designed to establish a Digital art education platform in the "Internet plus" environment. However, the above methods have a slower response speed for situations with a large number of online users.

Django is an excellent open source Web development framework of Python, and the bottom layer is built based on Python language. It has the advantages of simplicity, clarity, efficiency and security in Web development. It has a dynamic database access API with rich functions, and supports a variety of background databases, including Postgresql, MySql, Sqlite, and Oracle. Using Django's extensible built-in template, you can code the model layer control layer and the page template completely independently, and the structure is very clear. The Django framework has a strong background management function. Just use ORM to do simple object definitions, and you can automatically generate the database structure, as well as a full-featured management background [4, 5]. As the Django framework is a weak client, it is particularly suitable for website access and online course learning on mobile phones with less hardware configuration than PCs. Based on this, this paper will discuss the Django framework mobile terminal music education platform. Utilizing Apache servers to interact with users and using the Django framework to build a backend, a mobile terminal education platform architecture based on the Django framework is designed. Choose appropriate education platform hardware to increase platform stability. Improve the platform functions through the music education resource library of mobile terminals, construct a music teaching resource sharing service model, and achieve resource sharing.

2 Design of Education Platform Architecture Based on Django Framework

The bottom layer of the Django framework is built on Python, a scripting language that supports modules and packages, supports multiple platforms, is extensible, and has rich libraries. Because Django is a web framework written in Python, Python language is selected as the development language [6] when designing the mobile terminal music education platform. Django is an open source model, view, and controller style Web application framework driven by high-level Python programming language, which originated from the open source community. Using this architecture, programmers can easily and quickly create high-quality, easy to maintain, database driven applications. This is the main reason why OpenStack's Horizon component is designed with this architecture. In addition, the Django framework also contains many powerful third-party plug-ins, making Django highly extensible.

Apache is a widely used Web server, which supports Django better. The Apache server is used to interact with users, and the background is built by the Django framework to assist the MySQL database to provide resource storage and management services, which can meet the development needs of the education platform [7]. Based on the above contents, design the education platform architecture as shown in Fig. 1 below.

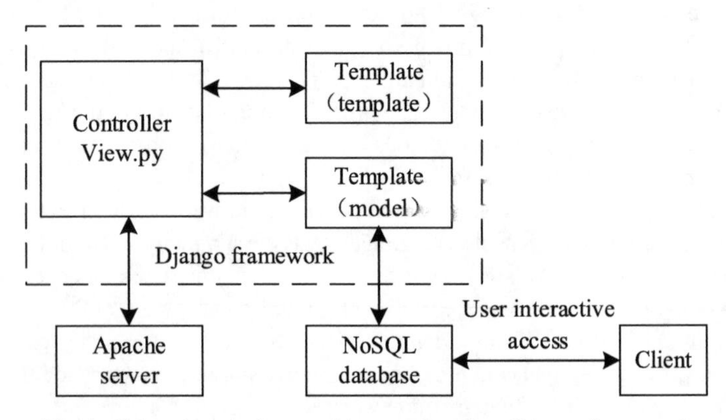

Fig. 1. Education platform architecture based on Django framework

This platform is developed using the Django framework+SpringMVC+MyBatis open source framework. The entire framework mainly uses Django+SpringMVC to implement the Web MVC design pattern, decouple the responsibilities of views and business objects, and simplify the development process [8]. Using MyBatis as the data persistence layer framework, compared with the traditional Sql statements with low reuse rate and the complex Hiberate framework, the Django framework has very strong flexibility and optimization, and is currently a highly used persistence framework.

The Django framework is a lightweight IOC and AOP container framework. On the basis of the Spring MVC and Django frameworks, Spring is used to integrate the Web layer and the persistence layer, and create the object relationship mapping of the model. It can design a better management interface for end users, implement URL design, and meet the education needs of mobile terminals.

3 Hardware Equipment Selection

On the basis of the above contents, the hardware structure of the education platform is designed as shown in Fig. 2 below.

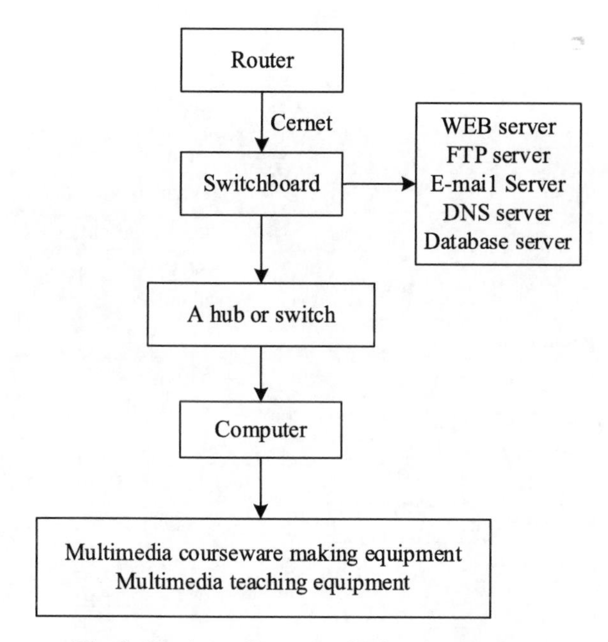

Fig. 2. Hardware Structure of Education Platform

From the content shown in Fig. 2 above, we can see that the education platform developed in this paper is composed of routers, switches, computers (and other mobile terminals), Web servers, FTP servers, E-mail servers, DNS servers, database servers, hubs, multimedia lesson production equipment, multimedia teaching equipment, etc. [9]. The following will take the Web server and hub as examples to select the hardware equipment in the education platform.

3.1 Web Server Selection

The Web server selected in this paper is an enterprise level website server with the model of Inspur NF5260M6. Inspur NF5260M6 is a 2U two-way rack server equipped with the third generation Intel Xeon expandable processor. Flexible modular design, deep module disassembly, highly flexible PCIe resources, and multiple front and rear window configurations to meet the customized needs of different customers. With innovative architecture, it is suitable for all kinds of Internet applications. The server technical parameters are shown in Table 1 below.

The NF5260M6 selected this time is based on the new generation of Intel Xeon's third-generation extensible processor. A single CPU has 40 cores and 80 threads, supports TDP 270W CPU, Remax 3.6 GHz, and three groups of 11.2 GT/s UPI interconnection links, so that the server has higher processing performance. It supports 32 3200 MT/s DDR4 ECC memories. The memory supports the RDIMM type and can provide excellent speed, high availability and up to 4T memory capacity [10]. It supports DCPMM type memory, with a memory capacity of 512 G and a bandwidth of 3200 MHz. While not

Table 1. Technical Parameters of Web Server

S/N	Project	Parameter
1	Brand	Inspur
2	Specifications	2U rack type
3	Weight	Full configuration <40 kg
4	Fan	6 6056 fans, supporting N + 1 redundancy
5	Working temperature	5 °C–45 °C
6	Outer packing box size	651 wide × 295 high × 1031 deep
7	Maximum temperature gradient (operation and storage)	20°C/H
8	Product model	INSPURNF5260M6 rack mounted
9	Internal hard disk bit	Optional 2 SATAM
10	Chipset	IntelC621A
11	Post I/O	2 USB3. 0, 1 RJ45, 1 VGA
12	Cpu/processor	Supports 1 to 2 Intel®xeon®Third generation scalable processing
13	Network card	One set of dual 10 Gigabit on-board network ports 10 Gb/s
14	Expansion Slots	Up to 6 standard PCIe

reducing the memory capacity and bandwidth, it can still save complete memory data when power is completely cut off.

At the same time, the server has the advantage of flexible expansion. "The front supports 12 3.5" hard disks or 24 2.5 "hard disks, the rear supports 4 2.5" hard disks, and the whole machine supports 16 hot swappable NVMe SSD full flash configurations. There is no direct outgoing PCIe slot on board, which flexibly meets customer customization requirements. It supports two optional notification hot plug OCP3.0 modules and provides 10 G, 25 G, 40 G, 100 G and 200 G network interface options, providing more flexible network structure for applications.

In the design, the server is based on the humanized design concept, and the whole system can realize tool free maintenance. Through the enhancement and optimization of some structural parts, rapid disassembly and assembly are realized, and the operation and maintenance time is greatly shortened. Through the unique intelligent control technology of Inspur and the advanced air cooling system, the whole machine can work in a better environment and ensure the stable operation of the system. The application of new BMC technology enables technicians to guide equipment through the Web management interface, fault diagnosis LED, etc., and mark the faulty machine through the UID indicator on the front panel, so as to quickly find the components that have failed (or are failing), thus simplifying maintenance work, speeding up problem solving, and improving system availability.

3.2 Hub Selection

The UTEK industrial 4-port RS485/422 UT-1304 intelligent hub is selected as the main hardware of the education platform. UT-1304 is a microprocessor (uninterruptible operation) designed internally. It is an RS485/422 bus dividing hub (HUB) designed to meet the requirements of RS485/422 large system in complex electromagnetic environment. The equipment supports a transmission rate of up to 115. 2KBPS. In order to ensure the safety and reliability of data communication, the RS-485/422 interface terminal uses the photoelectric isolation technology to prevent lightning surge from being introduced into the converter and equipment. The built-in photoelectric isolator and 8/20 us waveform 3000 A lightning surge protection circuit can provide 2500 V isolation voltage, and can effectively suppress lightning and ESD. At the same time, it can effectively prevent lightning strike and common ground interference. External switching power supply is used for power supply, which is safe and reliable. In the RS-485/422 working mode, the adopted discrimination circuit can automatically sense the direction of the data flow, and automatically switch the enabling control circuit to easily solve the problem of RS485/422 transceiver conversion delay. The RS-485/422 interface has a transmission distance of more than 1200 m and stable performance. It is widely used in the design and development of high-performance systems and platforms.

UT-1304RS-485/422HUB provides star RS-485/422 bus connection. Each port has short circuit and open circuit protection. With 2500 V photoelectric isolation, users can easily improve the RS-485/422 bus structure, divide the network segment, and improve the reliability of education resource transmission, communication, sharing and other processes. When equipment failure occurs, the network segment with the problem will be isolated to ensure the normal operation of other network segments [11]. This performance improves the reliability of the existing RS-485/422 network and effectively shortens the maintenance time of the network. Reasonable use of UT-1304RS-485/422HUB can design a reliable education platform.

In order to meet the access requirements of the hub, the technical parameters of the interface of the hub access education platform are designed according to the contents shown in Table 2 below.

On the basis of the above content, access the device to the music education platform of mobile terminal. The access method is shown in Fig. 3 below.

In the above way, RS-485/422 bus is extended to the application of four highly reliable RS-485/422 interfaces. 128 RS-485/422 hubs can be connected simultaneously on the RS-485/422 bus to complete the selection of hubs for music education platform.

Table 2. Technical Parameters of Hub Access to Education Platform Interface

S/N	Project	Parameter
1	Interface characteristics	The interface is compatible with EIA/TIA RS-485/422 standard
2	Electrical interface	RS-485/422 interface is the terminal block
3	Transmission medium	Twisted pair or shielded wire
4	Operation mode	Asynchronous half duplex, asynchronous full duplex
5	Signal indication	Seven signal indicators power supply (PWR), transmit (TXD), receive (RXD), fault (E1-E4)
6	Isolation	Isolation voltage 2500VRMS500VDC continuous, DC/DC isolation module
7	Transmission speed	300 bps–115. 2 Kbps
8	Working voltage	DC9-48 V
9	Working current	150 mA

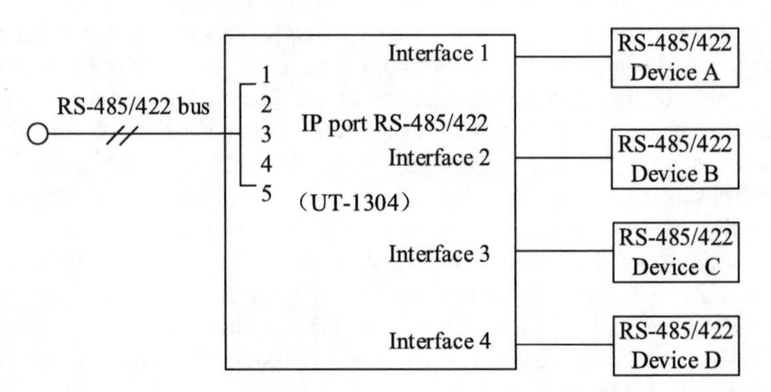

Fig. 3. Hub access mode on music education platform

4 Develop Music Education Resource Library of Mobile Terminal

On the basis of the above design content, in order to ensure a detailed understanding of the functions of the mobile terminal music education platform, it is necessary to design a complete education resource library as the support of the education platform.

The design of music education resource database includes three parts. In this design, the conceptual design and logical design of the education platform are described by the database conceptual model diagram, and the physical design of the database is described by the core table structure.

The database table design is a part of the database physical design. Through the description of the data table structure, the database physical design is realized. Select some tables to describe their physical structure.

(1) User information table: The user information table is used to store the system end user information and the system background desktop system, including user basic information and login related information. The fields, types, physical storage names and other information in this table are shown in Table 3.

Table 3. User Information

Field Name	Type and size	Field Notes	Physical name
User No	Character type (20)	Primary key cannot be empty	YHBH
User name	Character type (50)	Cannot be empty	YHMC
User Category	Character type (10)	Cannot be empty	YHLB
Login password	Character type (30)	Cannot be empty	DLKL
Contact information	Character type (50)	Can be empty	LXFS
Registration time	Date time type	Cannot be empty	ZCSJ
Notes	Character type (100)	Can be empty	BZSM

(2) Education information table: information information is used to store detailed information of information, including basic information, information publishing, browsing information, etc. The fields, types, physical storage names and other information in this table are shown in Table 4.

Table 4. Education Information

Field Name	Type and size	Field Notes	Physical name
Information No	Character type (20)	Primary key cannot be empty	ZXBH
Information Title	Character type (50)	Cannot be empty	ZXBT
Information category	Character type (10)	Cannot be empty	ZXLB
Release time	Date time type	Cannot be empty	FBSJ
Publisher	Character type (20)	Foreign key cannot be empty	FBR
Information content	Character type (2000)	Cannot be empty	ZXNR
Number of visitors	integer	Cannot be empty	ZLRS
Browse Class	Character type (100)	Cannot be empty	LLBJ

(3) Attachment information table: The attachment information table is used to store information and notification related document attachment information. The attachment information in the system is stored in the form of a file. The title, category, address, etc. of the document are stored in the attachment information table. The field, type, physical storage name, etc. of this table are shown in Table 5.

Table 5. Attachment Information

Field Name	Type and size	Field Notes	Physical name
Document No	Character type (20)	Primary key cannot be empty	WDBH
Document Title	Character type (50)	Cannot be empty	WIBT
Document category	Character type (10)	Cannot be empty	WILB
File address	Character type (50)	Cannot be empty	WIDZ
Upload time	Date time type	Cannot be empty	SCSJ
Uploader	Character type (20)	Foreign key cannot be empty	SCR
Number of visitors	Integer	Cannot be empty	ZLRS
Number of downloads	Integer	Cannot be empty	XZRS

According to the above method, design the daily notification information table, assignment publicity information table, student information table, famous works information table, and student practice score information table in the music education resource database of mobile terminal.

To ensure that the designed information table can provide a basis for the management of the education platform, the data entered in the information table shall be processed uniformly according to the following formula.

$$Y = Y_0 + \frac{Y_m - Y_0}{N_m - N_0}(X - N_0) \tag{1}$$

In formula (1): Y represents the teaching resource data after unified processing; Y_0 indicates the upper threshold of data unification processing; Y_m indicates the lower threshold of data unification processing; N_m represents the digital quantity of sampling data in the database; N_0 indicates the original format of the sampling data in the database; X represents a data transition scale. According to the above method, data conversion processing is carried out to complete the development of music education resource database of mobile terminal.

5 Build a Music Teaching Resource Sharing Service Model

College teaching resources are characterized by heterogeneity, dynamics, format diversity and cross organization. It is difficult to schedule resources among multiple users. In order to meet the interaction needs between resources, after completing the above design, it is necessary to build a music teaching resource sharing service model. The model expression is shown in the following formula.

$$K = \sum_{F>1}^{F} Y \|[A, L] - [A, L]_F\| \tag{2}$$

In formula (2): K represents the music teaching resource sharing service model; F linear expression of music teaching resources; A indicates the resource category; L Indicates the resource transmission mode. On this basis, starting from the current situation of the construction of teaching resources in colleges and universities, centering on the hot spots and key points of the current teaching reform, according to the characteristics of disciplines and majors in colleges and universities, through collaborative innovation, integrate massive teaching resource data in the existing environment of colleges and universities, build a music teaching resource sharing service system and model, and unify the codingData standards and database systems, etc., form a unified access data center [12, 13], and realize data sharing and unified management of teaching resources. The coding process of digital teaching resources is shown in the following formula.

$$v = \frac{a\sqrt{n_1^2 - n_2^2}}{\lambda \cdot K} \tag{3}$$

In formula (3): v indicates the digital teaching resource code; a represents the private cloud access interface; λ indicates the resource data transmission frequency parameter; n_1 indicates the data scheduling task; n_2 indicates a data transmission task. After ensuring that the resources in the education platform meet the needs of sharing and service, Universities should focus on "specialty group, specialty, curriculum, resources" and "With educational institutions, colleges, teachers and students as the main line, we will create a university professional teaching resource database that can be shared, shared and co built by colleges, enterprises and society, and a university teaching resource cloud that can meet the autonomous learning of teachers, students, teachers and social learners, and provide technical support for the optimal allocation and management of university teaching resources.

6 Music Practice Management and Online Classroom Management

On the basis of the above design content, in order to standardize the operation of the mobile terminal music education platform and give play to the higher value and efficiency of the education platform, it is necessary to further design the music practice management and online classroom management in the platform to improve the functions of the education platform.

Music exercises are used to realize the process of students completing music exercises through mobile phones, and exercise topics are made sound through mobile phones to reflect the effect of exercises, including accompaniment exercises, test exercises and famous works. The detailed class diagram of music practice is shown in Fig. 4 below.

In the above figure, the Works class is used for students to practice maintenance operations_ID attribute is student ID, work_Id is the work identification, the geWorksInfo method is to obtain the details of student exercises, the addWorks method is to add student exercises, the delWorks method is to delete student exercises, the saveWorks method is to save student exercises, and the searchWorks method is to query student exercises.

The FamousWorks class is an operation class for students to maintain online accompaniment exercises for famous works. The workType is the work category (Chinese,

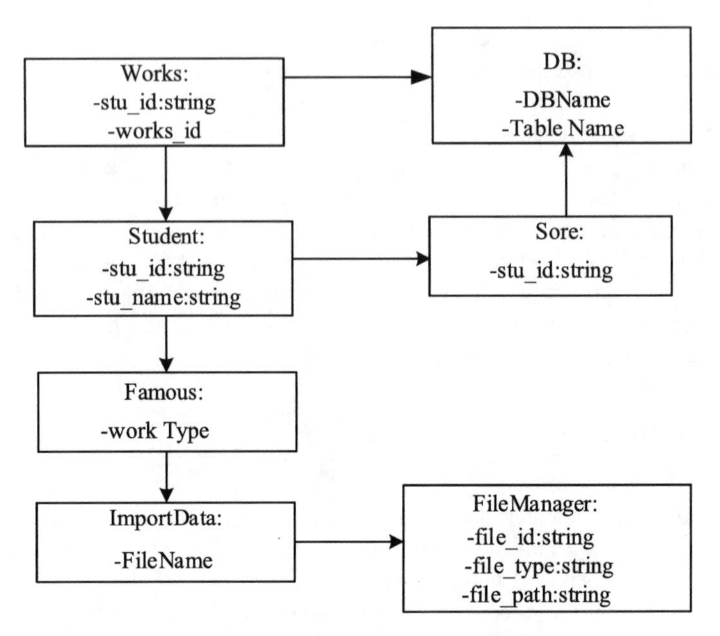

Fig. 4. Detailed Class Diagram of Music Practice

foreign, choral, etc.), the geWorksInfo method is to obtain the details of famous works, the addWork method is to add famous works, the delWorks method is to delete famous works, the saveWorks method is to save famous works, and the searchWorks method is to query famous works,the ImportData method implements the batch import of famous artists' works.

The Score class is used to maintain the practice scores. In addition to managing students' basic scores, it also needs to import the practice scores_The i attribute is the student ID, the getScore method is to obtain the details of the exercise scores, the saveScore method is to save the exercise scores, and the searchScore method is to query the exercise scores.

On this basis, online classroom management is carried out. Online classes include online classes, learning logs, learning reflection, online questions, resource browsing, etc. Online class is that learners can arrange to log in to the platform at any time for learning detection after they have formulated relevant target performance. In the process of online class, they can use the learning tools under this module to record notes, feelings, problems, etc. for future discussion and reference. After completing the learning task, the platform will automatically record the log of this learning for later evaluation score calculation. This module has designed the online class Independent Study, learning log StudyLogs, learning reflection record StudyNote and student information management student. When the online class needs music homework, get the latest music homework from the Resources class, start self-learning, and record the learning process (start time and end time) through StudyLogs. Through the above methods, the standardized management of music exercises and online classes in the education platform is realized, so as to complete the design of mobile terminal music education platform based on the Django framework.

7 Comparison Experiment

7.1 Experiment Preparation

The above research on the design of mobile terminal music education platform based on the Django framework has been completed from three aspects. In order to test the application effect of the platform, the following test will be carried out on the education platform by means of design comparison experiments, as shown below. Before the test, the technical parameters of the test environment are analyzed as shown in Table 6 below.

Table 6. Technical Parameters of Comparative Experimental Environment

S/N	Project	Parameter
1	Operating system	Windows 7
2	Development integration environment	MyEclipse6. 0
3	Development language	Java
4	Development package	JDK1. 6
5	Web client	HTML5+JavaScript
6	Client	Android 4. 0 or above
7	Data base	My SQL5. 6
8	Server operation	Windows Server 2019

Integrate ADT Bundle under MyEclipse to complete client development. On this basis, use the method designed in this paper to develop a mobile terminal music education platform based on the Django framework, integrate the education platform into the test environment according to the specifications, debug it, ensure that relevant work meets the technical specifications, and then introduce a mobile terminal music education platform based on cloud computing technology and a mobile terminal music education platform based on artificial intelligence response technology, as traditional platform 1 and traditional platform 2.

Stress test the music education platform of mobile terminal integrated in the test environment, simulate the number of virtual concurrent users, set the number of virtual users to 3000, 4000 and 5000, perform service access concurrently under different test conditions, view the stress test results of the music education platform under LoadRunner, and the test results are based on the platform response time. That is, the time from the front end submitting a data request to the service until the return result of the application system service can be received.

7.2 Analysis of Experimental Results

According to the above method, the experimental results are counted as the key indicators to test the application effect of the music education platform. The statistical experimental results are shown in Fig. 5 below.

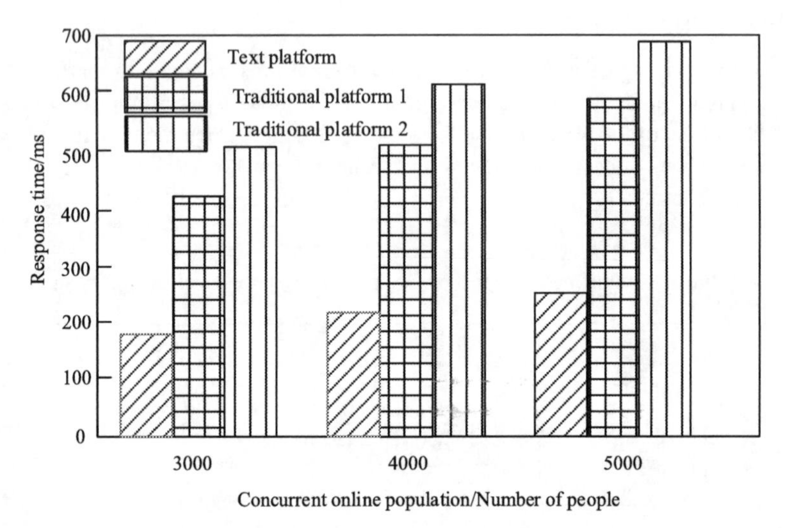

Fig. 5. Response duration of music education platform under different concurrent online population

According to the test, under the condition of 5000 concurrent users, the response time of the text platform in this paper does not exceed 3 s, which can meet the time requirements of the platform design. Compared with the platform in this paper, traditional platform 1 and traditional platform 2 have higher response times, and with the increase of the number of concurrent users, the platform response time continues to increase. Therefore, after completing the above experiment, we get the following conclusions: Compared with the traditional education platform, the mobile terminal music education platform based on the Django framework designed in this paper has a good application effect. This platform can pass the stress test in the comparison experiment. Under the condition of high concurrent users, the response time of this platform is still less than 3 s.

In order to better validate the application effect of the mobile terminal music education platform based on the Django framework, the error rate was used as the experimental indicator. The lower the error rate, the more effective the platform can handle requests, indicating the better stability of the platform. The error rate is the ratio of the number of errors that occur during the platform's request processing to the total number of requests. Set the number of operation requests to 100 and compare the error rates of different platforms, as shown in Table 7.

From the data in Table 7, it can be seen that the average error rates of Platform 1 and Platform 2 are 1. 32% and 1%, respectively, while the average error rate of this platform is only 0. 5%, indicating that the stability of this platform is good.

Table 7. Error rates for different platforms

Number of operation requests/time	Traditional platform 1/%	Traditional platform 2/%	Text platform/%
20	0	1	0
40	0	2	0
60	1.7	0	0
80	0	0	2.5
100	5	2	0

8 Conclusion

With the improvement of China's comprehensive national strength, the national government and families pay more and more attention to education. Education is a key factor that directly affects China's economic development and the healthy growth of the next generation of families. Therefore, more and more enterprises and individuals, whether in basic education or higher education, participate in the production of educational resources. At the same time, the corresponding educational resources are also increasingly rich. The enrichment of educational resources has improved people's learning efficiency, and people can find corresponding educational resources according to their own needs. However, the growing Internet resources make the content complex and muddied. In particular, people engaged in music education and learning will encounter many difficulties when searching and obtaining music education resources through the Internet. In order to solve this problem, this paper has completed this design by developing the music education resource database of mobile terminals, building the music teaching resource sharing service model, music teaching management and online classroom management. Design a mobile terminal music education platform architecture based on the Django framework, while establishing a complete education resource library to accelerate platform response speed. Build a music teaching resource sharing service model, increase music practice management and online classroom management, and improve platform functions. The design results have proved by comparative testing that it can improve the response speed of the platform, increase the stability of the platform, and provide users with a better and high-quality learning experience through this way, realize the overall optimization of quality education in China. In the future, with the development of the Django framework, the platform will be continuously improved to have better application effects.

References

1. Zafman, K.B., Riegel, M.L., Levine, L.D., Hamm, R.F.: An interactive childbirth education platform to improve pregnancy-related anxiety: A randomized trial. Am. J. Obst. Gynecol. **229**(1), 67.e1-67.e9 (2023). https://doi.org/10.1016/j.ajog.2023.04.007

2. Li, J., Wang, R.: Machine learning adoption in educational institutions: role of Internet of Things and digital educational platforms. Sustainability **15**(5), 4000 (2023). https://doi.org/10.3390/su15054000
3. Zhang, C., Li, X.: Construction of digital art education platform under the "Internet+" Environment. Mob. Inf. Syst. **19**, 1–13 (2023)
4. Doyle, A.A., Parameshwar, P.S., Tyson, N.: 60. @Gyn.Guide: A novel social media educational platform providing evidence-based and accessible information about gynecology. J. Ped. Adolesc. Gynecol. **36**(2), 198–199 (2023). https://doi.org/10.1016/j.jpag.2023.01.148
5. Wang, B., Qi, J., An, X., et al.: Research on four-dimensional innovative intelligent education platform based on cloud edge-end architecture. Comput. Intell. Neurosci. **2023**, 1–11 (2023). https://doi.org/10.1155/2023/2263033
6. Kaufmann, M.D., Steeb, T., Wessely, A., et al.: eImmunonkologie: development and launch of a virtual education platform for the immunotherapy of cutaneous neoplasms. Med. Sci. Educ. **33**(1) (2023)
7. Popescu, O., Leonte, N.: Development of Spatio-temporal orientation of children with down syndrome through educational platforms after Romanian pandemic lockdown. Sustainability **15**(2), 926 (2023). https://doi.org/10.3390/su15020926
8. Papadakis, S., Anastasaki, M., Gergianaki, I., et al.: Development and implementation of a continuing medical education program on non-alcoholic fatty liver disease for primary care practitioners in Europe. Front. Med. **10** (2023)
9. Nieto-Márquez, N.L., Baldominos, A., Soilán, M.I., et al.: Assessment of COVID-19's impact on EdTech: case study on an educational platform, architecture and teachers' experience. Educ. Sci. **12**(10), 681 (2022). https://doi.org/10.3390/educsci12100681
10. Shumilovskikh, L.S., Shumilovskikh, E.S., Schluetz, F., et al.: NPP-ID: Non-Pollen Palynomorph Image Database as a research and educational platform. Veg. Hist. Archaeobotany **31**(3), 323–328 (2022)
11. Liu, S., Gao, P., Li, Y., Weina, F., et al.: Multi-modal fusion network with complementarity and importance for emotion recognition. Inf. Sci. **619**, 679–694 (2023)
12. Zhang, Chao, Li, Huizi: Adoption of artificial intelligence along with gesture interactive robot in musical perception education based on deep learning method. Int. J. Hum. Robot. **19**(03) (2022). https://doi.org/10.1142/S0219843622400084
13. Zhang, Y.: Modern art design system based on the deep learning algorithm. J. Interconnect. Networks **22**(Supp05) (2022). https://doi.org/10.1142/S0219265921470149

Construction of Mobile Education Platform for Piano Tuning Course Based on LogicPro Software

Xiaojing Wu(✉)

Tianshui Normal University, Tianshui 741000, China
wuxiaojing2313@163.com

Abstract. In view of the problems of the current existing mobile education platform for piano tuning course in practical application, such as the poor practicability of the platform, the inability to ensure the smooth use of various functions, the long response time of the platform operation, and the inability to meet the requirements of users for high operating efficiency of the platform, LogicPro software was introduced to carry out the research on the construction of the mobile education platform for piano tuning course based on LogicPro software. First, the B/S architecture is taken as the core to design the overall structure of the platform. Secondly, complete the selection of the platform core switches, servers and other equipment, and realize the hardware design of the platform. Under the support of LogicPro software, optimize the upload of piano tuning teaching resources. Finally, through the application of mobile intelligent terminal equipment, the mobile time teaching of piano tuning course is realized. The experimental results prove that the new education platform can complete various tasks as expected in practical application, and has high practicability. At the same time, the platform has shorter operating response time, faster operating efficiency, and can achieve higher quality mobile education of piano tuning courses.

Keywords: Logicpro Software · Curriculum · Education Platform · Moving · Piano Tuning

1 Introduction

Piano tuning is an indispensable part of piano art. All piano players take a musical instrument with accurate melody and perfect sound as the premise of their artistic exertion. This is the basic principle that almost all people engaged in musical instrument playing know, and it is also the first step of their artistic pursuit. Behind the impressive sound of the piano, there is a hard discipline, namely the technology of tuning the sound of this big machine [1]. Therefore, the development and perfection of this subject is almost the premise of the development and perfection of piano art. This art was gradually established and developed since the reform and opening up in the 1980s. It was included in the curriculum of higher professional art schools, but it came too late.

L. Yun et al. (Eds.): ADHIP 2023, LNICST 549, pp. 283–297, 2024.
https://doi.org/10.1007/978-3-031-50549-2_20

"The modern piano was born in Europe in 1709." There are more than 220 strings and more than 8000 parts and keys inside the piano, which need the chord shaft, shaft plate and iron bone to tighten the strings. During the playing process, 88 sounds are made by hammer impact. Long term impact will cause the strings to loosen. As time goes by, the piano "runs out of tune" is an inevitable phenomenon. Not only that, the piano is mainly composed of wooden parts and metal materials, so it will be affected by temperature and humidity, resulting in inaccurate melody. Therefore, any piano in the world needs to be tuned regularly to ensure that the piano always keeps its best state. Modern piano has a history of more than 300 years. "Piano tuning has developed with the development of piano. As an independent technology and discipline, piano tuning has also existed for more than a century. It is documented that piano tuning started in Britain, and the British Piano Tuning Association was founded in 1913." Before the 19th century, the piano was generally owned by aristocrats. After the 19th century, the piano began to become popular, The increasing demand of ordinary families for piano has a direct impact on the demand for transfer lawyers.

Network teaching platform can enhance the openness of teaching space and enrich teaching resources to a certain extent. With its flexible teaching methods and convenient interactive teaching behavior, it has become one of the main ways of teaching application at present. Among them, the stability of the network teaching platform directly affects the overall teaching effect, and the real-time data transmission plays an important role in students' learning experience in class. Therefore, it is of great significance to ensure the stability of the long-term operation of the teaching platform and improve the real-time performance of data transmission.

Reference [2] established a teaching platform based on Hadoop cloud computing. Combining overall learning algorithms and statistical models. Calculate data scale. Hadoop cloud computing is used to integrate the Autocorrelation computing model to achieve efficient teaching platform research data processing. Simulate different operating conditions multiple times on the Hadoop platform to verify the accuracy of training capabilities. Utilize Hadoop based cloud computing to develop computing system performance and improve teaching platform research models. Reference [3] proposes an exploratory evaluation student learning system based on artificial intelligence, which enhances interactive learning experiences in nonlinear environments. Use concept mapping in Chatbot to improve students' learning ability. In addition, based on probability distribution analysis, students were validated through concept mapping, and their learning was evaluated using the system generated probability graph curve. However, the above methods have a slower response speed when there are many concurrent users.

LogicPro software is a music production software developed by Apple, which has the functions of composition, recording, editing, sound correction and mixing. The software has a large number of plug-ins and sound effects, and is a rich sound resource library. It is easy to input, store and change notes [4] with a mouse or an external MIDI keyboard. The input note signal can be used as a trigger to flexibly trigger various voices. Notes can be presented in multiple forms such as MIDI signal bar, staff, chord mark, etc.

The memory of piano tuning needs the cooperation of the player's hearing, vision and touch. Research shows that the human body can remember 15% of the content by hearing alone, 25% by vision alone, but 65% by combining hearing and vision. This software can fully mobilize the joint effect of multiple senses such as hearing, vision and touch, help students experience the actual sound effect of chord and melody matching, accurately understand and remember the structure of chord, and directly link the distance between chord sounds displayed on the computer screen with the distance between fingers when playing by themselves [5]. Through the synergy of multi-dimensional senses, improve the effectiveness of students' understanding and mastering knowledge points. Based on this, this paper will carry out the research on the construction of mobile education platform for piano tuning course based on LogicPro software.

2 Construction of Mobile Education Platform for Piano Tuning Course Based on LogicPro Software

2.1 Overall Structural Design of the Platform

Based on the above analysis, in terms of building the teaching platform, this paper creates a B/S architecture based on the above analysis. In terms of building the teaching platform, this paper creates a B/S architecture based piano tuning course teaching platform that uses the burst transmission mode to optimize the upload speed of teaching resources [6]. The platform covers information management, teaching management and user information management of piano tuning courses, and serves students to learn piano tuning skills from different aspects. The application of Logic Pro software can not only enhance the integration of harmony theory and actual tuning, but also optimize the teaching quality of piano tuning courses to the greatest extent while practicing listening, and provide sufficient conditions for students to effectively practice piano tuning.

The mobile education platform for the traditional piano tuning course has poor stability, and its teaching management function is not applicable. To solve this problem, this paper uses the B/S (Browser) Server, browser/server mode) structure to create a mobile teaching platform for piano tuning courses, which greatly improves the teaching level of piano tuning courses [7]. The platform covers three layers of architecture: presentation layer, business logic layer, and data access layer. The platform architecture is shown in Fig. 1.

The presentation layer is responsible for visual interface operation, allowing users to input their own information and transmit it to the background, outputting information requests, and realizing data sharing. It is the bridge between users and the platform.

The business logic layer is the key of the whole platform, which can realize the management function of the whole platform [8]. Encapsulate various function points to form atomic function points, so that the presentation layer can mobilize the information required by users. This layer includes three parts, namely, the information management part of the law regulation course, the teaching management part and the user information management part. Among them, the tuning course information management section can manage the piano tracks, such as adding new tuning course track information, deleting information, modifying information, querying specific information, playing tracks, etc. In the platform, only teachers have operation permissions for adding, deleting and modifying track information. Querying tracks is an operation permission item [9] that everyone using the platform has. Teaching management is an important part of the platform. After users click on the platform to enter the teaching management page, they

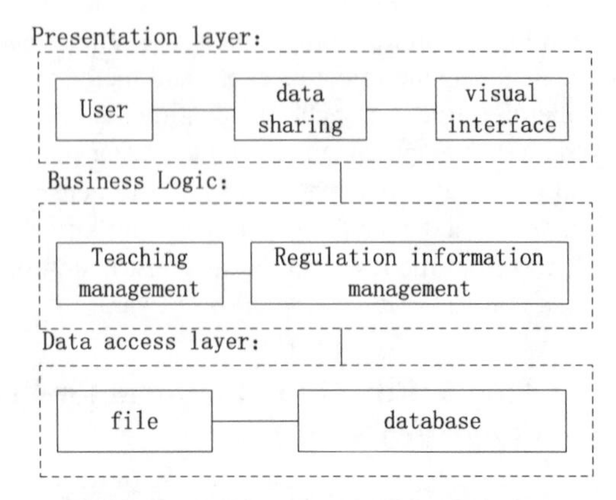

Fig. 1. Overall Structure Design of Platform

will conduct corresponding course management. Each sub module also covers different functions. The user information management section can uniformly manage the users in the platform and complete the platform operation and maintenance tasks. Only after the platform administrator's account is logged in, can you click from the main page of the platform to enter the user management page to complete user management, such as adding users, selecting designated users, modifying permissions, etc.

The data access layer is the entrance to access files and databases, which can be called by the business logic layer to complete the functions of querying, inserting, updating and deleting the piano tuning course resource data table.

2.2 Platform Hardware Design

The network construction of the mobile education platform for piano tuning course covers two aspects: the construction of the campus internal LAN and the construction of the WAN for teachers and students outside the campus. Figure 2 is the schematic diagram of the platform hardware connection structure.

The core switch selected in the platform is TP-LINK TL-SH5210PB 10 Gigabit uplink layer 3 network management PoE switch. The power consumption of the switch is 375W; RJ45 port is 8 * 2.5GE; Product size 440 × two hundred and twenty × 44 mm; The storage temperature is within the range of −40 °C–70 °C;T he working temperature is within the range of 0 °C–40 °C;The storage humidity is 5%–90% RH without condensation; The input power supply is 220VAC, 50 Hz [10]. TL-SH5210PB is a new 5-series 10 Gigabit uplink three-layer network management POE switch developed and launched by TP-LINK. It uses a new generation of high-performance hardware and software platform, provides flexible 2.5 G access and cost-effective 10 Gigabit uplink ports, and eight 2.5 G RJ45 ports support IEEE 802.3bt/at/af standard POE power supply. The maximum POE output power of the whole machine reaches 375 W. The device also supports three-layer routing protocol, complete security protection mechanism, perfect ACL/QoS policy and rich VLAN functions, which is easy to manage and maintain, and can meet the POE networking requirements of campus and enterprise campus [11]. TL-SH5210PB provides eight 100 M/1000 M/2.5 Gbps RJ45 ports and two 1 G/10 Gbps

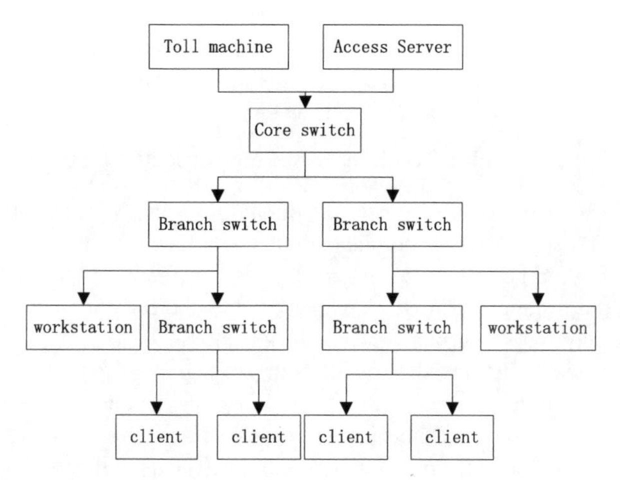

Fig. 2. Schematic diagram of platform hardware connection structure

SFP+ports. All ports support the wire speed forwarding function, which can easily meet the high-speed access requirements of terminal equipment such as WIFI 6 APs.

The SR868-460 type server is selected as the access server in the platform. The performance indicators of this type of access server are shown in Table 1.

Table 1. Access server performance indicators

S/N	performance index	numerical value
(1)	processor	Two or four second-generation Intel Xeon expandable processor platinum series, up to 205 W
(2)	Memory	Up to 6TB in 48 slots, using 128 GB DIMMs; 2666 MHz/2933 MHz (planned support) TruDDR4
(3)	Hard disk	Up to 24 2.5 inch storage bays supporting SAS/SATAHDD and SSDs, or 8 2.5 inch NVMe SSDs; And 2 images supporting startup M.2
(4)	network	Multiple options of 1 GbE, 10 GbE, 25 GbE, 32 GbE, 40 GbE or InfiniBand PCle adapters are provided;1 (2/4 ports) GbE or 10GbELOM card
(5)	Power Supply	2 hot pluggable/redundant: 750 W/1100 W/1600 W/2000AC 80 PLUS platinum power supply
(6)	RAID	Hardware RAID (16 ports at most), with flash memory cache; Up to 16 port HBAs

In the LAN, access to the mobile education platform for piano tuning courses can complete piano teaching, query and play teaching tracks, personnel management, etc. The online access platform for teachers and students on campus provides real-time homework upload, accompaniment download, online interaction, etc. At the same time, teachers and students outside the school can also use the network access platform to complete the same access operations as users inside the school.

2.3 Upload of Piano Tuning Teaching Resources Based on LogicPro Software

Logic Pro software can visually present the pitch, time value and strength of the notes through the MIDI signal bar in the piano shutter window, so that the three important musical elements can be visualized. After playing, the player can not only selectively replay the sound of the played music segment, but also clearly see the time value and strength of the played notes through the length and color of the MIDI signal bar [12]. The playing force is clearly distinguished by the multi-level gradient colors from green to red in the software, and is automatically quantified into numbers: the playing time value can be quantified to half a beat, quarter a beat, or even smaller time value units in the software through the enlarged grid in the piano shutter window. The quantification of strength and time value can help students fully understand the advantages and disadvantages of their own playing, and understand the gap between their own playing strength and time value and ideal playing effect, so as to effectively set short-term goals and improve practice efficiency. The synergy of touch, vision and hearing promotes the player's perception of musical elements such as chord structure, time value and strength.

Logic Pro software can input sound parts separately in separate tracks, and output or play one or more sound parts in combination. The sound source library provided with the software not only contains complete symphony, folk music, and electronic music timbres, but also can be implanted into other external sound sources to fully meet the user's auditory needs, achieve a richer auditory experience from the combination of musical instrument timbres, and intuitively show the sound effects when melodies appear independently or in combination. In teaching practice, the author found that when listening to the timbre of national instruments, students majoring in folk music have a higher recognition of single melody and chord than western orchestras; On the contrary, when listening to the performance of western orchestral instruments, students majoring in western music also have better recognition of single melody and chord than listening to folk music. Therefore, the author believes that in playing practice, students can set the music track as the type of instrument they are familiar with, for example, the students majoring in folk music set the melody track as bamboo flute, and the tuning course track as dulcimer. This rich and varied listening experience can not only enhance students' understanding and memory of the musical expression of chords, but also stimulate students' subjective initiative to try to create independently.

Hypothesis $x(t)$ is the transmission signal in the platform, $h(t)$ is the platform shock response function, $n(t)$ represents the Gaussian noise signal with an average value of 0 and a variance of 1. The received signal of the platform is recorded as:

$$y(t) = h(t)x(t) + n(t) \tag{1}$$

In the formula, $y(t)$ Indicates that the platform receives signals.

According to Shannon's theory, the platform channel capacity is:

$$C = \lg(1 + \rho v) \tag{2}$$

In the formula, v indicates the average signal power in the transmission period; ρ indicates the channel gain; C indicates the platform channel capacity. If the resource uploader can accurately obtain the quantization value of ρ, the platform will divide ρ into K quantization levels and define the quantity range as Ω_i. Set in sequence s_i, v_i is the quantitative range i effective transmission time and power, then the average expression of transmission rate is:

$$E(R) = \sum s_i \lg(1 + \rho v_i) \tag{3}$$

In the formula, $E(R)$ indicates the transmission rate.

The upload rate optimization of piano tuning course resources includes short-term power constraints and long-term power constraints. In the short-term power constraint, if it is in the period T, total power is P_c, power loss $P < P_c$, the constraint is transformed into:

$$s_i(v_i + \sigma) = P_c \forall i \in \{1, 2, \cdots, K\} \tag{4}$$

In the formula, σ represents the optimized upload rate of piano tuning course resources under short-term power constraints.

To get the best set ρ_i, calculated the Partial derivative of $E(R)$ to ρ_i, let the derivative be 0, and determine the best value $F(\rho_i + 1)$ of $F(\rho_i)$:

$$F(\rho_i + 1) = F(\rho_i) + \frac{s_i \lg(1 + \rho_i)}{\lg(1 + v_i \rho_i)} \tag{5}$$

Long term power constraints should be clear ρ_i and v_i. Ignoring the power condition, the long-term power constraints are transformed into two sub optimization problems. Hypothesis v_i in any known set, find a local optimal quantization boundary set ρ_i, through boundary set by water injection method ρ_i get the local optimal power distribution value of v_i. Under the premise of considering the calculated power, the alternative minimization method is used to clarify the power meter at the resource input and the calculation scheme under the constraint of the calculated power.

The power loss in the global resource transmission period is limited, and only the average power constraint is required to meet the conditions. If the power loss of a part of the quantization range is higher than the constrained power P_c, the following formula is obtained:

$$\sum_{i=1} (s_i P_i) \leq P_c \tag{6}$$

In the formula, P_i indicates assigned to quantification range of power Ω_i.

Use β All index sets representing the quantitative range of burst transmission, then the objective function is further obtained. set up λ And $\{\lambda_i\}$ As a nonnegative Lagrange multiplier Q The expression of the objective function is:

$$Q = \sum_{i \in \beta} s_i \lg(1 + v_i \rho_i) + \sum_{i \in \beta} \lg[1 + \rho_i (P_i - \sigma)] \tag{7}$$

Solve the partial derivative of the above formula in set β. IF $i \in \beta$, the transmission rate of piano tuning course resources meets the requirements. For the sudden transmission range in the teaching platform of piano tuning course, in order to eliminate the dual gap, the value P_i must be equal to 0 to prevent the problem of poor stability caused by the uneven distribution of power, so as to ensure the long-term stable operation of the teaching platform.

In the specific teaching practice, the following exercise steps can be used: in the piano tuning exercise, students can create two independent tracks, one for melody and the other for accompaniment, every time they finish improvisation with melody. Copy the melody MIDI signal bar in the first playing track to the new melody track, and then copy the accompaniment MIDI signal bar to the new tuning course track. When the melody and accompaniment are assigned to two different tracks, the timbre of the melody track can be arbitrarily changed to any other instrument, so that the timbre of the main melody voice part and the accompaniment voice part can be completely distinguished audibly, and the actual sound effect of the piano in the process of tuning can be felt. By changing the combination of musical instruments, a rich and varied auditory experience can be achieved. In the practice of melody free tuning, you can play the input melody part first and convert it into any suitable instrument timbre. While playing the melody part, play the melody free melody with both hands on the piano melody track. Through the simultaneous presentation of three voice parts, a more real-time, three-dimensional, realistic auditory perception can be achieved. Both of the above two methods have improved the drawbacks of boring and repetitive traditional exercises, increased students' interest in learning, and are beneficial to the improvement of practice effect. According to the law of sensory synergy, multiple senses work together on memory activities, which is far more effective than using a single sense alone. Logic Pro software can fully mobilize students' multiple sensory experiences, thus promoting the understanding and memory of knowledge, effectively making up for the shortcomings of the traditional teaching of theoretical harmony and keyboard harmony teaching that can not be effectively combined, and the theory and practice are disconnected, and achieving better teaching results than the traditional classroom repeated exercises using a single sense.

2.4 Mobile Education of Piano Tuning Course

The mobile intelligent terminal equipment is regarded as the necessary condition for realizing the mobile education of piano tuning course. Apply the platform software and programs designed above to the mobile intelligent terminal equipment [13, 14], and realize the connection between the mobile intelligent terminal equipment and the above platform hardware through WIFi, 4 G, 5 G and other communication networks. Learners can use mobile devices to log in to the platform anytime and anywhere for learning, which is of great significance for lifelong learning. Based on the operating characteristics of the mobile education platform for piano tuning courses and the needs of mobile learning, the following principles should be considered in the design of the mobile online education platform: first, friendly interface and simple operation; Second, good scalability and compatibility; Third, the main function of the mobile intelligent terminal device is to answer and make calls. The device application software should have the function of pausing, saving data or exiting when calling.

There are two development schemes for mobile front-end. The first is to use the browser as the sea client; The second is to develop client application software. The comparison between the two schemes is as follows: the first scheme can install the browser directly, but because the screen of the mobile terminal device is too small, the performance effect is affected and the interface jump is not very convenient. Moreover, due to the limited function of the browser, it is unable to provide various multimedia interactive learning capabilities in the online education platform. The second solution needs to develop front-end application software, but it can be customized to provide more complex interaction of various audio and video streams, so that learners can learn more easily and improve the learning effect.

According to the above analysis, the mobile education platform for piano tuning course designed in this paper uses B/C mode, the front-end, that is, the client, uses the way of developing application software, and the back-end uses the server. In order to make the system have good scalability and maintainability, the platform uses MVC mode to develop.

The client communicates with the server using HTTP protocol, and the data communication format is XML or JSON. The connection between the front end and the rear end uses the HttpURLConnection or HtoClient class. This connection is an important process of returning data from the server, and can obtain the inputStream byte stream object. The main advantage of HtoURLConnection is multi-threaded breakpoint upload and download, while the functions of HttpClient are more extensive. The server column intercepts the client's URL request and parses it, then obtains the data through HttpPost, generates XML or SON format data, returns it to the client using HttoResponse, converts it through nputStreamReader, and displays the returned data using BufferReader.

In order to ensure the smooth implementation of mobile education of piano tuning course, it is necessary to reserve sufficient educational resources and establish a mobile education database. First of all, we should establish a sub database of student education progress and each flow record to associate with identity information. The teacher sub database includes mobile phone number, alias, grade, student comments and remarks, including text real-time dialogue, related video text introduction, teaching time, teaching students (mobile phone number), teaching content coding, related videos, and current education evaluation.

The student sub database includes mobile phone number, alias, grade, teacher comments and notes, including real-time text dialogue, teaching time, teacher (mobile phone number), relevant education videos, and current education evaluation. The above two sub databases are used to record the content of one-on-one education information. The level content is refreshed according to the platform evaluation system calculated for each evaluation.

Establish the information database of each teacher's teaching content and schedule, which shall be edited by the teacher [15]. The next step is to select the time period and teaching content of the teacher. After the deposit is selected, the teacher's schedule can be selected by a student. Once the deposit is selected, other students can only see the unselected time period and select it. Once the students have chosen to pay the deposit, the teachers and students will enter a pair of continuous broadcast rooms directly according to the time agreement.

The mobile education of piano tuning course needs to build related technologies, including optical lens design, education server platform technology and terminal APP video service programming. Piano atlas preparation, photo file, put in the cloud disk shared directory, steel.

Piano track is to sort the file names and enter them into the piano knowledge base and sub base. If you omit the knowledge base and sub base, you can also directly call the file in the shared directory to call the music score file. Rent video cloud services from video service companies, such as establishing video connection services with Baidu, Tencent and Alibaba Cloud servers. The standard length of the piano lens is 1.5 m. Add a large field of view corner lens to the front lens of Ipad, and the lens can see 1.5–2 m. Put it on the piano spectrum shelf, you can see all the keys of the piano.

Write an app to transmit some videos of the piano key rotating in the video area in real time. According to the difference between the lens type and the display area, video processing needs to be performed again.

Write app programs to solve video rotation, up, down, left, right, video area selection function to adapt to the types of shots in various programs. Take the vertical placement of Ipad as an example. The upper (or lower) 1/5 area is used as the video area to display fingering, the lower 4/5 area (or upper) is used as the music score area, and the four corners of the video area can be marked with specific signs to indicate the direction status of the piano key video. Pairing practice requires pointing guidance in the same direction. When rotating the video, you should know whether your rotation is in the same direction or in the opposite direction.

According to the above settings, the teaching content of pre job regulation course can be more clearly displayed in the mobile intelligent terminal equipment, and the quality of mobile education can be improved.

3 Experimental Analysis

3.1 Experiment Preparation

In order to verify the practicability of the platform, the function and performance of the platform were simulated and tested. The stability of the designed platform is verified through different functional tests. Functional testing is to use test cases to verify the platform functions. In order to make the experimental results comparable, the education platform based on LogicPro software designed above in this paper is selected as the experimental group, the education platform based on education cloud is selected as the control group I, and the education platform based on big data is selected as the control group II. First, we tested the three management functions of the platform, and uploaded the attachment by selecting the tune exercise track; Modify a certain information in the tuning exercise track, and observe whether the modified information can be saved; Check whether the style can be deleted accurately; Check whether the music can be played quickly. The above four tests are used to test the information management function of modulation. On this basis, by judging whether the three education platforms can correctly find the required courseware; Whether to query the number of users effectively; Can you use keywords to browse jobs; Whether this function can be used to complete online problem solving and test the teaching management function of the course.

After testing the platform functions, use Loadlunner 9.5 to complete the performance simulation experiment for the running performance of the three education platforms, and use 300 concurrent virtual users to query at the same time to test the platform response time and throughput.

3.2 Result Analysis

Taking accompaniment information management and teaching management as an example, functional tests were carried out on two modules, and the results are shown in Tables 2 and 3.

Table 2. Test table of three platforms' regulation information management function

S/N	Test items	Results of experimental group	Results of control group I	Results of control group II
(1)	Uploading the information of tuning practice	adopt	adopt	Failed
(2)	Modify the tuning exercise information	adopt	Failed	adopt
(3)	Delete tuning practice information	adopt	Failed	Failed
(4)	Play the tune exercise track	adopt	adopt	adopt

Table 3. Test Table of Teaching Management Functions of Three Platform Regulating Courses

S/N	Test items	Results of experimental group	Results of control group I	Results of control group II
(1)	Courseware management	adopt	adopt	Failed
(2)	user management	adopt	adopt	adopt
(3)	job management	adopt	Failed	adopt
(4)	Online interaction	adopt	Failed	Failed

According to the results of various functional tests recorded in Tables 1 and 2, during the operation of the data function of the education platform based on LogicPro software proposed in this paper, the client has relatively complete verification actions. Without going through the server, the client can complete reliable information type calibration, prevent the input of dirty data, and ensure the quality of platform teaching resources, After inputting the complete piano improvisation accompaniment information, it can be quickly saved to the database, and the test results meet the expected goals. However, the other two education platforms fail to pass a test item in the application process, which cannot meet the actual needs of users.

According to the above experimental results, the data application of the education platform based on LogicPro software in this paper is more practical. On this basis, the operating performance of the three education platforms is compared with the results of the platform response time test. The response time test results of the three education platforms are shown in Figs. 3, 4 and 5.

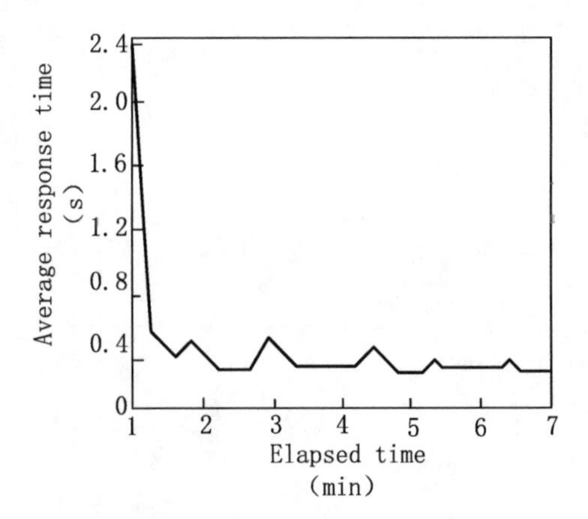

Fig. 3. Response time test results of the experimental group's education platform

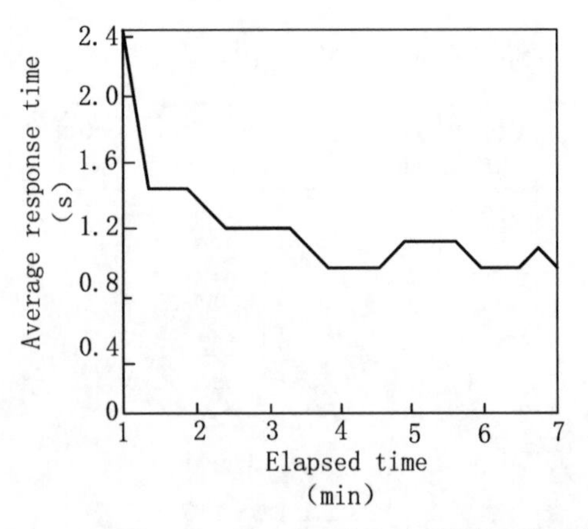

Fig. 4. Response Time Test Results of Group I Education Platform

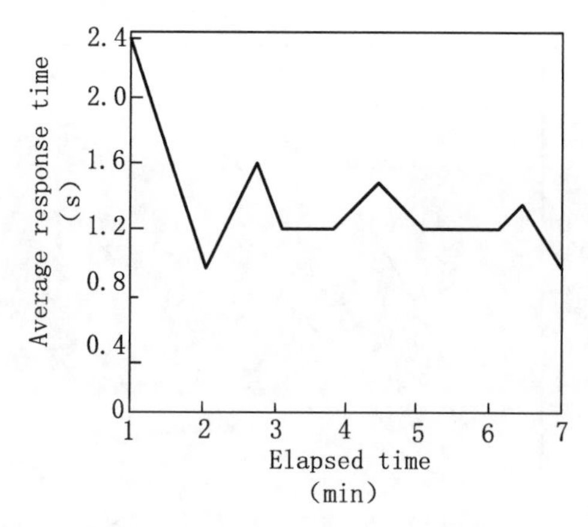

Fig. 5. Response Time Test Results of Group II Education Platform

By comparing the response time of the above three education platforms in the application process, it can be concluded that the average response time of the education platform platform in the experimental group is 0.384 s, the maximum response time is 2.398 s, and the minimum response time is 0.312 s under the condition of 300 concurrent virtual users. In general, when 300 concurrent users access, the average response time can meet the design goal within 1s, while the platform in this paper can meet the operational performance requirements within 1s, indicating the superiority of the actual application of the platform. The maximum response time of the other two education platforms is close to that of the education platform in this paper, but their minimum response time cannot reach the level of the experimental group, and the average response time cannot be controlled within 1s, which cannot meet the user's access requirements. Therefore, the above experimental results can prove that the education platform based on LogicPro software proposed in this paper has stronger practicability in practical applications, and the platform has good operational response performance, which can fully meet the needs of platform users.

Comparing the throughput of different platforms, the higher the throughput, the better the stability of the platform. The throughput comparison of different platforms is shown in Fig. 6.

From Fig. 6, it can be seen that Control Group I reached its maximum throughput of 135 bit/s in 7 s, while Control Group II reached its maximum throughput of 140 bit/s in 6.5 s, both of which showed a significant decrease. The maximum throughput of the experimental group is 149 bit/s, which is greater than the other two platforms. After reaching the maximum throughput, the throughput of the experimental group remains at a certain level, indicating that the processing ability of the experimental group is strong.

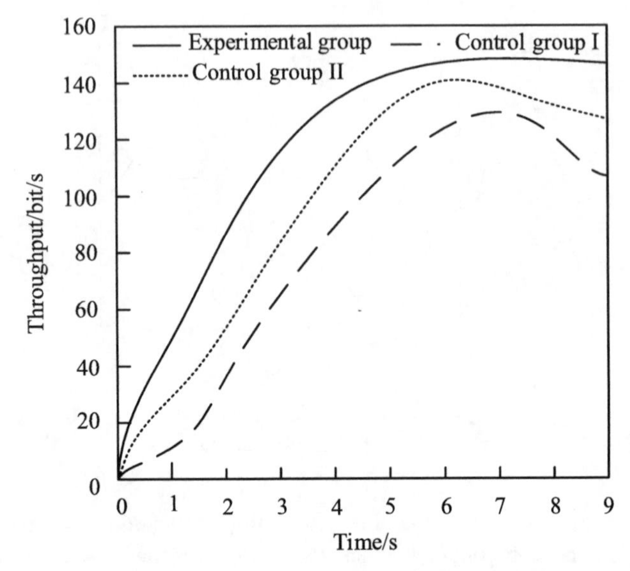

Fig. 6. Comparison of Throughput on Different Platforms

4 Conclusion

In order to improve the stability of the mobile education platform for piano tuning course and improve the teaching quality of the network platform, a mobile education platform for piano tuning course based on Logic Pro software was designed. The browser/server mode structure is used to build the teaching platform structure system architecture and network topology structure, to improve the stability of the teaching platform. Logic Pro software is used, combined with Shannon theory, through optimizing channel transmission, balancing power distribution, to create high-quality piano tuning exercises and teaching resources, so as to establish a good teaching platform for students and provide reference for the long-term stable operation of the teaching platform. Connect the above platform programs to mobile intelligent terminal devices using various communication networks, thereby achieving the construction of a mobile education platform for piano tuning courses based on LogicPro software. Through experiments, it has been proven that the mobile education platform for piano tuning courses based on LogicPro software can achieve its course teaching management function and has good stability. At the same time, the platform has good response performance and strong processing power.

Acknowledgement. Natural Science Fund Project in Shandong Province (2013ZRE27312).

References

1. Xie, T.: Design of automatic education classification management system in cognitive web services platforms using machine learning techniques. Int. J. e-Collabor. **19**(2), 1–19 (2023). https://doi.org/10.4018/ijec.316659

2. Liu, C.: Application of hadoop-based cloud computing in teaching platform research. J. Interconnect. Networks **22**(Supp05), 2147011 (2022)
3. Liu, L., Subbareddy, R., Raghavendra, C.G.: AI intelligence chatbot to improve students learning in the higher education platform. J. Interconnect. Networks **22**(Supp02), 2143032 (2022)
4. Qi, Y.: Retraction notice to "The role of mobile web platforms in the development of critical, strategic and lateral thinking skills of students in distance physical education courses" [Thinking Skills and Creativity Volume 42, December 2021, 100935]. Thinking Skills Creat. **46** (2022)
5. Silvera-Tawil, D., Bruck, S., Xiao, Y., et al.: Socially-assistive robots to support learning in students on the autism spectrum: investigating educator perspectives and a pilot trial of a mobile platform to remove barriers to implementation. Sensors **22**(16), 6125 (2022). https://doi.org/10.3390/s22166125
6. Ben, H., Wang, L., Ma, Z., et al.: Design of college physical education curriculum based on mobile app development platform. Wireless Commun. Mobile Comput. **2022**, 1–10 (2022). https://doi.org/10.1155/2022/9746549
7. Peng, Y., Zeng, Y.: A mobile teaching method of ideological and political education in colleges and universities based on android platform. Math. Probl. Eng. **2022**, 1–8 (2022). https://doi.org/10.1155/2022/9948451
8. Han, Y., Liao, J.: Design of feasibility analysis platform for college physical education based on mobile social network. J. Comput. Meth. Sci. Eng. **22**(2), 483–493 (2022). https://doi.org/10.3233/JCM-215767
9. Rácz, M., Noboa, E., Détár, B., Nemes, Á., Galambos, P., et al.: PlatypOUs—a mobile robot platform and demonstration tool supporting STEM education. Sensors **22**(6), 2284 (2022). https://doi.org/10.3390/s22062284
10. Zhang, Lizhe, He, Juan: Optimization of ideological and political education under the epidemic via mobile learning auxiliary platform in the era of digitization. Wireless Commun. Mobile Comput. **2022**, 1–9 (2022). https://doi.org/10.1155/2022/6149995
11. Wang, ., Jia, X., Cui, H., et al.: An interactive practice platform of English mobile teaching in colleges and universities based on open API. Int. J. Contin. Eng. Educ. Life-Long Learn. **32**(4), 418 (2022). https://doi.org/10.1504/IJCEELL.2022.124968
12. Leon-Paredes, G.A., Bravo-Quezada, O.G., Sacoto-Cabrera, E.J., Calle-Siavichay, W.F., Jimenez-Gonzalez, L.L., Aguirre-Benalcazar, J.: Virtual reality platform for sustainable road education among users of urban mobility in Cuenca, Ecuador. Int. J. Adv. Comput. Sci. Appl. **13**(6) (2022). https://doi.org/10.14569/IJACSA.2022.01306106
13. Wang, C.: Acute teaching method of college physical skills based on mobile intelligent terminal. J. Interconnect. Networks **22**(Supp05), 2147007 (2022)
14. Cheng, X., Fan, Y.: Research and design of intelligent speech equipment in smart english language lab based on Internet of Things technology. Procedia Comput. Sci. **198**, 505–511 (2022)
15. Liu, S., Xiyu, X., Zhang, Y., et al.: A reliable sample selection strategy for weakly-supervised visual tracking. IEEE Trans. Reliab. **72**(1), 15–26 (2023)

Temperature Control Technology in Heating Room Based on Multi-channel Temperature Signal Denoising

Li Liu[1,2], Riheng Chen[1,2], Jintian Yin[1,2](✉), and Qunfeng Zhu[1,2]

[1] Hunan Provincial Key Laboratory of Grids Operation and Control on Multi-Power Sources Area, Shaoyang University, Shaoyang 422000, China
yinjintian112@yeah.net

[2] School of Electrical Engineering, Shaoyang University, Shaoyang 422000, China

Abstract. In order to improve the accuracy of temperature control in the heating room and make the indoor temperature meet the comfort requirements of people for the living environment, the theory of multi-channel temperature signal denoising is introduced and the research on the technology of temperature control in the heating room based on multi-channel temperature signal denoising is carried out. First, the heating heat load is classified into three levels, namely, first level, second level and third level. On this basis, a heating indoor temperature model is built. Secondly, the pretreatment circuit design is completed to realize the noise removal of multi-channel temperature signals and improve the transmission quality of multi-channel temperature signals. Thirdly, the PID parameters are adjusted, and the temperature in the heating room is PID controlled. The fuzzy PID theory is introduced to design the fuzzy PID control. Finally, introducing fuzzy PID algorithm to make the steady-state error and maximum deviation of the heating temperature control system, the prediction and compensation control of the heating room temperature is carried out. The experimental analysis results show that the introduction of multi-channel temperature signal control technology can achieve high-precision control of indoor temperature in heating in practical application, so that the indoor temperature change can fully meet the comfort requirements of residents for indoor heating.

Keywords: Multi-Channel Temperature Signal Denoising · Indoor · Control · Temperature · Heating

1 Introduction

Heating can solve the basic living problems of people in cold regions in winter, and people's urgent demand for heating increases with the improvement of living standards [1]. With the proposal of national energy conservation and emission reduction, green development concept and the improvement of residents' heating demand, the traditional coal-fired boiler heating mode gradually withdrew from the heating stage due to environmental pollution, low efficiency and rising prices, and the heating modes such as

L. Yun et al. (Eds.): ADHIP 2023, LNICST 549, pp. 298–312, 2024.
https://doi.org/10.1007/978-3-031-50549-2_21

electricity instead of coal and electricity instead of oil have been widely used [2]. Electric energy has the characteristics of no noise, no waste gas, the most environmentally friendly and clean. In addition, the current situation of the country is that there is an excess of electricity, while water resources are scarce. In the future, the national power supply and demand will be unbalanced, and the state of serious oversupply will always be. The promotion of heating can increase the power load, which is not only conducive to solving the consumption of surplus electricity, but also can improve air quality, Reducing the consumption of water resources is of great significance for solving the current situation of national power surplus and long-term energy conservation and environmental protection [3].

Therefore, heating is the first choice of people in today's rising prices of other energy sources. The most important parameter in the process of heating is temperature. Room temperature comfort has always been one of the standards to measure the quality of heating [4]. Due to the different use nature of rooms, the required temperature will be different. If the unified constant temperature control strategy is adopted, the unity of heat supply and heat demand is not achieved, and a large amount of electric energy [5] is wasted. It is necessary to provide heating according to needs, change the disadvantages of unbalanced supply and demand in the past, and reduce energy consumption. Make heating more intelligent and humanized.

With the continuous increase of heating scale, the problem of dynamic imbalance in the heating process increases [6]. During static stable heating, heating will stop when the temperature reaches the set upper limit, and will be put into operation when it is lower than the lower limit of heating temperature. Because the thermal insulation of each room is affected by external factors such as height and orientation, all loads are not switched at the same time, which is easy to cause dynamic three-phase power imbalance. It will increase the loss of lines and transformers, affect the safe operation of electrical equipment, and reduce the heating quality in heating. Therefore, it is very important to improve the indoor temperature control performance.

Since the 1970s, Europe has implemented heating metering and charging, which attaches great importance to the energy saving measures of the Hydronics and users, especially in room temperature control, heat metering, and hydraulic balance regulation technology of the heating network. Compared with foreign countries, China's Hydronics lacks the use of heat metering devices and room temperature regulation methods, so it is difficult for users to achieve real-time regulation of heat consumption. Reference [7] proposed a mixed integer Linear programming (MILP) for short-term optimization of network temperature in 5GDHC system. This model includes air source heat pumps, compression coolers, and heat storage devices in central power generation units, as well as heat pumps, coolers, electric boilers, and heat storage devices in buildings. In addition, the model also considers the thermal inertia of the water masses in the network as additional heat storage devices. Reference [8] proposes the application of the variable universe idea in the temperature control system of variable air volume air conditioning. Based on traditional fuzzy PID control, the variable universe idea is introduced to design a controller with variable universe fuzzy PID algorithm. Finally, the fuzzy PID control algorithm and variable universe fuzzy PID control algorithm are respectively simulated to control the temperature of variable air volume air conditioning rooms Although different methods have been used in current research to achieve temperature control, in heating systems, temperature sensors may be subject to various noise interferences, such as electromagnetic interference and sensor noise. These noises can have an impact on temperature measurement results, leading to inaccurate indoor temperature control. In order to reduce or even avoid the waste of heat during heating, this paper will carry out the research on indoor temperature control technology based on multi-channel temperature signal denoising. In the research process, based on the establishment of heating load classification and indoor temperature model, a multi-channel temperature signal denoising preprocessing circuit was designed. Design a fuzzy PID controller for heating system temperature from two aspects: PID parameter tuning and indoor temperature PID control. Finally, achieve indoor temperature control through temperature estimation and compensation control in the heating room.

2 Temperature Control Technology in Heating Room Based on Multi-channel Temperature Signal Denoising

2.1 Classification of Heating Load and Establishment of Heating Indoor Temperature Model

It is assumed that the loads at all levels are evenly distributed on the three phases, that is, the three phase loads themselves are statically balanced [9]. In order to simplify the analysis, it is assumed that the number of loads at each level is the same, and the factory power of each load is different. The load temperature range at each level is set according to the thermal load level, and the temperature range is gradually reduced. Table 1 shows the classification of heating load.

Table 1. Classification of Heating Load

S/N	Load level	temperature range
(1)	Level I load	[22,25]°C
(2)	Secondary load	[18,22]°C
(3)	Level III load	[5, 10]°C

The establishment of the temperature model depends on the heat production of the heater and the heat dissipation and storage capacity of the room. Heat generation of electric heater W equal to the storage capacity of the room W_1 heat dissipation with the room W_2 and. Assuming that the room temperature distribution is uniform, the model can be expressed by the following formula:

$$W = W_1 + W_2 \tag{1}$$

In the formula, W_1 it can be calculated by the following formula:

$$W_1 = C_d T_{\text{int}}/dt \tag{2}$$

In the formula, C_d represents the specific heat fusion in air; T_{int} indicates the indoor temperature; t indicates the heating time. In the formula, W_2 it can be calculated by the following formula:

$$W_2 = (T_{\text{int}} - T_{out})/R = (T_{\text{int}} - T_{out})A\eta/\varepsilon \tag{3}$$

In the formula, T_{out} indicates the ambient temperature; R represents equivalent thermal resistance; A represents the wall area; η represents the heat transfer coefficient of the wall; ε indicates the thickness of the wall. Ignoring the influence of outdoor ambient temperature, the above formula is substituted into the formula above, and the following formula can be obtained by Laplace transformation:

$$W = (C * s * \varepsilon + A\eta)T_{\text{int}}(s)/\varepsilon \tag{4}$$

As the temperature change is accompanied by nonlinear and delayed phenomena, heat generation W and control quantity $U(s)$ is proportional, that is $W = KU(s)$. For delay time τ and the following relationship exists:

$$Y(s) = T_{\text{int}}(s) \tag{5}$$

Substitute the above formula to get:

$$G(s) = \frac{Ke^{-\tau s}}{Ts + 1} = \frac{0.6e^{-330s}}{1850s + 1} \tag{6}$$

In the formula, T represents the time constant; K represents static gain. In combination with the above formula calculation, formula (6) is taken as the transfer function of indoor temperature control for heating, and the subsequent design of indoor temperature control for heating is carried out on this basis.

2.2 Design of Preprocessing Circuit Based on Multi-channel Temperature Signal Denoising

After the classification of heating load and the establishment of indoor temperature model, the pretreatment circuit [10] is designed to denoise the multi-channel temperature signals. The preprocessing circuit mainly consists of input circuit, amplification circuit, multi-channel output modulation circuit, reference terminal (cold terminal) signal processing circuit, reference voltage generation circuit and manual test circuit. Its principle block diagram is shown in Fig. 1.

As mentioned above, the preprocessing circuit mainly consists of input circuit, amplification circuit, multi-channel output modulation circuit, reference terminal signal processing circuit, etc. The composition and working principle of the main circuit are briefly introduced below.

The input circuit is composed of 32 input unit circuits, and each input channel is equipped with an input unit circuit. The input unit circuit is specially designed for collecting and processing the thermal resistance signal. It is a bridge composed of precision resistors. The resistors on the three arms of the bridge are R1, R2 and R3 (where R1

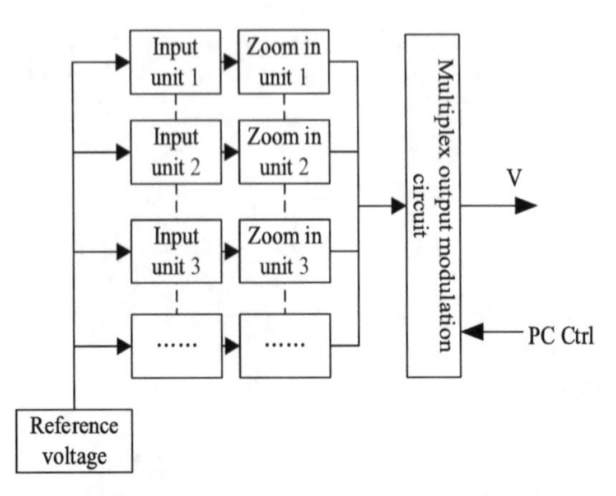

Fig. 1. Preprocessing circuit based on multi-channel temperature signal denoising

= R2 = 10 k Ω, R3 = 100 Ω, and the resistor on the other arm is thermistor Rt.R1 and R2 are respectively on the two upper arms of the bridge road, and Rt and R3 are respectively on the two lower arms of the bridge road. Voltage source of bridge circuit E = +10 V. Platinum thermistor Pt100 is used as thermistor, and its resistance value is R = 100.00 Ω at 0C. When the temperature is 2 °C, the output voltage of the bridge is 0 V. In practical applications, the temperature measurement object is often far away from the test circuit, that is, the temperature measurement object is separated from the bridge, which is to compensate the lead.

We adopt the three wire input connection mode. Since the lead is long on the side of the temperature measurement object, the R3 lead is also extended to one end of the temperature measurement object, connected with Rt to form a common point, and then from the common point lead back to the ground terminal of the test circuit, so that the reciprocating lead resistance is the same, and complete compensation of the lead resistance is obtained. As mentioned before, the input unit circuit is specially designed for collecting and processing the thermal resistance signal. When used to process thermocouple signals, the input unit circuit is bypassed. In this case, the two-wire input mode is adopted.

The amplification circuit is composed of 32 independent amplification unit circuits. Each input channel is equipped with an amplification unit circuit. The unit circuit is composed of ultra-low bias voltage VOS and ultra-low bias voltage drift operational amplifier ADOP07.The amplification unit circuit is a single-stage in-phase amplification circuit with a gain of ×10, ×100 Two gears. It can be selected and set through the gain adjustment jumper. The gain integration nonlinearity of the amplifier unit is better than 1.1%, and the long-term working stability (continuous 10 h of working test) is better than 0.18%. The gain of each amplifier unit has good consistency. The amplification unit has two modes of single end input and differential input, which can be selected by selecting input mode and setting jumper. The output of each amplification unit circuit is connected to a multi-channel output modulation circuit.

It is known from the working principle of the thermocouple that the thermoelectric potential of the thermocouple E_{AB} the size of is not only related to the temperature of the measuring end, but also related to the temperature of the reference end, namely:

$$E_{AB}(t, t_o) = f(t) - f(t_0) \tag{7}$$

In the formula, $E_{AB}(t, t_o)$ express t_o reach t thermocouple thermal potential at time; $f(t), f(t_0)$ represents a function. When using thermocouples, attention must be paid to the $E_{AB} - T$ (Thermoelectric potential and temperature) relationship curve or the thermoelectric potential value listed in its graduation table are given when the reference end is 0 °C (that is, t0 $=$ 0 °C).Therefore, in practical application, if the reference end is not 0 °C (that is, t0 $\neq$ 0 °C), it needs to be corrected, which is called cold end compensation. The correction can be carried out according to the following formula:

$$E_{AB}(t, 0) = E_{AB}(t, t_o) + E_{AB}(t_o, 0) \tag{8}$$

In the formula, $E_{AB}(t, 0)$ it indicates that the temperature at the measuring end of the thermocouple is t, the thermoelectric potential at the reference end at 0 °C; $E_{AB}(t, t_o)$ is the temperature at the measuring end of the thermocouple t, reference end is t_o the thermoelectric potential of, that is, the thermoelectric potential value measured by thermocouples; $E_{AB}(t, 0)$ it indicates that the reference terminal temperature of the thermocouple is t_o the correction value that should be added can be found from the scale of the thermocouple used.

According to this principle, the actual temperature value at the measuring end of the thermocouple, that is, the accurate value of the temperature, is $E_{AB}(t, 0)$ the corresponding value. Therefore, when using thermocouples to measure temperature, it is necessary to know the temperature of the reference end. Therefore, the temperature at the reference end must be measured at any time to correct the temperature measured by the thermocouple. We specially designed this circuit to collect and process the temperature signal at the reference end. This circuit is also composed of integrated operational amplifier ADOP07, whose circuit form is an inverse summation amplifier. There are two input branches, one from the static bias current generation circuit to generate a +273 μA's static bias current, the other is from the AD590 temperature sensor. The sensitivity of the device is 1 μA/K linear output current, that is, it will generate 1 μA Current. According to our design, when the temperature rises by 1 °C, it will provide −1 for the summation circuit μA current. Output voltage of reference terminal temperature signal processing circuit:

$$V_t = K(T - T_0) \tag{9}$$

In the formula, K represents the conversion coefficient; T represents the actual temperature at the reference end; T_0 represents the reference temperature, generally taken as $T = 0$ °C. After the circuit is tuned and calibrated, ensure that V t = 10 mV/K. The temperature sensor can be built-in (that is, installed on the circuit board and placed inside the chassis) or external (that is, placed outside the chassis). The machine is external and enters the circuit through the 32nd channel input connector of the instrument. The 32nd channel is designed as a special channel, which can be used as both thermocouple/thermal resistance signal processing channel and reference terminal signal processing channel, and can be preset through jumper.

2.3 PID Parameter Setting and PID Control of Heating Room Temperature

PID control is a classical technology widely used in the field of production control. Because the structure is simple and easy to understand, and the control effect is relatively good, it is widely used in industry. The core idea of PID control is to adjust the PID parameters according to the difference between the set value and the actual value of the controlled object. As long as the PID gain is reasonably adjusted to make the closed-loop control stable, the control goal can be achieved. Figure 2 is the block diagram of PID control principle of temperature in heating room.

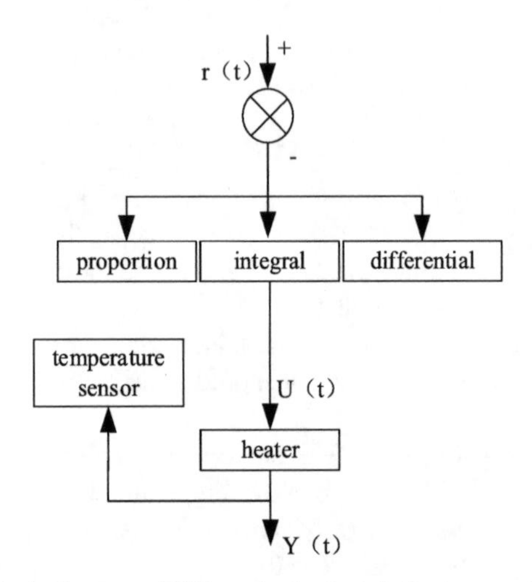

Fig. 2. Structure block diagram of PID control principle for temperature in heating room

When there is deviation in heating temperature control, the proportional link will reduce the temperature deviation in the form of proportion. The speed of adjustment is closely related to the size of the proportional coefficient. Of course, the proportional coefficient should also be properly selected. Too large or too small will have a certain impact on the stability of the control technology. The integration link can restrain the static error to a certain extent, and the effect of integration will vary with the size of the time constant. The larger the time constant is, the worse the effect will be. The smaller the time constant is, the opposite will happen. However, there are both advantages and disadvantages. When the time constant is larger, the overshoot decreases, and the stability effect is better. When the time constant is smaller, oscillation will occur. The differential link can reduce the time required for the system to reach stability. The magnitude of differential time constant will also affect the change of overshoot. Contrary to the integral part, the overshoot of the system will increase with the increase of the differential time. The smaller the differential time, the greater the time for the temperature to reach stability. Therefore, it is very important to choose an appropriate differential time constant.

As a determined system, the indoor temperature control system usually uses the rising curve method to measure the rising curve of the heating temperature control, and then obtain the parameters of the indoor temperature mathematical model, as shown in Fig. 3.

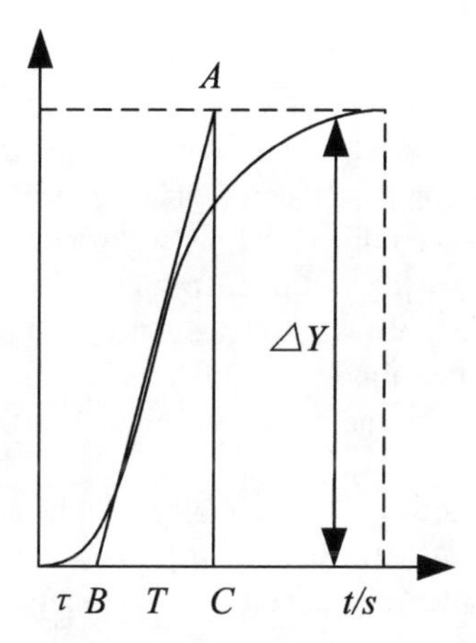

Fig. 3. Temperature Rise Curve in Heating Room

Make a tangent at the point where the temperature changes rapidly, and one end of the tangent is relative to the B point on the asymptote of the other end relative to the steady value A points, A the points mapped to the timeline are C point, origin to B the time period of the point is the lag time τ, for time constant T BC segment representation. In the measurement of the rise curve of the heating temperature control system, after adding the step signal, the heating temperature output will change correspondingly, and finally there will be a stable value, so the temperature control transfer function obtained according to this method is $\tau = 330$, $T = 1850$. This parameter is taken into the above heating load classification and heating room temperature model formula to obtain the temperature control transfer function $G(s)$:

$$G(s) = \frac{Ke^{-\tau s}}{Ts + 1} = \frac{0.6e^{-330s}}{1850s + 1} \tag{10}$$

The principle of PID is simple and easy to understand, and has strong operability. It has its own control mode and has been popularized in the engineering field. The scope of application is relatively wide and practical. As long as the three parameters of PID are set at appropriate values, the control system will be in a closed loop stable state, with small operating fluctuations, which is conducive to better completion of the work. PID control has strong robustness and does not depend on the traditional model. Especially when PID control is applied in such a complex environment as engineering, the parameters cannot be accurately known due to various factors, and the dynamic model is not easy to determine. In this case, the environment is usually considered first, and then the parameters are debugged with reference to the experience and knowledge of technicians. PID can play a good control role when other methods are not applicable.

2.4 Fuzzy PID Control Design of Temperature in Heating Room

In the actual control process, different control requirements need to be formulated for different working conditions, which requires the PDD controller to adjust its parameters accordingly with the change of working conditions, so as to reduce the steady-state error and meet the control requirements. However, this adaptive parameter adjustment strategy is difficult to be realized on the conventional PD controller for control systems that lack accurate identification models at present. The fuzzy controller can work out the corresponding fuzzy rule set according to human knowledge and experience, and realize the control of the control system that is difficult to be represented by mathematical model. Therefore, combining PID control with fuzzy control, a fuzzy PID control algorithm is proposed, which completes the online tuning of the three parameters of the PID controller through the process of fuzzification, fuzzy reasoning, and defuzzification, so that it can meet the corresponding control requirements. In order to further improve the control accuracy of indoor temperature control technology for heating, fuzzy control rules are introduced. The generation of fuzzy control rules needs to comprehensively consider whether they are cross, complete and consistent. The intersection here refers to whether rules are related and interact with each other; The sign that rules have good integrity is that each given input condition has a corresponding rule to execute; The rule consistency means that the conditions between control rules can be the same, but the conclusions cannot be very different, otherwise there will be contradictory situations. In practical application, it is necessary to avoid the occurrence of conflict control rules. Taking the deviation and deviation change rate of heating temperature as two input variables of fuzzy control rules, improving fuzzy control rules is to arrange the number of control rules reasonably and give them appropriate confidence. And optimizing the number of control rules is to ensure that there are neither too many nor too few rules, because when there are too many rules, the controller's processing becomes complex and time-consuming. When there are fewer rules, the controller's output may be incorrect or there may be no corresponding output due to insufficient consideration; The quality form of control rules is also very important, mainly to identify whether the conditions of control rules are reasonable and whether the rules contradict each other.

On the premise of obtaining the membership degree and membership degree value table of fuzzy control PID input and output, as well as relevant fuzzy rules, the parameter fuzzy adjustment matrix table can be obtained through fuzzy reasoning. In order to facilitate the use of the matrix table, the matrix table can be stored in the computer program memory. Fuzzy PID can detect its output through the computer at any time, and can also calculate the two inputs of fuzzy PID control in real time. Then the two inputs are fuzzed. Since the fuzzy matrix table has been stored in the computer, the PID parameter adjustment can be obtained according to the table.

The parameter tuning is mainly to consider the speed of response, the accuracy value under steady state, the size of overshoot, and whether the system control is stable. The specific rules of tuning need to be combined with the system output response curve. The tuning of the three parameters of PID needs to meet the requirements of the input of fuzzy PID control on the control parameters in different situations. Through the fuzzy PID parameter tuning principle, the corresponding regulation control rules of the three PID parameters can be summarized. Therefore, in this chapter, the deviation and deviation change rate of the heating temperature are used as the input of the fuzzy PID heating temperature control, and the output at this time is the adjustment of the three parameters of the proportional integral differential control. The relevant fuzzy subset based on fuzzy PID heating temperature control is expressed by these seven language variables {NB, NM, NS, ZO, PS, PM, PB}. The fuzzy universe of temperature error e and error change rate ec following the error change are {−2, 2}, {−1, 1}, respectively. Table 2 is the fuzzy rule table.

Table 2. Fuzzy rule table

	NB	NM	NS	ZO	PS	PM	PB
NB	PB	PB	PM	PM	PS	ZO	ZO
NM	PB	PB	PM	PS	PS	ZO	NS
NS	PM	PM	PM	PS	ZO	NS	NS
ZO	PM	PM	PS	ZO	NS	NM	NM
PS	PS	PS	ZO	NS	NS	NM	NM
PM	PS	ZO	NS	NM	NM	NM	NB
PB	ZO	ZO	NM	NM	NM	NB	NB

The contents in Table 2 are used as the basis for fuzzy PID control of the temperature in the heating room. Fuzzy reasoning is also called approximate reasoning. It is mainly based on the corresponding fuzzy rules to transform the relevant fuzzy variables from input to output, and can also be seen as a relationship function of fuzzy variables. There are many types of fuzzy reasoning methods, and different methods will lead to different conclusions. Of course, the length of reasoning will also be different. The most commonly used methods in application are Mamdani method, Zadeh method, Baldwin method, etc. Usually, a simple, convenient and effective reasoning method will be used, which can reduce the reasoning time.

In heating fuzzy PID temperature control, the temperature controllers designed for each load level are embedded into the simulink. The controller part of heating fuzzy PID temperature control uses the fuzzy logic toolbox Fuzzy Logic Toolbox to create a Mamdani type fuzzy inference controller, and uses the maximum minimum criterion (Max Min) as the approximate reasoning method of heating temperature control.

2.5 Temperature Prediction and Compensation Control in Heating Room

Although the introduction of fuzzy PID algorithm has reduced the steady-state error and maximum deviation of the heating temperature control system, the hysteresis characteristics of the temperature system have not changed significantly. Therefore, it is necessary to optimize the fuzzy PID. In this chapter, Smith fuzzy PID controller is obtained by combining Smith predictor on the basis of fuzzy PID. This controller combines the excellent characteristics of fuzzy PID and the compensation function of Smith predictor, which can improve the delay of the system very well. The main idea of Smith predictive compensation control is to calculate the exact transfer function.

A matched compensator is added to weaken or even eliminate the hysteresis effect in the feedback path. Smith compensation can also be called pure lag compensation control. Smith estimation compensation principle block diagram is shown in Fig. 4 below.

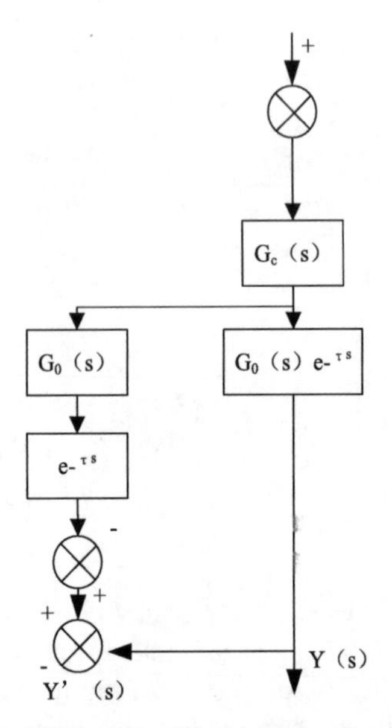

Fig. 4. Schematic Diagram of Smith Estimation Compensation

The transfer function of the controlled object is expressed as follows:

$$G(s) = G_0(s)e^{-\tau s} \tag{11}$$

In the formula, $G(s)$ represents the transfer function of the controlled object; G_0 represents an item without hysteresis. Add a compensator to eliminate the lag term $G_s(s)$ the compensator should be connected with the object in parallel, and can also be regarded as a new compensation object, and the controller should be designed according to the normal steps $G_c(s)$. In this article $G_c(s)$ it is a heating fuzzy PID controller, so the equivalent transfer function of Smith predictive compensation new compensation object can be expressed as:

$$G_r(s) = G_0(s)(1 - e^{-\tau s}) \tag{12}$$

In the formula, $G_r(s)$ indicates the compensation value. After Smith estimated compensation, due to $e^{-\tau s}$ it is outside the closed loop, so it does not affect the stability of the system. Smith predictor has the function of compensation, so its ultimate purpose is to use this function to minimize the influence of delay time on system stability. In the previous section, the fuzzy PID has shown good anti-interference ability in the temperature control of the heater, but its algorithm control effect still has room for improvement in the aspect of time delay in the temperature control. So the fuzzy PID control algorithm based on Smith predictor compensation is quoted, and this algorithm can make up for the disadvantage that Smith predictor is sensitive to the parameters of the controlled object model. When heating Smith fuzzy PID temperature control, the parameters of the predictor have been determined by the parameters of the controlled object model of heating temperature control, so the most important part to be solved is to design a fuzzy PID controller, and the fuzzy PID control has been designed in the previous section. According to the above discussion, the prediction compensation of the temperature control in the heating room is realized, so that the temperature control technology in this paper has higher control accuracy in practical application.

3 Experimental Analysis

3.1 Experiment Preparation

The traditional indoor temperature control technology for heating has the problems of inertia and delay lag in practical application. Although the temperature can be stabilized eventually, it will take a long time to adjust, and there will be some overshoot. To solve this problem, in order to verify whether the newly proposed control technology can be solved and further optimize the temperature control performance, the following comparative experiments are designed. Control the ventilation of the laboratory to maintain the stability of indoor temperature. Select semiconductor temperature sensors to collect temperature signals from different indoor locations, and use data acquisition equipment to convert the analog signals collected by the temperature sensors into digital signals and record them. Based on a residential building, select three rooms with the same area, indoor environment and other conditions in the residential building, and number the rooms as Room A, Room B and Room C. The control technology based on multi-channel temperature signal de-noising, the control technology based on fuzzy logic control and the control technology based on improved BP neural network proposed in this paper are respectively used and set as the experimental group, control group A and control group B to control the temperature in the heating room of the three rooms. Figure 5 shows the temperature change curve of the heating room according to the requirements of the residential building.

It can be seen from Fig. 5 that the temperature of the standard curve of temperature change in the heating room is controlled above 22 °C to meet the needs of indoor people for environmental temperature.

3.2 Result Analysis

Taking the curve in Fig. 3 as the basis, compare the curve with the temperature change curve in the heating room obtained after the application of each control technology, and get the results as shown in Figs. 6, 7 and 8.

According to the analysis of the experimental results obtained, only the temperature change curve in the heating room under the application of the experimental group control technology is consistent with the change of the standard curve, and the difference is not more than ±3 °C. Compared with the application of Group A control technology, the change trend of the temperature in the heating room is basically the same, but the

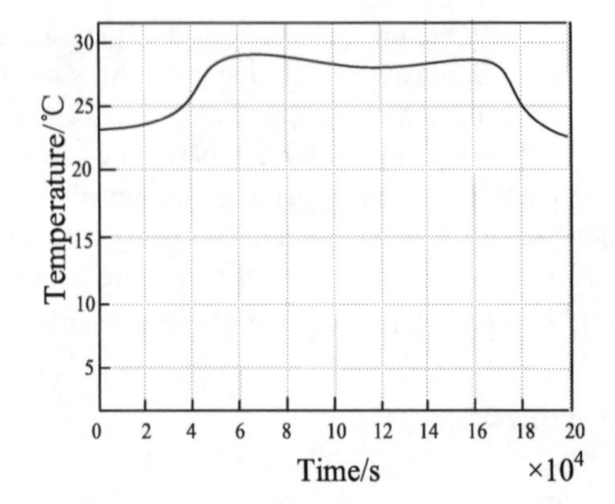

Fig. 5. Standard Curve of Temperature Change in Heating Room

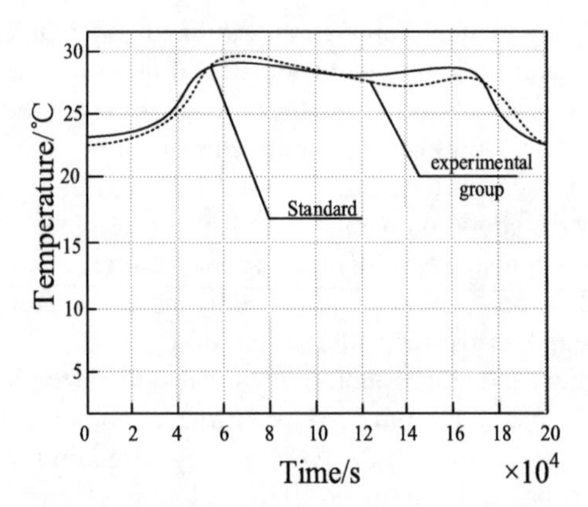

Fig. 6. Indoor temperature change of experimental group control technology heating

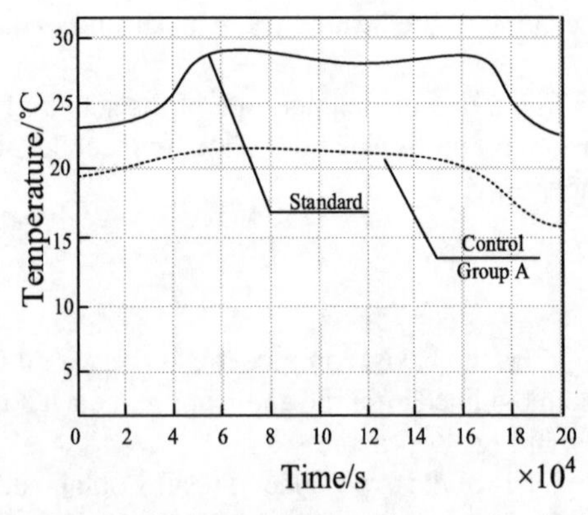

Fig. 7. Temperature change in heating room with control technology of Group A

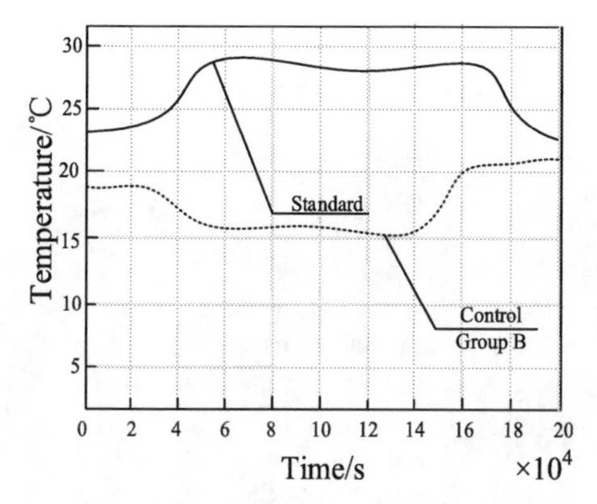

Fig. 8. Temperature change in heating room with control technology of Group B

specific temperature values differ greatly, and the temperature is generally lower than 20 °C, which cannot meet the residential requirements of indoor residents. Under the application of control technology of Group B, the change trend of temperature in the heating room is inconsistent with the standard curve, and the temperature difference at the same time exceeds 10 °C. Therefore, the above experimental results can prove that the control technology designed in this paper based on multi-channel temperature signal denoising has higher control accuracy in practical applications, can fully meet the needs of indoor temperature changes in heating, and provide residents with a more comfortable indoor environment.

4 Conclusion

During heating, temperature, as an important indicator of heating comfort, needs to be controlled in different ranges according to the needs of room temperature. Based on this, this paper proposes a heating room temperature control technology based on multi-channel temperature signal de-noising, and uses comparative experiments to verify the feasibility of this control technology. In practical application, the temperature control technology in this paper can classify the heat load according to the room properties and control the temperature in different ranges. With the development of artificial intelligence and the Internet of Things technology, intelligent algorithms can be applied to indoor temperature control of heating based on multi-channel temperature signals in the future. By analyzing a large amount of temperature data and other environmental parameters, the system can automatically learn and optimize temperature control strategies, achieving more intelligent and adaptive indoor temperature control. In addition, it is believed that in the future, technology based on multi-channel temperature signal denoising can be integrated with other fields of technology, such as machine learning, fuzzy control, optimization algorithms, etc. This can further improve the accuracy and stability of temperature control, and achieve more efficient management of heating systems.

Acknowledgement. Shaoyang City Science and Technology Plan Project (2022GZ3034); Hunan Provincial Natural Science Foundation of China (2023JJ50270, 2023JJ50267); Hunan Provincial

Department of Education Youth Fund Project (21B0690); Hunan Provincial Department of Science and Technology Science and Technology Plan Project (2016TP1023).

References

1. Gao, Y., Shohei, M., Yasunori, A.: Energy saving and indoor temperature control for an office building using tube-based robust model predictive control. Appl. Energy **341** (2023)
2. Chen, C., An, J., Wang, C., et al.: Deep reinforcement learning-based joint optimization control of indoor temperature and relative humidity in office buildings. Buildings **13**(2) (2023)
3. Yang, Z., Zhang, W., Lin, X., et al.: Optimization of minor-LiCl-modified gypsum as an effective indoor moisture buffering material for sensitive and long-term humidity control. Build. Environ. **229** (2023)
4. Cho, S., Nam, H.J., Shi, C., et al.: Wireless, AI-enabled wearable thermal comfort sensor for energy-efficient, human-in-the-loop control of indoor temperature. Biosens. Bioelectron. **223** (2023)
5. Cheng, J., Kang, M., Lin, W., et al.: Preparation and characterization of phase change material microcapsules with modified halloysite nanotube for controlling temperature in the building. Construct. Build. Mater. **362** (2023)
6. Sun, C., Liu, Y., Cao, S., et al.: Integrated control strategy of district heating system based on load forecasting and indoor temperature measurement. Energy Reports **8** (2022)
7. Wirtz, M., Neumaier, L., Remmen, P.,et al.: Temperature control in 5th generation district heating and cooling networks: An MILP-based operation optimization. Appl. Energy **288**(24), 116608.1–116608.13 (2021)
8. Yang, S., Li, S., Yu, H.: Study on room temperature control of VAV air conditioning system. Comput. Simul. **39**(4), 284–289 (2022)
9. Juan, B.V., Luis,C.L., David, M.V.: Indoor temperature and relative humidity dataset of controlled and uncontrolled environments. Data **7**(6) (2022)
10. Hatef, H., Kaiser, A., Jarek, K.: Dynamic heating control measured and simulated effects on power reduction, energy and indoor air temperature in an old apartment building with district heating. Energy Build. **268** (2022)

Research on Pedestrian Tracking in Urban Rail Transit Stations Based on Adaptive Kalman Filtering

Bo Li(✉)

Wuhan Railway Vocational College of Technology, Wuhan 430000, China
libo95123@sina.com

Abstract. Traditional methods mainly use kernel-weighted feature histograms as tracking models, which are easily influenced by the similarity of tracking distributions, resulting in lower mean average precision (mAP) for tracking. In order to effectively address the issues of traditional methods, a new pedestrian tracking method based on adaptive Kalman filtering for urban rail transit stations is proposed. By combining pedestrian micro-walking state analysis with urban rail transit station pedestrian tracking features, a pedestrian tracking model is constructed. The urban rail transit station pedestrian tracking algorithm is designed using adaptive Kalman filtering, and pedestrian tracking is achieved based on the tracking model. Experimental results show that the designed pedestrian tracking method based on adaptive Kalman filtering for urban rail transit stations has a higher mAP for tracking and has certain practical value.

Keywords: Adaptive Kalman filter · City · Track · Traffic station · Pedestrian tracking

1 Introduction

Pedestrians are the main participants in the traffic system. Ensuring smooth pedestrian traffic and pedestrian safety is an important goal of urban traffic system construction [1]. However, in the current traffic system research, vehicles are often the focus of the research. In the actual urban traffic system [2], especially in a typical mixed traffic system such as China, detailed pedestrian information on urban roads is the basis for achieving urban traffic safety and efficient traffic. From the following analysis of a set of traffic data, we can understand the importance of the auxiliary driving system. Every year, more than 39000 pedestrians die in traffic accidents around the world, and more than 430000 people are injured in traffic accidents. In traffic accidents [3], the number of pedestrians injured is only second to the number of passengers injured in accidents. Therefore, people are eager to strengthen the protection of pedestrians in traffic activities and hope that some automatic detection systems can appear to protect pedestrians.

The most basic step of pedestrian research is to extract the traveller target [4] from the target. After extracting the targets in the scene, the next thing to deal with is to

L. Yun et al. (Eds.): ADHIP 2023, LNICST 549, pp. 313–327, 2024.
https://doi.org/10.1007/978-3-031-50549-2_22

distinguish pedestrians from other targets. There are many pedestrian discrimination methods available. The basic way is to extract some pedestrian intrinsic features [5], and then compare them based on these features. These methods solve the problem of pedestrian detection in specific applications. However, in complex environments, due to the inability to solve various difficulties in human shape and appearance [6], as well as different human motion modes, existing pedestrian discrimination algorithms need to be improved in robustness and accuracy. In practical applications, it is usually necessary to collect and analyze video images in real time, which requires that the algorithm can extract the size, position, shape, contour and other information of passengers in a very short time [7], which is the real-time requirement of pedestrian detection algorithm. At present, research on pedestrian tracking has made certain progress. For example, reference [8] proposed a frame difference pedestrian tracking method, which mainly uses improved three frame difference method combined with morphological technology to detect and track pedestrian targets in surveillance videos. It can effectively fill some "voids" in pedestrian targets and ensure the quality of pedestrian tracking. Reference [9] proposes a pedestrian tracking method based on sample learning. By specifying the target to be detected in the first frame of the surveillance video, a hybrid classification model can be autonomously generated for object detection. Online progressive learning algorithms are used to learn the changes in the target's posture and update the model. Combined with color based object tracking algorithms, a high-precision object detection and tracking system is automatically constructed to ensure tracking quality.

The detection of pedestrians in complex environments is very difficult. For example, installing a camera in an application environment of a freely moving platform, such as an automobile driving aid system, due to the complexity of pedestrians, some conventional pedestrian detection methods in static background, such as the pedestrian detection method based on frame difference, are no longer applicable in complex environments. The pedestrian detection method based on sample learning has a high error rate in complex background, so pedestrian detection needs to be optimized in some ways to ensure the accuracy of detection. From the detection algorithm itself, the effect of improving detection is limited after all. In recent years, how to use some intelligent algorithms to improve the visual processing process has become a research hotspot. Under the above background, this paper analyzes the traffic characteristics of pedestrians, and uses adaptive Kalman filter to design an effective pedestrian tracking method for urban rail transit stations. The innovative points of effective pedestrian tracking methods for urban rail transit stations designed using adaptive Kalman filtering are mainly reflected in adaptability, features based on micro walking states, consideration of station environment, and evaluation of tracking effects. These innovative points enable this method to better adapt to complex station environments and provide accurate and robust pedestrian tracking results, which has practical application potential. The main contributions of this study include technological innovation, analysis of pedestrian micro walking states, application practices, and improvement of traffic safety. These research results provide important theoretical basis and practical application value for the safety management, operation optimization and Transportation planning of urban rail transit stations.

2 Pedestrian Tracking Method for Urban Rail Transit Stations Based on Adaptive Kalman Filter

2.1 Analysis of Pedestrian Traffic Characteristics of Urban Rail Transit Stations

Influenced by the dynamics of urban traffic, pedestrian traffic characteristics will change dynamically, affecting the final tracking effect. Therefore, this paper first analyzes the pedestrian traffic characteristics of urban rail transit stations. Pedestrian traffic is a very important part of urban traffic. Pedestrian traffic is not as orderly as vehicle queuing. Pedestrians are affected to varying degrees by different factors such as pedestrians themselves, other pedestrians, and the environment, showing a random, complex, and changeable feature. Therefore, the study of pedestrian traffic must have a comprehensive understanding of the characteristics of pedestrian traffic. To understand the characteristics of pedestrian traffic, we can start from the micro and macro aspects.

The micro aspect is aimed at the movement of a single pedestrian. It mainly studies the space size, stride length, stride frequency, stride speed, and the role of other pedestrians and the environment when walking. The macro aspect refers to the group pedestrian movement, which mainly studies the relationship between the three important elements of pedestrian flow speed, density and flow, and pedestrian self-organization behavior. At the same time, environmental factors also have a greater impact on pedestrian walking characteristics [10]. In order to make the research more practical, it is necessary to conduct field research on the platform area of urban rail transit stations. Based on the analysis of the layout of platform facilities and pedestrian flow lines, specific research plans and locations are determined. According to the data obtained from the research and the results of observation videos, the characteristics of pedestrian walking behavior and following behavior in the platform area are analyzed to provide a basis for subsequent research. According to the characteristics of pedestrian stride, a feature map is drawn in this paper, as shown in Fig. 1 below.

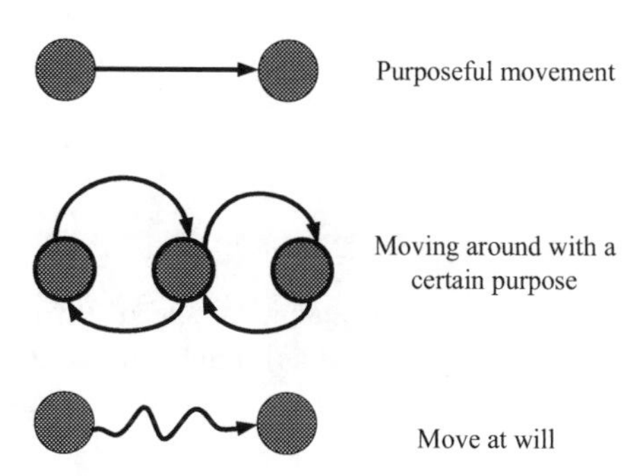

Fig. 1. Characteristic Diagram of Pedestrian Moving Stride

It can be seen from Fig. 1 that the purpose of pedestrian travel is the highest walking speed when commuting, and the walking speed when commuting can better represent the walking speed of pedestrians in urban rail transit stations.

Pedestrian static space demand refers to the space required by the body when the pedestrian is still, including the space actually occupied by the pedestrian and the safe space required by the pedestrian to maintain a certain distance from the surrounding environment psychologically. The actual occupied space depends on the shoulder width and chest thickness of pedestrians. The psychological exclusion of pedestrians should be considered for the psychological safety distance. The schematic diagram of pedestrian's static space demand is shown in Fig. 2 below.

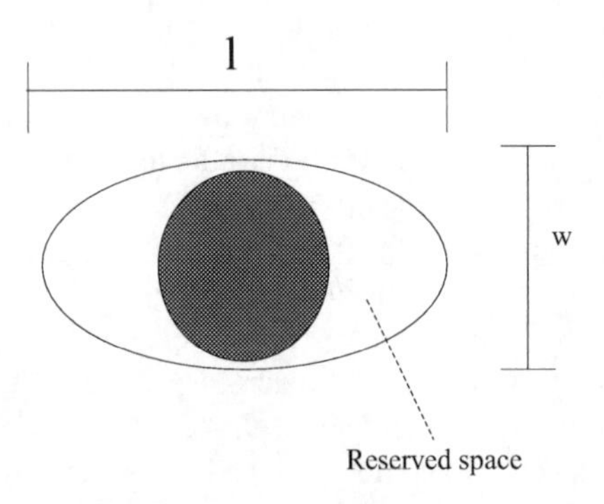

Fig. 2. Pedestrian Static Space Demand

It can be seen from Fig. 2 that the safe space of pedestrians is mainly related to their own gender, customs, personality, status, interpersonal relationships, environment and other factors. The safety space between acquaintances is smaller than that between strangers. When the environment is comfortable, women will generally reserve 0.37 square meters of safety space, while men will reserve 0.74 m^2 of safety space, and the safety zone is vulnerable to environmental impact. The safety space required by pedestrians will be greatly reduced when the surrounding pedestrian density is high. Pedestrian speed refers to the average speed of pedestrians in the study area in a certain period of time, which is related to the walking speed of each person $v(t)$ as shown in (1) below.

$$v(t) = \frac{\sum_{i=1}^{N} v_i(t)}{N} \tag{1}$$

In formula (1), $v_i(t)$ represents the instantaneous speed at which pedestrians pass, N represents the total number of people passing through, and the formula for calculating the average speed of pedestrians in the space at this time v as shown in (2) below.

$$v = \frac{L}{\sum_{i=1}^{N} \frac{t_i}{N}} \tag{2}$$

In formula (2), L represents the walking distance, t_i it represents the time consumed by pedestrians to complete the walking distance. Pedestrian density refers to the average number of pedestrians per unit area, which is calculated by k as shown in (3) below.

$$k = \frac{N}{L \times W} \tag{3}$$

In formula (3), W it represents the width of sidewalk. According to the above pedestrian density characteristics, the traffic class relationship can be set to generate the pedestrian speed, density, and flow tracking characteristic formula, as shown in (4) below.

$$q = v \times k \tag{4}$$

In formula (4), q represents pedestrian flow, v represents pedestrian speed, k represent pedestrian density. According to the above pedestrian tracking characteristics, pedestrian tracking parameters can be effectively calculated, which is the basis for subsequent tracking model construction.

2.2 Build Pedestrian Tracking Model of Urban Rail Transit Station

In order to solve the problem that the average tracking accuracy mAP is low due to the tracking distribution similarity when the kernel weighted feature histogram is used as the target tracking model, this paper constructs a pedestrian tracking social model for urban rail transit stations based on the pedestrian tracking feature parameters of (1)–(4).The social force model is a microscopic simulation mechanical model, which describes the movement of pedestrians under the action of social forces according to the principle of Newtonian mechanics. The social force in the model is based on the interaction results between pedestrians themselves, between pedestrians, and between pedestrians and the environment. It emphasizes more on the subjective initiative of pedestrians and can more realistically display the characteristics of pedestrian traffic. The social force pedestrian tracking model is shown in Fig. 3 below.

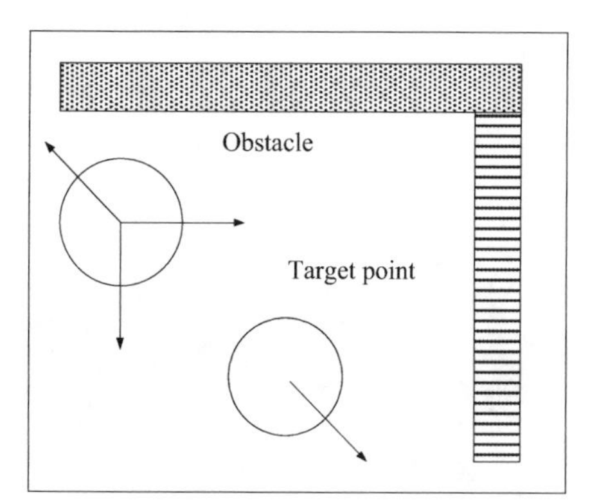

Fig. 3. Pedestrian tracking model of traffic social forces

It can be seen from Fig. 3 that in the traditional social force model, pedestrians are mainly affected by three forces: pedestrian self driving force refers to the power from the target point, and the direction points to the target point; Interaction between pedestrians. The force surface of the environment on the pedestrian is the force from the obstacle that the pedestrian receives, and the direction is perpendicular to the obstacle and points to the pedestrian. At this time, the final force on the pedestrian F As shown in (5) below.

$$F = m_i \frac{dv(t)}{dt} \tag{5}$$

In formula (5), m_i represents the mass of pedestrians, $dv(t)$ represents the force exerted by the environment on pedestrians, dt represents equivalent reaction. The moving acceleration of pedestrians can be calculated according to the force relationship of pedestrians $v_i(t)$, as shown in (6) below.

$$v_i(t) = at + v_0(t) \tag{6}$$

In formula (6), at represents the actual speed of pedestrians, $v_0(t)$ it represents the repulsive force of pedestrians against obstacles.

Self driving force is the force generated by pedestrians to reach the target point as soon as possible, maintain the desired speed, and select the shortest path subjectively without external environmental interference. When pedestrians are disturbed and the actual speed does not reach the expected speed, self driving force will be generated to accelerate pedestrians. The direction of self driving force is determined by the pedestrian pointing to the target point, and the size is determined by the pedestrian's expected speed and actual speed F_{ID} the calculation formula is as follows (7).

$$F_{ID} = m_i \frac{1}{\tau_i}(ve - v_i) \tag{7}$$

In formula (7), τ_i represents pedestrian response time, ve represents the actual speed of pedestrians, v_i represents the expected speed of pedestrians.

Pedestrians are affected by four forces when walking, including pedestrian self driving force, pedestrian force, pedestrian and obstacle force, and pedestrian gravity. When walking, pedestrians will select the following condition 4 in the field of vision.

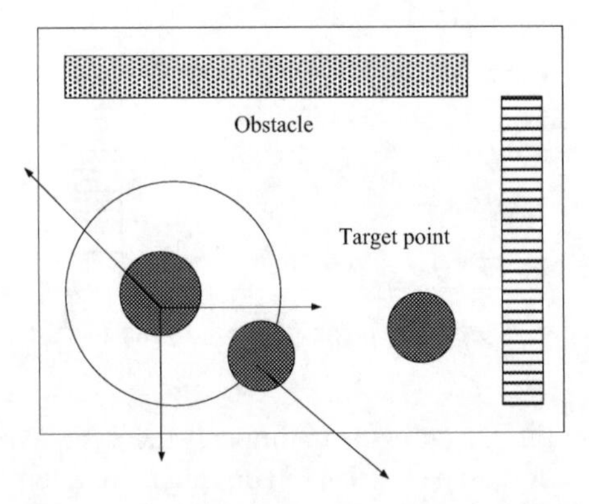

Fig. 4. Follow force of pedestrians in the model

It can be seen from Fig. 4 that, combined with the influence of various directional factors, this paper optimizes the pedestrian tracking model above, and constructs the pedestrian tracking optimization model for traffic stations $f_{\max}$ as shown in (8) below.

$$f_{\max} = \varphi m_i \frac{v_i}{\tau_i} \tag{8}$$

The above model studies the relationship between the following behavior revision coefficient and pedestrian traffic efficiency. Through research, it is found that pedestrian traffic efficiency increases with the increase of the following behavior revision coefficient. This shows that pedestrians follow the same direction of movement. The stronger the willingness of other pedestrians on the bus, the faster the pedestrians adjust their speed in the direction of movement, the faster the pedestrian flow will be formed, and the traffic efficiency will be improved. By studying the relationship between pedestrian speed and pedestrian density, it is found that when the revision factor is 0.2, the simulation results are more realistic, so the revision factor of the model is set to 0.2.

2.3 Design of Pedestrian Tracking Algorithm for Traffic Stations Based on Adaptive Kalman Filter

Only using the pedestrian tracking model of the traffic station built above for tracking may lead to the problem of limited tracking efficiency. Therefore, this paper designs the pedestrian tracking algorithm of the traffic station based on adaptive Kalman filter. Adaptive Kalman filter means that while filtering with measured data, the filter itself constantly judges whether there is any change in the system dynamics, estimates and modifies the model parameters and noise statistical characteristics of the above design, so as to improve the filter design and reduce the actual error of the filter. Therefore, foreground extraction is needed first, which can be realized by background subtraction, frame subtraction, optical flow and other methods. Considering the application environment and real-time nature of the algorithm, this paper uses the Gaussian mixture model with high adaptability to the background and moderate operating efficiency to extract foreground.

The Gaussian mixture model uses K Gaussian distributions to represent the characteristics of each pixel in the image, and matches each pixel in the current image with the Gaussian mixture model. If the matching is successful, the point is determined as the background point; otherwise, the point is the front spot, and at this time, each pixel X becomes the probability of tracking the background point $P(X_i)$ as shown in (9) below.

$$P(X_i) = \sum_{k=1}^{k} w \times \eta \tag{9}$$

In formula (9), w represents the high-speed distribution weight at different times, η represents the mean variance of Gaussian distribution, and the probability density function can be generated according to the tracking probability calculated above $\eta(x)$, as shown in (10) below.

$$\eta(x) = \frac{1}{(2\pi)\sigma} e \tag{10}$$

In formula (10), σ represents the matching coefficient, e representing the pixel matching value, the Gaussian mixture model can accurately evaluate the background model and carry out multi-target detection. In this paper, the EM algorithm is used to train the Gaussian mixture distribution. Through recursive calculation, a pedestrian tracking image with good adaptability can be obtained.

After the above steps are completed, this paper uses Faster RCNN network as the pedestrian detection network, which is divided into two processes: training and testing. The two processes are similar. Before the image sequence is input into the network, the data set is first preprocessed. Therefore, it is necessary to manually calibrate the Bounding Box for pedestrians in the training data set, obtain the minimum coordinates surrounding pedestrians, and save the coordinate information as an xml file for Faster RCNN to read. Since the Faster RCNN network has no fixed requirements for the size of the input image, it does not need to normalize the image sequence. When the image is input into the network for feature extraction, it will unify the image normalization scale according to the set scale.

Input the whole image into CNN for feature extraction. CNN has 13 relu layers and 4 pooling layers. The convolution core of each convolution layer is 3, and all convolutions are expanded outward for one circle. At the same time, the convolution core of each pooling layer is 2, and the step is 2, to obtain the feature map of the image. Think of the extracted feature as a 51 * 39 256 channel image, and input it into the RPN network to generate the region proposal. Essentially, the sliding window is made on the feature map finally extracted by CNN to obtain the multi-scale aspect ratio of the region proposal, that is, for each position of the feature image, three kinds of scale and aspect ratio are considered, and a total of nine possible candidate windows (anchors) are considered.

The anchors with the largest overlap ratio with the ground truth in these candidate regions are recorded as foreground samples, and then those with the overlap ratio with the bounding box greater than 0.7 are selected from these foreground anchors as foreground samples, and those with the overlap ratio less than 0.3 are recorded as background samples. Finally, the probability belonging to foreground and background is output from these 256 dimensional features. The proposed feature maps will be calculated and extracted at the ROI pooling layer, sent to the full connection layer to determine the target category, and the bounding box region will be used to obtain the target position detection box with the final accurate position.

The research of pedestrian re recognition began with multi vision tracking. In the re recognition algorithm, there are two very important parts: image description and distance measurement. A good distance measurement is very important because when the sample variances are consistent, high-dimensional visual features usually do not extract invariants. Among them, distance measurement methods are classified into supervised learning and unsupervised learning, global learning and local learning. In the re recognition algorithm, most of the work is mainly supervised global distance measurement learning. The general idea of global metric learning is to make vectors of the same class closer and those not belonging to the same class farther. The most commonly used is Markov distance, which is an extension of Euclidean distance using linear scale and rotation methods in the feature space. The algorithm designed in this paper tracks pedestrians based on Euclidean distance, so as to quickly obtain the tracking distribution relationship of pedestrians and maximize the tracking accuracy of pedestrians at stations.

3 Experiment

In order to verify the tracking performance of the designed pedestrian tracking method for urban rail transit stations based on adaptive Kalman filter, this paper selects a research area and compiles a simulation experiment program to compare it with the conventional frame difference pedestrian tracking method and the pedestrian tracking method based on sample learning. Experiments are carried out as follows.

3.1 Experiment Preparation

This chapter takes the platform investigation area of a subway station line 2 as the research background, applies the pedestrian following behavior simulation model program compiled by MATLAB software, selects reasonable parameters, and conducts simulation experiments on pedestrian tracking behavior in the platform area of urban rail transit stations. First, we need to build the functional structure of the experimental platform according to the characteristics of MATLAB software, as shown in Fig. 5 below.

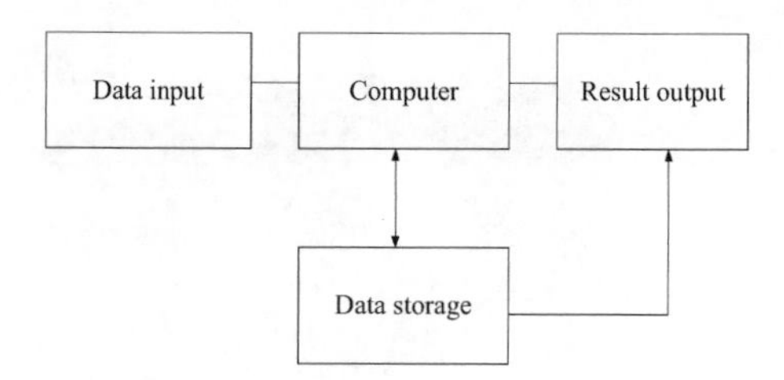

Fig. 5. Functional structure of the experimental platform

It can be seen from Fig. 5 that this experiment will divide the simulation experiment platform into four parts: data input part, data operation part, data storage part and result output part. The experimental platform built in this paper is divided into two parts: local computer side configuration and remote server side configuration. The specific configuration of the local side and server side is shown in Table 1.

This experiment selects two data sets, namely Crowd Human pedestrian data set and the traffic hub pedestrian data set collected and produced in this paper. Crowd Human pedestrian data set is a benchmark data set for pedestrian detection in the crowd, and the data set collected and produced in this paper is the pedestrian data set under various environmental conditions in each traffic hub.

There are 15000 training pictures, 4370 verification pictures and 5000 test pictures in Crowd Human pedestrian data set. The pedestrian individuals in each picture use three kinds of label frames, namely head frame, body frame and visible area frame. In order to intuitively display each label frame in the data set, draw a rectangular box for the pedestrian in the picture with the cv2.rectangle function at the end of the data set, The

Table 1. Configuration of Experimental Platform

Experimental platform		detailed parameters
Local end	CPU	i5-6500 CPU
	operating system	64 bit window7 operating system
	Memory	8G
	operating system	Linux centos7-1
Server side	CPU	40*Intel(R) Xeon(R) Silver 4210 CPU @ 2.20 GHz
	GPU	TITAN RTX (24G VRAM)
	Memory	8* 16 GB TruDDR4 2933 MHz
	other	CUDA10.2, cuDNN7

pedestrian individuals in each picture use three kinds of annotation frames, namely head frame, body frame and visible area frame. The annotation status of different annotation frames is shown in Fig. 6 below.

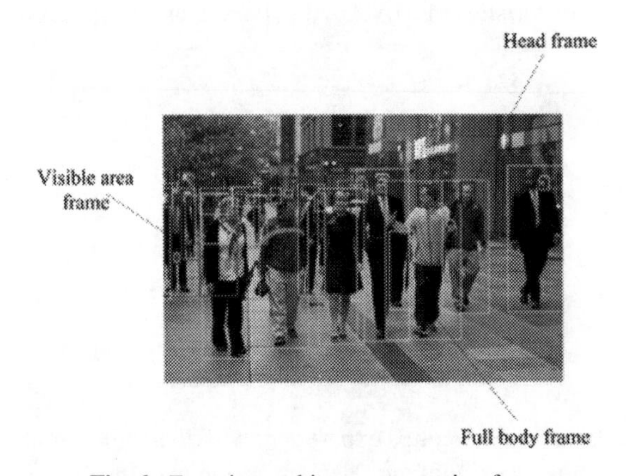

Fig. 6. Experimental image annotation frame

The data structure of this dataset is shown in Fig. 7 below.

As shown in Fig. 7, first use python programming to process the odgt annotation file of Crowd Human pedestrian dataset. Each ID of the odgt file is a picture. Divide it according to the ID, and extract the information under each ID. The targets tested in this paper are bbox and hbox. Therefore, the overall length and width of the picture, the coordinate information of the bbox box, and the coordinate information of the hbox box are extracted, Each image extraction information is saved as a separate xml file, and the xml file is named with the corresponding image name, which corresponds to the images stored in the JPEGImages folder one by one, and all are stored under the Annotations folder. Combining the above dataset data structure, this paper has produced an effective experimental dataset image, as shown in Fig. 8 below.

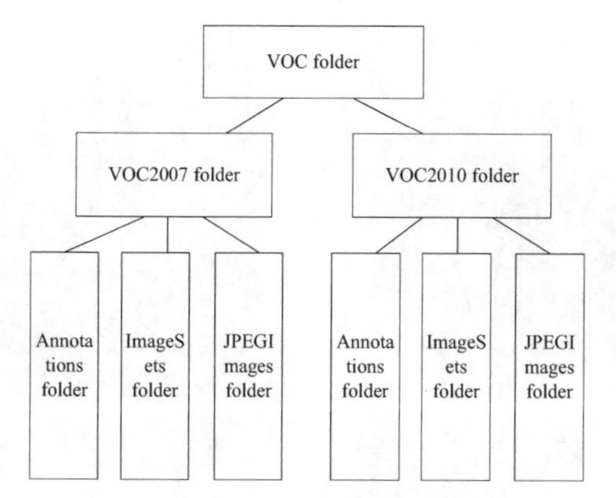

Fig. 7. Dataset Data Structure Diagram

It can be seen from Fig. 8 that this experiment detects pedestrians in the traffic hub. Due to the large number of detected targets, the target features are seriously lost in the case of serious occlusion. Therefore, this paper adopts the idea of head and shoulder detection to collect and produce the pedestrian head and shoulder data set of the traffic hub. There are 6196 pictures in total, with a total of 83072 effective numbers marked. The pictures are mainly from the two ways of photography and network, including the train exit Traffic intersection, station square and other scenes.

The Crowd Human pedestrian data set is made into a standard VOC data set format to prepare for algorithm training and testing. After the experimental data set is selected, the experimental indicators can be selected. According to the requirements of pedestrian tracking at the station, this paper selects the average accuracy rate mAP as the evaluation indicator, and its accuracy rate P and recall rate R as shown in (11) and (12) below.

$$P = \frac{TP}{TP + FP} \tag{11}$$

$$R = \frac{TP}{TP + FN} \tag{12}$$

In formula (11)–(12), TP in order to correctly detect the quantity, FP is the number of false checks, FN The number of pedestrians detected as the background is the number of pedestrians detected as the background by mistake. At this time, take the pedestrian detection at the traffic hub as an example, take the recall rate as the abscissa, and the accuracy rate as the ordinate, draw the P-R curve, and use the integral to calculate the value of mAP, as shown in (13) and (14) below.

$$F1 = \frac{2TP}{2TP + FN + FN} \tag{13}$$

$$mAP = \int_0^1 P(R)d(R) \tag{14}$$

At this time, F1 can be used as a comprehensive measure of accuracy and recall. The closer to 1, the better the effect will be.

Pictures of pedestrians at the exit

Pictures of pedestrians at the entrance

Bus stop pedestrian map

Pictures of pedestrians in station square

Pedestrian map at traffic intersections

Pedestrian Map at Intersection

Fig. 8. Picture of experimental data set

In the simulation in this paper, the pedestrian is represented by a circle, and the K-means algorithm is used in Crowd Human dataset and transportation_hub_The experimental clustering parameters obtained by clustering on the Human dataset are shown in Table 2 below.

Table 2. Experimental clustering parameters

data set	Input size	Anchor Box
Crowd Human(fbox)	608	(4,10), (7,19), (10,35), (14,53), (20,27), (20,76), (28,114)
	416	(3, 8) (5,17) (8, 29)
	320	(2, 8) (3, 15) (6, 19)
Crowd Human(hbox)	608	(2, 3) (4, 5) (5, 8)
	416	(3, 8) (5, 17) (8, 29)
	320	(1, 2) (2, 3) (5, 8)
Transportation_ hub_Human	608	(1, 1) (1, 2)
	416	(6, 11) (9, 15) (12, 19)
	320	(5, 9) (12,19)

Table 2 shows that K-means++ can be used in Crowd Human dataset and transportation_hub_The clustering effect on the Human dataset can determine the number of experimental samples and effectively carry out subsequent pedestrian tracking experiments.

3.2 Experimental Results and Discussion

Combined with the above experimental preparations, pedestrian tracking experiments at urban rail transit stations can be carried out in the selected experimental data set, that is, five experimental samples are preset, and the pedestrian tracking methods at urban rail transit stations designed in this paper based on adaptive Kalman filter, conventional frame difference pedestrian tracking methods, and pedestrian tracking methods based on sample learning are used for tracking, Use formula (11)–(14) to calculate the average accuracy mAP of the three methods in different samples. The experimental results are shown in Table 3 below.

Table 3 shows that the average accuracy mAP of the pedestrian tracking method for urban rail transit stations designed in this paper based on adaptive Kalman filtering in different samples is close to 1.0, while the average accuracy mAP of the conventional frame difference pedestrian tracking method and the pedestrian tracking method based on sample learning is quite different from 1.0.The above results prove that the pedestrian tracking method of urban rail transit station designed in this paper based on adaptive Kalman filter has good tracking effect, accuracy and certain application value.

Table 3. Experimental Results

The pedestrian tracking method designed in this article for urban rail transit stations based on adaptive Kalman filtering	
Sample number	mAP
CA01	0.954
CA02	0.957
CA03	0.986
CA04	0.974
CA05	0.936
Frame difference pedestrian tracking method	
CA01	0.654
CA02	0.639
CA03	0.665
CA04	0.659
CA05	0.598
Pedestrian tracking method based on sample learning	
CA01	0.745
CA02	0.569
CA03	0.623
CA04	0.558
CA05	0.628

4 Conclusion

Accurate detection of pedestrians is fundamental to ensuring the safety of transportation hubs. With the rapid development of urbanization, cities are facing increasing traffic pressure, congestion, and frequent traffic accidents. In this context, effective positioning and tracking of pedestrians are necessary to improve traffic safety. Therefore, this paper determines the characteristics of pedestrian tracking in urban rail transit stations based on micro-walking states. A pedestrian tracking model is constructed, and an adaptive Kalman filtering algorithm is designed for urban rail transit station pedestrian tracking. The traditional Kalman filtering method is usually based on static models to estimate the position of pedestrians, but in practical situations, the motion state of pedestrians may change. This method introduces self-adaptability and adjusts the parameters of the filter based on real-time observation data to adapt to changes in pedestrian motion status, thereby improving the accuracy and robustness of tracking. The experimental results demonstrate that the designed method for pedestrian tracking in urban rail transit stations has good tracking performance, accuracy, and practical value. Pedestrian tracking in urban rail transit stations has wide application prospects, as it can enhance safety, operational management efficiency, and urban planning levels. As technology continues to advance and be applied, pedestrian tracking will play an increasingly important role in the future. By integrating technologies such as artificial intelligence, computer vision, and deep learning, the accuracy and efficiency of pedestrian tracking can be improved. It is also important to integrate pedestrian tracking technology with urban planning to provide scientific and accurate data support for city design and transportation planning. Through proper pedestrian flow guidance and planning, the urban traffic conditions and living environment can be improved.

References

1. Yang, S., Chen, Z., Ma, X., et al.: Real-time high-precision pedestrian tracking: a detection-tracking-correction strategy based on improved SSD and Cascade R-CNN. J. Real-Time Image Proc. **2**, 19 (2022)
2. Xu, Y., Chan, H.Y., Chen, A., et al.: Walk this way: Visualizing accessibility and mobility in metro station areas on a 3D pedestrian network. Environ. Plan. B Urban Anal. City Sci. **49**(4), 1331–1335 (2022)
3. Xu, X., Li, X., Zhao, H., et al.: A real-time, continuous pedestrian tracking and positioning method with multiple coordinated overhead-view cameras. Measurement **178**, 109386 (2021)
4. Xu, Y., Shmaliy, Y.S., Hua, L., et al.: Decision tree-extended finite impulse response filtering for pedestrian tracking over tightly integrated inertial navigation system/ultra wide band data. Meas. Sci. Technol. **3**, 32 (2021)
5. Zhou, Y., Yang, W., Shen, Y.: Scale-adaptive KCF mixed with deep feature for pedestrian tracking. Electronics **10**(5), 536 (2021)
6. Zhang, X., Wang, X., Gu, C.: Online multi-object tracking with pedestrian re-identification and occlusion processing. Vis. Comput. **5**, 37 (2021)
7. Wong, K.Y., Luo, H., Wang, M., et al.: Recognition of pedestrian trajectories and attributes with computer vision and deep learning techniques. Adv. Eng. Inform. **49**, 101356 (2021)
8. Huang, Y., Li, D., Cheng, J.: Simulation of pedestrian-vehicle interference in railway station drop-off area based on cellular automata. Phys. A **579**(4), 126142 (2021)
9. Mo, D.L., Yang, R.H., Wu, W., et al.: Pedestrian target detection in surveillance video based on improved three-frame difference method. Sci. Technol. Innov. Guide **18**(6), 138–142 (2021)
10. Shao, W., Zhao, F., Luo, H., et al.: Particle filter reinforcement via context-sensing for smartphone-based pedestrian dead reckoning. IEEE Commun. Let. PP(99), 1–1 (2021)

Badminton Flight Trajectory Location and Tracking Algorithm Based on Particle Filter

Zhiyong Huang[1(✉)] and Yuansheng Chen[2]

[1] Guangzhou Huashang Vocational College, Guangzhou 511300, China
13560455687@163.com
[2] Guangzhou Huali College, Guangzhou 511325, China

Abstract. In order to meet the accuracy requirements of badminton robot, a badminton flight trajectory location and tracking algorithm based on particle filter is proposed. Collect badminton flight images for morphological and filtering processing. Using image features and matching to detect badminton flying targets. Predict the flight trajectory of a badminton ball considering its force situation. Using particle filter algorithm to determine the flight position of badminton, and through real-time updates, achieve the positioning and tracking of badminton flight trajectory. The experimental results show that the trajectory positioning and length tracking errors of the designed algorithm are 4.12 m and 0.13 m, respectively. The tracking update delay is only 7.6 s, and the tracking success rate is as high as 97%. The design method effectively solves the problems of large trajectory positioning tracking errors, low stability, and efficiency.

Keywords: Particle filter · Badminton · Flight path · Location tracking

1 Introduction

With the improvement of people's living standards, sports have been integrated into people's daily life. Among them, badminton is a sport suitable for all ages. In order to capture the dynamic path of badminton quickly and accurately, a badminton robot needs to be designed to track the high-speed badminton in real time. Target tracking can be defined as the process of obtaining the motion state information of the target continuously in the video sequence [1] based on the prior template information of the known target. Target tracking is a challenging technical difficulty in the field of computer vision. Target tracking technology is closely related to target detection and recognition technology. Generally, in practical engineering applications, it is necessary to detect the target of interest first, and then switch to automatic tracking mode after accurately capturing the target. During the tracking process, it constantly analyzes the information such as target shape, scale, and motion rules, and evaluates, classifies, and recognizes the target attributes. Target tracking technology plays an important role in both military and civilian fields.

L. Yun et al. (Eds.): ADHIP 2023, LNICST 549, pp. 328–344, 2024.
https://doi.org/10.1007/978-3-031-50549-2_23

After a long time of research, the current track location and tracking algorithm development is more mature, including: track location and tracking algorithm based on multi-sensor, track location and tracking algorithm based on dynamic Snake model, and track location and tracking algorithm based on network computing. The track location and tracking algorithm based on multi-sensor uses mobile robots as tracking targets, According to its kinematic model, analyze the key parameters of motion control in the trajectory tracking process, collect and fuse multi-sensor data, process the signals of each sensor respectively, and fuse the angle information on this basis, so as to reduce the positioning error. The fuzzy control method is studied to realize the application of fuzzy control in the trajectory tracking of mobile robots, and good positioning and tracking results are obtained. The track location and tracking algorithm based on dynamic Snake model combines KLT optical flow method, selects the strong feature points in the contour points obtained in the current frame to estimate the optical flow, takes the estimation result as the initial contour of the next frame Snake, and determines the target location and tracking results according to the contour features of the target. However, the above flight path location and tracking algorithm is still very complex and faces many challenges in the actual operation process, including: the information loss caused by the projection from the three-dimensional world to the two-dimensional image plane makes tracking difficult; Noise in the image itself; Tracking the light changes in the scene, mutual occlusion between objects, etc. Applying the traditional location and tracking algorithm to the location and tracking of badminton flight trajectory, there is an obvious problem of large location and tracking errors. Therefore, particle filter algorithm is introduced.

The so-called particle filter is to find a group of Random sample To approximate probability density function, using Sample mean Instead of integral operation system state Of Minimum variance estimation These samples are vividly called "particles", so they are called particle filtering. The idea of particle filter is based on Monte Carlo method, which uses particle set to represent probability, and can be used in any form of state space model. Its core idea is to learn from Posterior probability It is a sequential importance sampling method to express the distribution of random state particles extracted from. The particle filter algorithm is used to optimize the location and tracking algorithm of badminton flight trajectory, in order to improve the location and tracking accuracy of badminton flight trajectory. Based on the above analysis, this article proposes a badminton flight trajectory localization and tracking algorithm based on particle filtering. Collect badminton flight images, preprocess the initial flight images through morphology and filtering to obtain the correct moving target. By using image features and matching, detect badminton flying targets, consider the force on the badminton, and predict the badminton flight trajectory. Using particle filter algorithm to determine the flight position of badminton, real-time updates are used to achieve badminton flight trajectory positioning and tracking, thereby improving the positioning and tracking accuracy of badminton flight trajectory.

2 Design of Badminton Flight Trajectory Location and Tracking Algorithm

2.1 Collecting Badminton Flight Images

Through stereo vision to identify and track badminton, reconstruct the three-dimensional information of badminton in the badminton court. This requires the visual system to cover the court completely and without blocking. At the same time, in order to find badminton earlier, the visual height is generally higher than the net height. The fixed focus industrial camera is selected as the stereo vision equipment. The measurement accuracy of the stereo vision equipment in the Z-axis direction is required to be:

$$c_Z = \frac{h^2}{d_{\text{optic axis}} * f} W_{xy} \tag{1}$$

Among them, h is the depth of field, $d_{\text{optic axis}}$ and f are the optical axis distance and camera focal length, W_{xy} is the pixel size of the visual device [2]. The measurement accuracy c_X and c_Y of stereo vision equipment in the X-axis and Y-axis directions are:

$$c_X = c_Y = \frac{h}{f} W_{xy} \tag{2}$$

According to Formula 1 and Formula 2, the measurement error in the depth direction is proportional to the square of the measurement distance, and the measurement error in the X and Y directions is proportional to the measurement distance. Use the stereo vision equipment that meets the accuracy requirements to obtain the acquisition results of badminton flight images according to the representation principle in Fig. 1.

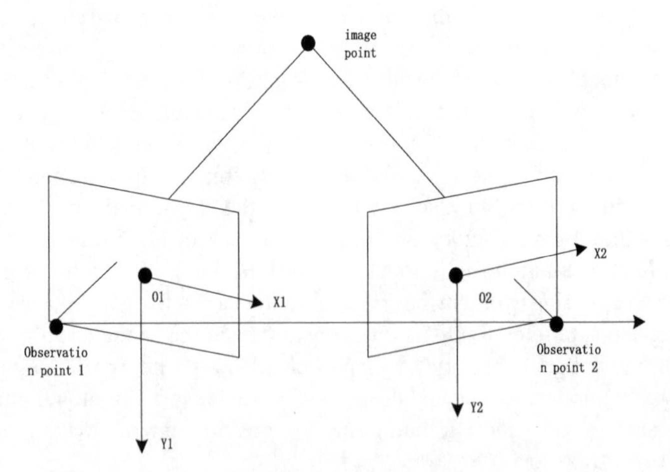

Fig. 1. Schematic diagram of badminton flight image acquisition

In order to avoid measurement distortion caused by wide-angle lens distortion during badminton flight image acquisition, a 100° field of view lens is selected. Set the image acquisition frequency, and get the image acquisition results at any time during the badminton flight.

2.2 Initial Badminton Flight Image Pre-processing

Generally, because the camera has shifted or rotated, the image background in the image sequence will produce motion, and the motion of the background will lead to the change of the coordinate system where the target is located. A series of operations against the target will produce errors, resulting in the final failure to obtain the moving target correctly. When this happens, in order to obtain the moving target in the image sequence using the frame difference method and other methods, it is necessary to first register the image sequence and convert it to the same coordinate system. This method is called global motion compensation [3, 4]. The essence of global motion compensation is to find the corresponding relationship between two adjacent images, and then convert the two images to the same coordinate system through this relationship. The main steps of the algorithm are described as follows: first, find the corresponding feature points in the two images, and then obtain the geometric transformation model between their position coordinates according to the position information of the two feature points in their respective images, so that other pixel points in the two images can be transformed into the same coordinate system through this transformation model, Then the image on the moving background can be further processed as the image on the fixed background. Morphological filtering uses the information of the image contained in the structure element to carry out logical operations on the corresponding area pixels, obtain new pixels after filtering, and constantly update the pixels in the image to achieve the purpose of changing the overall structure of the image. The morphological filtering processing results of the initial badminton flight image can be expressed as:

$$\begin{cases} A \odot B = \left\{ z_{\text{corrosion}} \middle| (B)_{z_{\text{corrosion}}} \subseteq A \right\} \\ A \oplus B = \left\{ z_{\text{expansion}} \middle| (B)_{z_{\text{expansion}}} \cap A \neq \emptyset \right\} \end{cases} \tag{3}$$

In the above formula, A and B they respectively represent the badminton flight image and structural elements initially collected, $z_{\text{corrosion}}$ and $z_{\text{expansion}}$ corresponds to the results of image erosion and expansion processing. $(B)_{z_{\text{corrosion}}}$ and $(B)_{z_{\text{expansion}}}$ are the corrosion and expansion treatment results of structural elements, respectively. Due to jitter, external light, wind and other factors in the image acquisition process, there will be some noise in the image generation process, which will reduce the image quality and have a negative impact on the further processing of later images. Therefore, image filtering algorithm is often used for image denoising. In order to ensure the image preprocessing effect, the processing method of combining mean filtering, Gaussian filtering and median filtering is adopted. The mean filtering is to replace the pixel value of each pixel in the local neighborhood of the original image with the mean value of all pixel values in the neighborhood. The process of mean filtering $I_{\text{Mean filtering}}(x, y)$ can be quantified as:

$$I_{\text{Mean filtering}}(x, y) = \frac{1}{m} \sum I(x, y) \tag{4}$$

Among them, m refers to the number of pixels in the badminton flight image, $I(x, y)$ represents the flying image of badminton. In addition, the basic principle of Gaussian filtering is that the weight value is selected according to the shape of the Gaussian function, and the median filtering is to assign the pixel value of the original image through the processing of the template [5]. The process of Gaussian filtering $I_{\text{Gaussian filter}}(x, y)$ and median filtering $I_{\text{median filtering}}(x, y)$ is as follows:

$$\begin{cases} I_{\text{Gaussian filter}}(x, y) = \dfrac{1}{2\pi\chi^2} \exp^{\frac{-I(x+y)^2}{2\chi^2}} \\ I_{\text{median filtering}}(x, y) = \text{med}\{I(x - m_x, y - m_y)\} \end{cases} \tag{5}$$

In Formula 5, χ^2 is the variance of image pixels, m_x and m_y respectively represents the number of pixels in the horizontal and vertical directions of the badminton flight image, med{} is the median calculation function. Repeat the above operations, filter all the pixels in the collected badminton flight images, and obtain the preprocessing results of the initial badminton flight images.

2.3 Detection of Badminton Flying Targets

The detection principle of badminton flying target is: according to the characteristics of badminton flying environment, realize the separation of foreground and background, and extract the features of foreground image. Through the matching of foreground image features and badminton features, determine whether there are badminton flying targets in the current image, and then obtain the detection results of badminton flying targets. Figure 2 shows the basic detection process of badminton flying targets.

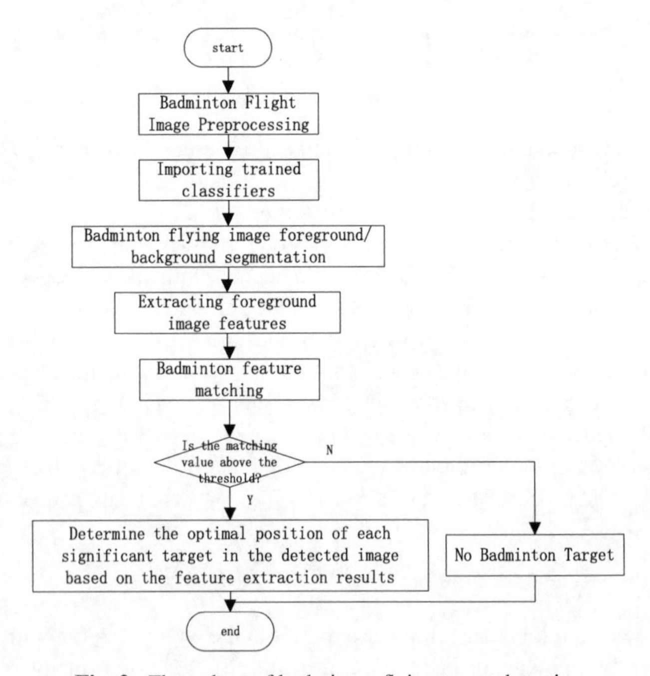

Fig. 2. Flow chart of badminton flying target detection

Let two single channel gray images of the same size for difference operation be $I_1(x, y)$ and $I_2(x, y)$, the image after difference operation is $I_{\text{prospect}}(x, y)$, we can get:

$$I_{\text{prospect}}(x, y) = |I_2(x, y) - I_1(x, y)| \tag{6}$$

It can be seen from Formula 6 that the target can be enhanced by calculating the absolute value of the pixel difference at the corresponding position of the two images. There are light, wind and other interferences in the external environment, resulting in a lot of noise in the real scene. Therefore, in order to obtain a stable background, background modeling [6] is required. Mathematically, it can be realized through a single Gaussian model, in which the iterative process of the mean and covariance of image pixels represented by the model is:

$$\begin{cases} \mu_{t+1} = (1 - \omega)\mu_t \omega I_t \\ A_{t+1} = (1 - \omega)A_t + \omega(I_t - \mu_t) \end{cases} \tag{7}$$

Among them, ω is the experience threshold, A_t and A_{t+1} are t and $t + 1$ covariance matrix of time, μ_t and μ_{t+1} are the pixel mean values of the badminton flight images at time t and $t + 1$, respectively. The background part of the image can be obtained by Formula 7, and the image difference can be performed by Formula 6 to realize the division of foreground and background. On this basis, color, contour, texture, gradient and other features are extracted from the foreground image in the image, and color features and gradient direction features are fused to generate a comprehensive histogram feature. The color histogram feature has little change in the case of plane rotation and partial occlusion, which is suitable for describing the target and is widely used. In order to reduce the influence of light changes during tracking, HSV color space [7, 8] is used. The kernel function representing spatial information is used to calculate the weighted histogram, which improves the robustness of the color histogram to describe the target features; The edge area may contain background, with small contribution and small weight allocation. Each channel of H and S components is assigned a quantization level of 8. Assume that the target state is (i, j) is the center, r is the radius, and the position of the pixel in the target area is X_i, then the normalized color histogram in the target area is:

$$Q_{\text{colour}} = \frac{\sum_{i=1}^{m} X_i f_{\text{Weight}}\left(\left\|\frac{1}{r}\right\|\right) f_{delta}[\varphi(X_i) - \lambda]}{\sum_{i=1}^{m} X_i f_{\text{Weight}}\left(\left\|\frac{1}{r}\right\|\right)} \tag{8}$$

Among them, $f_{\text{Weight}}()$ and $f_{delta}()$ are weight kernel function and delta function respectively, $\varphi()$ represents the image of the color level index on the histogram, λ is the index of the color level in the histogram. Similarly, the extraction results of contour features, texture features, and gradient features can be obtained. Finally, the feature extraction results are fused and marked as Q. Match the image foreground image features with badminton standard features, and the matching results are as follows:

$$s = \frac{Q(t) \cdot Q_{\text{badminton}}}{\|Q(t)\| \cdot \|Q_{\text{badminton}}\|} \tag{9}$$

Among them, $Q_{\text{badminton}}$ is a standard feature of badminton, $Q(t)$ express t the feature extraction results [9] of badminton flight images were collected at all times. If the result of Formula 9 is higher than the threshold s_0, indicating that there is a badminton flying target in the current image, otherwise it is considered that there is no tracking target in the current image.

2.4 Predetermine Badminton Flight Path

Through the analysis of the real-time force on badminton, the basic trend of badminton flight trajectory is judged, and the prediction results of badminton flight trajectory are obtained. Figure 3 shows the stress of badminton in flight.

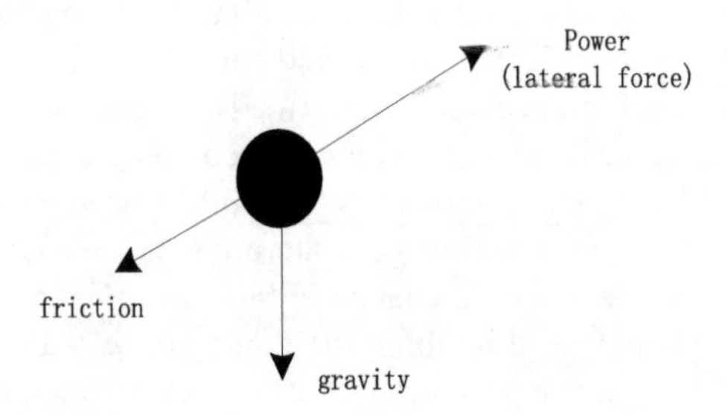

Fig. 3. Schematic diagram of force analysis of flying badminton

Badminton is mainly affected by aerodynamics and gravity during flight. Gravity is always vertically downward. Aerodynamics can be divided into three components, including the resistance along the badminton flight path, the lift that prevents the badminton from falling, and the lateral force that makes the badminton track yaw sideways [10]. The aerodynamic force on the badminton is in direct proportion to the maximum cross-sectional area of the badminton and the incoming flow pressure. The following three equations can be obtained:

$$\begin{cases} X = \kappa_{\text{resistance}} P \cdot S \\ Y = \kappa_{\text{lift}} P \cdot S \\ Z = \kappa_{\text{lateral force}} P \cdot S \end{cases} \tag{10}$$

Among them, $\kappa_{\text{resistance}}$, κ_{lift} and $\kappa_{\text{lateral force}}$ represent drag coefficient, lift coefficient and lateral force coefficient respectively, and the above coefficients are dimensionless proportional coefficients. P represents the actual flow pressure, S represents the maximum cross-sectional area of badminton. When the badminton flies at a certain initial speed in the air, the badminton will hit the air, and the air will make the badminton blades converge to the axis, and part of the air will also be compressed, so that the air pressure at the front end of the badminton cork will increase, and the air will produce a force opposite to the movement direction of the badminton, which is called air resistance. Generally speaking, badminton can be divided into two parts, the front part composed of cork, which gathers most of the weight of the badminton and the skirt part composed of 16 pieces of fur. The weight is light, but it will produce air resistance during flight. Based on this mass arrangement, the gravity gravity center of the badminton should be located at the head of the badminton. The badminton skirt forms a large wind resistance plane in the badminton flight direction. Compared with the ball holder, it will suffer greater air resistance, so the force application point of air resistance is concentrated at the tail of the badminton. Considering the force characteristics of badminton, the prediction results of badminton flight trajectory are obtained.

2.5 Using Particle Filter Algorithm to Determine Badminton Flight Position

Combined with the prediction results of badminton flight trajectory, the accurate flight position of badminton in the image is determined using particle filter algorithm. Figure 4 shows the operation principle of particle filter algorithm.

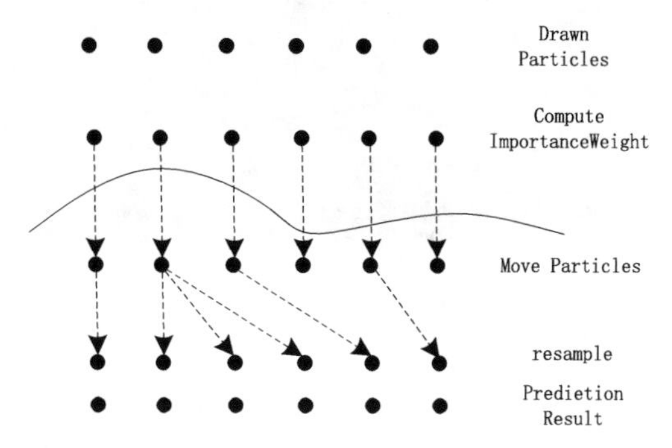

Fig. 4. Schematic diagram of particle filter algorithm

Following the principle shown in Fig. 4, the particle filter algorithm is executed through the steps shown in Fig. 5.

In the actual badminton flight positioning process, first initialize the particles in the badminton flight environment, and get the initial probability density distribution n_{particle} particle set of points, initialize all particle weights to:

$$\varpi_0 = \frac{1}{n_{\text{particle}}} \tag{11}$$

At time k, the dynamic model of object motion is used to predict the particle state at time k and generate the particle set at time k [11]. Formula 12 is used to update the weight of each particle in the particle set at time k:

$$\varpi_k = \varpi_0 \rho\left(x_k^i\right) \tag{12}$$

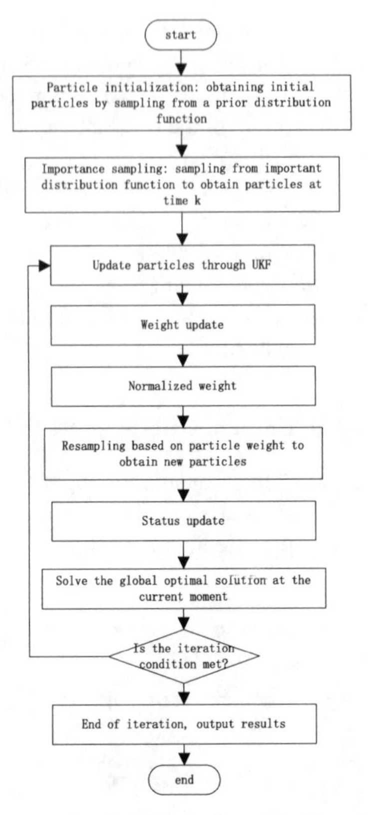

Fig. 5. Operation flow chart of particle filter algorithm

Among them, $\rho()$ represents the probability density distribution of particles, x_k^i represents the state variable of the particle at time k. The effective sampling scale is defined as:

$$L_{eff} = \frac{1}{\sum_{k=1}^{n_{\text{particle}}} \varpi_k} \tag{13}$$

If L_{eff} is less than the threshold value, resampling will be performed. In the process of iteration, the weight of some particles will become smaller and smaller. These particles do not help state estimation much. Instead, they will occupy a lot of computing resources. Therefore, it is necessary to evaluate the weight of all particles to resample particles. That is, the particles with large weight value are copied and the particles with small weight value are reduced, and the small weight particles are replaced by the new sample sequence with the same weight value. Because the residual resampling algorithm has relatively high calculation efficiency and small calculation variance, the residual resampling algorithm [12] is adopted. Repeat the above operations to obtain the global optimal solution at the current time. After the iteration conditions are met, the output particle filter result is the positioning result of the badminton flight path, marked as $(u(t), v(t))$.

2.6 Realize Badminton Flight Track Positioning and Tracking

According to the above method, the badminton flight track positioning results at each time are fused to obtain the positioning and tracking results of the badminton flight track. The tracking results can be quantified as:

$$G = \sum_{t=1}^{T} (u(t), v(t)) \tag{14}$$

Among them, T is the flight time of badminton. Finally, the positioning and tracking results of badminton flight trajectory are visualized and marked with real-time flight trajectory position information to complete the positioning and tracking of badminton trajectory.

3 Experimental Analysis of Track Location and Tracking Performance Test

For the purpose of verifying the trajectory location and tracking accuracy of the badminton flight trajectory location and tracking algorithm based on particle filter, a performance test experiment was designed by combining white box and comparison. The basic principle of this experiment is to install a micro position sensor on the badminton equipment to be located and tracked, obtain the actual value of the real-time position of the badminton, and use this as the comparison standard to judge the accuracy of the location and tracking results of the badminton flight trajectory based on particle filter. In the process of badminton flight, start the location tracking algorithm of optimal design, get the corresponding output results, compare with the actual values obtained, and get the test results that reflect the tracking accuracy of the optimal design algorithm. The principle of comparative testing is to compare the accuracy test results of the design algorithm with the accuracy performance of the traditional algorithm, reflecting the advantages of the design method in performance.

3.1 Choose Badminton and Its Flight Environment

In this experiment, the standard badminton is selected as the target of location and tracking. The length of the standard badminton is between 64 mm and 70 mm. For the same badminton, all the hair pieces are the same length. The top opening of the badminton is arranged in a circular pattern, and the opening diameter is between 58 mm and 68 mm. The smaller the diameter of the badminton, the smaller the flight resistance in the air, the farther the flight distance, and the heavier the ball head will produce the same situation, and vice versa. The badminton feather flying at high speed in the air will be subjected to lateral force, which will lead to the spin of the badminton. The faster the spin speed is, the slower the flight speed of the badminton will be. On the contrary, the faster the flight speed of the badminton will be. Generally, it is better to keep the spin speed at 370–400 rpm. The standard badminton court is selected as the flight environment of badminton, as shown in Fig. 6.

Fig. 6. Actual View of Badminton Flight Field

The standard badminton court is arranged in a rectangle. The sidelines of the court are drawn with white paint with a width of 40 mm. The sidelines are divided into singles and doubles. The width of the singles sideline is 5.18 m, the length is 13.40 m, and the width of the doubles sideline is 6.10 m, the length is 13.40 m. The middle height of the court is 1.524 m. The court is divided into two halves. Players from both sides hit the ball in two halves of the net. According to the regulations of international competitions, the height of the standard badminton court should be at least 9 m. There should be no interfering objects in the entire three-dimensional space where the badminton court is located, so as to ensure that the feather apparatus can fly in the air without interference. The color of the court wall should be dark, and there should be no reflectors or reflective surfaces, so as to avoid interference to athletes. The wind speed in the court should not be greater than 0.2 m/s. Before the experiment, a micro position sensor with a weight of no more than 1g shall be embedded inside the badminton court, and the real-time collected data of the sensor shall be transmitted to the host computer by using the wireless communication network covered inside the badminton court.

3.2 Configure the Flight Path Location and Tracking Algorithm Operating Environment

In order to make the platform and experimental data more accurate, the Intel model i74790 central processor with better hardware facilities is selected. The size of the memory card is 8 GB, which can store the data to be used. The hard disk requirements are relatively high, and 36 GB can be selected to avoid insufficient memory. The network card is set within the range of 1000 mb/s. In order to realize the visual output and measurement of the badminton flight track positioning and tracking results, SONY FCB-EX480CP series new high pixel DSP integrated color analog camera is adopted. It has super light sensitivity and intelligent automatic night vision function, and can remotely set all camera parameters without manual operation, The camera image parameters can be adjusted in time according to the effect of the monitor image to present the image optimization. The minimum illumination of the camera is color 0.7 lx, and the power consumption range is [1.5 W, 2.5 W]. The MV-200 industrial image acquisition card is used as the image acquisition card. This industrial image acquisition card can capture high-precision images in real time, with stable hardware structure design and underlying functions, good hardware compatibility, and can work stably and reliably in a harsh working environment. MV-200 industrial image acquisition card can capture images with high definition, high resolution and high fidelity. In terms of software environment, software development is carried out under the Windows operating system, and the integrated development environment is Visual Studio 2013.

3.3 Input Operation Parameters of Particle Filter Algorithm

In order to meet the call requirements of the badminton flight path location and tracking algorithm to the particle filter algorithm in the running process, the relevant operation coefficient of the particle filter algorithm is set. Set the total number of initial particles to 200, the movement time to 15 s, and assume that one action is taken every second. In addition, the size of the sports field is consistent with the size of the badminton flight field selected.

3.4 Describe the Performance Test Process

This experiment was conducted in the singles scene, and the number of target shuttlecocks tracked in the scene was 1. Athletes with more than 3 years of badminton experience were selected as the athletes for this experiment to ensure that the three movements of high distance ball, flat shot and pick ball were completed in badminton. With the support of sensors, the actual position information of badminton was obtained, as shown in Table 1.

Take the data in Table 1 as the actual position data of badminton flight track. In the configured experimental environment, the optimized badminton flight path location and tracking algorithm based on particle filter is converted into program code that can be run directly by the computer by using development tools, and the set running parameters of particle filter algorithm are input into it. Start the positioning and tracking procedure before the start of badminton movement, and get the corresponding flight path positioning and tracking results with the badminton movement, as shown in Fig. 7.

Therefore, the tracking position information of badminton flight track at any time can be obtained. In order to reflect the performance advantages of the design algorithm, the traditional track location and tracking algorithm based on multi-sensor and the track location and tracking algorithm based on dynamic Snake model are set as the experimental comparison algorithm. The development and operation of the comparison algorithm are completed according to the above process, and the corresponding flight track location and tracking results are obtained.

Table 1. Information of actual position of badminton flight track

Time/s	Lob		Flat shot		Pick the ball	
	x	y	x	y	x	y
1	10	1.6	10	1.1	10	0.6
2	15	1.8	15	1.1	10	0.8
3	20	2.0	20	1.1	12	1.0
4	25	2.2	25	1.1	15	1.6
5	30	2.5	30	1.1	20	1.7
6	35	2.3	35	1.2	25	1.8
7	40	2.2	40	1.2	30	1.9
8	45	2.0	45	1.1	35	2.0
9	50	1.8	50	1.1	40	2.2
10	55	1.6	55	1.1	45	2.4

3.5 Setting Performance Test Indicators for Track Positioning and Tracking

According to the purpose of the experiment, the tracking performance of the algorithm is reflected from the tracking accuracy and tracking timeliness of the track location. The test indicators of the tracking accuracy are tracking error and track length error. The numerical results are as follows:

$$\varepsilon = \sum_{t=1}^{t_{\text{experiment}}} \left|x_{\text{track}}(t) - x_{\text{reality}}(t)\right| + \left|y_{\text{track}}(t) - y_{\text{reality}}(t)\right| \tag{15}$$

In the above formula, $(x_{\text{track}}(t), y_{\text{track}}(t))$ and $\left(x_{\text{reality}}(t), y_{\text{reality}}(t)\right)$ represent the badminton position tracking result and actual position coordinates respectively. The test results of badminton flight path length tracking error are as follows:

$$\varepsilon_l = \left|l_{\text{track}} - l_{\text{reality}}\right| \tag{16}$$

Among them, l_{track} and l_{reality} are the tracking value and the actual value of the badminton flight path length. In addition, the test index of track tracking timeliness is tracking update delay, and its numerical result is:

$$\Delta t = t_{out} - t_{\text{sense}} \tag{17}$$

where, t_{out} indicates the output time of badminton flight track positioning and tracking results, t_{sense} the time of data generation is sensed by the built-in sensor for badminton. Finally, the smaller the tracking error and track length error are, which proves that the higher the tracking accuracy of the corresponding algorithm is, the smaller the tracking update delay is, which indicates that the better the track tracking timeliness is.

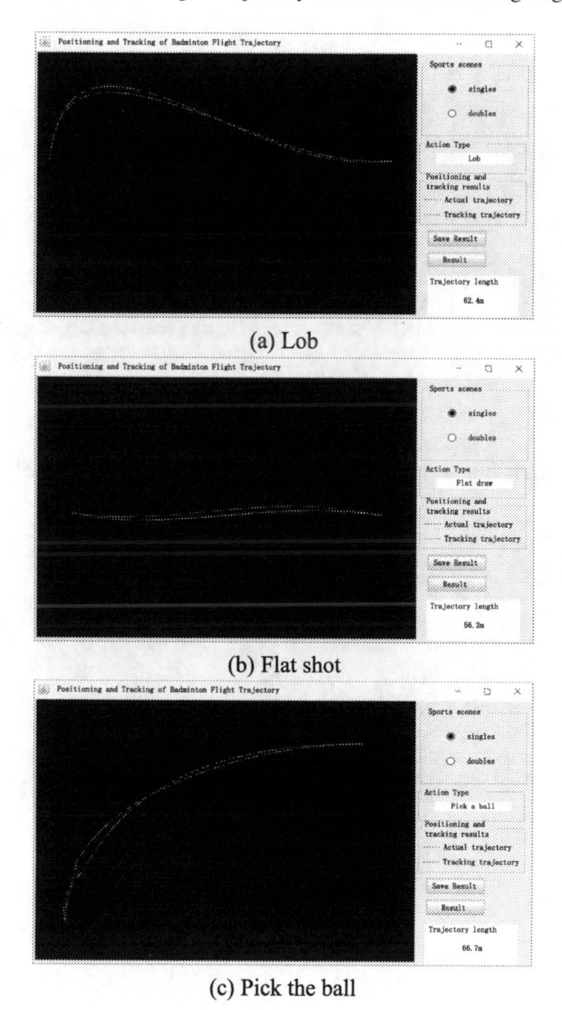

Fig. 7. Badminton Flight Path Positioning and Tracking Results

3.6 Performance Test Results and Analysis

Through the statistics of relevant data and the calculation of Formula 15, the test results of badminton flight track positioning and tracking error are obtained, as shown in Fig. 8.

It can be seen intuitively from Fig. 8 that the tracking error of the optimized design algorithm is smaller than that of the traditional positioning and tracking algorithm. Because the design method collects badminton flight images and preprocesses the initial flight images through morphology and filtering, effectively reducing the positioning and tracking error of badminton flight trajectory. In addition, the test results of track length error are shown in Table 2.

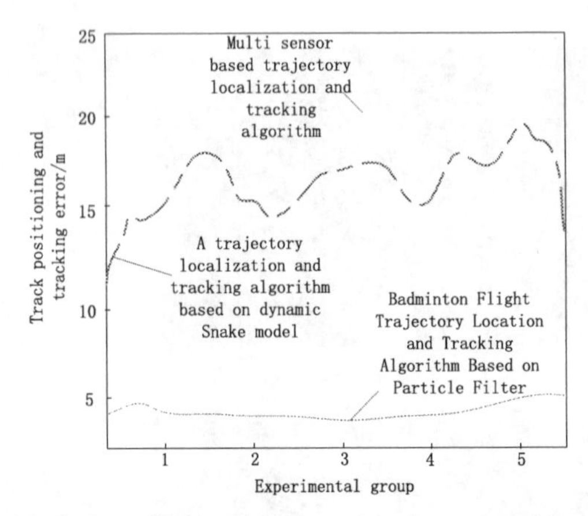

Fig. 8. Test results of badminton flight trajectory positioning and tracking errors using different methods

Table 2. Test Data Table for Length Error of Badminton Flight Trajectory by Different Methods

Experiment No	Actual length of badminton flight path/m	Multi sensor based track location and tracking algorithm output track length/m	Track location and tracking algorithm based on dynamic Snake model output track length/m	Badminton flight trajectory location and tracking algorithm based on particle filter output trajectory length/m
1	62.4	62.9	62.0	62.3
2	56.2	56.9	56.8	56.0
3	66.7	65.8	66.2	66.5
4	61.8	61.2	61.4	61.7
5	56.8	56.0	56.3	56.7
6	66.5	65.7	66.1	66.4

By substituting the data in Table 2 into Formula 16, it is calculated that the average track length tracking error of the two comparison algorithms is 0.72 m and 0.47 m respectively, and the average track length tracking error of the badminton flight track location and tracking algorithm based on particle filter is 0.13 m. In addition, the test results of track tracking timeliness are obtained through the calculation of Formula 17, as shown in Fig. 9.

As can be seen from Fig. 9, the average tracking update delay of the two comparison algorithms is 16.8 s and 17.2 s respectively, while the average tracking update delay of the badminton flight path location and tracking algorithm based on particle filter optimized design is 7.6 s.

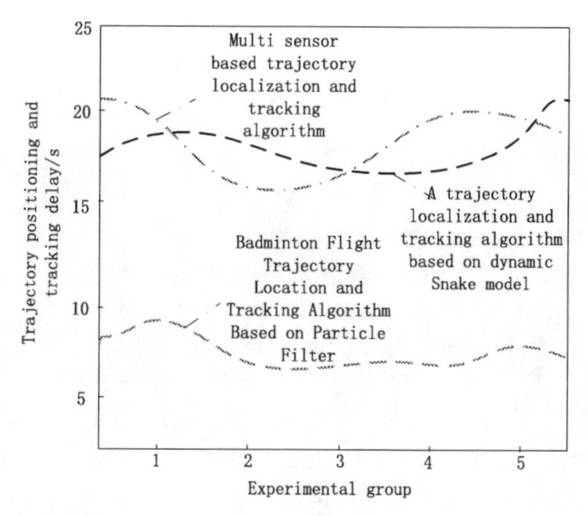

Fig. 9. Comparison results of time-effectiveness testing for badminton flight trajectory positioning and tracking using different methods

In order to verify the stability of badminton flight trajectory positioning and tracking in the design method, the success rate of tracking is used as an evaluation indicator. The higher the value, the higher the stability of badminton flight trajectory positioning and tracking in the method. Three methods were used for comparison, and the success rates of badminton flight trajectory localization and tracking using different methods are shown in Table 3.

Table 3. Success rate of badminton flight trajectory localization and tracking using different methods

Experiment No	Tracking success rate of trajectory positioning and tracking algorithm based on multiple sensors/%	Tracking success rate of trajectory positioning and tracking algorithm based on dynamic Snake model/%	Tracking success rate of badminton flight trajectory localization and tracking algorithm based on particle filter/%
1	92.0	89.3	98.9
2	91.8	87.0	96.7
3	90.2	86.5	95.8
4	91.4	88.7	97.2
5	92.3	86.7	96.5

According to Table 3, when the experimental number is 5, the tracking success rates of the multi-sensor based trajectory positioning and tracking algorithm and the dynamic Snake model based trajectory positioning and tracking algorithm are 91.5% and 87.6%, respectively. The tracking success rate of the badminton flight trajectory localization and tracking algorithm based on particle filtering is as high as 97%. From this, it can be seen that the particle filter based badminton flight trajectory localization and tracking algorithm has high stability.

4 Conclusion

With the development of computer image processing technology and the advent of camera timely return visit system, visual tracking and detection technology has become increasingly mature, making target tracking and monitoring not only applied to artificial intelligence and other fields, but also launched a new chapter in sports. In sports, for the detection of balls, especially for the sphere moving in the air, because of its fast motion speed, ordinary cameras can not track its motion state in real time for the rapid movement of the sphere, and the image capture during the movement of the sphere will also have drag. The optimized badminton flight trajectory location and tracking algorithm based on particle filter effectively solves the problems existing in the traditional tracking algorithm, and realizes the accurate location and tracking of badminton flight trajectory. The experimental results show that the optimized design algorithm has good effectiveness and feasibility.

References

1. Song, D., Gan, W., Yao, P., et al.: Guidance and control of autonomous surface underwater vehicles for target tracking in ocean environment by deep reinforcement learning. Ocean Eng. **250**, 110947 (2022)
2. Zuo, R., Li, Y., Lv, M., et al.: Realization of trajectory precise tracking for hypersonic flight vehicles with prescribed performances. Aerosp. Sci. Technol. **111**(2), 106554 (2021)
3. Lv, Z., Li, F., Qiu, X., et al.: Effects of motion compensation residual error and polarization distortion on UAV-borne PolInSAR. Remote Sens. **13**(4), 618 (2021)
4. Gishkori, S., Daniel, L., Gashinova, M., et al.: Imaging moving targets for a forward scanning SAR without radar motion compensation. Signal Process. **185**, 108110 (2021)
5. Pereira, S.C., Conde, E.R., Luz-de-Almeida, L.A., et al.: A power-line communication system employing median filtering for power control enhancement of wind generators. Int. J. Commun. Syst. **35**(12), e5203 (2022)
6. Wang, X.-W., Peng, H.-J., Liu, J., Dong, X.-Z., Zhao, X.-D., Lu, C.: Optimal control based coordinated taxiing path planning and tracking for multiple carrier aircraft on flight deck. Defence Technol. **18**(2), 238–248 (2022)
7. Lv, C., Li, J., Kou, Q., et al.: Stereo matching algorithm based on HSV color space and improved census transform. Math. Probl. Eng. **2021**, 1–17 (2021)
8. Waldamichael, F.G., Debelee, T.G., Ayano, Y.M.: Coffee disease detection using a robust HSV color-based segmentation and transfer learning for use on smartphones. Int. J. Intell. Syst. **37**(8), 4967–4993 (2022)
9. Dong, C., Guo, Z., Chen, X.: Robust trajectory planning for hypersonic glide vehicle with parametric uncertainties. Math. Probl. Eng. **2021**(5), 1–19 (2021)
10. Wang, B., Zhang, Y., Zhang, W.: Integrated path planning and trajectory tracking control for quadrotor UAVs with obstacle avoidance in the presence of environmental and systematic uncertainties: Theory and experiment. Aerospace Sci. Technol. **120**, 107277 (2022)
11. Buelta, A., Olivares, A., Staffetti, E., et al.: A gaussian process iterative learning control for aircraft trajectory tracking. IEEE Trans. Aerospace Electron. Syst. **57** (2021)
12. Cardoso, D.N., Esteban, S., Raffo, G.V.: A new robust adaptive mixing control for trajectory tracking with improved forward flight of a tilt-rotor UAV. ISA Trans. **2021**(110-), 110 (2021)

Design of Substation Battery Condition Monitoring System Based on SDH Network

Feng Xu(✉), Quan Zi, Chen Zhao, Nannan Wang, and Yan Wang

Suzhou Power Supply Company, Suzhou 23400, China
19166586244@163.com

Abstract. In the process of substation battery condition monitoring, because the status parameters are real-time fluctuations, resulting in relatively large errors in the monitoring results, this paper proposes the design and research of substation battery condition monitoring system based on SDH network. TW-SDH6000 is used as the hardware device of the substation battery condition monitoring system. In the software design phase, the SDH network is used to simulate and process the battery resources, so that when the relevant equipment in the substation is abnormal and the battery status fluctuates, the self-healing function can be used to filter this part of data. In the condition monitoring phase, the PCA method is used to reduce the dimension of the original data, and Pearson correlation coefficient is used to analyze the relationship between the original data and the simulation processing results. Realize accurate monitoring. In the test results, the error of the design method for the monitoring results of battery voltage amplitude, input frequency and coil proportion is significantly lower than that of the control group.

Keywords: SDH Network · Substation Battery · Condition Monitoring · TW-SDH6000 · Simulation Processing · Self-Healing Function · Pca Method · Pearson Correlation Coefficient

1 Introduction

The maximum usable capacity that the battery can release as a backup power supply is the most important indicator. It can be seen from the definition of the state of health (SOH) of the battery that under certain discharge conditions, its value is the ratio [1] of the capacity released by the battery from the full state at a certain rate to the cut-off voltage and its corresponding nominal capacity. Therefore, the maximum usable capacity of the battery is directly related to its SOH value. However, SOH can be directly measured and displayed [2], unlike the amount of fuel in a household car, because there is no exact parameter that is proportional to the health status of the battery, and the health status depends on the comprehensive evaluation and estimation of multiple time-varying parameters of the battery [3]. Research on battery SOH estimation methods can be roughly divided into three categories: traditional offline method, data driven method and fusion model method. The traditional off-line method is to measure the terminal

L. Yun et al. (Eds.): ADHIP 2023, LNICST 549, pp. 345–359, 2024.
https://doi.org/10.1007/978-3-031-50549-2_24

voltage, discharge current, temperature and other basic physical electrical quantities [4] of the battery for off-line large current discharge, and establish a mathematical model based on various physical parameters using statistical model identification method [5, 6]. The offline method mainly includes internal resistance folding algorithm, capacity attenuation method, partial discharge method, impedance analysis method [7], etc. Therefore, the off-line method is not widely used at this stage and is only applicable to theoretical research in the laboratory. The data driven method can be adaptive to meet the changing system parameters, with excellent real-time performance, good robustness and high precision. Data driven methods mainly include Kalman filter, support vector machine (SVM), correlation vector (RVM) and artificial neural network (ANN) [8]. The estimation principle of the data-driven method is relatively simple, and it does not need to pay attention to the complex degradation mechanism inside the battery. The estimation principle of the fusion model method is relatively complex. The circuit model and digital drive fusion method not only focus on the complex degradation mechanism inside the battery [9], but also need to learn the historical data from the perspective of the historical degradation data of the battery in the subway power supply system with the help of machine learning algorithm to estimate the parameters of the circuit degradation model. Combined with the high accuracy of the circuit model, the battery SOH is evaluated [10]. At present, the fusion model method, which is a new SOH estimation method integrating several battery models or estimation methods, is more and more important in practical situations. Its purpose is to maximize the use of the advantages of two different models to overcome their shortcomings, so as to improve the estimation accuracy [11].

At this stage, there have been some achievements in the estimation of battery SOH, but there are still many problems, because most of the time when the battery pack is in standby, it is connected in parallel with the charging device for trickle charging. The battery with a service life of less than five years has only been discharged for two years. If the power supply system is not powered off, the battery will never be discharged. However, most of the existing online monitoring systems rely on floating charge information for health assessment, which has a large error or even no reference. Or the battery needs to be discharged in depth offline, but the reliability and safety of power supply cannot be guaranteed when the battery capacity is less than 80%.So how to use shallow discharge or even floating charge to evaluate the battery performance is the difficulty of the current online detection technology.

On this basis, this paper proposes the design and research of the substation battery condition monitoring system based on SDH network, and carries out targeted design from both hardware and software perspectives, and analyzes the application performance of the design system. To solve the problem of low reliability of monitoring data, TW-SDH6000 is used as the main hardware equipment for data transmission, SDH network is used to filter battery resources, PCA method is used to reduce the dimension of original data, and the problem of large errors in monitoring results caused by high data complexity is solved. Accurate monitoring of battery status parameters is completed through Pearson correlation coefficient.

2 Hardware Design

The substation battery condition monitoring system designed in this paper is based on SDH network. Therefore, it is necessary to set up a reliable SDH optical transceiver device according to the actual needs. Therefore, TW-SDH6000 is used as the SDH optical transceiver of the system in this paper.TW-SDH6000 is a new generation open MSAP integrated service access platform. SDH group channel optical port rate supports STM-1, STM-4 levels, and can be smoothly upgraded to STM-16 level. It has the bidirectional bearing capacity of SDH/MSTP transmission network and IP MAN, supports SDH and Ethernet independent bus architecture, and can truly achieve high reliability integrated service access and transmission.

The TW-SDH6000 has powerful slot crossing and IP data exchange capabilities, which can realize flexible scheduling of various services, and perfectly realize the access and convergence of traditional E1 services, optical branch services, EOS/EOP services and other services; Provide complete service protection capability and provide customers with stable transmission lines; It supports multiple network topologies and can meet the application requirements of complex networks; It has a comprehensive and intelligent management platform, which can realize end-to-end monitoring and management of business [12]. It is widely used in the field of large customer dedicated line access, base station interconnection and video monitoring. After analyzing the characteristics of TW-SDH6000, its structure configuration is shown in Table 1.

Table 1. TW-SDH6000 Equipment Structure Configuration

Category	Board name	Model	Brief description	Insertable slot
SDH uplink	STM-4 Optical cluster road panel	CU622	Two × STM-4, built-in cross connection and SET functions	5, 8
	STM-1 Optical cluster road panel	CU155	Two × STM-1, built-in cross connection and SET functions	5, 8
Ethernet Core convergence	Ethernet Core aggregation disk	CUGE4A	Two × GX + 2 × GE supports port VLAN and 802.1QVLAN and link aggregation Trunk function. The 1000M Ethernet data of up to 9 service slots can be aggregated	6, 7

The overall structure of TW-SDH6000 is displayed as a 19 inch 7U plug-in box with 16 card slots; LVDS has no clock wiring backplane, SDH and GE independent bus structure. It can be seen from the information shown in Table 12 that the TW-SDH6000 has a dual SDH cluster board, dual Ethernet core convergence board, dual network management board, and dual power supply board structure, so it is safer and more reliable in the operation phase.

The configuration of SDH group service of TW-SDH6000 is analyzed, as shown in Table 2.

Table 2. SDH Group Service Configuration of TW-SDH6000 Equipment

Board name	Model	Brief description	Insertable slot
E1 mapping disk	E1-16	16 channel E1, optional 75 Ω or 120 Ω	1–4, 6–7, 9–11
STM-1 Optical branching board	TU155-8	Eight × STM-1, built-in cross connection function.VC12 full crossing is supported in the board	1–4, 6–7, 9–11
EoS Optical access panel	EoS-8FX8	Eight × FX, 8VCG physical isolation, EoS, GFP/LCAS/VCAT, flow control, TS1000 protocol	1–4, 6–7, 9–11
EoS Electric access panel	EoS-8FE8	Eight × FE, 8VCG physical isolation, EOS, GFP/LCAS/VCAT, flow control	1–4, 6–7, 9–11
EoP Optical access panel	EoP-8FX8	Eight × FX, 8VCG physical isolation, EoP, GFP/LCAS/VCAT, multiple E1 converter mode; HDLC package, single E1 converter mode;TS1000 protocol	1–4, 6–7, 9–11
EoP Electric access panel	EoP-8FE8	Eight × FE, 8VCG physical isolation, EoP, GFP/LCAS/VCAT, multiple E1 converter mode; HDLC package, single E1 converter mode;	1–4, 6–7, 9–11

According to Table 2, the TW-SDH6000 equipment can provide up to four STM-4 or STM-1 group optical interfaces; LC type SFP optical transceiver module is used for group optical ports, which can support hot plug; Maximum cross capacity is 32VC4 × 32VC4、96TU-3 × 96TU-3、2016VC12 × 2016VC12. In addition, the TW-SDH6000 device also supports 1 + 1 cluster board card protection to achieve hot backup of cluster, crossover, and clock modules, as well as single/double end 1 + 1 linear multiplex section protection and SNCP protection, which can maximize the connection requirements of point-to-point, chain, ring and other topologies.

The configuration of Ethernet core switching service of TW-SDH6000 equipment is analyzed, as shown in Table 3.

It can be seen from Table 3 that the TW-SDH6000 device supports up to 4 gigabit optical ports and 4 gigabit electrical ports in terms of Ethernet core switching service configuration; Each disk can independently aggregate data from nine gigabit Ethernet ports on the backplane.

Table 3. Ethernet core switching service configuration of TW-SDH6000 equipment

Board name	Model	Brief description	Insertable slot
EoP optical access convergence disk	EoP-8FX8A	Eight × FX, 8VCG, EoP, GFP/LCAS/VCAT, multiple E1 converter mode; HDLC package, single E1 converter mode; Support switching function	1–4, 6–7, 9–11
EoP power access convergence panel	EoP-8FE8A	Eight × FE, 8VCG, EoP, GFP/LCAS/VCAT, multiple E1 converter mode; HDLC package, single E1 converter mode; Support switching function	1–4, 6–7, 9–11
EoS service convergence disk	EoS-16GE2A	One × Combo port + 1 × GE port + 1 × FE test port.16 VCGs; Board convergence ratio: 16:1;Port VLAN and 802.1QVLAN support QinQ. Support GFP/LCAS/VCAT	1–4, 6–7, 9–11
EoP business Convergence disk	EoP-16GE2A	One × Combo port + 1 × GE port + 1 × FE test port.16 VCGs; Board convergence ratio: 16:1;Port VLAN and 802.1QVLAN support QinQ. Support multiple E1 or single E1 conversion mode. It can support GFP and HDLC packaging structures	1–4, 6–7, 9–11

In addition, in terms of multiple branch service configurations, the TW-SDH6000 device supports the access and convergence of downlink SDH optical branch, E1, EOS, EOP and other branch services. With such settings, it can be interconnected with remote Ethernet optical transceiver, protocol converter, SDH optical termination and other products. At the same time, the TW-SDH6000 equipment also supports 1 + 1 channel protection and built-in E1 bit error tester test function, which maximizes the stability and reliability in the operation phase. Combine application requirements in different environments. The TW-SDH6000 device is designed specifically for the timing function. It can track the timing source of group optical ports, external synchronous timing source and branch timing source, provide it with external synchronous timing source output, and support automatic or forced selection of timing source, as well as four timing modes of free oscillation, tracking, locking and maintaining.

Finally, the configuration of TW-SDH6000 equipment management is shown in Table 4.

Table 4. TW-SDH6000 Device Management Configuration

Category	Board name	Model	Brief description	Insertable slot
Snap in	Network management panel	NMU7000	The dual network management disks are backup to each other, providing system management functions and supporting network management cascading	12, 13
Power Supply	DC power panel	RPW300DC	Dual power redundant backup	14, 15
	AC power panel	RPW300AC		14, 15

With the help of the management configuration shown in Table 4, the TW-SDH6000 device can provide equipment rack top alarm output, and support the DCC management channel and the built-in DCN network management channel. In SNMP_V1 and SNMP_Under the V2 protocol, it can be used with the RayView network management platform of C/S architecture to realize the online software upgrade function of local device boards and remote devices.

3 Software Design

3.1 Simulation Processing of Battery Resources Based on SDH Network

In order to reasonably show the structure of the power SDH communication network to users, observe the impact of faults on the SDH network, the impact on the circuit carrying service, and the recovery process of the circuit to the service, the author determines the functional requirements, network resource simulation requirements, and network behavior requirements of the power SDH communication network self-healing simulation system through the research on the self-healing mode of the power SDH communication network. SDH network operation and maintenance personnel apply self-healing simulation in the simulation system in the following ways, as shown in Fig. 1.

(1) Configuring the SDH network includes setting and removing the connection of devices and logical ports; Configuration of cross connection in the circuit; The configuration of the power protection group and the configuration of the circuit carrying the services.
(2) Simulate and repair the accident, observe the impact of the accident on the business in the SDH network, and get the alarm reported by the corresponding equipment after the accident. Realize the repair of the accident, eliminate the impact of the accident on the SDH network, and eliminate the corresponding alarm after the accident is repaired.
(3) SDH network self-healing application: the system receives the incentive of service failure and network failure, and completes network self-healing through protection, so as to recover services.

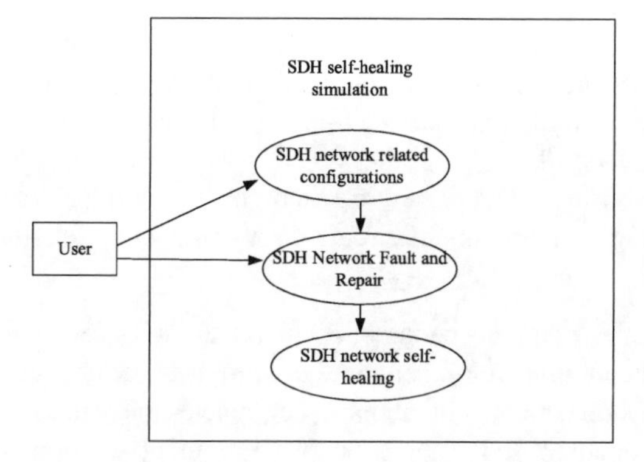

Fig. 1. SDH network self-healing simulation function implementation mode

Figure 2 is the structure diagram of battery resource model based on SDH network.

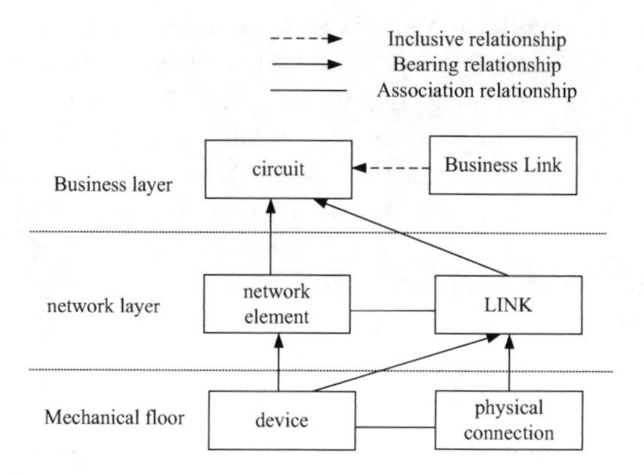

Fig. 2. Structure of battery resource model based on SDH network

As shown in Fig. 2, the network resource model required for self-healing simulation of power SDH communication network can be divided into three levels of equipment, network and service for modeling. The device layer mainly models the information of the physical network composed of physical devices and physical connections; The network layer mainly models the information of the topological network composed of logical network elements and logical links; The service layer is mainly used to model the information of the service network composed of the service link and the circuit carrying the service link.

Based on the condition monitoring of the battery in the substation, the self-healing simulation function of the power SDH communication network is mainly divided into the following three parts:

(1) Simulation of equipment layer resources and their status changes caused by faults, including resource simulation of infrastructure sites in SDH network and resource simulation of physical connecting optical cables in SDH network.

(2) The simulation of network layer resources and the simulation of their state changes caused by faults, including the resource simulation of transmission system, SDH equipment, logical port, multiplex section, time slot, cross connection and protection group in SDH network.
(3) Simulation of business layer resources and corresponding alarm generated due to accident; Resource simulation of circuit, circuit protection relationship, service and service group.

On this basis, in the resource management phase, the self-healing simulation function of the power SDH communication network needs to meet the demand for SDH network simulation. The resource management mainly completes the creation and deletion of the topological network composed of logical network elements and logical links. The basic behavior settings of SDH transmission can be expressed as.

$$s(x) = \sum [d_i(s), c_i(s)] \tag{1}$$

Among them, $s(x)$ Indicates the basic behavior of SDH transmission, $d_i(s), c_i(s)$ Indicates deletion and modification actions with transmission SDH as the core, including adding transmission system and setting transmission system related information; Modify the information related to the transmission system and delete the transmission system, and delete the station, device, port, cross connection, time slot and other information in the transmission system in cascade. The basic behaviors of the site include adding sites and setting site related information; Modify the relevant information of the site and delete the site, and cascade delete the device port, cross connection, time slot, connection with other sites and other information in the site.

For the basic behavior of optical cables, the specific settings can be expressed as.

$$l(x) = \sum [d_i(l), c_i(l)] \tag{2}$$

Among them, $l(x)$ Represents the basic behavior of the optical cable, $d_i(l), c_i(l)$ Indicates the deletion and modification actions with optical cable as the core, including adding optical cable, and setting the information such as the initial site and type of optical cable; Modify the optical cable, set the initial site and type information of the optical cable, and delete the optical cable.

For the basic behavior of the SDH equipment TW-SDH6000, the specific settings can be expressed as.

$$S(x) = \sum [d_i(S), c_i(S)] \tag{3}$$

Among them, $S(x)$ Represents the basic behavior of TW-SDH6000, $d_i(S), c_i(S)$ Indicates the deletion and modification actions with TW-SDH6000 as the core, including adding SDH equipment and setting the corresponding information of SDH equipment; Modify the corresponding information of the SDH device, delete the SDH device, delete the connection information between the SDH device and the power supply device, and cascade delete the port, cross connection, time slot and other information on the SDH device.

According to the way shown above, the design of battery resource simulation processing operation is realized based on SDH network.

3.2 Substation Battery Condition Monitoring

PCA (Principal Component Analysis) is the most commonly used linear dimension reduction method [13, 14]. The main measure of PCA dimension reduction is the maximum value of data variance after coordinate transformation. When the square difference is the largest, the dimension of the transformed coordinate axis is lower, but the original data information is relatively perfect. Principal component analysis (PCA) is a dimensionality reduction method that retains the most original data information. In the stage of substation battery condition monitoring using battery resource simulation processing data in SDH network, the data dimension is reduced first. Let the n-dimensional vector be a coordinate axis direction of the target subspace (called the mapping vector), maximize the variance after data mapping, and have

$$\max_{W} \frac{1}{m-1}\left(\sum W^T (X_i - \overline{X})\right)^2 \tag{4}$$

Among them, m Is the number of data instances of battery resource simulation processing in SDH network, X_i Is the vector representation of data instance i, $\overline{X}$ Is the average vector of all data instances. It is defined as a matrix containing all mapping vectors as column vectors. After linear algebraic transformation, the following optimization objective function can be obtained

$$\min_{W} tr(W^T A W), s.t. W^T W = I \tag{5}$$

Among them, tr Represents the trace of the matrix, A Is the data covariance matrix, which can be expressed as

$$A = \frac{1}{m-1}\sum (X_i - \overline{X})(X_i - \overline{X})^T \tag{6}$$

Easy to get the best W Is determined by the data covariance matrix k It is composed of the eigenvectors corresponding to the largest eigenvalues. These eigenvectors form a set of orthogonal bases and best retain the information in the data. PCA output Y Can be expressed as

$$Y = W'X \tag{7}$$

from X The original dimension of is reduced to k Dimension. PCA seeks to maximize the internal information of the data after dimensionality reduction, and measure the importance of the projection direction by the size of the data variance in the projection direction. Principal component analysis is to condense data and condense multiple indicators into several unrelated general indicators (principal components), so as to achieve the purpose of dimension reduction.

The relationship between the characteristic parameters obtained after dimensionality reduction by PCA and the status data after simulation processing of battery resources in SDH network is unknown, so the dimensionality reduction is directly used.

It is unreasonable to use the parameters of as the input of the health evaluation model to evaluate the state data, so the dimension reduced feature.

The correlation coefficient between the characteristic data and the state data is determined, and the characteristic parameters that meet the linear relationship are extracted as the health factor input of the model to obtain the state data value. In this paper, Pearson correlation coefficient is used as an evaluation index to measure the degree of correlation between each group of data, which is calculated by the product difference method. Based on the deviation between the two groups of data and their respective average values, the correlation between the two variables is reflected by multiplying the two deviations. Pearson correlation coefficient is defined as the quotient of covariance and standard deviation between two variables, which can be expressed as

$$\rho_{X,Y} = \frac{\text{cov}(X, Y)}{\sigma_X \sigma_Y} = \frac{E[(X - \mu_X)(Y - \mu_Y)]}{\sigma_X \sigma_Y} \tag{8}$$

Among them, $\rho_{X,Y}$ The Pearson correlation coefficient between the status data after simulation processing of battery resources and the ideal value, σ_X and σ_Y Represent the covariance of characteristic parameters that meet the linear relationship, μ_X and μ_Y Represent the covariance of the characteristic parameter covariance satisfying the linear relationship respectively. The Pearson correlation coefficient is [–1,1], and the degree of correlation between data is determined by ρ the absolute value of reflects that the larger the absolute value is, the higher the correlation between data X and Y. It should be noted that the correlation coefficient can be defined only when the standard deviation of two variables is not zero.

In this way, according to the correlation between X and Y, combined with the simulation results of battery resources in SDH network, the monitoring of battery status parameters is realized.

4 Test Analysis

4.1 Experimental Platform Setting

In the experimental demonstration phase, simulation experiments are mainly used to determine the feasibility of the scheme. The simulation experiments are mainly based on MATLAB tools. MATLAB is mainly used for data analysis, calculation and model simulation. Its main feature is that it has strong data operation and processing capabilities, and can realize matrix operation, fourier transform and optimization algorithms. Simulink is a visual simulation tool in MATLAB, which supports system design simulation and automatic code generation, while providing graphics editor, result export and other functions, its main function modules are as follows:

(1) Continuous module: mainly including calculus kernel function model;
(2) Discrete module: mainly including integrator filter;
(3) Function & Tables: mainly including function calls and user-defined functions;
(4) Math module: mainly including common mathematical operation module;
(5) Receiver module (Sinks): mainly including oscilloscope and other graphic display modules;

(6) Input source module (Sources): mainly includes some signal generation modules, such as sine signal generator.

The specific experimental process is as follows:

Step 1: Use TW-SDH6000 as the hardware equipment of the substation battery condition monitoring system;

Step 2: Use SDH network to simulate and process battery resources, and use self-healing function to filter some data;

Step 3: Reduce the dimension of the original data by PCA method;

Step 4: Use Pearson correlation coefficient analysis and process the relationship between the results to achieve accurate monitoring.

The experimental platform is mainly based on the signal generator, sensor, lock-in amplifier and battery. The core material of the sensor is nanocrystalline. As a new soft magnetic material, nanocrystalline material has high permeability and wide frequency characteristics, and has high inductance, small size, good filtering effect, and good thermal stability. It is widely used in high-performance magnetic cores, inductive components transformer, etc., nanocrystals have become high-quality materials for common mode inductance cores, and their characteristics are shown in Table 5.

Table 5. Common Mode Inductor Core Parameters

S/N	Characteristic	Index content
1	Texture of material	Fe based nanocrystals
2	Saturated magnetic induction	1.25T
3	Permeability (10 kHz)	80000
4	Permeability (100 kHz)	20000
5	Resistivity	115
6	Lamination coefficient	0.78
7	Core shape	Toroidal core

In order to ensure that the equipment has strong anti-interference capability and is easy to install, the nanocrystalline coil is often equipped with a steel shell. During the test, the data-driven monitoring method and the fusion model monitoring method are set as the control group for the test, and the performance of the designed system is objectively evaluated by analyzing the specific test results.

4.2 Induction Measurement Experiment

The experiment mainly adopts the control variable method, and the sensor material is nanocrystalline. Finally, the relationship between the output voltage and the resistance to be measured is calculated. The output voltage is mainly related to the amplitude and frequency of the excitation and the turns ratio of the original and auxiliary coils. The control variable experiment is designed at the loading end to explore the impact of each influencing element on the monitoring results of the data system and analyze the relevant errors.

(1) Amplitude experiment.

During the experiment, first determine the frequency range. Under the 0 Ω resistance test environment, the frequency changes from 10 Hz to 1 MHz. It is found that at a low frequency, the environmental interference is obvious, and at high frequency, the interference can be significantly reduced.

The data is stable, and the frequency of 500 kHz is the best. Therefore, 500 kHz frequency is selected to measure the change of output voltage caused by the change of voltage amplitude, and the average error of output voltage under different voltage amplitude is obtained by transforming different resistance values. See Table 6.

Table 6. Average error of test results under different voltage conditions/%

Amplitude/V	Data driven monitoring method	Fusion model monitoring method	This paper designs the monitoring system
1.0	16.56	17.26	16.41
2.0	16.33	16.30	15.14
3.0	15.85	15.99	15.41
4.0	13.02	14.30	11.76
5.0	2.45	3.35	1.72
6.0	2.28	3.24	1.01
7.0	2.20	3.20	1.00
8.0	1.96	2.15	1.21

According to the test results shown in Table 6, in the three groups of test results, when the voltage amplitude is within the range of 6.0–7.0V, the average error of the corresponding monitoring results is the smallest. Among them, the error corresponding to the data driven monitoring method is within the range of 2.20–2.30%, the error corresponding to the fusion model monitoring method is within the range of 3.20–3.25%, and the error corresponding to the monitoring system designed in this paper is within the range of 1.00–1.05%. In contrast, the monitoring results of the monitoring system designed in this paper are more reliable. Not only that, the data information of the overall test results is analyzed. Among them, the maximum error of the data driven monitoring method is 16.56%, the maximum error of the fusion model monitoring method is 17.26%, and the maximum error of the monitoring system designed in this paper is 16.41%. In contrast, the monitoring system designed in this paper also shows obvious advantages.

(2) Frequency experiment.

According to the amplitude experiment results, when the voltage amplitude is between 6.0 and 7.0 V, the corresponding monitoring error is small. For this reason, the voltage amplitude is selected as 6.6 V, and the average error of different monitoring methods and system outputs under different frequency inputs is tested. The data results are shown in Table 7.

Table 7. Average Error of Test Results at Different Frequencies/%

Frequency (Hz)	Data driven monitoring method	Fusion model monitoring method	This paper designs the monitoring system
5000	33.01	33.71	20.86
10000	22.77	22.74	11.58
15000	16.15	16.29	9.71
20000	9.46	9.74	7.2
50000	1.78	2.68	1.15
100000	1.56	2.52	1.09
500000	1.51	2.51	1.03
1000000	2.02	2.21	1.27

According to the test results shown in Table 7, among the three groups of test results, when the input voltage frequency is 500 kHz, the average error of the corresponding monitoring results is the smallest. Among them, the error corresponding to the data-driven monitoring method is 1.51%, the error corresponding to the fusion model monitoring method is 2.51%, and the error corresponding to the monitoring system designed in this paper is 1.03%. In contrast, the monitoring system designed in this paper is more reliable for the monitoring results of testing battery frequency status parameters. Similarly, the data information of the overall test results is analyzed. Among them, the maximum error of the data-driven monitoring method is 33.01%, the maximum error of the fusion model monitoring method is 33.71%, and the maximum error of the monitoring system designed in this paper is 20.86%. The battery condition monitoring system designed in this paper also shows obvious advantages.

(3) Original and auxiliary coil experiment.

Based on the above experimental results, when testing the coil, set the amplitude and frequency of the excitation signal as 6.6 V and 500 kHz respectively. On this basis, under the condition of different primary and secondary coil ratios, the output average error under different coil ratios is obtained by transforming different resistances. The specific test results are shown in Table 8.

According to the test results shown in Table 8, when the coil ratio is 1:2, the average error of the corresponding monitoring results is the smallest among the three groups of test results. Among them, the error corresponding to the data-driven monitoring method is 0.42%, the error corresponding to the fusion model monitoring method is 0.70%, and the error corresponding to the monitoring system designed in this paper is 0.01%. In contrast, the monitoring results of the monitoring system designed in this paper for testing the proportional state parameters of the battery coil are more reliable. Similarly, the data information of the overall test results is analyzed. Among them, the maximum error of the data-driven monitoring method is 1.65%, the maximum error of the fusion model monitoring method is 2.00%, and the maximum error of the monitoring system designed in this paper is 1.01%. The battery condition monitoring system designed in this paper also shows obvious advantages.

Table 8. Average error of test results under different coil ratios/%

Coil ratio	Data driven monitoring method	Fusion model monitoring method	This paper designs the monitoring system
4:1	1.65	1.76	0.91
3:1	1.49	1.54	0.02
2:1	1.48	1.49	0.01
1:1	1.18	2.00	1.01
1:2	0.42	0.70	0.01
1:3	1.25	0.99	0.19
1:4	1.17	0.98	0.85

Based on the above test results, it can be concluded that the substation battery condition monitoring system designed in this paper can achieve accurate monitoring of different state parameters of the battery. This is because the design system uses TW-SDH6000 as the hardware device of the battery, uses SDH network to filter some battery resources, combines PCA method to reduce the dimension of the original data, and uses Pearson correlation coefficient to achieve accurate monitoring of the status of the substation battery. The minimum monitoring error of the designed system under different voltage conditions, frequency conditions, and coil ratios can reach 1.00%, 1.03%, and 0.01%, resulting in the best monitoring effect.

5 Conclusion

As an electrochemical product, the health status of battery is affected by the actual charging and discharging characteristics, ambient temperature, discharge depth state of charge and other factors, especially in the application of series batteries gradually expanding, the performance of the entire battery pack will deteriorate rapidly, so it is of great significance to realize online SOH monitoring of batteries. In addition, SOC provides the basis for battery energy management and charging and discharging control strategies. Accurate estimation of battery SOC can improve the utilization rate of energy storage batteries and facilitate system energy management and scheduling. Therefore, it is of great significance for the management, maintenance and fault prevention of the energy storage system to realize the real-time data acquisition of the battery and the estimation of the SOC and the SOH of the battery state of charge. The TW-SDH6000 is used as the main hardware equipment to improve the reliability of battery condition monitoring. The SDH network is used to filter battery resources. Aiming at the problem of high dimensions of the original data, PCA method is used to reduce the dimensions of the data, and Pearson correlation coefficient is used to achieve accurate monitoring of different state parameters of the battery. The system has good practical application value. With the help of the monitoring system designed in this paper, it is also hoped that it can provide valuable help for the management of the substation battery and help the battery to maximize its role in the operation of the power system.

References

1. Kim, D.H., Kim, M.S., Prabakar, K., et al.: Efficient management of fast charging systems based on a real-time monitoring system. Electronics **11**(4), 520 (2022)
2. Si, Y., Korada, N., Ayyanar, R., et al.: A high performance communication architecture for a smart micro-grid testbed using customized Edge Intelligent Devices (EIDs) with SPI and Modbus TCP/IP communication protocols. IEEE Open J. Power Electron. **PP**(99), 1 (2021)
3. He, H., Sun, F., Wang, Z., et al.: China's battery electric vehicles lead the world: achievements in technology system architecture and technological breakthroughs. Green Energy Intell. Transp. **1**(1), 24 (2022)
4. Salterio, N., Foti, D., Bogod, N., et al.: P.214 Improved cognition after endoscopic third ventriculostomy in adult obstructive hydrocephalus using repeatable battery for the assessment of neuropsychological status. Can. J. Neurol. Sci. / J. Canadien des Sciences Neurologiques **48**(s3), S81 (2021)
5. Fan, X., Yan, H.: Research on anti-interference screening of on-line measurement of battery ageing under smart grid big data. IOP Conf. Ser. Earth Environ. Sci. **692**(2), 022003–022010 (2021)
6. Masatsugu, O., Goki, S., Megumi, K., et al.: Computed tomography, not bioelectrical impedance analysis, is the proper method for evaluating changes in skeletal muscle mass in liver disease. JCSM Rapid Commun. **3**(2), 103–114 (2020)
7. Li, B., Jones, C.M., Adams, T.E., et al.: Sensor based in-operando lithium-ion battery monitoring in dynamic service environment. J. Power. Sour. **486**(13), 229349 (2021)
8. Pradeep, K.G.M., et al.: Energy efficient scheduling algorithm for structural health building monitoring system (SHBM) to increase the battery lifetime. Turkish J. Comput. Math. Educ. (TURCOMAT) **12**(3), 5005–5012 (2021)
9. Juhi, J.E., Rajeswari, R.: Implementation of ABMS with Cuk converter for enhanced battery life using Internet of Things. Int. J. Mod. Trends Sci. Technol. **7**(5), 107–111 (2021)
10. Lee, S.H., Yang, T., Kim, T.S.: Mobile-based sensing scheme to minimize battery power consumption for urban monitoring systems. Electronics **10**(2), 198 (2021)
11. Zhan, Y.Y., Lu, X.H.: Big data anomaly extraction algorithm based on uncorrelation test. Comput. Simul. **38**(3), 245–248460 (2021)
12. Liu, S., Wang, S., Liu, X., Gandomi, M.D., Muhammad, K., De Albuquerque, V.H.C., et al.: Human memory update strategy: a multi-layer template update mechanism for remote visual monitoring.IEEE Trans. Multimedia **23**, 2188–2198 (2021)
13. Pan, T.: Intrusion detection method of Internet of Things based on Multi GBDT feature dimensionality reduction and hierarchical traffic detection. J. Q. Comput. **3**(4), 161–171 (2021)
14. Zhao, P.: Analysis of PCA method in image recognition with MATALAB. (2014-4), 124–126 (2021)

Dynamic Tracking Method for Train Number of Rail Transit Signal System

Liwen Liu[1](✉), Chao Cai[2], Yulong Wang[2], and Zhiwen Chen[3]

[1] Wuhan Railway Vocational College of Technology, Wuhan 430205, China
LiwenLiu613@126.com
[2] Wuhan Metro Operation Co., Ltd., Wuhan 430205, China
[3] Department of Mechanics and Electronics, Wuhan Railway Vocational College of Technology, Wuhan 430205, China

Abstract. The conventional dynamic tracking method of train number of rail transit signal system mainly uses the ZG (Zone Controller) regional controller to report the train position, which is vulnerable to the influence of the occupation of the logical section, resulting in the mismatch of the tracking display position. Therefore, a new dynamic tracking method of train number of rail transit signal system needs to be designed. Namely, the GPRS train number dynamic tracking server is installed, the dynamic tracking module of this number on the side of the rail transit signal system is developed, and the dynamic tracking algorithm of the train number of the rail transit signal system is designed combined with the artificial neural network, thus realizing the dynamic tracking of the train number. The experimental results show that the designed dynamic tracking method for train number of rail transit signal system has good tracking effect, and the matching between the tracking display position and the actual position is reliable and has certain application value, which has made certain contributions to improving the safety of rail transit.

Keywords: Rail Transit Signal · System · Train Number · Dynamic Tracking · Method

1 Introduction

With the rapid development of national economy and the acceleration of urbanization process, the number of urban population and vehicles are increasing [1], and the huge flow of people and vehicles often lead to traffic congestion on urban roads [2–4]. In order to alleviate the pressure of urban traffic, rail transit has been built [5–7]. With the continuous progress of science and technology and the large-scale development of urban rail transit, the urban rail transit network will certainly become more developed [8]. From manual driving under the supervision of ATP to semi-automatic driving in ATO mode [9], and then to fully automatic driverless driving with the highest level of automation, the automation level of urban rail transit system has gradually improved. In

L. Yun et al. (Eds.): ADHIP 2023, LNICST 549, pp. 360–376, 2024.
https://doi.org/10.1007/978-3-031-50549-2_25

a fully automated driving system, the work of the train driver is transferred to a highly integrated automated control system, which enables fully automated operation from the depot to the main track. For the safety of vehicle operation, it is necessary to track the train number, but the recognition effect is not good at this stage, and relevant scholars have carried out research.

For example, literature [10] puts forward the design method of neural network PID controller for high-speed train speed tracking, which introduces neural network and builds a PID controller based on it to track trains. But the tracking effect of this method is not good. Literature [11] proposes a fault tolerant tracking control method for high-speed trains that considers actuator performance constraints. Based on the auxiliary system constructed by hyperbolic tangent function, this method constructs an augmented speed tracking control model for high-speed trains. In order to avoid the first derivative of virtual control signal in the controller, a fault tolerant tracking controller for high-speed train is designed by using dynamic surface method and adaptive control technology. The stability of the controller is analyzed based on Lyapunov function, and the train tracking is realized. However, the error between the tracking position and the actual position is large and the fitting is poor.

Under the above background, in order to improve the dynamic tracking effect of rail transit signal system, this paper designs a new dynamic tracking method of rail transit signal system based on the operation characteristics of rail transit signal system, which makes a certain contribution to improving the operational reliability of rail transit.

2 Design of Dynamic Tracking Method for Train Number of Rail Transit Signal System

2.1 Installing GPRS Train Number Dynamic Tracking Server

In order to solve the problem of tracking display position mismatch caused by the occupation effect of logic section[9]when the ZG regional controller reports the train position, this paper applies GPRS technology to install a GPRS train number dynamic tracking server. GPRS network adds GPRS service support node and GPRS gateway support node, two functional entities, on the basis of GSM-R network. The actual installed network structure is shown in Fig. 1 below.

It can be seen from Fig. 1 that GPRS shares the base station of the GSM-R system, but the BSC needs to add a packet control unit for processing packet data and wireless packet channel management, as well as SGSN and GGSN.Functions of the PCU: This function entity can be set up with the BSC or used as a separate network element [10]. The Gb interface between the PCU and the SGSN is a standard interface defined in the specification, and the interface between the PCU and the BSC is an internal interface. It is responsible for processing the data service of the wireless channel Y As shown in (1) below.

$$Y = \frac{DLTBFPEDCH + DLTBFPBDCH}{TRAFFCI, GPRSSCAN} \quad (1)$$

Where, $DLTBFPEDCH$ Represents the number of allocated businesses, $DLTBFPBDCH$ Represents the amount of contribution data, $TRAFFCI$ Represents the degree of reuse,

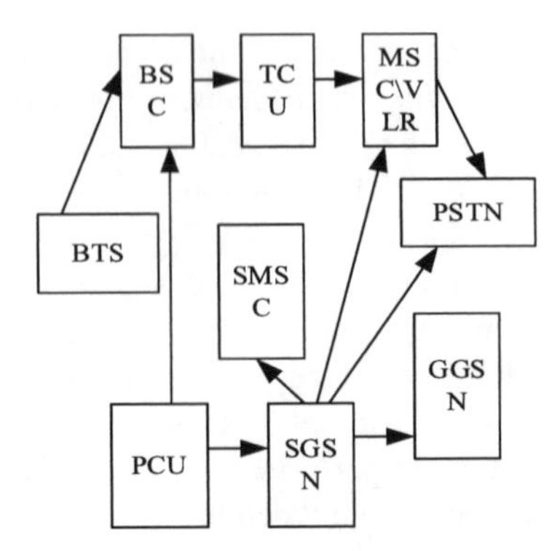

Fig. 1. Dynamic Tracking Server Installation Network Structure

GPRSSCAN It represents the total amount of services. When managing and allocating the wireless data channel, the above processing mode allows multiple users to access the same wireless resource, compress, encrypt and forward user data, and also has the functions of power control, quality control and channel coding scheme selection.

SGSN is an important part of GPRS backbone network and the core part of packet switching. It is connected to PCU through frame relay. The function is similar to the MSC/NLR function of the GSM system. The SGSN not only processes signaling transmission in packet switching, but also processes and transmits data packets. The number of channels at this time *PS* As shown in (2) below.

$$PS = TOTAL - ERLANG \tag{2}$$

Where, *TOTAL* Represents the data rate, *ERLANG* Representing the share of coding services, SGSN can perform mobility management, security management, access control and routing functions for MS. That is to record the relevant information of the mobile data user currently active in the SGSN area, such as location information, which can be modified, deleted, etc., and is responsible for the attach and detach, location update, paging, authentication, encryption, etc. of the data user; Be responsible for the establishment, maintenance and release of the logical link between MS and SGSN; The selection of negative expensive routes and the storage and forwarding of information; Generate original billing data *TBF* As shown in (3) below.

$$TBF = (TOTAL - ERLANG) \times TBFD \tag{3}$$

Where, *ERLANG* Represents the maximum value of data service, *TBFD* It represents the tracking loss rate. GGSN maintains the GPRS backbone network internally, and can connect multiple data networks externally, such as Internet, Enterprise Network, X.25 Network, etc. It is the gateway between the GPRS backbone network and the external data network; Its position in GPRS data network is similar to that of GMSC in traditional GSM network; Be responsible for generating the original billing data of data services.

The GPRS interface server can complete the data protocol conversion between the vehicle application entity and the ground application entity, as well as the data storage and forwarding service, which plays a key role. The structure of GPRS interface server is shown in Fig. 2 below.

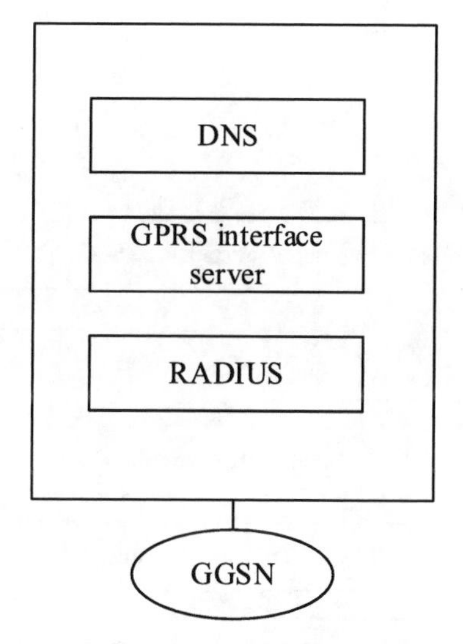

Fig. 2. Structure of GPRS interface server

As shown in Fig. 2, the GPRS interface server is composed of data forwarding unit, data recording unit, external DNS unit, Radius unit, disk array unit, three-layer Ethernet switch and management and maintenance terminal in hardware. The GPRS interface server is functionally composed of the following logical units: GGSN interface unit, CTC/TDCS interface unit, TCDS interface unit, GIS information service interface unit, mobile ticket system interface unit, data forwarding unit, etc. By connecting the above logical units, the tracking logical path can be effectively run to improve the sensitivity of dynamic tracking.

2.2 Develop Dynamic Tracking Module for Train Number of Rail Transit Signal System

For an urban rail transit [14, 15], it is composed of many stations. According to the deployment of CI, these stations can be divided into different interlocking areas. In the train identification and tracking module, an interlocking area and a station in the interlocking area adjacent to the interlocking area are regarded as a tracking area. In each tracking area, a set of train identification and tracking module processes can be placed on the ATS extension in the interlocking area, and each interlocking area can have one or more stations, so a tracking process can be responsible for tracking one or more stations, and the processes are connected through the communication platform, so as to realize the identification and tracking of trains in different areas and between different interlocking areas, the dynamic tracking process of train number of rail transit signal system is shown in Fig. 3 below.

It can be seen from Fig. 3 that in the same station in the same tracking area, trains are identified and tracked according to the status and relationship of equipment in the station; For different stations in the agreed tracking area and between different stations in different tracking areas, the train shall be identified and tracked according to the

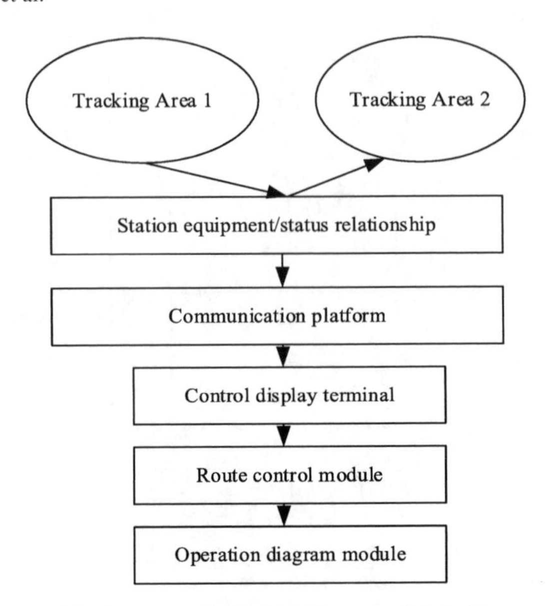

Fig. 3. Dynamic Tracking Process of Train No

equipment relationship between stations. The relationship between the station equipment and the inter station equipment is configured in the ATS system data according to the interlocking table information, and the dynamic tracking module of train number can be designed according to the dynamic tracking process of train number.

The first is the train identification module. In the ATS system, the train identification number plays a very important role. It is not only the unique identification of the train, but also the premise for the ATS system to realize its train route arrangement, automatic operation adjustment, train tracking and other functions. Therefore, the design of the train identification number is particularly important. The data structure of train identification module is shown in Fig. 4.

It can be seen from Fig. 4 that the change identification of train number is the mark for adding, deleting, modifying and moving the train identification number. The train ID is the number of the train, the table number, the number of the running line in the running chart, the train number and the train operation; Train set number is the train number; Destination number is the number of the destination where the train will arrive; The office code/line number is the number of the line to which the current train belongs; Station code, the number of the station where the train is located; Train number window number, the number of train number window where the train is located; Train number refers to the number of trains to distinguish multiple trains in the same train number window.

The train information maintenance module mainly includes adding, deleting, modifying and moving the train identification number. Its information processing structure is shown in Fig. 5.

It can be seen from Fig. 5 that the train information maintenance module receives the commands from the ATS control and display terminal to add, delete, modify and

Train number change sign
Train ID
Table number
Train number
Train unit number
Destination number
Office Code
Station code
Train number window number
Train serial number

Fig. 4. Data Structure of Train Identification Module

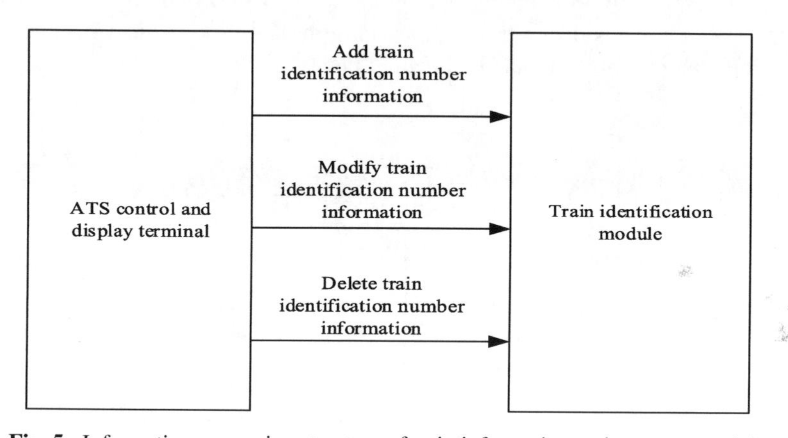

Fig. 5. Information processing structure of train information maintenance module

move the train identification number, then compares and judges the train identification number in the whole ATS system, finally maintains the train identification number, and then feeds back the modified train identification number information to the control and display terminal, and gets the correct display. When the train position is updated. This module can check the information C As shown in (4) below.

$$C = \sum_{i=1}^{N} R_i - d \tag{4}$$

Where, R_i The representative modifies the information group number, d It represents the standard information group number, i stands for field number. After information verification, subsequent information processing can be carried out.

Aiming at the requirement of ATO train number information verification in the train identification [16, 17] requirement, the ATO train number information processing plug-in is designed. The basic functions of the plug-in include: identifying new trains according to ATO train number information, checking logical train number and generating alarm. In CBTC mode, track and check the train according to the CBTC train position information. When a track is occupied by a train, but there is no train identification number in the train number window corresponding to the track, and the track is not represented by a red light band, an alarm will be sent and the alarm information will be sent to the control and display terminal.

When the train arrives at the platform, compare the train set number of the train in the ATS system and the ATO system. If the number is the same, update the train start, stop, door status and other information. If different, the ATS system needs to send an alarm, and the dispatcher can modify it according to the actual situation. When a track is occupied by a train, and there is no train identification number in the train number window corresponding to the track, but the track is indicated by a red light band, an alarm will be given, and ATO train number information will be used to generate logic tracking train number, and the train identification number will be displayed in the train number window corresponding to the track. According to ATO stop stability information, set the stop stability status of the train at the platform, and the structure diagram of the processing plug-in of this module is shown in Fig. 6 below.

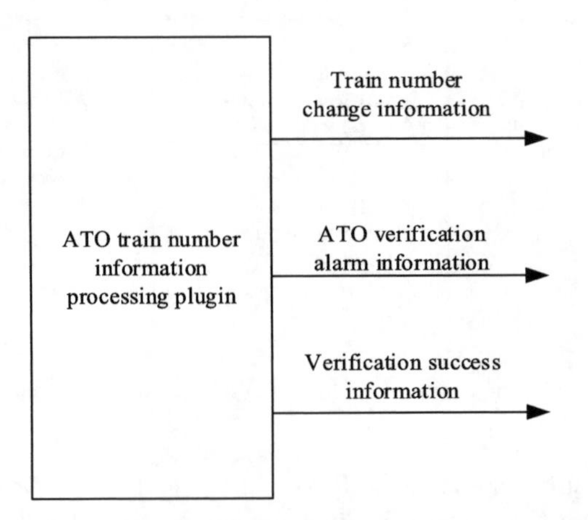

Fig. 6. Structure of processing plug-in

It can be seen from Fig. 6 that the above plug-in can obtain the position and change information of the train for dynamic tracking. The process of tracking train number according to ATO information generation logic is: search train information in the plan according to train set number, if the corresponding operation line is found, update train information according to the operation line; otherwise, generate the first train with only train set number and destination number of 0, and send an alarm.

The plan information management module mainly aims at the plan information management requirements in the train tracking requirements, and designs the plan management plug-in and the automatic point reporting plug-in. The basic functions of the plan management plug-in include: receiving and analyzing the train operation plan from the train diagram; Operation adjustment mode and its changes received from train diagram; Find the operation line information for the train according to the train set information; Provide turn back information for trains requiring turn back according to the plan and adjustment mode, and use the above functional modules to effectively track and report points and determine the train number of the rail transit signal system.

2.3 Design Dynamic Tracking Algorithm of Train Number

In the process of train number dynamic tracking, the influence of actual tracking environment changes on the tracking results is often ignored. Therefore, this paper uses artificial neural network to design an effective train number dynamic tracking method for rail transit signal system. Artificial neural network (ANN) is a mathematical model of information processing using a structure similar to the synaptic connection of brain nerves. It is a nonlinear system composed of a large number of simple computing units. The steepest descent method can be used to continuously adjust the weighted value and threshold of the neural network, so that the sum of squares of the network errors can be minimized.

BP neural network [18–20] consists of a large number of simple basic elements, namely neurons, which are interconnected to form a nonlinear dynamic system. Although the structure and function of each neuron are relatively simple, the dynamic system behavior generated by a large number of neuron combinations is very complex. In the process of calculation, the use of artificial neural network can greatly improve the speed of calculation.

Train number tracking refers to the automatic calculation of the train number of the running or stopped train at a certain time and place according to the adjustment plan information and actual operation of the train. Train number tracking technology mainly includes three aspects: acquisition of original train number, logical tracking processing of train number, and train number verification. The original train number is generally obtained according to the train adjustment plan. The verification of train number is to check the calculated train number with the planned train number, wireless train number, etc., and judge whether the calculated train number is correct according to certain rules to improve the accuracy of the output train number. Among them, the logic tracking is the most complex, including the tracking of the train in the station and the tracking of the section. The essence of logic tracking train number is that only one train can run or stop at the same time in the same block section, and the dynamic tracking strategy function generated at this time a_t, as shown in (5) below.

$$a_t = u(s_t|\theta^n) + N_0 \tag{5}$$

Where, u Represents regression tracking noise, s_t Represents the tracking target weight, θ^n Represents the moving angle, N_0 Represents the tracking parameters. Under different tracking states, the dynamic tracking effect of train number will change. Therefore, updated follow-up review L, as shown in (6) below.

$$L = \frac{1}{n}\sum_{j}(y_i - Q(s_t, \theta^n))^2 \tag{6}$$

Where, y_i Represents the tracking gradient, n Represents the train number update parameters, $Q(s_t, \theta^n)$ Represents the update coordinate, and a reasonable tracking reward function can be obtained for the above update d, as shown in (7) below.

$$d = \sqrt{\sum_{j=1}^{6}(a_j - c_j)^2} \tag{7}$$

Where, a_j Represents the current train number of the traffic signal system, c_j Represents the actual train number, and the artificial neural network tracking model is constructed according to the above reward function x_i, as shown in (8) below.

$$x_i = f(t) = \begin{cases} 0, t = f_i \\ 1, t \neq f_i \end{cases} \tag{8}$$

Where, $f(t)$ Represents the tracking state information function, f_i Representing the condition function, the train number tracking objects are divided into four categories: station tracking objects, interval tracking objects, train number verification objects and original train number objects. First, these four elements are set as the first layer, and the tracking objects in the station and in the section are divided into the second layer, namely, a single object, including turnout object, track object, turnout free object, approach departure object, indicator object, block section object, section annunciator object, etc., as the input layer. The output of tracking hidden layer can be further generated according to the check relationship of train number z_k, as shown in (9) below.

$$z_k = f_i(\sum_{i=1}^{n} vk) \tag{9}$$

Where, vk It represents the check and matching parameters of the train number. At this time, train number dynamic tracking y_j can be obtained according to the above output, as shown in (10) below.

$$y_j = z_k f(t)\left(\sum_{i=1}^{n} w\right) \tag{10}$$

Where, w It represents the weighted value of the tracking transfer function. The above tracking algorithm can effectively determine the dynamic tracking position to ensure the accuracy of the final tracking display.

3 Experiment

In order to verify the tracking effect of the designed dynamic tracking method for train number of rail transit signal system, this paper builds a simulation tracking ATS topology structure, and compares it with the conventional dynamic tracking method for train number of rail transit signal system based on machine learning and the dynamic tracking method for train number of rail transit signal system based on reinforcement learning algorithm. Experiments are carried out as follows.

3.1 Experiment Preparation

In combination with the experimental requirements, this paper designs an effective ATS line topology based on a rail transit signal system and in combination with the characteristics of the station yard diagram. The topology is mainly composed of the starting point coordinates, the ending point coordinates, the specified logical direction, the length of the side and other parameters. The experimental ATS line topology is shown in Fig. 7 below.

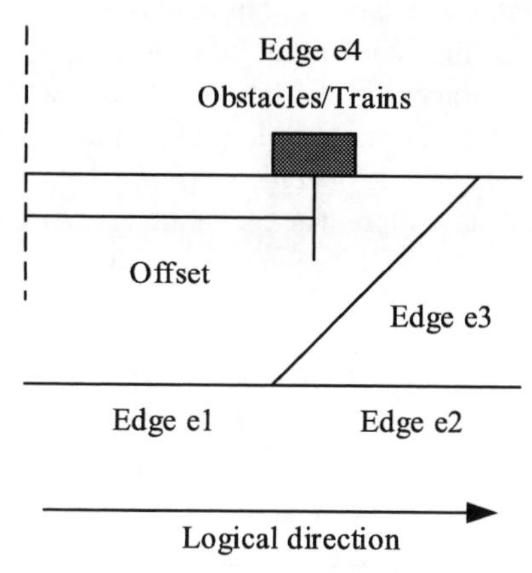

Fig. 7. Experimental ATS Topology

It can be seen from Fig. 7 that any point on the edge is uniquely determined by the ID value number of the edge plus the offset, that is, the offset relative to the starting point of the edge. In the ATS simulation system, it is not necessary to carry out absolute positioning for the train. By reading the train information data received by ATS, obtaining the train occupation ID value and the offset value, the specific position of the train in the station yard can be calculated, so as to achieve accurate positioning of the train. This paper determines the logical direction of the side from left to right according to the signal plane layout, as shown in Fig. 8 below.

It can be seen from Fig. 8 that in reading train information, the offset of the occupied section of the train relative to the logical direction is read in turn according to the logical direction of the side, and then the accurate position of the train is calculated.

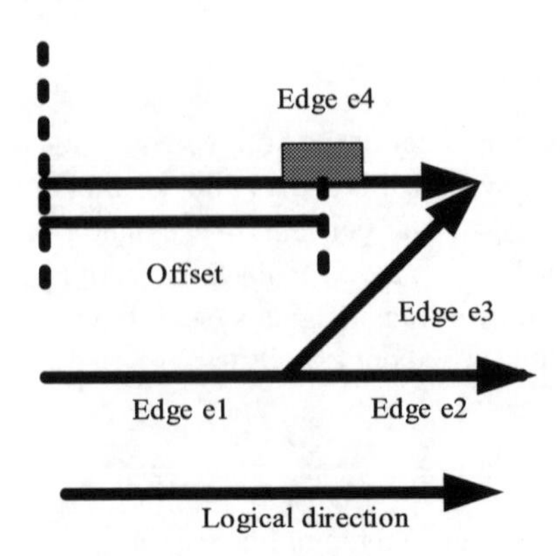

Fig. 8. Logical Direction of Side

The data in the station includes static data and dynamic data. Static data refers to the data that will not change with the change of signal equipment status or train position, that is, fixed data, which can only be modified when designing the station diagram.In this experiment, static data is configured and stored in the format of *. Xml file. By reading the *. Xml file, station entity instances are generated. Each instance corresponds to a signal equipment in the floor plan. The static data does not change with time. During the experiment initialization, the static data will be used as the line representation and basic data of train operation. After the experiment, the experimental platform will automatically save these data for future use. At this time, the data flow of signal equipment in the experimental station is shown in the following Fig. 9.

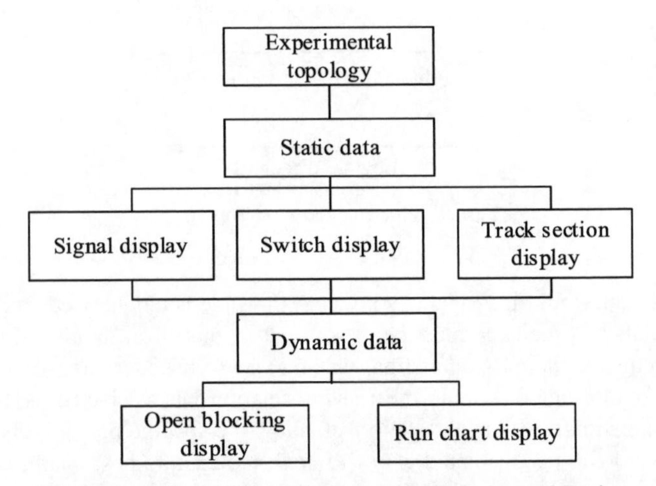

Fig. 9. Data Flow of Signal Equipment in Experimental Station

It can be seen from Fig. 9 that dynamic data refers to the data that changes in real time with the operation of the train. With the operation of the train, the status of the signal equipment changes constantly. Dynamic data is used to control the refresh of the dynamic status of the signal equipment. According to the above signal equipment data flow, the static data structure table of the experimental signal can be generated, as shown in Table 1 below.

Table 1. Static Data Structure of Experimental Signal

Data Attribute Name	Data type
Sig Name	String
Sig_ PointX	Int
Sig_ Pointy	Int
Sig_ Id	Int
Sig_Leftes QD Name	String
Sig_Right QD Name	String
Sig_DefaultStatus	Bit
Sig_ ProDir	Bit
Type	Int

It can be seen from Table 1 that the above static data structure meets the tracking experiment requirements, but it is difficult to track the dynamic attributes of train number using only the static data structure, so it is necessary to update the dynamic data structure table of the experiment, as shown in Table 2 below.

Table 2. Dynamic Data Structure of Experimental Signal

Data Attribute Name	Data type
Sig-G	TINYINT
Sig-H	TINYINT
Sig-U	TINYINT
Sig-HU	TINYINT
Sig-Broken	TINYINT
Sig-Off	TINYINT
Sig-AR	TINYINT
Sig-FS	TINYINT

It can be seen from Table 2 that the dynamic properties of dynamic signals mainly include the current light on state, route enabling state, blocking state, etc. of the annunciator. Obtain the dynamic data value of each annunciator through data interaction with the server, and update the display in real time. Five segment representation method is adopted for single turnout, as shown in Figs. 3 and 4 below. In order to facilitate the experimental division, this paper uses the five section method to distinguish, so that the train can be accurately positioned. The schematic diagram of experimental division is shown in Fig. 10 below.

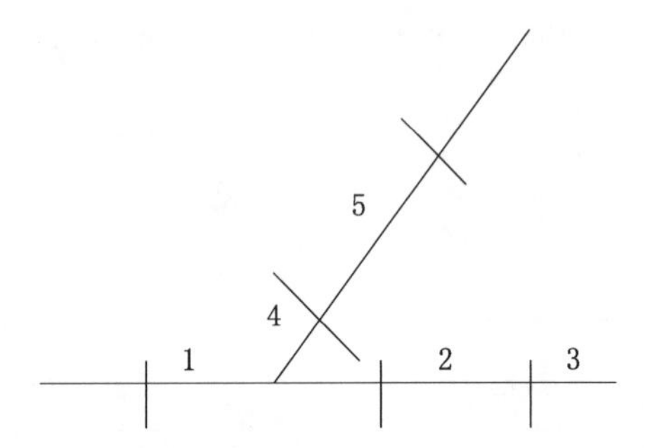

Fig. 10. Schematic Diagram of Experimental Section Division

It can be seen from Fig. 10 that during the design, the secondary number is set to 0 for the parking track and the section without turnout, and the secondary number is assigned to the section with turnout according to different sections. When positioning, accurate positioning can be achieved by using the primary number and secondary number.

In order to effectively carry out experiment interaction, this experiment uses the data information interaction between the area controller/on-board computer/computer interlocking/ATS system to enable ATS to complete train tracking identification and display, station diagram display, temporary speed limit control and other functions. All data interactions between the system and other subsystems are simulated and designed by the device simulation server. The browser only needs to communicate with the server to obtain relevant experimental data. At this time, the definition of experimental communication information is shown in Table 3 below.

Table 3. Definition of Experimental Communication Information

Name	Mess Type	Sending direction	Period or not
ATO control command	0 x 0201	Browser ~ Server	Aperiodic
ATO status information	0 x 0202	Server ~ Browser	Cycle
Train information	0 x 0204	Server ~ Browser	Cycle
Vehicle mounted equipment alarm information	0 x 0206	Server ~ Browser	Aperiodic

It can be seen from Table 3 that the train information is tracked and identified by using the train information sent by the server. Control commands can send dispatching control commands to the server according to operation requirements. ATS system monitors all trains in the station through ATO status information sent by the server. On board equipment alarm information is the fault alarm information of train equipment sent by the server to ATS system. Since all data interactions in the experiment are conducted with the server, the browser client only needs to send command related data, and the server sends the processed data to the device simulation server after logical processing, so all data formats involved in this paper are interactions with the server rather than real signal subsystems.

In order to effectively improve the network transmission efficiency of data, this paper defines the data format between the Web Socket server and the browser client as JSON format. The command information in the browser client is in JS information format, while the command information transmitted in the network must be in JSON string format; So first, you need to use the JSON function to convert the JS object into a JSON object for transmission in the network, that is, to serialize the JS object. Similarly, the JSON string information sent by the server must also be deserialized. That is, the front end can analyze the data only after you use the JSON function to convert the JSON string transmitted from the back end into a JS object. After the above experiment preparation is completed, the dynamic tracking experiment of train number of rail transit signal system can be carried out.

3.2 Experimental Results and Discussion

On the basis of the above experimental preparation, the preset train number dynamic tracking command is used for tracking experiments, that is, the train number dynamic tracking method of rail transit signal system designed in this paper, the train number dynamic tracking method of rail transit signal system based on machine learning, and the train number dynamic tracking method of rail transit signal system based on reinforcement learning algorithm are used for tracking. Record the dynamic tracking results of the three methods under different train positions, as shown in Table 4 below.

Table 4. Experimental Results

This paper			
Train position	Running direction	Actual train number	Track and display train number
1G	←	D1	D1
2G\3G	→	D2	D2
4G	←	D3	D3
5G	→	D4	D4
6G	←	D5	D5
A Dynamic Tracking Method for Train Number in Rail Transit Signal System Based on Machine Learning			
1G	←	D1	D1
2G\3G	←	D2	D1
4G	→	D3	D2

(*continued*)

Table 4. (*continued*)

This paper			
Train position	Running direction	Actual train number	Track and display train number
5G	←	D4	D2
6G	→	D5	D3
A Dynamic Tracking Method for Train Number in Rail Transit Signal System Based on Reinforcement Learning Algorithm			
1G	←	D1	D1
2G\3G	←	D2	D2
4G	→	D3	D2
5G	←	D4	D2
6G	→	D5	D3

It can be seen from Table 4 that under the preset operating environment, the dynamic tracking position of the train number dynamic tracking method of the rail transit signal system designed in this paper is matched with the actual train position, and the train number dynamic tracking method of the rail transit signal system based on machine learning. The dynamic tracking position of the train number dynamic tracking method of the rail transit signal system based on the reinforcement learning algorithm is quite different from the actual train position. The above experimental results prove that the train number dynamic tracking method of the rail transit signal system designed in this paper has good tracking effect, reliability and certain application value.

4 Conclusion

The development of rail transit technology has put forward higher demands on passenger volume, efficiency, comfort, energy conservation, etc. In areas with mature subway development, many cities consider upgrading existing lines to fully automated lines to improve operation efficiency and automation level. In the metro project, ATS system can obtain the occupation and clearing information of all physical tracks and logical tracks beside the line through the main line interlocking, and obtain the occupation and clearing information of physical axle counters or track circuits in the depot/parking lot through the depot interlocking, respectively realizing the tracking of train number based on the interlocking occupation information of the main line. Tracking of train set number based on interlocking occupancy information in depot.In the tram project, the operation dispatching system can only obtain the occupancy and clearing information of the section with turnout through the main line turnout controller, and the occupancy and clearing information of the trackside section without turnout can not be obtained, which makes it impossible to track the train number completely according to the occupancy and clearing information reported by the turnout controller. According to the characteristics of the rail transit signal system. An effective train number dynamic tracking method is designed, and experiments are carried out. The results show that the designed dynamic tracking method has good tracking effect, reliability and certain application value, and has made certain contributions to improving the operation reliability of the rail transit signal system.

References

1. Hu, J., Ma, T., Ma, K., et al.: Three-dimensional discrete element simulation on degradation of air voids in double-layer porous asphalt pavement under traffic loading. Constr. Build. Mater. **313**(5), 570–582 (2021)
2. Guan, L., Wang, D., Shao, H., et al.: Understanding the topology of the road network and identifying key bayonet nodes to avoid traffic congestion. Int. J. Mod. Phys. C **34**(03), 316–328 (2023)
3. Wei, K., Vaze, V., Alexandre, J.: Transit planning optimization under ride-hailing competition and traffic congestion. Transp. Sci. **56**(3), 725–749 (2022)
4. Li, J., Ma, M., Xia, X., et al.: The spatial effect of shared mobility on urban traffic congestion: evidence from Chinese cities. Sustainability **13**(24), 1–10 (2021)
5. Suryakala, V., Rajalakshmi, T., Kolangiammal, S., et al.: A novel vision based embedded framework system to detect and track dynamic vehicles. IOP Conf. Ser. Mater. Sci. Eng. **1130**(1), 51–60 (2021)
6. Wen, Y., Wu, R., Zhou, Z., et al.: A data-driven method of traffic emissions mapping with land use random forest models. Appl. Energy **305** (2022)
7. Lee, S., Jain, S., Ginsbach, K.: Dynamic-data-driven agent-based modeling for the prediction of evacuation behavior during hurricanes. Simul. Model. Pract. Theor. Int. J. Fed. Euro. Simul. Soc. **106**(1), 193–207 (2021)
8. Hu, J., Xiao, F., Mei, B., et al.: Optimal energy efficient control of pure electric vehicle power system based on dynamic traffic information flow. IEEE Trans. Transp. Elec. **8**(1), 510–526 (2021)
9. Xu, D., Zhou, D., Wang, Y., et al.: Temporal and spatial heterogeneity research of urban anthropogenic heat emissions based on multi-source spatial big data fusion for Xi'an, China. Energy Build. **240**(4), 110–123 (2021)
10. Liang, X., Xiao, L., Wang, X., et al.: Design of neural network PID controller for high-speed train speed tracking. Comput. Eng. Appl. **57**(10), 252–258 (2021)
11. Xu, C., Gu, X., Wang, L.: Fault tolerant tracking control of high speed train considering actuator performance constraints. J. Chin. Inertial Technol. **30**(4), 545–552 (2022)
12. Wu, Q., Cheng, S., Li, L., et al.: A fuzzy-inference-based reinforcement learning method of overtaking decision making for automated vehicles. Proc. Inst. Mech. Eng. Part D. J. Automobile Eng. **236**(1), 75–83 (2022)
13. Wang, Z., Zhao, X., Chen, Z.: A dynamic cooperative lane-changing model for connected and autonomous vehicles with possible accelerations of a preceding vehicle. Expert Syst. Appl. **173**(2), 1–18 (2021)
14. Afrin, T., Yodo, N.: A probabilistic estimation of traffic congestion using Bayesian network – ScienceDirect. Measurement **174**(1), 1–13 (2021)
15. Jiang, P., Liu, Z., Zhang, L., et al.: Advanced traffic congestion early warning system based on traffic flow forecasting and extenics evaluation. Appl. Soft Comput. **118**(8), 108544–108570 (2022)
16. Zhang, X., Nie, X., Sun, Z., et al.: Re-ranking vehicle re-identification with orientation-guide query expansion. Int. J. Distrib. Sens. Netw. **18**(3), 205–214 (2022)

17. Yang, Z., Zhu, Y., Zhang, H., et al.: Moving-vehicle identification based on hierarchical detection algorithm. Sustainability **14**(1), 211–224 (2021)
18. Luo, Q., Li, J., Zhang, H.: Drag coefficient modeling of heterogeneous connected platooning vehicles via BP neural network and PSO algorithm. Neurocomputing **18**(1), 484–896 (2022)
19. Song, S., Xiong, X., Wu, X., et al.: Modeling the SOFC by BP neural network algorithm. Int. J. Hydro. Energy **46**(38), 65–77 (2021)
20. Liu, L.: Lute acoustic quality evaluation and note recognition based on the softmax regression BP neural network. Math. Prob. Eng. **22**(7), 46–59 (2022)

Centralized Monitoring System of Rail Transit Multiple Signals Based on Bus Technology

Bo Li(✉)

Wuhan Railway Vocational College of Technology, Wuhan 430000, China
libo95123@sina.com

Abstract. Conventional rail transit signal monitoring systems are prone to being affected by pooling aggregation during image downsampling processing, resulting in abnormal monitoring functions. Therefore, this study designs a new centralized monitoring system for multi-channel signals in rail transit based on bus technology. In the hardware part of the system, SBMA RF receiver, SZ45XIT magnetic random access memory, and SPACECOM electric zoom monitoring camera are installed to support smooth operation of the system. In the system software section, based on the design of traffic multi-channel monitoring signal processing algorithms, a signal centralized monitoring function module was generated based on bus technology. The test results indicate that the various functions of the system operate in an orderly manner, with reliability and application value. In addition, compared to traditional systems, the signal monitoring delay of this system is lower.

Keywords: Bus technology · Rail transit · Multiple · Signal · Centralized monitoring system

1 Introduction

In recent years, China's economy has risen rapidly and developed into the second largest economy in the world. With the continuous improvement of people's living standards and the upgrading of resident consumption [1] in China, the automobile market is experiencing rapid growth. The popularity of cars and the number of cars owned by thousands of people are increasing year by year. According to the data of the National Bureau of Statistics, China's civil automobile ownership in 2018 was about 240 million [2], an increase of 10.5% over last year. The development of the automobile era has led to an explosive growth in the number of cars, which has led to traffic accidents and congestion. Both developed and developing countries are facing many traffic problems, which greatly reduces the efficiency of people in production and life [3]. In order to solve these problems, traditional means such as increasing the number of roads, widening the original roads, and establishing overpasses have been used to alleviate the contradiction between cars and roads. However, due to the huge population of China [4], the development of urban roads is limited, and this method cannot fundamentally improve the traffic

L. Yun et al. (Eds.): ADHIP 2023, LNICST 549, pp. 377–392, 2024.
https://doi.org/10.1007/978-3-031-50549-2_26

conditions of cities. The introduction of intelligent transportation system is a change of traditional thinking mode, which conforms to the development of the automobile era and can effectively improve the acute traffic problems [5].

The intelligent transportation system integrates the intelligent achievements in scientific information technology, sensor technology, data communication technology, artificial intelligence and other fields, and is committed to creating a comprehensive, multi-functional, efficient and intelligent traffic monitoring system [6]. Since the intelligent transportation system was put forward to the gradual improvement of the system, the level of road management in China has been greatly improved. The number of deaths caused by traffic accidents can be reduced by 30% every year, and the utilization rate of vehicles can be increased by 50%. The intelligent transportation system has achieved effective results. The proposed intelligent transportation system conforms to the development trend of informatization in the world today [7], can effectively coordinate the relationship between roads, people and vehicles, and improve the tense traffic situation, which is the mainstream direction of future transportation system development.

The intelligent transportation system is composed of various subsystems, each of which is responsible for different fields and processes different information. These information are collected and correlated through the established intelligent information processing platform, in which the acquisition of vehicle information is crucial. At present, the acquisition of vehicle information mainly includes loop detection, infrared detection, video detection and radar detection. Compared with foreign countries [8], domestic involvement in the field of intelligent transportation started relatively late. Before the start of the research on intelligent transportation, the detection of various violations of vehicles is mainly to judge and distinguish according to the situation, and in most cases still rely on manual judgment and decision, while the traffic videos recorded on urban roads are only used as evidence after the occurrence of traffic accidents, and there is no subsequent data. Under the above background [9], in order to improve the safety of China's rail transit, this paper designs a new centralized monitoring system for rail transit multi-channel signals based on bus technology.

2 Hardware Design

2.1 SBMA RF Receiver

The traffic multi-channel signal centralized monitoring system mainly works in the 315 MHz and 433 MHz Sub GHz frequency bands, and adopts ASK and FSK coding methods. At present, there are mature RKE schemes in the market, such as RF transceivers produced by Infineon, Cypress and other companies, and separate RF transmitters and receivers [10]. This design focuses on signal reception, so a separate SBMA RF receiver with more friendly cost will be selected. The SBMA RF receiver has added a microcontroller. The development process is more complex, and the receiver integration is high. All circuits of the RF receiver are configured by external circuits, so the flexibility is high [11].

The SBMA RF receiver does not need to configure more external components in actual operation. The operating ISM frequency band is between 300 MHz and 450 MHz. The current popular 32 pin thin QFN package is selected, and the applicable temperature range is –40 ~ +125 °C. The temperature range meets the environmental requirements of the centralized signal monitoring system. The SBMA RF receiver is equipped with various active components to maintain the normal operation process of superheterodyne, such as a low-noise amplifier supporting AGC function, a fully integrated phase-locked loop, local oscillator, etc., to better adapt to the low power consumption operation mode. The SBMA RF receiver is integrated with a discontinuous reception mode, which can be configured scientifically and reasonably using the serial interface bus. The requirements for centralized monitoring of signals are met. The composition of the RF receiver is shown in Fig. 1 below.

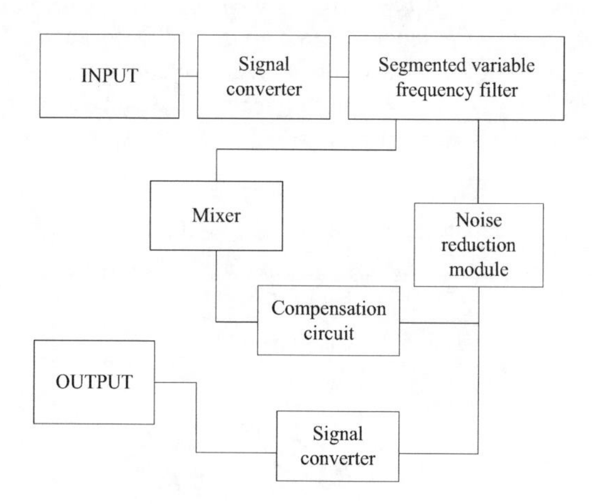

Fig. 1. Structure of SBMA RF Receiver

It can be seen from Fig. 1 that the peripheral components of the SBMA RF receiver are in rich demand. In addition to ensuring the matching network design of the antenna input, it can also better suppress the adjacent band interference and improve the sensitivity of the receiver. The operating frequency can be fine tuned by adjusting the system clock, which is theoretically within 300–450 MHz [12]. In view of the important role of the clock in frequency selection, the clock used in this system is replaced by a temperature compensated crystal oscillator clock.

2.2 SZ45XIT Magnetic Random Access Memory

In the process of centralized monitoring of multiple signals of rail transit, some emergencies may occur, and the reliability problem is mainly manifested in emergencies such as power failure. Dynamic random access memory.

The capacitance can only be refreshed through current during power on to store the information operation program. All data transmitted from the camera at the time of power off cannot be retained in case of emergency power off. In practical applications, the data at the moment of power failure are often important evidence. Dynamic random access memory (DRAM) has a simple and efficient structure. Each bit requires only one transistor and one capacitor. However, the DRAM capacitor inevitably has leakage.

As long as the charge is insufficient, data errors will occur, and the capacitor must be refreshed periodically. Therefore, using dynamic random access memory as program operation and data cache is easy to lose important data or cause system paralysis with low reliability.

Each time the system restarts, the master microprocessor needs to find the startup code from the flash memory and load the system program into the dynamic random access memory [13]. Moreover, the process of charging and discharging capacitors in dynamic random access memory takes time, and the refresh frequency can easily reach the upper limit in advanced technology. The above process and the limitations of flash memory's read/write speed and read/write times affect the efficiency of information interaction, reduce the system's smoothness, and make the system unable to run at high speed. In view of the storage defects of the conventional monitoring system, this paper selects SZ45XIT memory as the core memory of the system. The operating parameters of this memory are shown in Table 1 below.

Table 1. Operation parameters of magnetic random access memory

Name	Parameter
Package	Cut Tape (CT)
Life cycle	Active
Overall dimensions	4 mm\5 mm\1.5 mm (Dimensions)
Package	SOIC-8
Pin count	8
Installation method	Surface Mount
Power supply current	5 mA
Supply voltage	4.5~5.5 V
Output voltage	0.0~5.3 V
Output current	1.6 mA
Supply voltage	4.50 V
Operation temperature	125 °C
ECC Ncode	EAR99

It can be seen from Table 1 that the nonvolatility of the above memory improves the reliability of the video monitoring system. When the video monitoring system is powered down suddenly, the audio data just imported into the system cache can be saved, so there is no need to worry about the loss of important real-time data, and the reliability of RAM in the same environment is higher than that of dynamic RAM. This is very important for the video monitoring system in important places [14].

Thanks to the non-volatile nature of the magnetic random access memory, the memory connected to the main control microprocessor does not need to carry out code input and manned operation data after each power on after the first power on. Moreover, the magnetic random access memory itself has the advantage of high-speed reading and writing, which can greatly save the startup time and enable high-speed data transmission of the video monitoring system. The magnetic random access memory (MRAM) can be infinitely rewritten, which reduces the number of times and costs of maintenance of the system's storage part. The low power consumption also saves costs. The nonvolatile and low-power nature of magnetic random access memory eliminates the trouble of traditional video surveillance systems matching backup batteries or super capacitors, and greatly reduces the cost.

2.3 SPACECOM Electric Zoom Monitoring Camera

In the process of centralized monitoring of multiple signals in rail transit, it is necessary to continuously collect monitoring images for multi-target analysis. Therefore, SPACECOM electric zoom monitoring camera is selected as the acquisition hardware of the system. SPACECOM camera belongs to the day and night lens, which is used in forest fire prevention, coastal defense, border defense and other industries. It belongs to the long focus electric zoom lens with good comprehensive performance. The performance parameters of the electric zoom monitoring camera are shown in Table 2 below.

Table 2. SPACECOM Surveillance Camera Parameters

Name	Parameter
Focal length	15~500 mm
Canvas Size	1\1.8
Aperture ratio	1:4.0
Aperture Range	F4.0~360
Installation interface	C
Viewing angle	23.19° × 17.49°
Operating mode	Electric
Close object distance	2.9 m
Filter size	105 mm
Overall dimensions (mm)	122\125\267
Weight (g)	2900

It can be seen from Table 2 that the image quality and color of SPACECOM electric zoom monitoring camera are clear. The built-in auto focus function of the lens makes the operation more convenient. The fog penetration function can penetrate the fog, which solves the problem that objects cannot be seen in the heavy fog environment. Multiple sets of filters are used, suitable for different application scenarios. It has 24-h monitoring capability. Provide high definition color images in the daytime; Fine black and white images are available at night. The built-in temperature sensing device can sense the thermal expansion and contraction caused by the temperature information of the lens, which can automatically compensate the back focus position and prevent the image from being blurred. That is, when used in an environment with large temperature difference, the image can always be kept clear without refocusing. The lens adopts advanced standards, which can work normally in bad weather and has good environmental adaptability. It is used in monitoring scenarios such as railway, forest fire prevention, border and coastal defense, urban commanding heights, and environmental monitoring, and meets the monitoring requirements of the centralized signal monitoring system designed in this paper.

3 Software Design

3.1 Design Traffic Multi-channel Monitoring Signal Processing Algorithm

As mentioned earlier, conventional rail transit signal monitoring systems are susceptible to the influence of pooling aggregation during image downsampling processing, which affects the final monitoring effect. In response to this issue, this study utilized Tiny-YOLOv3 technology to design an effective centralized monitoring algorithm for traffic multi-channel signals.

In Tiny-YOLOv3 technology, the polarized beam is placed according to the antenna, and the three-dimensional information of the rail transit target is solved using the spectrum pairing ratio method.

Phase comparison angle measurement is based on the phase error of the echo signal between two antennas. When the radar antenna radiates electromagnetic waves, the direction of the target is determined according to the characteristics that the echo signal is strongest when the antenna beam axis is aligned with the target, and the echo signal becomes weaker when the antenna beam axis deviates from the target. The schematic diagram of the phase angle test at this time is shown in Fig. 2 below.

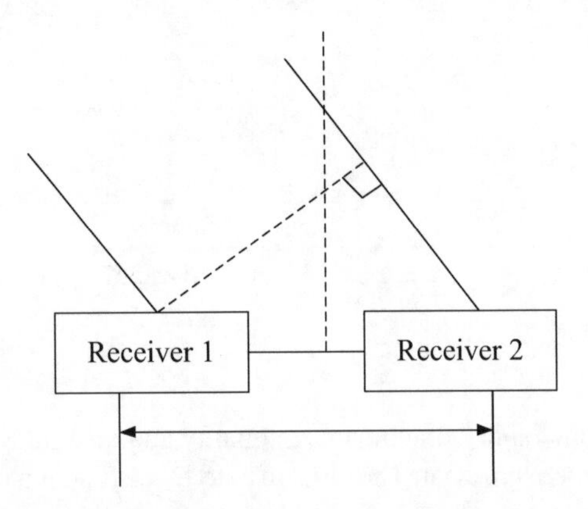

Fig. 2. Schematic diagram of monitoring phase angle test

In the environment of Fig. 2, the processing phase difference $\Delta\varphi(\theta)$ of the multi-channel monitoring signal for rail transit is:

$$\Delta\varphi(\theta) = \varphi(1) - \varphi(2) \tag{1}$$

In formula (1), $\varphi(1)$ Represents the phase angle of receiver 1, $\varphi(2)$ It represents the phase angle of receiver 2. When the electromagnetic wave encounters an object, it will bounce back and be received by the receiver. At this time, the monitoring wave path difference of the collected signal can be calculated according to the included angle between the echo and the normal direction of receiver 1 and receiver 2, as shown in (2) below.

$$\Delta R = d \cdot \sin \theta \tag{2}$$

In formula (2), d Represents the distance between receiver 1 and receiver 2, θ Represents the included angle of the monitoring receiver. Under the condition that the distance d between the radar receiving antenna 1 and receiving antenna 2 is fixed, the incident angle of the echo signal is related to the phase difference. Determine the phase difference corresponding to the receiving antenna 1 and receiving antenna 2 to determine the incident angle of the echo signal, find the location of the target, and then monitor the average value of the echo signal a As shown in (3) below.

$$a = \frac{\sum_{i=1}^{N} \theta_i}{N} \tag{3}$$

In formula (3), θ_i Represents the average signal acquisition angle, N It represents the incident amount of echo signal. According to the above calculated signal monitoring parameters, signal processing steps can be designed.

Step 1: AD sampling is performed on the multi-channel signals of the horizontal polarization beam and the vertical polarization beam receiving channels, converting the analog signal into a digital signal, and reordering the sampled data to obtain multiple sets of beat signal data.

Step 2: FFT transform each group of LFSK beat signals. In order to solve the problem of spectrum leakage, this paper introduces Blank man window to get N spectrum graphs when the signal is transformed, and the expression of Blank man window at this time $w[n]$ As shown in (4) below.

$$w[n] = S_i^k \left(4\pi \frac{n}{N-1} \right) \tag{4}$$

In formula (4), S_i^k Represents spectrum leakage parameters, n Represent the number of spectrum graphs, and calculate the modulus square of the spectrum graph obtained in step 2 to obtain the amplitude frequency response. Then sum the amplitude frequency responses of horizontal receiving channel 1, horizontal receiving channel 2, vertical receiving channel 1 and vertical receiving channel 2 respectively. Finally, the frequency spectrum of the horizontally polarized beam is cumulated into periodic patterns, and the corresponding amplitude frequency response D As shown in (5) below.

$$D = \frac{w[n]}{a} \cdot s_a \tag{5}$$

In formula (5), s_a Represents the periodic pattern accumulation of the spectrum of the vertically polarized beam, and the corresponding amplitude frequency response S_S^A

As shown in (6) below.

$$S_S^A = S_F + D\backslash s_a \tag{6}$$

In formula (6), S_F It represents polarization accumulation response. According to the results obtained from formula (6), the design of a multi-channel monitoring signal processing algorithm for rail transit is completed using the following process:

Process 1: Understand the amplitude frequency response of the signal. The amplitude frequency response describes the amplitude variation of a signal at different frequencies. By analyzing the amplitude frequency response of a signal, the frequency domain characteristics and frequency distribution of the signal can be understood.

Process 2: Preprocess the signal. Before applying the algorithm, it is necessary to preprocess the signal, including filtering, denoising, enhancement, and other operations. Select an appropriate filter based on the amplitude frequency response of the signal to remove unnecessary frequency components or enhance the frequency components of interest.

Process 3: Feature extraction. Feature extraction is the process of converting signals into feature vectors or feature matrices with recognition and representation capabilities. Frequency domain characteristics (such as spectrum characteristics, power Spectral density), time domain characteristics (such as mean, variance, energy) or other characteristics (such as wavelet transform coefficients) can be used.

Process 4: Algorithm design. Based on the results of feature extraction, signal processing and analysis are realized through Time–frequency analysis.

3.2 Generate Centralized Monitoring Structure Based on Bus Technology

The bus technology can effectively classify data and improve the comprehensive performance of the monitoring system. It can connect each monitoring component to the computer processing unit to form an effective centralized monitoring structure. Fieldbus is a decentralized, digital, intelligent, bidirectional, multipoint, multistation, and multivariable communication system used between field control terminals and control centers. It provides network services according to standards, and is characterized by high reliability, good stability, strong anti-interference ability, high communication speed, and low maintenance cost. Based on the above characteristics, this paper designs a centralized monitoring structure, as shown in Fig. 3 below.

It can be seen from Fig. 3 that the traditional CAN bus is a single-layer structure system. The bus controller and the bus terminal module are both attached to a pair of buses, and all bus devices are at the same level. In long-distance communication, if the single-layer bus structure is adopted, the work of the whole bus will be affected when the line fault occurs. In the new multi-layer network structure, the bus centralization module is used as the middle layer. The bus controller is responsible for managing the bus centralized module, while the centralized module manages the terminal module, reducing the number of nodes on the primary bus and reducing bus conflicts.

The system divides the stations of the whole line into several blocks according to the proximity principle. Each block is equipped with CAN bus centralization station. The control host is responsible for managing each bus centralization module, and each bus

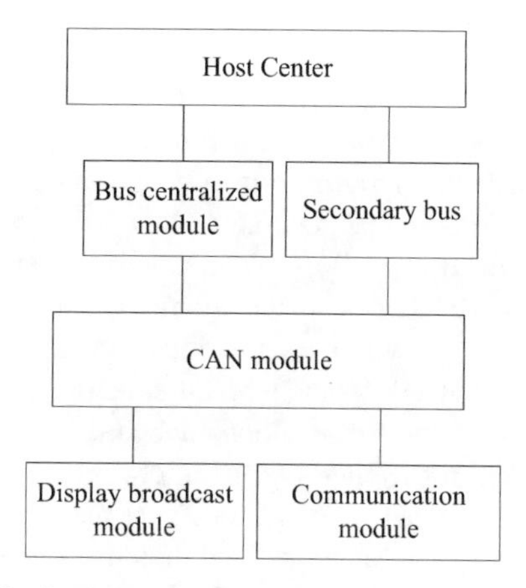

Fig. 3. Bus Centralized Monitoring Structure

centralization module manages the corresponding terminals attached to its subordinate bus.

Here, each centralization station is equipped with two bus centralization modules, of which module 1 is in the main working state, responsible for patrolling each communicator and issuing various commands sent by the control host to the communicator, train number display and automatic broadcast control module. Module 2 is in the hot standby state. It receives the command from the control host and detects whether module 1 is in the working state. When module 1 is found to be faulty or receives the start command sent by the control host, it turns to the main working state.

The two-layer structure reduces the number of nodes on the bus, and reduces the transmission of redundant information, as well as the information round-trip between the field and the control room. The CAN bus adopts the method of closing the node for the node with serious bus conflict, so as to reduce the network failure caused by the conflict. However, after the node is closed, the communication between the field equipment and the center is also interrupted. The reduction of the number of nodes can reduce conflicts and improve the reliability of the bus *MTBF* As shown in (7) below.

$$MTBF = \frac{1}{\lambda}(1 - R) \tag{7}$$

In formula (7), λ Represents the round-trip reliability coefficient of information, R It represents the communication integral. If the reliability calculated at this time is qualified, it proves that the performance of the above structure is good. Otherwise, it needs to detect relevant problems and make reasonable optimization.

3.3 Design Signal Centralized Monitoring Function Module

Combined with the above structure, we can use the drawing class SurfaceView of Android system to write the upper application program and realize the preview of video image on the LED of the development board. There are two ways to install hardware drivers in the kernel: one is static compilation.

When compiling the kernel, it is directly selected and compiled into the kernel, and can be used directly after the kernel is started; The other is dynamic loading. This article creates the spca5xx entry in the driver/usb/media directory.Mkdirspca5xx, return to the/driver/usb grid, place the portal in this directory, and run path-p1.2.6.12.path.If correct, the corresponding file will be generated in the usb/media/spca5xx directory.

Then add and modify the drivers file. Compile the kernel and modules. In the source root directory, execute make, that is, make modules, specify the cross compilation environment, and make changes within the previous level of Makeage.

Copy the module file to the development board directory, insert the USB camera into the development board, and restart the development board. This video monitoring system uses a portable Usb camera for testing, that is, to collect video data. When collecting video, the Android system can achieve real-time automatic capture of video, and then call each frame through interface functions, and then use surface view classes to draw, so as to complete the real-time preview function of the image. The video data collection process is shown in Fig. 4 below.

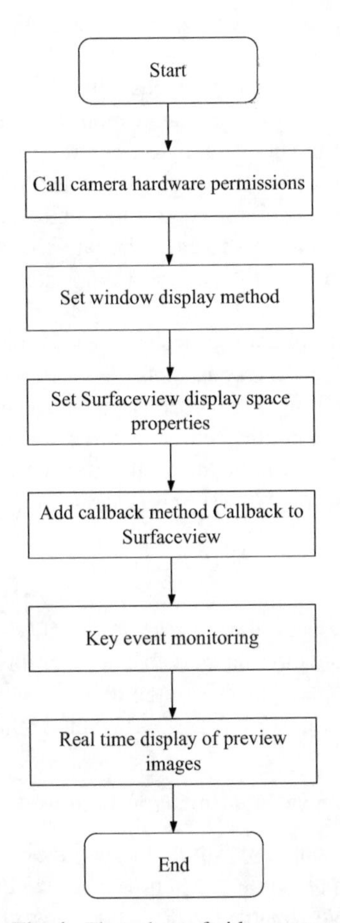

Fig. 4. Flow chart of video capture

As shown in Fig. 4, this article adds hardware permissions to call the camera, adds the following code to the Android manifest file, and sets the window display mode. When performing operations such as modifying, creating, and destroying data, this function is realized by setting a callback function.

After the video information is collected, the video monitoring system will preview the video in real time through the LED display, which involves the knowledge of Android interface design. Android user interface is based on GUI system and programmed in Java language. The Android SDK supports a large part of the functionality of the standard Java Runtime Environment (JRE).The Java platform supports the use of XML through many different formulas, and a large number of XML related Java APIs are fully supported on Android.

For the interface design based on the Android platform, including the basic components of building the screen, even if you use XML to define the screen and load it into the code, you also need to handle various tasks for the user interface, and then realize the design of the program UI through a single or multiple active applications. Thus, an effective user interface can be formed to allow users to easily access the periphery and realize the preview of real-time video information.

4 System Test

In order to verify the monitoring effect of the designed rail transit multi-channel signal centralized acquisition and monitoring system based on bus technology, this paper built an experimental platform to verify the operation function of the designed system, and carried out system testing, as follows.

4.1 Test Preparation

During the research and development of the video monitoring system, testing is an indispensable means to measure the correctness, stability and effectiveness of video stream information acquisition, algorithm coding and RTP video transmission protocol in the system. Through the theoretical research and specific development of the system, reasonable and rigorous testing can better test the functionality of the system, It can also reflect the problems existing in the system through testing, provide the direction for improvement, and then optimize the system design to improve the software application effect. The test and development platform is 1.7.0 of Java Development Kit 7, which is a relatively stable version of JDK_The SR2 version of Eclipse software is used to provide the development environment of the system. The Android SDK is. The hardware of the experiment includes a camera, a radar main control board, a radar slave control board and a PC. The relative positions of the camera and the radar master and slave control boards are fixed, which can be reinforced with brackets. The experimental hardware placement diagram is shown in Fig. 5 below.

It can be seen from Fig. 5 that the design and processing of the radar master and slave control boards are the same. The Rogers 5880 material is used as the dielectric board, and XMC4500F100K1024 is used as the microprocessor chip and BGT24MTR12 is used as the RF chip. The thickness is about 10mil. The camera used for shooting is Xinanshi,

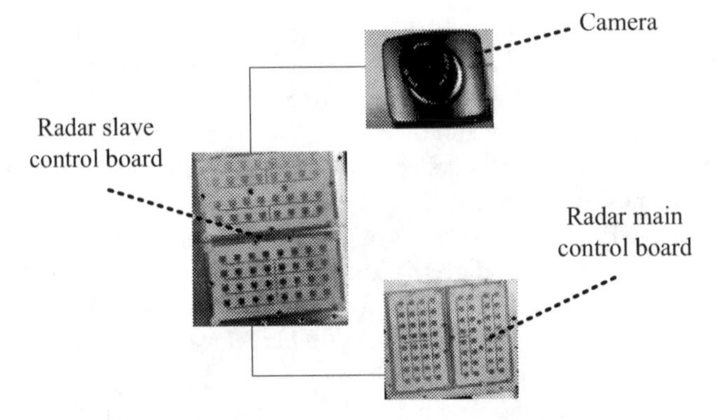

Fig. 5. Schematic diagram of experimental hardware placement

with a pixel size of 720 × 576. This paper uses the convolution neural network target detection model based on YOLO, which has a large number of convolution operations, and has certain requirements for the computing ability of the computer. The software information used in the experimental platform is shown in Table 3.

Table 3. Software Information of Experimental Platform

Device Name	Model
Operating system	Windows7
Processor	Intel Xeon CPU E5-2620 v3
Memory	6 GB
Graphics card	NVIDIA GeForce GTX 1060
Software platform	Visual Studio2015
Programming Language	C/C++
CUDA	10.1
CUDNN	7.0

It can be seen from Table 3 that the above parameters meet the system test requirements. In order to improve the reliability of the system test, the test monitoring radar is preset in this paper. The parameter indicators are shown in Table 4 below.

It can be seen from Table 4 that before the system test, it is necessary to set the target to move 40 m away from the radar and record the speed of the target detected by radar signal processing.

The purpose of the test is to check whether the program can complete the specified task according to the process set by the requirements during the running process, and whether it can handle correctly in the case of errors or exceptions, so as to avoid downtime or system paralysis. Since the system is based on the BS framework, it is necessary to test the function and performance of the front end, as well as the compatibility under

Table 4. Parameters and Indexes of Experimental Test Monitoring Radar

Name	Parameter indicators
Frequency	24 GHz
Working wavelength	0.0125 m
Maximum detection distance	120 m
Maximum detection speed	±240 km/h
FM bandwidth	200 MHz
Coherent processing cycle	7.68 ms
FFT points	512
LFSK frequency difference	f1:554\f2:13.03
Distance between receiving antennas of horizontally polarized beams	24.15 mm
Distance between receiving antennas of vertically polarized beams	24.15 mm

various browsers. The most important thing is to check the checksum processing on the server side. The requirement is implemented as a program, and then each requirement process is accepted through the program. This analysis method is usually drawn as a V-type test model, as shown in Fig. 6.

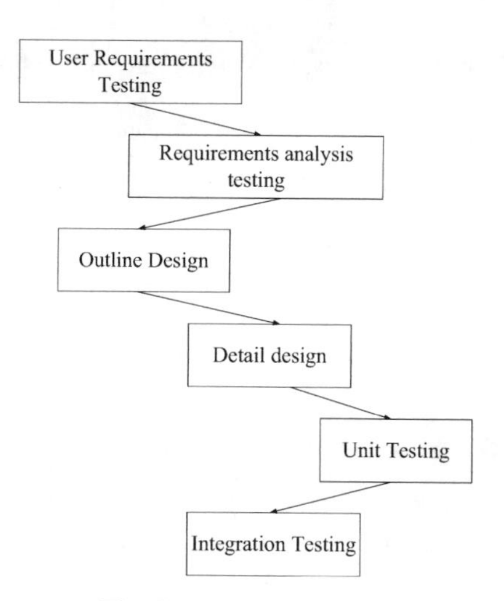

Fig. 6. System test model

As shown in Fig. 6, from the model diagram, the whole system test consists of four parts. They are unit test, integration test, confirmation test system test and acceptance test. Before the unit test of the system, each function test of the system needs to be carried out first when developing each function of the system.

4.2 Test Results and Discussion

Under the above test preparation, the system test can be carried out, that is, the debugging test environment. Run the rail transit multi-channel signal centralized monitoring system designed in this paper based on bus technology. At this time, the test results of each functional module are shown in Table 5 below.

Table 5. System Test Results

TEST	Expected results	Test result
Login Module	Enter the username and password, confirm login, display successful login and enter the system main interface	Test passed
Traffic signal monitoring module	Enter the username and password, confirm login, enter the traffic signal control module, click on manual control of traffic signals, and control the display of lights	Test passed
Monitoring information collection module	Enter the username and password, confirm login, enter the video monitoring module, click on the video screen to display the video information of each intersection normally	Test passed
Information modification module	Log in to the information modification interface, click Modify, and prompt for successful modification	Test passed

It can be seen from Table 5 that each functional module of the rail transit multi-channel signal centralized monitoring system designed in this paper based on bus technology operates orderly. The above test results prove that the performance of the designed signal centralized monitoring system is reliable and has certain application value.

4.3 Comparison of Test Results and Discussion

To further highlight the application advantages of the system in this article, the traditional rail transit signal monitoring system based on heterogeneous Internet of Things (Traditional System 1) and the embedded technology based rail transit signal monitoring system (Traditional System 2) are compared. The application performance of the three systems is analyzed from the perspective of signal monitoring delay, and the results are shown in Table 6.

By analyzing the results shown in Table 6, it can be seen that as the number of experiments increases, there are certain differences in signal monitoring delays among different systems. Among them, the maximum signal monitoring delay for Traditional system 1 is 2952 ms, the maximum signal monitoring delay for Traditional system 2 is 2053 ms, and the maximum signal monitoring delay for System of this paper is only 984 ms. In contrast, the monitoring timeliness of the System of this paper is higher.

Table 6. Comparison of Signal Monitoring Delay in Different Systems (ms)

Number of experiments	System of this paper	Traditional system 1	Traditional system 2
10	980	2952	1763
20	967	2948	1869
30	964	2921	1826
40	981	2945	1955
50	984	2950	2053

5 Conclusion

This article designs an effective centralized monitoring system for traffic multi-channel signals based on bus technology, which improves the application performance of the system from two perspectives of software and hardware design. The test results indicate that the system has good application performance and is effective in reducing signal monitoring latency.

References

1. Velmurugan, P., Ashok, B.: Improving the quality of service by continuous traffic monitoring using reinforcement learning model in VANET. Int. J. Model. Simul. Sci. Comput. **13**(06), 310–327 (2022)
2. Meng, B., Damanhuri, N.S., Othman, N.A.: Smart traffic light control system using image processing. IOP Conf. Ser. Mater. Sci. Eng. **1088**(1), 01–08 (2021)
3. Saldivar-Carranza, E.D., Hunter, M., Li, H., et al.: Longitudinal performance assessment of traffic signal system impacted by long-term interstate construction diversion using connected vehicle data. J. Transp. Technol. **4**, 11–20 (2021)
4. Yao, J., Qiu, J.: Research on road traffic flow prediction based on SSA-BP algorithm. J. Southwest Univ. (Natural Science Edition) **44**(10), 193–201 (2022)
5. Liu, Z., Cao, Y., Sha, A., et al.: Energy harvesting array materials with thin piezoelectric plates for traffic data monitoring. Constr. Build. Mater. **302**(4), 124–136 (2021)
6. Sofuoglu, S.E., Aviyente, S.: GLOSS: tensor-based anomaly detection in spatiotemporal urban traffic data. Sig. Process. Off. Publ. Euro. Assoc. Sig. Process. (EURASIP) **192** (2022)
7. Singleton, P.A., Runa, F.: Pedestrian traffic signal data accurately estimates pedestrian crossing volumes. Transp. Res. Rec. **2675**(6), 429–440 (2021)
8. Nawaz, A., Zafar, N.A., Alkhammash, E.H.: Formal modeling of responsive traffic signaling system using graph theory and VDM-SL. Sustainability **13** (2021)
9. Cao, B., Liu, W., Zhang, L., et al.: Simulation analysis of signal coverage of ADS-B base station based on DEM. J. Phys. Conf. Ser. **1865**(4), 1–8 (2021)
10. Zhu, J., et al.: Lightweight web visualization of massive road traffic data. J. Southwest Jiaotong Univ. **56**(05), 905–912 (2021)
11. Pan, C.: Frequencyshift rail transit signal detection method based on time-frequency analysis. Mach. Elec. **40**(01), 71–75 (2022)
12. Sun, M., Wei, H., Li, X., Yu, J., Xu, L.: Refined traffic state detection of road based on multidimensional density clustering. J. Geomat. Sci. Technol. **36**(04), 412–417 (2019)

13. Zhu, L.: Intelligent wireless signal monitoring of urban rail transit CBTC system. Urban Mass Transit **24**(S1), 117–121 (2021)
14. Yu, J.: Research on architecture of urban rail transit signaling system. Urban Mass Trans. **25**(10), 131–135+143 (2022)

Equilibrium Scheduling of Dynamic Supply Chain Network Resources Under Carbon Tax Policy

Hao Zhu(✉)

Business School, Shanghai Sanda University, Shanghai 201209, China
zhu_scmglobal@163.com

Abstract. The supply chain network is usually composed of multiple nodes and paths, and the energy consumption and carbon emissions of each node will change, leading to dynamic changes in the network structure and resource demand. In order to improve the utilization rate and accuracy of dynamic supply chain network resource balanced scheduling, under the current carbon tax policy, a comprehensive study on dynamic supply chain network resource balanced scheduling was carried out. Firstly, the RSUC dynamic supply chain network model is constructed to reflect the operation changes of the dynamic supply chain network under the carbon tax policy in real time through the iterative operation of the model. Secondly, a multi-path link routing algorithm for dynamic supply chain network resource balanced scheduling is designed, and the one with the highest success rate is selected as the transmission path for dynamic supply chain network resource balanced scheduling. Calculate the use of resources in dynamic supply chain network nodes, obtain the use and surplus of dynamic supply chain network resources under the carbon tax policy, and make every node's resources be used as far as possible. On this basis, according to the actual situation and characteristics of the supply chain network operation, the dynamic quota of network resource balanced scheduling is designed in an all-round way, so as to achieve the goal of dynamic supply chain network resource balanced scheduling under the carbon tax policy. Experimental analysis shows that after the application of the new method, the network resource utilization rate is high under different network workload data segments.

Keywords: Carbon Tax Policy · Dynamic · Supply Chain · Network · Resources · Equilibrium · Dispatch

1 Introduction

Carbon tax refers to the tax levied on carbon dioxide emissions. It aims at environmental protection and hopes to slow down global warming [1] by reducing carbon dioxide emissions. The carbon tax reduces fossil fuel consumption and carbon dioxide emissions [2] by taxing fossil fuel products such as gasoline, aviation fuel and natural gas

L. Yun et al. (Eds.): ADHIP 2023, LNICST 549, pp. 393–407, 2024.
https://doi.org/10.1007/978-3-031-50549-2_27

downstream of coal and oil in proportion to their carbon content. Different from the greenhouse gas emission reduction mechanism based on market competition such as total emission control and emission trade, carbon tax can be achieved by adding very little additional management cost [3]. Because of its connection with global climate change, carbon tax is theoretically set to require a global international management system to achieve optimal output, but this is not inevitable. When a country or region determines emission limits and emission reduction targets, it has considerable advantages to implement carbon taxes at the national or regional level. The impact of carbon tax is extensive and far-reaching, involving many aspects of social economy and people's life. The carbon tax should not only consider environmental effects and economic efficiency, but also consider social benefits and international competitiveness. The implementation effect of carbon tax varies greatly in different countries and regions at different stages of economic and social development. But in the long run, carbon tax is an effective environmental and economic policy tool, which can effectively reduce CO2 emissions. Reducing energy consumption, changing the energy consumption structure, and inhibiting economic growth in the short term will be conducive to the healthy development of the economy in the medium and long term. However, it will expand the income distribution gap between capital and labor and intensify social injustice [4].

Resource scheduling schemes have different classification methods according to different focuses. In specific applications, different schemes can be selected according to different needs and optimization objectives of the system [5]. According to the number of available channels, it can be divided into single channel and multi-channel resource scheduling;According to whether the network topology information needs to be obtained, it can be divided into topology transparent and topology related resource scheduling;According to whether the centralized controller is set to assist resource allocation, it can be divided into centralized, distributed and centralized/distributed hybrid resource scheduling. According to the way of resource allocation, it can be divided into fixed, competitive and mixed resource scheduling. As the premise and foundation of data transmission and management of the Internet of Things system, resource balanced scheduling plays an important role in ensuring the performance of the dynamic supply chain network and effectively avoiding transmission interference [6]. However, many IoT resource scheduling schemes do not consider the "energy hole" problem of nodes near the base station, and ignore the problem of timely recovery and reuse of network resources when nodes fail. Generally, priority based resource scheduling sets the priority of data in advance, and cannot detect multiple types of emergency data at the same time. In some data monitoring applications, data priority is difficult to set in advance. In some heterogeneous IoT resource scheduling, slot reuse rate is difficult to improve, and network resource utilization rate is low. The clustering technology of the Internet of Things divides the network into different clusters. By managing other member nodes through cluster head nodes, it can effectively reduce the energy consumption of the Internet of Things and extend its life cycle. Affected by the carbon tax policy, there are certain risks and hidden dangers in the balanced scheduling process of dynamic supply chain network resources, and the scheduling timeliness is poor, which cannot ensure the efficient operation of the dynamic supply chain network [7].

In view of the above problems, this paper has carried out an in-depth study on the balanced scheduling of dynamic supply chain network resources under the current carbon tax policy, to ensure the efficient and high-quality operation of the dynamic supply chain network.

2 Research on Network Resource Equilibrium Scheduling of Dynamic Supply Chain

2.1 Build RSUC Dynamic Supply Chain Network Model

In the dynamic supply chain network resource balanced scheduling method under the carbon tax policy designed in this paper, first, the power supply problem of dynamic supply chain network data acquisition equipment needs to be solved.

One of the obstacles to the application of dynamic supply chain network is the power supply of data acquisition equipment. For many applications of the Internet of Things, wired power supply is impractical, and regular battery replacement will cause more complex operation and maintenance. In the multi hop communication of the Internet of Things, nodes near the base station undertake more data forwarding work, which will lead to the rapid depletion of battery energy and the "energy hole" phenomenon. This problem can be alleviated by non-uniform clustering strategy. Many resource scheduling schemes do not consider the "energy hole" problem, and ignore the problem of resource recovery and reuse of failed nodes. This chapter proposes IoTResourceScheduling based on Unequal Clustering (RSUC) [8]. The structural framework of RSUC dynamic supply chain network model is shown in Fig. 1.

As shown in Fig. 1, a non-uniform clustering strategy based on dynamic topology is proposed to alleviate the "energy hole" problem in the multi hop IoT;Then we study the routing selection and layering mechanism based on non-uniform clustering, establish the multi hop routing constraint criteria, find the optimal relay node, and layer the cluster heads according to the data transmission chain routing. Finally, we propose resource scheduling for intra cluster communication and inter cluster communication based on non-uniform clustering. In the resource scheduling of intra cluster communication, the cluster head node dynamically reclaims the resources of the failed node according to the threshold of the number of members of the cluster. In the resource scheduling of inter cluster communication, the cluster head node obtains different sending and receiving timeslots according to different layers of the transmission chain. In RSUC based resource scheduling, clusters near the base station are allocated fewer intra cluster communication slots and more inter cluster communication slots. The farther the cluster is from the base station, the earlier the inter cluster communication ends and enters the intra cluster communication, rather than waiting for all cluster heads to complete the inter cluster communication, thus improving the utilization of resources. Experiments show that through reasonable joint resource allocation of channels and timeslots for intra cluster and inter cluster communication, data can be transmitted in parallel as far as possible to reduce network energy consumption and improve network throughput [9].

In the dynamic supply chain network model, the dynamic heterogeneous clustering and multi hop routing are comprehensively considered. IoT equipment mainly includes:

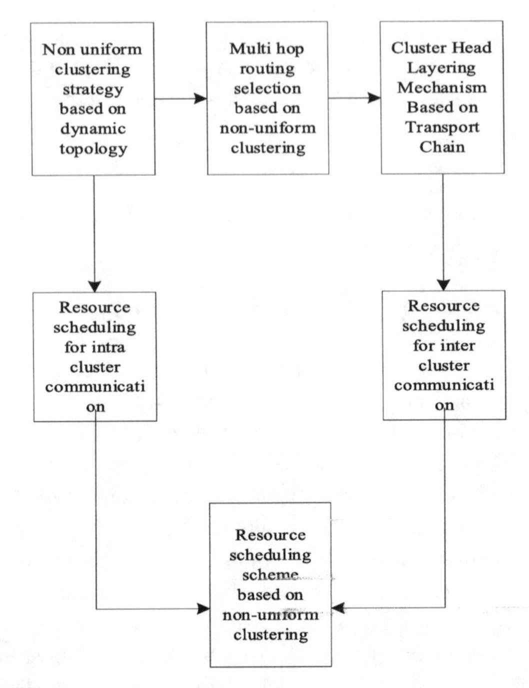

Fig. 1. Structure framework of RSUC dynamic supply chain network model

data collection node, cluster head node, base station and master control equipment. Each cluster has a cluster head node, which collects the data of nodes in the cluster and transmits the data to the base station in a multi hop manner. Set the specific functions of each node as follows:

1. Data collection node

   The data acquisition node is mainly responsible for sensing and collecting environmental and equipment information in the monitoring area, such as humidity, temperature, noise, vibration, infrared sensing and other data. After clustering, the data collection node will become a cluster member (CM) node and send the collected data to the cluster head node [10] in one hop mode.
2. Cluster head node

   The cluster head node (CH) is mainly responsible for generating the resource scheduling table for intra cluster communication, receiving the data from the data collection node in the cluster, merging the data, forwarding the data to the base station in a multi hop manner, and broadcasting time synchronization information, channel and slot scheduling table updates and other information in the cluster.
3. Base station

   The base station (BS) is mainly responsible for generating the resource scheduling table of inter cluster communication, collecting the data information of cluster head nodes, finally transmitting the data to the master control equipment, and broadcasting time synchronization information, channel allocation table, inter cluster timeslot scheduling table and other information throughout the network.

4. Main control equipment

All data is finally transmitted to the master control equipment. The master control equipment manages and processes the collected data uniformly, and provides real-time information for users.

The RSUC dynamic supply chain network model established in this paper needs to meet the following constraints, as shown in Table 1.

Table 1. Constraints of RSUC dynamic supply chain network model

number	constraint condition
1	Each node of the Internet of Things can indicate the received signal strength according to the received signal strength Indicator, RSSI) to calculate the distance
2	The transmission power of each node of the Internet of Things is adjustable
3	Generated by periodic data collection of data collection node
4	The cluster head node collects the data of the cluster members in the communication within the cluster and fuses them However, in the process of inter cluster communication, the received information is not fused considering the differences of data in each cluster
5	The base station supports multi interface and multi-channel communication
6	In the intra cluster communication, the data collection node can complete the current week of the node in one unit time slot Data transmission of periodic collection. The size of the unit timeslot can be adjusted according to the application environment
7	In a time slot, a node can only communicate with one neighbor node at most. Node in one The time slot is only in the sending state or receiving state

According to the constraints described in Table 1, the RSUC dynamic supply chain network model is established. Through the iterative operation of the model, the operation changes of the dynamic supply chain network under the carbon tax policy are reflected in real time, laying the foundation for the balanced scheduling of network resources in the following text.

2.2 Design a Dynamic Supply Chain Network Resource Balanced Scheduling Multi-path Link Routing Algorithm

The reasonable scheduling of multi-path links can effectively improve the reliability of data transmission, and the more multi-path links, the more obvious the improvement of reliability. Therefore, the number of multi-path links can be reasonably increased when generating data flow graph routes. It should be noted that the existence of multi-path links will lead to more communication conflicts and larger data transmission delay. In order to ensure effective data transmission, while increasing multi-path links to improve reliability, it is necessary to ensure that the data transmission delay is within the superframe range. In addition, the existence of multipath links can improve the reliability of data transmission. For a single data stream, the overlap of paths between different data streams will not affect the transmission success rate.

Therefore, the multi-path link routing algorithm designed in this chapter generates routes for each data stream in turn according to the priority order of the data stream. The design idea of the algorithm is described in detail below.

(1) Generation of the main path: When generating the main path, the data stream will not generate multi-path links, so only the length of the path needs to be considered at this time. You can directly use the single source shortest path algorithm to generate the main path of the data stream. If multiple shortest paths with the same length appear, select the path with the least overlap with other paths as the final main path.
(2) Generation of secondary path: when generating the secondary path, it is necessary to consider both the path length and the number of multi-path links, and there is no unified relationship between the two factors. You can use the method of generating alternative paths, and then compare the transmission success rate to select the final secondary path. The schematic diagram of generation of secondary path for balanced scheduling of dynamic supply chain network resources is shown in Fig. 2.

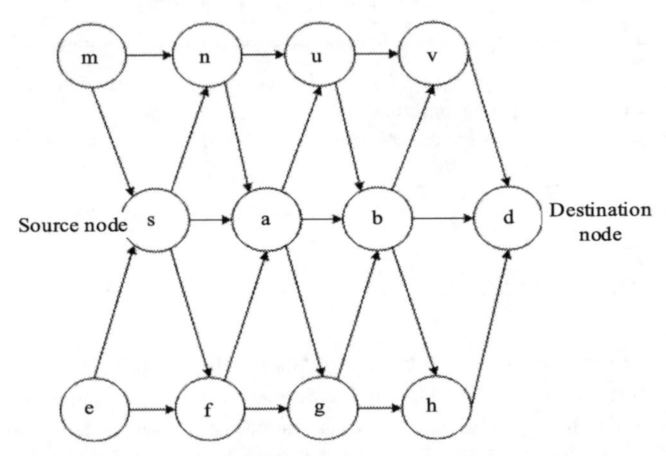

Fig. 2. Schematic diagram of dynamic supply chain network resource balanced scheduling secondary path generation

As shown in Fig. 2, it is the schematic diagram of generation of secondary path for dynamic supply chain network resource balanced scheduling designed in this paper. The specific generation process is as follows:

1. Determine the source node of the secondary path according to the primary path, generate the first alternative path through the shortest path algorithm, and represent it with C, representing the jth alternative secondary path of the ith secondary path of the data flow f. If there are shortest paths of equal length, they are all regarded as alternative paths.

2. Generate the shortest path containing multi-path links as an alternative path. The main path of the data stream is s → a → b → d, so it can be determined that the source node of its first secondary path is s and the destination node is d. The shortest path from node s to node a is generated through the shortest path algorithm, and then it reaches the destination node through links (a, b) and (b, d) to generate the shortest path containing two multi-path links: similarly, the shortest path is generated from node s to node b, Then it reaches the destination node through links (b, d), and generates the shortest path containing a multipath link. Note that the primary path and the secondary path cannot be identical.
3. Calculate the transmission success rate under different alternative paths. The formula is:

$$g(n, s[n]) = \prod_{i=1}^{i=n} \left[1 - (1 - p)^{s[i]}\right] \tag{1}$$

Among them, $g(n, s[n])$ Indicates the success rate of path transmission; p Represents the average transmission success rate of data stream sets in the entire network; $s[i]$ It indicates the number of link transmissions of data transmission. Through calculation, the one with the highest success rate is selected as the final secondary transmission path for balanced scheduling of dynamic supply chain network resources.

2.3 Calculating the Resource Usage of Network Nodes

After completing the design of multi-path link routing algorithm for balanced scheduling of dynamic supply chain network resources, next, calculate the resource usage of dynamic supply chain network nodes.

For the resource scheduling problem of network intensive applications, this paper combines the ideas of NodeResources LeastAllocated and NodeResour cesB balanced Allocation algorithms in the Kubernetes default scheduler, and proposes a optimal scheduling algorithm BWNA. When scheduling pods, this algorithm not only considers whether the resource scheduling among multiple nodes in the cluster is balanced,And considering whether the use of different resources on a single node is balanced, the most important thing is that this algorithm can comprehensively consider the three indicators of CPU, memory and network bandwidth, and then calculate the resource usage of network nodes to meet the scheduling requirements of network intensive applications in all directions.

The algorithm design is as follows: Assume there is k Nodes $N_1, N_2, N_3, ..., N_k$; node N_i The CPU capacity of is recorded as C_{cpu}^i, the memory capacity is recorded as C_{mem}^i, the network capacity is C_{net}^i Capacity represents the total amount of resources that a node can provide. Resource usage of nodes U The CPU usage is recorded as U_{cpu}^i, the memory usage is recorded as U_{mem}^i, network usage is U_{net}^i, and we set the weights of CPU, Memory, and Net to W_c 、 W_m 、 W_n. When a Pod needs to be scheduled, assume that the CPU, Memory, and Net requests of the Pod are R_{cpu} 、 R_{mem} 、 R_{net} The scheduling process is as follows:

1) Calculate each node N_i Usage of Pod resources running on U_{cpu}^i 、 U_{mem}^i 、 U_{net}^i.

2) When the Pod to be scheduled is scheduled to the node N_i Calculate the resource usage of the node after T^i_{cpu}, T^i_{mem}, T^i_{net}, the formula is as follows:

$$T^i_{cpu} = U^i_{cpu} + R_{cpu} \tag{2}$$

$$T^i_{mem} = U^i_{mem} + R_{mem} \tag{3}$$

$$T^i_{net} = U^i_{net} + R_{net} \tag{4}$$

3) Calculate the remaining available resource of the node E^i_{cpu}, E^i_{mem}, E^i_{net}, the formula is as follows:

$$E^i_{cpu} = C^i_{cpu} - T^i_{cpu} \tag{5}$$

$$E^i_{mem} = C^i_{mem} - T^i_{mem} \tag{6}$$

$$E^i_{net} = C^i_{net} - T^i_{net} \tag{7}$$

Through calculation, we can obtain the use and surplus of network resources in the dynamic supply chain under the carbon tax policy, so that the resources of each node can be used as far as possible, and lay a good foundation for the balanced scheduling of resources among multiple nodes in the subsequent network cluster.

2.4 Design Dynamic Quota for Balanced Scheduling of Network Resources

After obtaining the resource usage of dynamic supply chain network nodes, based on this, and according to the actual situation and characteristics of the supply chain network operation, the dynamic quota for balanced scheduling of network resources is comprehensively designed, so as to achieve the goal of balanced scheduling of dynamic supply chain network resources under the carbon tax policy.

To solve the problem that the supply chain network resource quota used by multiple users cannot change dynamically, this paper designs a DQ controller. This controller starts from the cluster resource change, listens to the increase and deletion of nodes in the cluster, calculates the proportion of resource quota that needs to be dynamically adjusted according to the ratio of cluster resource change, and finally modifies the resource quota data under multiple users. DQ is designed as follows:

1) DQ counts the total amount of CPU and memory resources of all available nodes of the cluster TotalCpu and TotalMem, and saves the NodeList of the node list of the current cluster.
2) Monitor the changes of resources such as nodes in the cluster through NodeInformer. When adding or deleting nodes in the cluster, perform step 3.
3) Update the node list in the NodeList. If the CPU and memory resources of the newly added/deleted nodes are NodeCpu and NodeMem, respectively, calculate the change rate of CPU and memory resources in the cluster. The calculation formulas are:

$$P_{cpu} = \frac{M_{cpu1} + M_{cpu}}{M_{cpu}} \tag{8}$$

$$P_{mem} = \frac{M_{mem1} + M_{mem}}{M_{mem}} \tag{9}$$

Among them, P_{cpu} Indicates the CPU change rate in the dynamic supply chain network cluster; P_{mem} Indicates the change rate of memory resources in the dynamic supply chain network cluster; M_{cpu1} Indicates the CPU of the new/deleted node; M_{mem1} Indicates the memory resources of new/deleted nodes; M_{cpu} Represents the total change of network CPU; M_{mem} Total changes in network memory resources.

4) Calculate the change rate of cluster resources according to different strategies. The DQ controller supports the mean and maximum values
(Max) and minimum (Min) are calculated as follows.

$$P_{Mean} = \frac{P_{cpu} + P_{mem}}{2} \tag{10}$$

$$P_{Max} = \max(P_{cpu}, P_{mem}) \tag{11}$$

$$P_{Min} = \min(P_{cpu}, P_{mem}) \tag{12}$$

Through calculation, the change rate of dynamic supply chain network cluster resources under different strategies is obtained.

5) According to the change policy configured in the DQ, select one of Formula 3 to Formula 5 to get the final rate of change ratio, then traverse the namespaces in the configuration that need to dynamically change the resource quota, and finally dynamically change the resource quota items in the DQ under these namespaces according to the rate of change ratio.

To sum up, DQ calculates the change rate of the node's CPU and memory resources, and then averages or compares them to get the final change rate. Finally, it adjusts the quota of the ResourceQuota under NamepSpace to achieve the goal that the ResourceQuota changes with the changes of cluster resources, and then achieves the goal of balanced scheduling of dynamic supply chain network resources under the carbon tax policy.

3 Experimental Analysis

3.1 Experiment Preparation

In order to objectively test the feasibility of the dynamic supply chain network resource balanced scheduling method proposed in this paper, the experimental analysis is carried out as shown below. Simulate the collaborative work of five cloud services, and take five data segments with the same length from the load tracking of Google dataset to evaluate the algorithm in this chapter.

Google cluster is a group of physical machines connected through a high bandwidth network. All clusters share a common cluster management system, which distributes work to machines and loads the cluster in the form of jobs. A job consists of one or more tasks. Each task is accompanied by a set of resource requirements, which are used to schedule the task to the appropriate server. The resource requirements and usage data of the task come from the information provided by the cluster management system. The data set used in this article is Google's one month data tracking of Google clusters. Data appears in the form of a table and contains six data attributes: machine attributes, machine events, job events, task constraints, task events, and task attributes. Data is generated by various physical machines with different processing capacities in the cluster, and the workload in the data shows a high degree of heterogeneity and variability. It consists of jobs with different resource requests and processing priorities. Each job is divided into one or more tasks, and the data contains the demand information of each task and the physical machine information of processing tasks. There are difficulties in processing tracked big data. The amount of tracked data after compression is about 40G, and the size of all tables after decompression is about 300GB. It should be noted that all the data provided by Google cannot reflect the real value of tracking. For privacy purposes, Google standardizes the data so that researchers can use these data for analysis and experiments.

The use of real load tracking makes the results more realistic and reliable, and can better reflect the randomness and volatility of user requests. The data segment includes 200 5-min intervals, lasting more than 16 h in total. The selected workload data segments correspond to cloud service 0, cloud service 1, cloud service 2, cloud service 3, and cloud service 4, respectively. In addition, this chapter assumes that the general scheduling strategy is first come, first serve. IaaS providers provide four types of virtual machines: small, medium, large, and extra large. They have different capacities and costs. The specific parameter settings are shown in Table 2.

Table 2. Parameter Settings of Virtual Machines Used in Experiments

VM Type	Large	Medium	Small
Core	4	2	1
CPU (MIPS)	10000	5000	2500
RAM (GB)	75	37.5	17
Disk (GB)	8500	4100	1600
VM price ($/h)	14	7	3.75
VM initiation price ($)	0.13	0.06	0.025

Set the virtual machine according to the parameters shown in Table 2. Secondly, build the test environment for this experiment. For task data of different scales, this chapter simulates the establishment of different number of servers and server farms: 30 servers and 10 server farms are set. For each server, it is assumed that there are 21 CPU resources and 21 RAM resources, and some servers are connected together as a server farm. In fact, the number and structure of server farms are determined by CSPs. In each server, suppose there are three virtual machines, and each task can run on one of them. In the resource balanced scheduling experiment designed in this paper, the non-uniform clustering strategy and resource scheduling of intra cluster and inter cluster communication are used to reduce network energy consumption. The initial battery voltage of nodes in the dynamic supply chain network is 3.3 V. Figure 3 shows the measured voltage values of 10 randomly selected nodes in the first and third layers after uneven clustering recorded on June 20, 2021 and July 20, 2021, respectively.

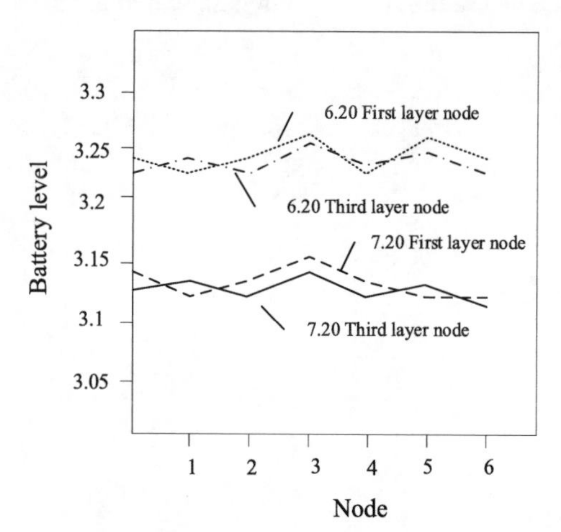

Fig. 3. Battery power corresponding to nodes at different layers in dynamic supply chain network resource balanced scheduling

It can be seen from the monitoring values that the remaining battery power of the nodes on the first layer and the third layer is similar. This experiment shows that the non-uniform clustering strategy makes the energy consumption of nodes at different levels relatively balanced, thus avoiding the "energy hole" problem caused by the frequent forwarding of data by the first layer nodes near the base station. In the resource scheduling experiment, the convex optimization theory is used to obtain the regular data change rate that reflects the normal operation of the dynamic supply chain network. Based on the change rate of conventional data, real-time monitoring the temperature change of the supply chain network equipment bearing.

3.2 Result Analysis

After the dynamic supply chain network resource balanced scheduling is selected, the network resource utilization rate is used as the evaluation index of this experiment. Network resource utilization refers to any time interval Δt Internal, the utilization rate of the supply chain network cloud service CPU. The calculation expression is:

$$P(\Delta t) = \frac{M(\Delta t)}{T(\Delta t)} \tag{13}$$

Among them, $P(\Delta t)$ Represents any time interval Δt Internal, the utilization rate of the supply chain network cloud service CPU; $M(\Delta t)$ Represents any time interval Δt

The cloud service requests the allocated machine commands per second for users; $T(\Delta t)$ Represents any time interval Δt The total MIPS of the leased server. The higher the utilization rate of network resources, the better the balanced scheduling effect of dynamic supply chain network resources under the carbon tax policy, the higher the CPU utilization rate, and vice versa. Dynamic supply chain network delay includes transmission delay, propagation delay, processing delay and queuing delay. In this experimental test, the average network delay of the supply chain is shown in Fig. 4.

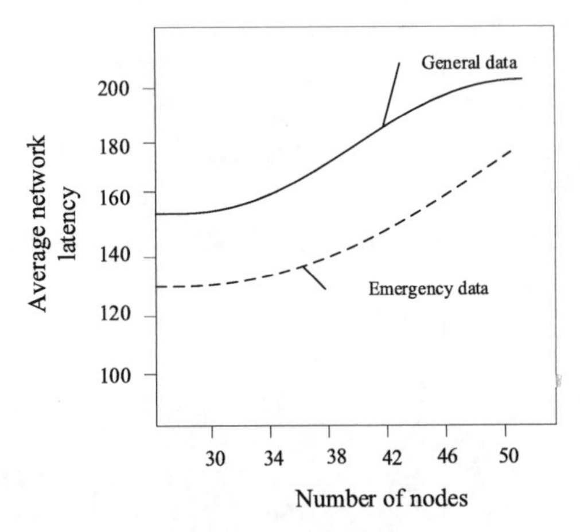

Fig. 4. Dynamic supply chain evaluation network delay

From Fig. 4, it can be seen that the supply chain network data monitoring system adopts the UCPDR scheduling scheme, which prioritizes the processing of emergency temperature data by allocating time slots and channels. The application of this scheduling scheme effectively reduces the average network delay of emergency data, keeping it within 160, which is significantly lower than the average network delay of general data. This scheduling strategy is of great significance for improving the real-time, reliability, and emergency response capabilities of data monitoring systems.

In order to make the experimental results more intuitive and convincing, the experimental method and principle of comparative analysis are introduced. Set the dynamic supply chain network resource balanced scheduling method proposed in this paper as the experimental group, and set the conventional network resource balanced scheduling method as the control group for comparative analysis. Under the selected supply chain network workload data segment, namely cloud service 0, cloud service 1, cloud service 2, cloud service 3, and cloud service 4, respectively measure the utilization rate of dynamic supply chain network resource balanced scheduling after the application of the two methods, and draw the comparison chart of evaluation indicators as shown in Fig. 5.

The comparison results of the evaluation indicators in Fig. 5 show that the two network resource balanced scheduling methods show different performance effects after application. Among them, after the application of the dynamic supply chain network resource balance scheduling method under the carbon tax policy proposed in this article,

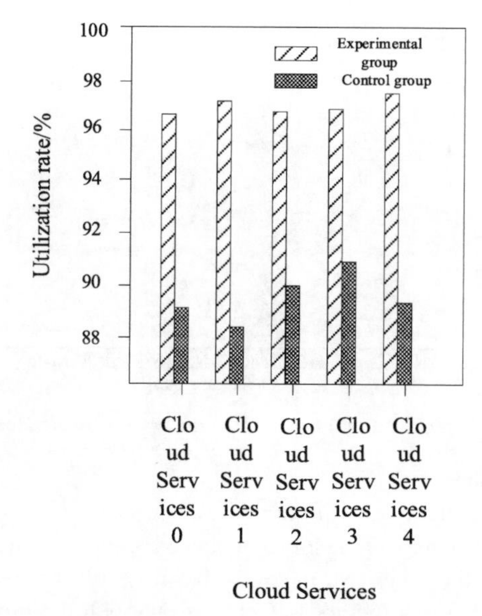

Fig. 5. Comparison Diagram of Experimental Evaluation Indicators

the resource utilization rate remains above 96% under different network workload data segments, while the overall resource utilization rate of the conventional method remains between 80% and 92%. Through comparison, it can be seen that the dynamic supply chain network resource balance scheduling method proposed in this article has always been higher than the conventional method under the carbon tax policy. It is not difficult to see that the resource balanced scheduling method proposed in this paper has high feasibility, can effectively reduce the resource waste generated in the operation of dynamic supply chain network, improve CPU utilization, and has significant advantages in the balanced scheduling effect.

In order to further validate the effectiveness of the method proposed in this paper, the methods of reference [6] and reference [7] in the introduction were used to compare with the method proposed in this paper. A random transmission path was randomly selected for testing in this method.

The network resource Transmission delay is selected as the evaluation index, which is one of the main indicators to show the load balance performance of each channel of the network interface. The smaller the Transmission delay of network resources, the faster the calculation speed of channel load, that is, the shortest time for channel load balancing, the shorter the waiting time of data queue, and the better the resource transmission performance. The experimental results are shown in Fig. 6.

From the experimental results in Fig. 6, we can see that the Transmission delay of network resources in this method is kept within 15 ms, while the Transmission delay of the other two methods is significantly higher than that in this method. This is because this method constructs the RSUC dynamic supply chain network model, and improves and optimizes it. This method can more accurately predict and schedule network resources to meet the needs of different nodes and reduce Transmission delay. Although this

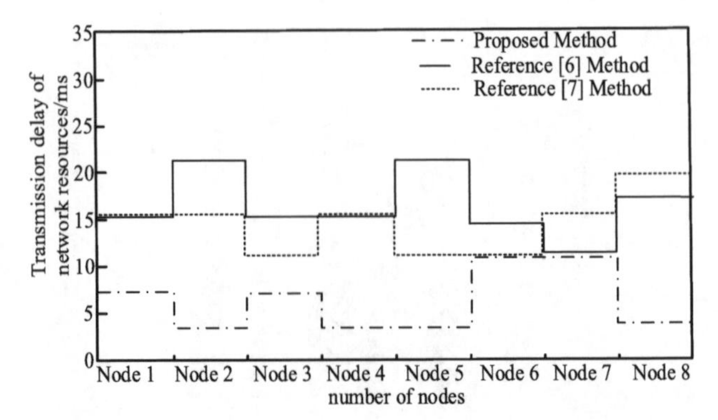

Fig. 6. Comparison of Transmission delay of Network Resources

method does not select the one with the highest success rate as the transmission path for dynamic supply chain network resource balanced scheduling, its Transmission delay is slightly lower than the other two methods, but it is still the lowest of all methods. It can be concluded that this method selects the one with the highest success rate as the transmission path for dynamic supply chain network resource balanced scheduling, which can ensure the rapid transmission and processing of data in the network, thus reducing Transmission delay.

4 Conclusion

To improve the overall performance of data transmission and resource scheduling in dynamic supply chain networks, this article studies the key technologies for optimizing resource balance scheduling in dynamic supply chain networks oriented towards carbon tax policies. Build an RSUC dynamic supply chain network model and design a dynamic supply chain network resource balance scheduling multi-path link routing algorithm to reflect the operational changes of the supply chain network under carbon tax policies in real time. By iteratively running the model, the usage of node resources in the dynamic supply chain network is calculated, and dynamic quotas are designed to achieve the goal of balanced scheduling of supply chain network resources under the carbon tax policy. The experimental results show that under different network workload data segments, the proposed method has a high utilization rate of network resources and can effectively solve task scheduling problems. However, in actual cloud environments, relying solely on task scheduling to achieve resource balance is not ideal and should be improved in future research.

References

1. Yang, Y.: Balanced scheduling method for big data of network traffic based on set-pair analysis strategy. J. Physics: Conference Series **2187**(1), 012065 (2022)
2. Stipanovic, I., Bukhsh, Z.A., Reale, C., et al.: A multiobjective decision-making model for risk-based maintenance scheduling of railway earthworks. Appl. Sci. **11**(3), 965 (2021)

3. Bu, B.: Multi-task equilibrium scheduling of Internet of Things: a rough set genetic algorithm. Comput. Commun. **184**, 42–55 (2022)
4. Wang, C., Wang, Z., Xu, X., et al.: A balanced sensor scheduling for multitarget localization in a distributed multiple-input multiple-output radar network. Int. J. Distrib. Sens. Netw. **17**(7), 71–78 (2021)
5. Yan, H., Liu, P., Lu, X.: Two-agent vehicle scheduling problem on a line-shaped network. J. Industrial and Manage. Optimization **19**(7), 4874–4892 (2023)
6. Wang, J.B., Lv, D.Y., Wang, S.Y., et al.: Resource allocation scheduling with deteriorating jobs and position-dependent workloads. J. Industrial and Manage. Optimization **19**(3), 1658–1669 (2023)
7. Cao, B., Zhang, J., Liu, X., et al.: Edge-cloud resource scheduling in space-air-ground integrated networks for internet of vehicles. IEEE Internet of Things J. 9(8), 5765–5772 (2021)
8. Alcaraz, J., Anton-Sanchez, L., Saldanha-Da-Gama, F.: Bi-objective resource-constrained project scheduling problem with time-dependent resource costs. J. Manufacturing Syst. **63**, 506523 (2022)
9. Wang, Z., Yao, N., Liu, Z.: Research on key technology of edge-node resource scheduling based on linear programming. J. Adv. Manuf. Syst. **22**(01), 85–96 (2023)
10. Li, F., Wang, C.: Simulation of blockchain resource allocation algorithm based on mobile edge computing. Computer Simulation **39**(9), 420–424 (2022)

Information Theory and Coding for Social Information Processing, Civilian Industry Technology Tracks

Incremental Update Algorithm of Athlete Physical Training Information Under Dynamic Iterative Sampling

Yuansheng Chen[1](✉) and Zhiyong Huang[2]

[1] Guangzhou Huali College, Guangzhou 511325, China
Chenys01030@163.com

[2] Guangzhou Huashang Vocational College, Guangzhou 511300, China

Abstract. An incremental update algorithm of athlete physical training information based on dynamic iterative sampling is proposed to address the problems of lack of real-time and low computational efficiency in the process of athlete physical training information analysis. The dynamic iterative sampling technique is combined to collect large-scale athlete fitness data, obtain athlete fitness training information based on the incremental update framework, map the existing athlete fitness training data input values into the high-dimensional feature space of the informational network, and combine with the incremental learning algorithm to perform fast updates to better understand the athletes' fitness status. The experimental results show that the root mean square error, the average relative error and the correlation coefficient of the samples after the application of this algorithm are better. It reflects the athletes' physical training situation more accurately and has certain application value.

Keywords: Athletes · Physical Training Information · Dynamic Iterative Sampling · Incremental Learning · Update Algorithm

1 Introduction

An important feature of High-performance sport is the need to constantly explore the athletes' sports potential, so as to develop athletes' sports skills. From the first Olympic Games to the present, sports training has roughly gone through four stages: the natural development stage, the new technology stage, the high-sport stage, the multidisciplinary integrated utilization stage, and the scientific training stage. The goal of each stage of progress and training is to promote the improvement of athletes' physical fitness level to reach the best level. With the continuous improvement of modern sports, the development of athletic ability reaches its maximum. As one of the main components of athletic ability, the level of physical development is increasingly prominent in modern sports. In this context, exploring and studying the basic theory and methods of developing athletes' physical fitness, seeking the best theoretical model of physical fitness training, making physical fitness training more scientific, athletes' physical fitness training information

L. Yun et al. (Eds.): ADHIP 2023, LNICST 549, pp. 411–423, 2024.
https://doi.org/10.1007/978-3-031-50549-2_28

network and optimization, has become the goal of modern sports training pursuit. Physical fitness is the basic athletic ability of the human body expressed through physical abilities such as strength, speed, endurance, coordination, flexibility and agility. It is an important part of athletes' competitive ability. The purpose of physical training is to meet the needs of all kinds of sports competitions. Through adopting targeted training methods, we strive to maximize the physical fitness of athletes and achieve the goal of physical training. In recent years, many scholars at home and abroad have conducted various studies on physical fitness and have made many research results. However, in general, they still lag behind the existing research on training practices and have not yet formed a complete understanding of the information network of physical fitness training for athletes.

For the whole body, physical training helps to improve the coordination of the central nervous athlete physical training information network, so that the body can be more coordinated between the active muscle synergistic muscle and the antagonistic muscle during contraction during exercise. More and more scholars, researchers and physical training coaches gradually pay attention to the important value of high-tech achievements in physical training and have made many useful attempts, but there are some problems such as lack of real-time, low computational efficiency, etc. [1–3] for the update of information on athletes' physical training, for example.

In order to solve these problems, this paper discusses the incremental updating algorithm of athletes' physical training information under dynamic iterative sampling, which provides an effective application method for improving athletes' training level. By introducing dynamic iterative sampling technology, online learning and real-time response can be achieved by continuously sampling and updating data during the training process; combined with incremental learning algorithms, athletes' physical training data can be updated more effectively to improve the training effect.

2 Database Update Algorithm Based on Incremental Learning Algorithm

Athletic physical fitness training refers to the process of implementing scientific and systematic physical fitness training for different projects, requirements, and individual characteristics in sports training, and improving athletes' physical, technical, and psychological qualities through certain methods and means. The necessity of athlete physical fitness training is that athletes need to display their best state and performance in competitions, and the improvement of good physical fitness, technical ability, and psychological quality is the key to ensuring that athletes achieve excellent results.

Strong stamina and good endurance performance is required as athletes are constantly running, jumping, turning and changing direction during the race. Physical fitness training can help athletes continuously improve their endurance and endurance performance, allowing them to better adapt to the intensity and rhythm of the competition, and thus achieve better results. In addition, physical fitness training can also help improve the physical fitness of athletes, such as stability, flexibility, coordination, etc., thereby improving their performance level; Reducing muscle soreness and strain, preventing minor and severe injuries to athletes. Physical training can reduce muscle tension and reduce the tension of the body's muscles through activities. Professional physical fitness coaches establish scientific methods to comprehensively improve athletes' physical fitness, strengthen the strength and flexibility of ligaments, tendons, and muscles, while avoiding excessive injuries to athletes, thereby maintaining health and reducing injuries. Therefore, athletes' physical fitness training is necessary to ensure their physical health.

Physical training can help athletes improve their technical abilities. Physical coordination is a crucial factor in athletes' athletic skills. Physical fitness training can enhance athletes' physical coordination, enabling them to better grasp and apply technical movements, thereby better exerting their technical level in competition and improving their performance. From this, it can be seen that athlete physical fitness training information is constantly increasing and there is redundancy. At the same time, incremental information updating refers to comparing newly obtained information with known information and effectively updating it based on the newly obtained information to improve the efficiency and accuracy of the information system. In today's rapidly developing information society, information systems need to be constantly updated with changes in information to ensure their effectiveness and accuracy. Therefore, incremental information updates have the following necessity:

(1) Improving information update efficiency: Incremental information updates can reduce the time and cost of information updates, while quickly comparing and updating new information with existing information, thereby improving the efficiency and speed of information updates.
(2) Higher accuracy: Traditional batch update methods can only compare and update new information with all known information, which consumes time and effort. However, incremental update of information can quickly compare and update new information with known information, reduce information redundancy, and improve information accuracy and accuracy.
(3) Improving system security: Incremental information updates can regularly compare and update the security policies in the system with the latest security updates, effectively improving the security of the information system.
(4) Improving user experience: Incremental information updates can reduce the time users need to wait for information updates, thereby improving user experience and satisfaction.

In summary, incremental information updates can improve the efficiency, accuracy, and security of information systems, while also improving user experience. Therefore, it is of great necessity in information systems.

2.1 Incremental Update Framework

Due to the lack of an informative and intelligent physical training system to command athletes' physical training information network. Therefore the relevant theoretical research work is slowly updated, the relevant disciplines do not reveal enough about the basic laws of physical fitness training, and some researchers lack accurate understanding and testing of the characteristics of professional sports. Many factors that constitute training and competition are dynamically changing. Exploring the dynamic balance between various factors involves a wide range of aspects and is really difficult to achieve; In addition, the cultivation of outstanding athletes is a long-term and complex process, including the content of basic theories related to body science, as well as the profound connotations of sports humanities and sports sociology; Some factors are controllable and predictable, but there are also many unforeseeable changes; There is measurable data, but there are also hidden unknown factors. These uncontrollable, non quantitative, and unpredictable factors need to be studied, and there is currently no landmark research. At the same time, there is a problem of a large amount of information overload and a large amount of duplicate and invalid information in the original network information. Directly selecting information related to users from network information can easily lead to query errors [4, 5]. Therefore, a complete incremental update framework is needed, Meet the requirement of utilizing dynamic iterative sampling.

(1) In order to maintain the normal operation of the athlete physical training information network, it is necessary to use the permission management module to manage and maintain the information network of athlete physical training. The specific functions include user information management (such as user registration, login, user permission settings, etc.). Authorization management involves assigning permissions to roles, and there are many corresponding relationships between accounts, roles, and permissions. Figure 1 shows the permission management module

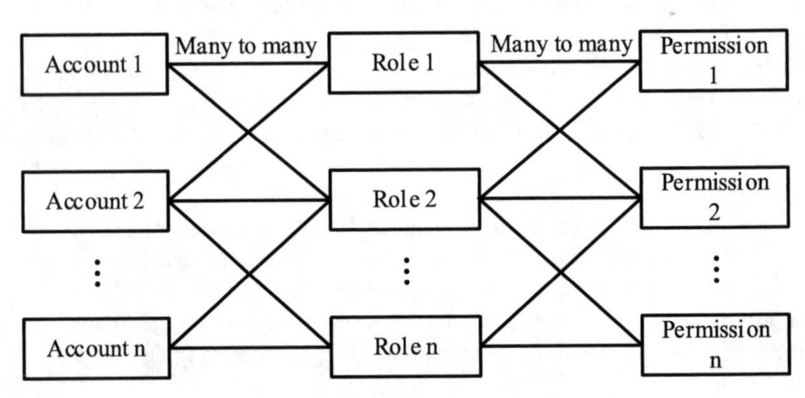

Fig. 1. Permission Management Module

In Fig. 1, the work of adding, deleting, and modifying user information in the athlete physical training information network is mainly completed, and the list mainly shows important fields such as number, real name, user name, department, role, creation time, and account status [7].

A pop-up window displays the new user interface function, and if the interface information is too much, the page can be reopened to display it. In addition to creating a new account, it is also necessary to make changes to the existing account (editing function in the operation bar), but the user's real name, username, and other information cannot be changed during the change process. Instead, it is necessary to change the real name and username to editable status according to the actual needs of the project.

Creating a new role is the process of describing that role and giving it permissions. If changes are allowed, select them from the drop-down list. If the number of permissions is large and the classification is complex, you can display it as a list of groups and let the user select it via the selection box. It is recommended to add a select all/reverse feature to simplify the operation. The edit feature in the toolbar is used to modify the existing roles, you can change the role name, description, status and permissions. After each modification, the update time is recorded.

In the case of a large number of permissions and complex types of permissions (page permissions, operation permissions, data permissions) in the athlete physical training information network, the permissions management page is presented in the form of a list with the number of permissions, name, type, description, creation time, etc., in order to ensure the convenience of the administrator and reduce the possibility of errors [8, 9].

The entry page for new permissions has different requirements for the athlete physical training informatics network, some of which require the development of the input code, some of which require the product manager to enter the URL of the new permission, and some of which do not show the new permission entry in the athlete physical training informatics network.

(2) The indicator management module is mainly used for browsing, searching, adding and deleting evaluation indicators, and experts or managers can adjust the evaluation indicator system through this module.

(3) The collaboration and communication module provides a variety of communication methods for evaluators, and the forum communication is used to provide feedback on evaluation results and evaluation problems so that managers or experts can make adjustments to the evaluation index system, and this module can be implemented through the forum. Fancy aliases are replaced by human names and displayed in a tree shape, making it easy for users to find. Self-developed multiple communication methods to ensure customer needs and secondary development, helping enterprises to realize QQ/MSN-like communication methods. Timely processing of foreign emails to ensure normal communication with customers. During the outbound period, you can notify related matters through SMS to ensure the normal work of relevant personnel.

2.2 Data Information Abstraction Based on Dynamic Iterative Sampling Technology

In each module, a series of data information is often abstracted into a mathematical theory-based information network for description. In the process of abstracting data information into information network graphs, the information nodes are mainly abstracted as points, while the relationships between different information nodes are abstracted as network edges. In the process of studying data mining and learning. Athlete physical training information network as a kind of special network that can more reflect heterogeneous relationships [10, 11],mainly consists of network nodes, node relationships and node relationship attributes, etc., for which dynamic iterative sampling techniques are introduced to represent the given spatial dimensional set and spatial data point set settings as follows, respectively:

$$H = \{h_1, h_2, \cdots, h_n\} \tag{1}$$

$$K = \{k_1, k_2, \cdots k_m\} \tag{2}$$

In the formula, H represents the given spatial dimension set, K represents the spatial data point set, h represents the spatial dimension, k represents the spatial data point, n represents the total number of spatial dimensions within the set, and m represents the total number of spatial data points within the set. According to the maximum value of the data space boundary, perform a query on the data contour of the real estate registration space, and calculate the length of the envelope in each spatial dimension at the initial time:

$$\begin{cases} C(h_a) \cdot \varpi = L \\ C(h_b) \cdot \varpi = 0 \end{cases} \tag{3}$$

In the formula, C represents the envelope, ϖ represents the initial time, L represents the maximum value of the data space boundary, h_a, h_b represents two data space dimensions, and a, b represents the spatial dimension. Among them:

$$1 \le a \le n,\ b \in \{1, \ldots, n\},\ b \ne a \tag{4}$$

When a contour point is found, the remaining contour points may only exist in the remaining query space. On a long-term dimension, time series reconstruction can be randomly simulated through temporal order. Combining the given time, perform multi-step prediction on the model:

$$f_{i+p} = G^h(X_d) \tag{5}$$

In the equation, f_{i+p} represents the multi-step advance prediction value of step i and step p forward; X_d represents the error value of a single prediction. Follow the principle of minimizing structural risk to process the given sample set and establish a regression function equation:

$$y(x) = w^T p(x) + b_i \tag{6}$$

where, r $y(x)$ epresents the loss function value of the sample data set; w^T represents the operational weight vector; And $p(x)$ represents a nonlinear spatial mapping function; b_i represents the input offset. To solve the goal planning problem, map the existing input values to a high-dimensional feature space:

$$\begin{cases} \min G(w_d, p_i) = \frac{\|w_d\|^2}{2} + \frac{\xi_m \sum_{i=1}^{n} p_i^2}{n} \\ s.t. y_i = w^T p(x) + p_i \end{cases} \tag{7}$$

where, $G(w_d, p_i)$ represents the optimization objective function of regularization, where w_d and p_i represent the input and output vectors of samples respectively; ξ_m is the relaxation factor of the sensitive loss function; y_i represents the value of the regression function. Therefore, during the contour point query process, the space that has already been queried is represented as:

$$S_1 = \{1 \leq h_a \leq L, 0 \leq h_b \leq C(h_b)\} \tag{8}$$

In the formula, S_1 represents the queried space. The remaining query space can be calculated based on the entire data space:

$$\begin{aligned} S_2 &= S - S_1 \\ &= \{1 \leq h_a \leq L, 0 \leq h_b \leq L\} - \{1 \leq h_a \leq L, 0 \leq h_b \leq C(h_b)\} \\ &= \{1 \leq h_a \leq C(h_a), C(h_b) \leq h_b \leq L\} + \\ &\quad \{C(h_a) \leq h_a \leq L, C(h_b) \leq h_b \leq L\} \end{aligned} \tag{9}$$

In the formula, S represents the entire data space, and S_2 represents the remaining query space.

3 Storage and Filtering of Athlete Physical Training Information

The specific storage structure is shown in Fig. 2.

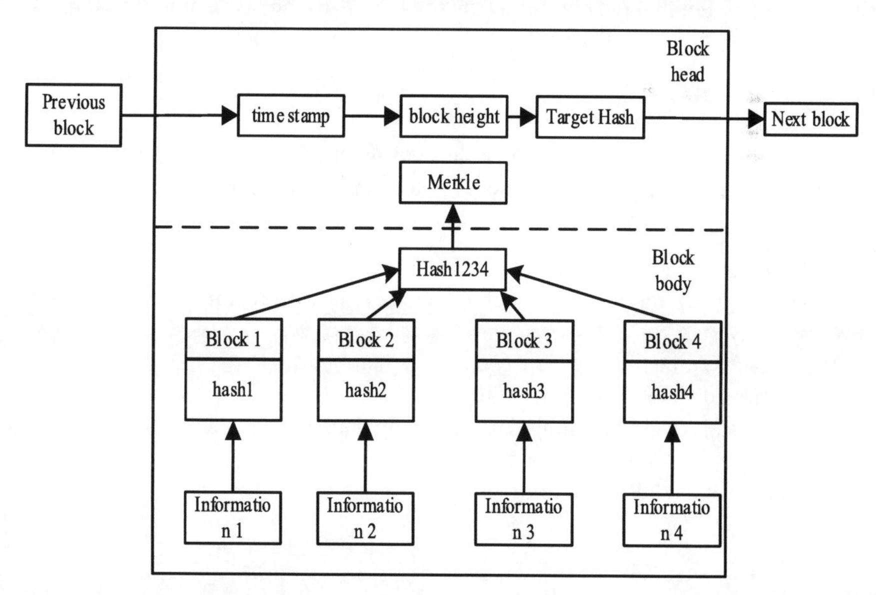

Fig. 2. Blockchain storage structure diagram of athlete physical fitness training information

Each block in the blockchain contains a block and its head, and the blockbody is the foundation, forming the entire blockchain through a large blockbody. At the same time, the target block is generated by combining the hash value of the previous block with a random number. Genhash verified the authenticity of the transaction and confirmed that there were no false transactions. Each block will store each transaction data separately and determine the structure of the transaction data based on the functionality of the blockchain. Blockchain contains a lot of information about transactions. A large amount of data is displayed in the Merkle tree, and these data are saved in the tree structure. In blockchain storage, an incentive mechanism is used to enable each node to effectively reach a consensus. In order to achieve maximum benefits, both parties will not change the authentication results of other blocks. In the actual storage process, when power companies request storage from edge devices, they can choose to store data on any node and only send storage requests to that node for a certain period of time. After receiving the request, the storage node sends back the response from the storage and provides the service sequence number of the storage to the storage node. When the storage node is received, the storage node can perform data filtering. After filtering the updated data through a temporary database, it is also necessary to check the location of the updated data. Only after the graphic data is checked and found to be correct can the real estate registration database be considered updated. Import accurate updated data directly into the current database and store it uniformly with historical data. Unlike temporary databases, the implementation of current database updates requires two steps: one is to replace the updated layer with data, and the other is to adjust the replaced data to historical data for subsequent calls. In addition, the update methods for the two databases are the same, except that the current database no longer requires feature filtering.

4 Experiments

A certain team has 20 athletes, and their physical and training data are shown in Table 1:

During the experiment, the algorithm related parameters were:

(1) Training Plan:
   1) Training time: 5 days per week;
   2) Training content: dribbling, shooting, defense, rebounding, physical fitness training, etc.;
   3) Intensity: Adjust according to personal situation;

(2) Incremental update algorithm for information:
   1) Input data: Weekly training data and body data, which need to be adjusted and modified according to actual situations;
   2) Output data: adjusted training plan;

Table 1. Body Data and Training Data

Sportsman	age	Height (cm)	Weight (kg)	Body fat percentage (%)	Weekly training frequency	Training intensity	Physical fitness score
1	20	185	75	15	5	8	80
2	22	180	80	12	6	7	70
3	25	175	70	18	4	6	60
4	23	190	90	13	5	9	90
5	19	170	65	16	4	6	70
6	21	185	80	10	6	8	80
7	24	195	100	20	5	9	90
8	20	175	70	14	4	6	60
9	22	180	75	12	5	7	70
10	23	185	80	16	6	8	80
11	19	170	65	13	4	6	70
12	20	185	75	15	5	8	80
13	22	180	80	12	6	7	70
14	25	185	70	18	4	6	60
15	23	190	90	13	5	9	90
16	19	180	65	16	4	6	70
17	21	185	80	10	6	8	80
18	24	195	100	20	5	9	90
19	20	185	70	14	4	6	60
20	22	180	75	12	5	7	70

Experimental steps:

(1) Collect basic information about athletes, including age, height, weight, body fat percentage, etc., and divide them into different groups according to their characteristics.

(2) Set a training plan, including weekly training time, training content, and intensity, and guide athletes in groups.

(3) Adopting information incremental update algorithm and Reference 3 Deterministic learning based gain recognition algorithm, the training plan is dynamically adjusted based on the training data of each athlete every month to adapt to changes in their body and needs.
(4) Using root mean square error, average relative error, and correlation coefficient as evaluation indicators for the algorithm, the effectiveness and superiority of the algorithm are evaluated by comparing the data before and after updates. The calculation formula is:

$$E_{RMSE} = \sqrt{\frac{1}{N_m}\sum_{i=1}^{n}\left(p_i - p_j\right)^2} \tag{10}$$

$$E_{MAPE} = \sqrt{\frac{1}{N_m}\sum_{i=1}^{n}\left|\frac{p_i - p_j}{p_i}\right|} \tag{11}$$

$$R_k = \sqrt{\frac{\left(N_m \sum_{i=1,j=1}^{n} p_i p_j - \sum_{i=1}^{n} p_i \sum_{j=1}^{n} p_j\right)^2}{\left(N_m \sum_{i=1}^{n} p_i - \left(\sum_{i=1}^{n} p_i\right)^2\right)\left(N_m \sum_{j=1}^{n} p_j - \left(\sum_{j=1}^{n} p_j\right)^2\right)}} \tag{12}$$

In the formula, E_{RMSE} represents the root mean square error and E_{MAPE} represents the average relative error. The smaller the values of the two, the higher the accuracy of the algorithm; N_m represents the total amount of data; p_i and p_j represent actual data and monitoring data, respectively; R_k represents the correlation coefficient, and the closer this value is to 1, the better the fitting effect of the algorithm. Using the above formula, obtain the error of the algorithm to determine the performance of different algorithms.

The experimental comparison results are shown in Fig. 3:

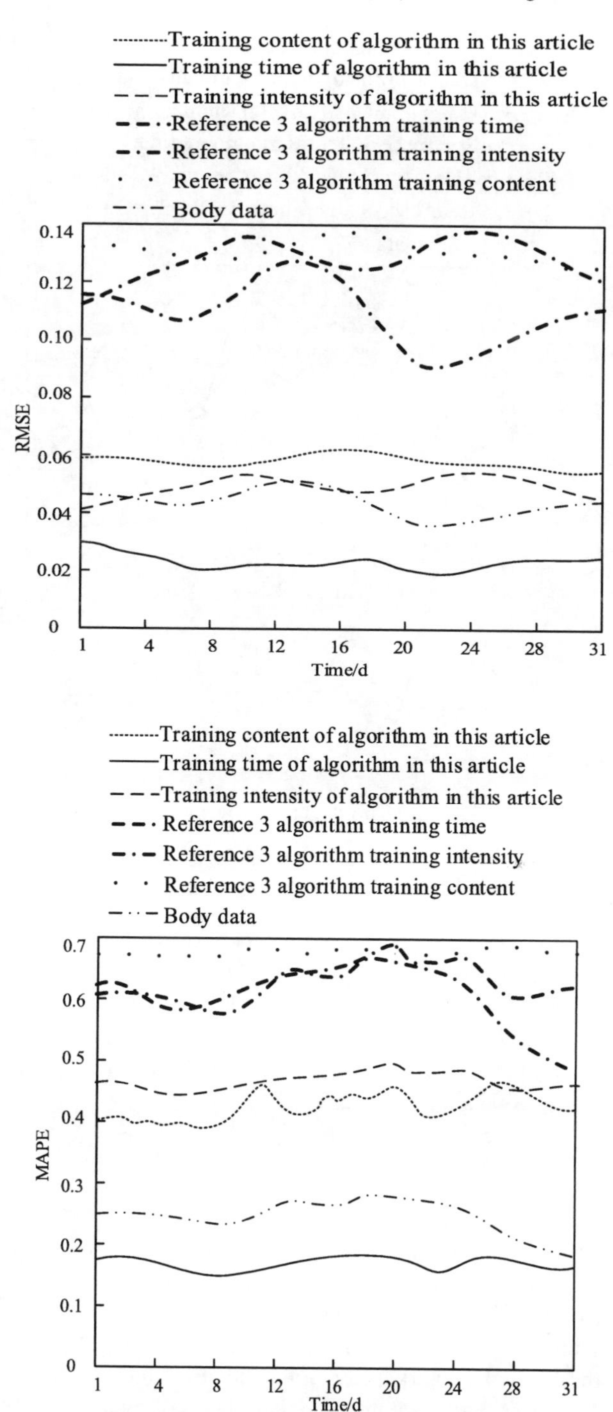
Training content of algorithm in this article
Training time of algorithm in this article
Training intensity of algorithm in this article
Reference 3 algorithm training time
Reference 3 algorithm training intensity
Reference 3 algorithm training content
Body data
RMSE
0.14
0.12
0.10
0.08
0.06
0.04
0.02
0
1
4
8
12
16
20
24
28
31
Time/d
Training content of algorithm in this article
Training time of algorithm in this article
Training intensity of algorithm in this article
Reference 3 algorithm training time
Reference 3 algorithm training intensity
Reference 3 algorithm training content
Body data
MAPE
0.7
0.6
0.5
0.4
0.3
0.2
0.1
0
1
4
8
12
16
20
24
28
31
Time/d

Fig. 3. Algorithm Performance Test Results

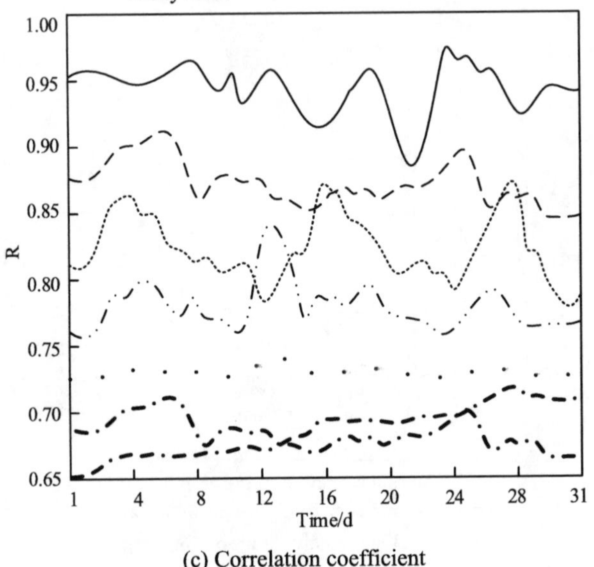

(c) Correlation coefficient

Fig. 3. *(continued)*

In Fig. 3, the difference between the predicted physical performance and actual performance of athletes using information incremental update algorithms is relatively small, with smaller RMSE and MAPE, and a larger correlation coefficient. This indicates that the updated and filtered information in the database meets the requirements of training plan adjustment, and the adjustment results are good, indicating that the incremental update algorithm has good application effects.

5 Conclusion

A dynamic iterative sampling based incremental update algorithm for athlete physical fitness training information is proposed. Build a multi-module incremental update framework, based on dynamic iterative sampling technology to abstract analyze data information, effectively filter out a large amount of information overload and repeated invalid information in the original network information, thereby avoiding the occurrence of user information query errors, and meeting the requirements of dynamic iterative sampling. Apply a database update algorithm based on incremental learning algorithm, combined with a blockchain storage structure for athlete sports training information, to store and filter athlete sports training information. After filtering the updated data through a temporary database, Huqiu updates the location of the data to accurately predict the future physical performance of athletes and achieve dynamic updates of training information.

In the future, we will further explore the application of incremental update algorithms for athletes' physical fitness training information in more fields. For example, it can be applied to the training of other sports events, or applied to the monitoring and management of athletes' physical condition, which will further improve the algorithm, improve its prediction accuracy and training effectiveness, and achieve better results in practical applications.

Aknowledgement. Research project of higher education in 2022 under the "Fourteenth Five-Year Plan" of Guangdong Higher Education Association. The title of the project is "Research on the Teaching Innovation of Experiential Expansion Training Infiltration into Public Physical Education Courses in Higher Vocational Education", the host of the project is Chen Yuansheng, and the project approval number is 22GQN63.

References

1. Zhong, Y., Liang, X.: A Hybrid Evaluation of Information Entropy Meta-Heuristic Model and Unascertained Measurement Theory for Tennis Motion Tracking **12**(3), 263–279 (2022)
2. Ong, P., Chong, T.K., Ong, K.M., et al.: Tracking of Moving Athlete from Video Sequences using Flower Pollination Algorithm **38**(3), 939–962 (2022)
3. He, C., Ye, L., Sulaimani, H.J., et al.: Training Method of Sports Athletes Using the Nonlinear System of Moving Human Body Competitive Ability **30**(2), 2240093 (2022)
4. Qiu, S., Hao, Z., Wang, Z., et al.: Sensor Combination Selection Strategy for Kayak Cycle Phase Segmentation Based on Body Sensor Networks **9**(6), 4190–4201 (2022)
5. Liu, S., Li, Y., Fu, W.: Human-centered attention-aware networks for action recognition. Int. J. Intell. Syst. **37**(12), 10968–10987 (2022)
6. Zheng, H., Li, P., He, J.: A novel association rule mining method for streaming temporal data. Annals of Data Sci. **9**(4), 863–883 (2022)
7. Zhang, C., Du, Z., Yang, Y., et al.: On-shelf utility mining of sequence data. ACM Transactions on Knowledge Discovery from Data **16**(2), 21.1–21.31 (2022)
8. Munshi, M., Shrimali, T., Gaur, S.: A Review of Enhancing Online Learning using Graph-Based Data Mining Techniques **26**(12), 5539–5552 (2022)
9. Feng, G., Fan, M., Chen, Y.: Analysis and Prediction of Students' Academic Performance Based on Educational Data Mining **10**, 19558–19571 (2022)
10. Li, D., Xiao, F., Zheng, Y.: Research on the Inheritance and Protection of Data Mining Technology in National Sports **34**(13), e5893.1-e5893.10 (2022)
11. Fan, Y., Zheng, Z., Zheng, W.: Prediction of college students' athletic performance based on improved SSA-LSSVM model. Computer Simulation **40**(1), 8 (2023)

Design of Mobile Education Platform for University Network Law Popularization Based on Streaming Media Technology

Yu Zhao(✉) and Liang Zhang

Changchun University of Finance and Economics, Changchun 130000, China
zhaoyy4100@163.com

Abstract. In order to make the function and performance of the university network law popularization mobile education platform meet the needs of users, the design of university network law popularization mobile education platform based on streaming media technology is proposed. Using the education platform client and streaming media server, this paper designs the structure of the university network law popularization mobile education platform. Combined with the design of sign in incentive module, legal knowledge integration module and legal knowledge extraction module, this paper designs the functional module of university network law popularization mobile education platform, and realizes the design of university network law popularization mobile education platform. The test results show that the platform in this paper can meet the user's requirements for the platform functions through the sign in incentive function test. In terms of the storage space and memory occupancy of the platform, it can also meet the user's requirements for the platform performance.

Keywords: Streaming Media Technology · Education Platform · Legal Education · Sign In Incentive · Knowledge Integration

1 Introduction

At present, there is a certain gap in the cultivation of rule of law awareness and rule of law thinking among most college students in China. On the one hand, the lack of cultivation of rule of law awareness in childhood education and family education results in the lack of college students' own grasp of legal common sense, and on the other hand, the lack of surrounding rule of law atmosphere and environment results in the deficiency of the guidance of the force of abiding by the law, Many college students are not only lack of rule of law education in their knowledge learning experience, but also lack of proper rule of law atmosphere edification [1] on campus. Therefore, the construction of the legal education platform in colleges and universities is particularly important. It can not only make up for the lack of legal knowledge of most students, but also reshape the legal personality of college students and improve their legal literacy.

L. Yun et al. (Eds.): ADHIP 2023, LNICST 549, pp. 424–438, 2024.
https://doi.org/10.1007/978-3-031-50549-2_29

In domestic research, Shi Wanli et al. [2] designed a smart education platform based on big data analysis technology, including data resource database, big data analysis layer, smart education information cloud service layer, application service layer and presentation layer, to achieve one-stop online services in the education process. The big data analysis layer uses the Hbase database combined with the SQL computing execution engine to analyze the smart education data, and transmits the analyzed student, teacher and resource information to the smart education information cloud service layer. The intelligent education information cloud service layer enables platform users to enjoy services such as platform storage files, course management and course publishing by verifying users, service binding and service provision. The administrator of the online learning module in the intelligent education information cloud service layer reviews teachers' lesson preparation materials and students' learning resources, and realizes the functions of teachers' online or video teaching and students' curriculum selection and determination, so as to realize online information exchange among teachers, administrators and students. The platform test results show that the platform can provide personalized teaching and management according to students' personalized learning behavior, realize online communication between teachers, administrators and students, and the platform education resource storage service and portal service can be realized, with high application value. Gao Xue et al. [3]. First of all, the characteristics and advantages of distributed cognitive theory in the experience design of online education platform are analyzed through literature, and user research is conducted by questionnaire and interview methods to clarify the influencing factors of online learning from the perspective of distributed cognition. Secondly, the Coursera platform is analyzed as a case, and each element is distributed in its own context for analysis. Finally, the experience design strategy of online education platform from the perspective of distributed cognition is extracted. In the proposed design strategy, design methods such as visual knowledge map, collaborative tools to promote deep interaction, and social behavior to establish mutual benefit and sharing can meet the learning needs of online learning users, and solve the problems such as low learning autonomy and low communication and collaboration efficiency.

In foreign research, Song Y et al. [4] proposed an application method of mobile education in the intelligent campus assisted autonomous learning platform in order to solve the problems of low teaching effect and poor student performance in traditional intelligent campus construction methods. By analyzing the connotation and characteristics of mobile learning, this paper studies the influence of mobile education on assisted autonomous learning in smart campus. By means of questionnaire survey, the teachers who have carried out mobile education and those who have not carried out mobile education were investigated. And compare the test scores of mobile education classes and non mobile education classes. The survey results show that the application of mobile education in the intelligent autonomous learning platform can diversify teaching methods, improve teachers' teaching effects, improve students' learning achievements, and thus improve the teaching level of intelligent campus on the premise of improving students' enthusiasm for class.

At present, digital multimedia teaching has already entered the learning of various courses in colleges and universities, which is certainly essential for the development of law popularization education for college students. However, there are still many shortcomings in how to build and improve the existing digital media law popularization platform. On the one hand, the multimedia infrastructure of some colleges and universities is not perfect. On the other hand, the participation and initiative of college students are also lacking. It is necessary to play the role of digital media law popularization education platform. First of all, colleges and universities should improve relevant infrastructure and supporting equipment, so that students can easily and quickly access the latest laws and regulations, relevant legal cases and legal advisory information. Also, teachers should learn to effectively use the existing digital media for law popularization teaching, so that law popularization education is no longer limited to boring textbooks, but the latest and fastest integration with social life, explain the latest legal social hot spots, and let the law popularization education keep pace with the development of the times.

Based on the above research background, this paper applies streaming media technology to the design of university network law popularization mobile education platform, so as to improve the ability of university students to popularize legal knowledge.

2 Structure Design of University Network Law Popularization Mobile Education Platform

The mobile education platform for online legal popularization in universities mainly consists of an education platform client and a streaming media server. The education platform client includes four modules: data transmission control module, partner management module, media data buffer, and media playback. The streaming media server utilizes a handshake protocol to build a network connection and transmit client streaming information to the streaming media server, achieving playback of audio and video files. This completes the structural design of the mobile education platform for university network legal education.

2.1 Design Education Platform Client

The education platform client is composed of four parts: data transmission control module, partner management module, media data buffer and media play. Its structure is shown in Fig. 1.

The partner management module is mainly responsible for the management and maintenance of the relationship between this node and other nodes. When a node joins a P2P network to carry out legal education, this module is responsible for sending requests to the server, obtaining the resource information of the parent node, establishing a resource information table, and then establishing a connection with the parent node. The partner management module stores the information of the parent node and child node, and carries out message communication with the parent node and child node every two seconds to detect whether the other party is still on the network and remains available, so that when a node suddenly leaves, the connection can be quickly restored in the P2P network.

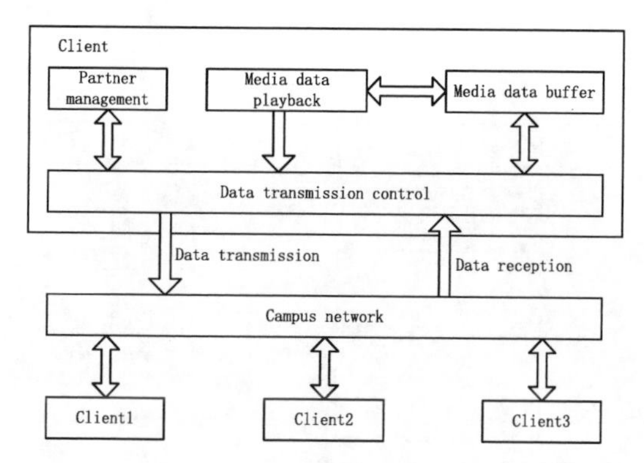

Fig. 1. Client Structure

The media data buffer is mainly responsible for receiving the streaming media data transmitted from the server or other clients, updating the streaming media data of this node and storing it.

The function of the media data playback module is to play the data in the buffer by calling the Media Player [5] that comes with Windows.

The data transmission control module is used to open a data buffer in the client host memory, store the received streaming media data to the buffer, and when the buffer is full, submit the data that first entered the buffer to the player at a certain rate for playing [6]. While caching data streams, it also distributes data streams to its child nodes.

2.2 Design Streaming Media Server

The process of streaming media server transmitting encapsulated media data to the player through streaming media protocol requires several steps: handshake protocol, network connection, network stream, and playback. The details are as follows:

Step 1: Handshake protocol

The handshake starts with the client passing C0 and C1 blocks. After the server obtains C0 and C1, it will pass S0 and S1. Once the client obtains S0 and S1, it will pass C2. After obtaining C0 and C1, S2 will be passed. Once the client and server acquire S2 and C2, the handshake ends. The schematic diagram of handshake protocol is shown in Fig. 2.

Step 2: Build network connection

The client passes the "connection" in the command message to the server. After the server obtains the connection command, it will send the protocol message of selecting the window size to the client, which is the same as the application in the connection command. The server transmits the message about the bandwidth protocol to the client. After the client processes the set bandwidth protocol message, it will send the confirmation window size protocol message to the server.

Step 3: Build network flow

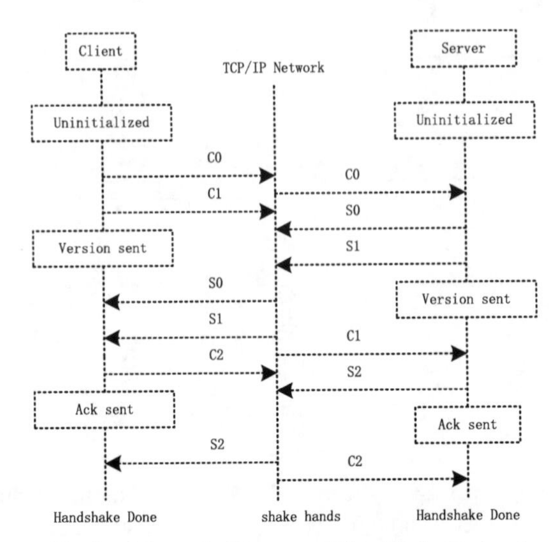

Fig. 2. Schematic Diagram of Handshake Protocol

The client transmits the "Create Stream" message in the command message to the server. After the server gets the "Create Stream" message, it sends the "Result" in it, and then transmits the client stream information.

Step 4: Play

The client transmits the "Play" in the command message to the server. After getting the play message, the server transmits the block size protocol message. The server passes the "streambegin" of the client to urge the client to obtain the stream ID. After the playback command is sent, the server will transmit the "response status" in it to facilitate better transmission of the client's "playback" command. Then, the server transmits the audio, video and other files to be played by the client.

3 Design the Function Module of the University Network Law Popularization Mobile Education Platform

After designing the structure of the mobile education platform for online legal education in universities, construct the functional modules of the platform. This module includes a check-in incentive module, a legal knowledge integration module, and a legal knowledge extraction module. The check-in incentive module is mainly aimed at students, and completing check-in can earn rewards. Design student check-in and re signing processes, distribute incentive coins to users upon completion of the event, and achieve check-in incentives. The Jacobi method is used to establish the characteristic equation of legal knowledge resources, and the template features are Linear map to the low dimensional feature space, so that the Bhattacharyya Distance between categories is maximized, and the legal knowledge integration module is completed. Extract legal knowledge based on the objectives of legal education tasks and the characteristics of course content in mobile education. This enables the design of functional modules for the mobile education platform for online legal popularization in universities.

3.1 Design the Sign in Incentive Module

The main users of sign in incentive are students. Students complete sign in before class to get rewards. If they can't sign in five minutes after class, they need to contact the tutor for re signing. The url address of sign in incentive is/signIn/episodes/{epidedId} [7], the request method is post, and the parameters required by the request method are shown in Table 1.

Table 1. Interface Parameters of Sign in Incentive Request

Field	Must	Type	Remarks
Userid	No	Int	User ID, no need to display the transmission, and then check and obtain the permission according to the requested context
Version	No	String	The version information requested by the user. The low version application does not support the function of continuous sign in incentive, does not return the corresponding results, does not need to be transmitted, and is obtained in the context
Epirsod	Yes	Int	Course id
Teamid	Yes	Int	Class ID, which is the class of the user in this course

The check-in process of students in the network law popularization mobile education platform is shown in Fig. 3.

There are two levels of check-in activity: student dimension and course dimension. In the curriculum dimension, each class of students in a series of courses belongs to the sign in range of the series of courses, and the sign in progress is different in different courses;In the student dimension, the sign in activity changes with the change of students. The sign in of students in different courses is unified progress. After signing in in one series of courses, the sign in activity in another series of courses is continuous.

In order that students will not be confused by the different progress of signing on different courses, the student dimension of signing in activity is selected here. However, this will bring some other complications. Since the current check-in times and previous check-in status are not in a series of courses, the student's check-in order cannot be obtained directly, so it is necessary to add the student's check-in order in the check-in table to indicate the student's position in this round of check-in, and the previous check-in statuses can also be obtained. In addition, a flag bit is required to mark whether the previous check in status is complete or partially complete, so that the client can display the final different award status.

There are two steps to obtain the status of students' current round of check-in. The first step is to obtain the latest check-in position, and the second step is to obtain the previous check-in status in the check-in record table according to the current check-in position. In addition, it is also necessary to record the user's non sign in status when the user has not signed in, so that it will not be lost when obtaining it later. The flow chart of the supplementary signing module is shown in Fig. 4.

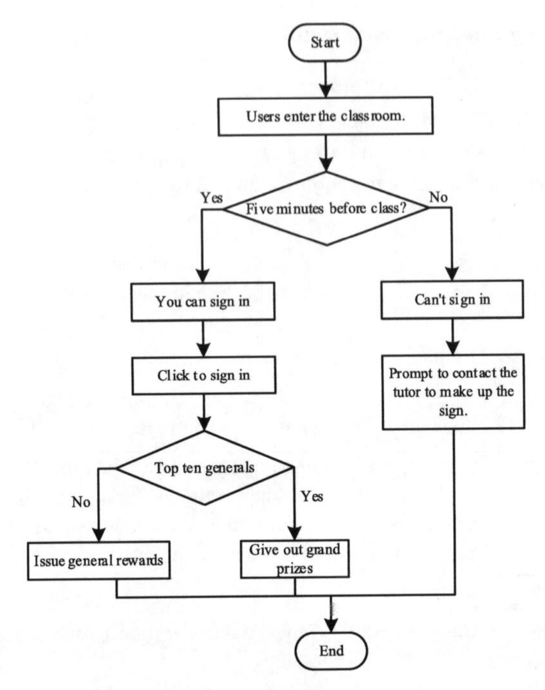

Fig. 3. Flow Chart of Student Sign in Incentive Module

The teacher needs to guide the teacher to select courses and students, generate corresponding supplementary signing records for students after judgment, and issue gold coins. If the student does not purchase, does not attend class, has too many times to sign in, and does not need to sign in for the course, the signing will fail and the corresponding information will be returned.

The time to generate an unsigned record is when the class is over, and the time to receive the bill for the class is when the unsigned user who should sign for the class creates an unsigned record. This will add some redundant records to the sign in table, because theoretically, the sign in table should only record the courses that users have signed in. Here, redundant storage of courses that users have not signed in is required, but it is necessary to achieve incentives for continuous sign in. In addition, it is also necessary to judge whether it is the last time to sign in for this round and whether it is necessary to send the grand prize. After signing in, users need to issue incentive gold coins. This is done through events, instead of blocking the return of the user's sign in results [8]. The reward service will consume after receiving the gold coin issue event, issue the corresponding gold coin to the user's account and add the related gold coin record.

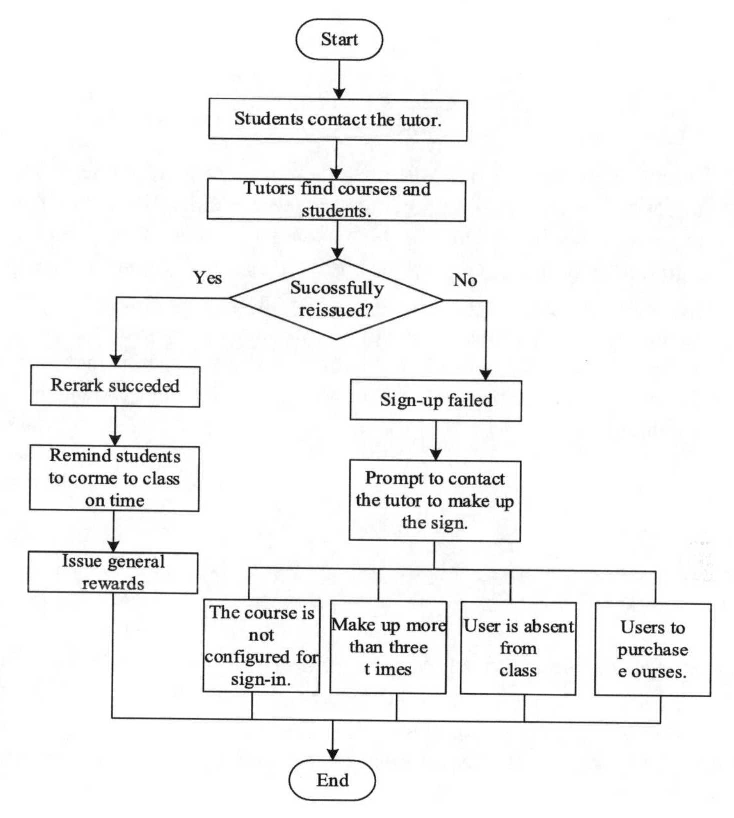

Fig. 4. Flow chart of supplementary signing of sign in incentive module

3.2 Design Legal Knowledge Integration Module

Assumed by n represents the number of sample characteristic variables in the original legal knowledge resource integration module, X representing the original sample n characteristic variables, satisfied with $X = x_1, x_2 \cdots x_n$. After orthogonal transformation u^* comprehensive variables (y_1, y_2, y_3), R representative sample X characteristic equation [9] of the integration module of legal knowledge resources is constructed using Jacobi's method and expressed by formula (1):

$$\lambda^n(i) = \frac{\{R \times X\}^n}{\{(y_1, y_2, y_3)\}} \cdot \frac{(x_1, x_2, \cdots, x_n)}{u^*} \tag{1}$$

Assuming λ_i represents the number of non negative eigenvalues of the sample correlation coefficient matrix, and rank them to satisfy the $\lambda_1 \geq \lambda_2 \geq \lambda_n \geq 0$. Extract the features of the first m legal knowledge resource integration modules through the ratio of λ_i to feature noise interference.

Assuming α before the representative m, if the variance of the characteristics of the legal knowledge resource integration module and its proportion in the total variance, then use formula (2) to calculate the characteristic variance contribution rate of the legal knowledge resource integration module. The formula is:

$$\beta(p) = \frac{m + \alpha}{\mu(R)} \times \nu(e)\left(\sigma^* + \kappa\right) \tag{2}$$

where, $\mu(R)$ represents the weight of the characteristic sample of the legal knowledge resource integration module, information entropy representing different characteristics $\nu(e)$, σ^* represents the best threshold of the behavior variables of the legal knowledge resource integration module, κ represents the observed variable of student characteristics.

Assuming $X = x_1, x_2 \cdots x_n$ is defined as the characteristic random vector of the observable legal knowledge resource integration module, a_{ij} represents the factor load, representing the i variable vs j correlation coefficient of the factor, then use formula (3) to give the characteristic observable random vector of the legal knowledge resource integration module:

$$\partial(X) = \frac{X \times F}{\left(a_{ij}\right)_{n \mp m}} \times c_i + \varepsilon_i \times X_i \tag{3}$$

where, F represents an unobservable vector, c_i represents the load of the integration strategy factor of legal knowledge resources, ε_i representative influence c_i Unique factor of.

Integrate legal knowledge resources into information sources X^* The uncertainty degree of is defined as shannon entropy [10], calculate using the source space of source X^* and the probability of integration module events [11].

Assuming $I(\lambda_i)$ represents the information function of the integration module of legal knowledge resources, we use formula (4) to define the behavior probability of the integration of legal knowledge resources:

$$\xi(w) = \frac{I(\lambda_i) \times H(X^*)}{\zeta(k)} \times \vartheta(R) \tag{4}$$

where, the amount of information $\zeta(k)$ representing legal knowledge, $\vartheta(R)$ represents the cumulative information contribution rate of the legal knowledge resource integration module.

Integrate the module behavior probability with legal knowledge resources $\xi(w)$. For the basis, the feature vector corresponding to the non-zero eigenvalue of the legal knowledge resource integration module is defined as the classification feature vector, and the samples of the knowledge resource integration module are linearly mapped to the low dimensional feature space to maximize the Bhattacharyya Distance between its categories. The legal knowledge resource integration module is given. The specific steps are detailed as follows:

Assuming $\Phi(^\circ)$ represents a nonlinear mapping function, w represents a vector in the kernel space, integrating module probability with legal knowledge resources $\xi(w)$ based on, the feature vector corresponding to the non-zero eigenvalue of the legal knowledge resource integration module is defined as the classification feature vector, and the formula (5) is used to express:

$$J_B = \frac{(\mu_2 - \mu_1)}{\xi(w)} \cdot \frac{(W, Z_i) \times \omega_i}{\Phi(^\circ)} \tag{5}$$

where, μ_1 and μ_2 respectively represent a vector in the kernel space, W and Z_i mean and variance of the sample representing the category after projection to the one-dimensional space, ω_i represents a kernel function.

Assuming $o(k)$ represents that the sum of variances between the features of the kernel space is the smallest, and use formula (6) to linearly map the sample of the legal knowledge resource integration module to the low dimensional feature space, and promote the Bhattacharyya Distance between its categories to be the largest, namely:

$$\iota(u) = \frac{o(k) \times i_{(\nu)} \times J_B}{E(p)} \tag{6}$$

where, $E(p)$ represents the effective feature vector of the behavior classification of the legal knowledge resource integration module, $i_{(\nu)}$ represents the isoline value between two eigenvectors.

Based on the calculation result of formula (6), the integration module of legal education knowledge resources is established by using formula (7), namely:

$$\partial^*(y) = \frac{\iota(u) \times J_B}{\xi(w)} + H\left(X^*\right) \tag{7}$$

According to the above process, complete the design of legal knowledge integration module.

3.3 Design Legal Knowledge Extraction Module

In the field of legal education knowledge provided by mobile education platforms, the goal of legal education in mobile education is the product of learners' distance education learning needs, task activity needs, and legal course content.

Hypothesis $g(U)$ on behalf of the content design process of law popularization courses in the mobile education platform, $\mu(W)$ represents the teaching activities of learners, through the digital integration of legal education resources [12], formula (8) is used to give the characteristics of the content of legal education courses:

$$\psi(\kappa) = \frac{g(U) \times \mu(W)}{v(o)} \times \mu(j) \times \varpi(b) \tag{8}$$

where, $\varpi(b)$ represents the knowledge field of law popularization and teaching provided by the mobile education platform, $v(o)$ and $\mu(j)$ is the performance characteristics of the teaching content of popularizing the law, which represents the linear learning and leap learning of learners, will be based on the needs of the teaching environment of popularizing the law and the teaching objectives of popularizing the law $\xi(c)$ decomposes and reorganize. According to the organizational form of mobile education learners, use formula (9) to provide the extraction structure of legal resources suitable for learners' legal education and teaching mode:

$$\chi(y) = \frac{v(o) \times \mu(j)}{\xi(c)} \times \frac{k(\rho) \times z(i)}{\ell(a)/\partial(n)} \tag{9}$$

where, $k(\rho)$ represents learners' learning motivation, $z(i)$ represents the performance characteristics of learners in the process of popularizing the law. The internal psychological activity process of learners is related to the optimization and selection of legal course content by mobile education teachers, as well as the ratio of the performance characteristics of legal teaching content. Cognitive laws are related to the learning state of learners and the ratio of the performance characteristics of jumping learning legal teaching content.

Use formula (10) to build the extraction module of mobile education quality legal resources, namely:

$$m(g) = \frac{\iota(\vartheta)}{B(i)} \times \varsigma(\Omega) \times \alpha(d) \tag{10}$$

where, $\iota(\vartheta)$ represents the learning state of learners in the process of online law popularization and mobile education [13], $B(i)$ represents learner autonomy learning strategies provided in the extraction of common law resources, $\varsigma(\Omega)$ represents the learners' attention to the teaching content of the law course, $\alpha(d)$ represents the satisfaction of the content of the course for different types of learners.

To sum up, through the design of sign in incentive module, legal knowledge integration module and legal knowledge extraction module, complete the functional module design of university network law popularization mobile education platform.

4 Test Analysis

4.1 Test Environment

In order to improve the ability of college students to popularize legal knowledge and test the function and performance of the education platform, it is necessary to clarify the use environment of the mobile education platform before the test, so as to simulate the request in the real situation and achieve the same test effect as online. The test environment configuration of this article is as follows:

Hardware environment:
A PC with Windows operating system
MBP of a Mac OS operating system
A mobile phone on the Android platform
A mobile phone on the ios platform
Software environment:
Chrome browser
Mobile student app
Teacher app
Tutor following tool app
Other required software such as Paytm, Facebook, Whatsapp, etc.

4.2 Functional Test

In order to test the various functions of the university network legal popularization mobile education platform, this paper takes the sign in incentive module as the test object. The test steps are: after the development is completed, other members of the team will review the code. After the review and modification is completed, the self test will be conducted. The self test will be released in the test environment for front-end joint debugging, and then it will be proposed to the testers. After the testers pass the test, they will go online, finally, carry out online backtesting. For example, Table 2 is used for sign in incentive function test.

Table 2. Sign in incentive function test cases

Preconditions	Execution steps	Expected results	Actual results
The user has purchased the corresponding course	Enter the classroom within five minutes from the opening of the classroom to the beginning of the class, and click the sign in button;	You can sign in and complete the animation of common sign in, and send basic gold coins to users;	Consistent with expected results
	Sign in at the last class break that won the grand prize;	You can sign in, complete the sign in animation of the grand prize, and calculate the random proportion of gold coins to send to users;	
	If the previous courses are not fully signed, the last time of signing in is required;	You can sign in, complete the animation of ordinary sign in, and send basic gold coins;	
	Enter the classroom five minutes after the beginning of the class and click Sign In;	It is not allowed to sign in, and prompts that the tutor needs to be contacted for re signing;	
	The tutor will sign up for the students who have not signed in	Can be countersigned	

According to the results in Table 2, the platform in the article can sign in and prompt users to contact tutors, and tutors can also make up signatures, which meets user needs.

4.3 Performance Test

In the performance test, in order to highlight the advantages of the platform in this paper, the education platform based on big data analysis technology, the education platform based on distributed cognition and the education platform based on intelligent assistance are introduced for comparison. The storage space and memory occupancy of the test platform are as follows.

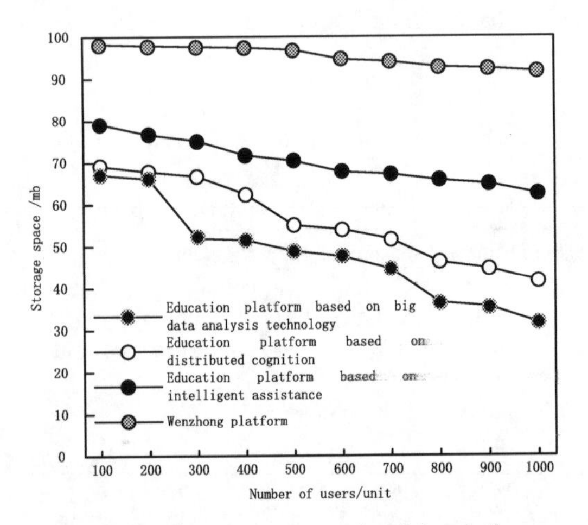

Fig. 5. Storage space test results of the platform

It can be seen from the results in Fig. 5 that when the intelligent assisted education platform is adopted, the storage space decreases significantly with the increase of the number of users. When the number of users in the platform increases from 100 to 1000, the storage space decreases from 67.6 mb to 32.1 mb. When using the education platform based on distributed cognition, the storage space is between 40 mb and 70 mb, and the range of storage space is basically the same as that of the education platform based on intelligent assistance, but it is larger than that of the education platform based on intelligent assistance. When the education platform based on big data analysis technology is adopted, the range of storage space changes is reduced, between 60 mb and 80 mb. When using the platform in this article, with the increase of the number of users, the storage space is still more than 90 mb.

According to the results in Fig. 6, with the increase in the number of user requests, the change trend of the memory occupancy of the education platform based on big data analysis technology, the education platform based on distributed cognition, and the education platform based on intelligent assistance is basically the same, all more than 20%. When using the platform in this article, with the increase of the number of user requests, the memory occupancy is always within 10%, which has high performance.

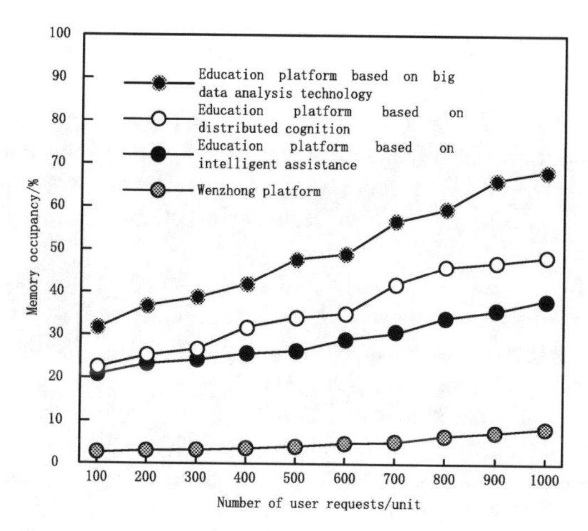

Fig. 6. Memory occupancy test results of the platform

5 Conclusion

This paper proposes the design and research of university network legal popularization mobile education platform based on streaming media technology. Design a mobile education platform client for university network legal popularization through four modules: data transmission control module, partner management module, media data buffer, and media playback. Design a streaming server based on streaming technology to achieve audio and video file playback. Utilizing the post request method and Jacobian method to establish platform functional modules, reducing the use of storage space, and thus achieving the design of a mobile education platform for university network law popularization. Through experiments, it has been proven that the storage space required by the platform in the article is small and the memory occupancy rate is always low, it is found that the function and performance of this platform can meet the needs of users. Although this research has achieved some results, there are still many shortcomings. Although streaming media technology has been applied to the design and development of actual projects, it has not optimized the configuration of various details of streaming media parameters, such as the response time of services. The next step is to optimize the parameter configuration of streaming media from this perspective.

Aknowledgement. 2023 Jilin Province Higher Education Research Project: Research on University Telecom Internet fraud from the Perspective of Collaborative Governance (JGJX2023D685).

References

1. Lin, H.: The construction of mobile education in cloud computing. Procedia Computer Sci. **183**(8), 14–17 (2021)
2. Shi, W., Zhang, Y.: Intelligent education platform design based on big data analysis technology. Modern Electronics Technique **43**(9), 150–153 (2020)

3. Gao, X., Jiang, X.: Design strategy of online education platform from the perspective of distributed cognition. Packaging Eng. **43**(12), 365–371 (2022)
4. Song, Y.: Application of mobile education in assisted autonomous learning platform in intelligent campus. Int. J. Continuing Eng. Educ. Life-Long Learning **30**(2), 104–119 (2020)
5. Liu, L., Subbareddy, R., Raghavendra, C.G.: AI intelligence chatbot to improve students learning in the higher education platform. J. Interconnection Networks **22**(Supp 2), 2143032 (2022)
6. Liu, S., He, T., Li, J., et al.: An effective learning evaluation method based on text data with real-time attribution - a case study for mathematical class with students of junior middle school in China. ACM Trans. Asian and Low-Resource Language Information Processing **22**(3), 63 (2023)
7. He, J., Zhao, H.: Mobile-based education design for teaching and learning platform based on virtual reality. Int. J. Electrical Engineering Educ. 002072092092854 (2020)
8. Qiao, X.: Optimization of college students' educational resources integration. Computer Simulation **34**(8), 239–242 (2017)
9. Xu, L., Zhou, Q.: App design of distance art education platform under internet ecological environment. International J. Electrical Eng. Educ. 002072092098352 (2021)
10. Balenzuela, M.P., Wills, A.G., Renton, C., et al.: A new smoothing algorithm for jump Markov linear systems. Automatica **140**, 110218–110228 (2022)
11. Feng, M., Zhang, H.: Application of Baidu Apollo open platform in a course of control simulation experiments. Comput. Appl. Eng. Educ. **30**(3), 892–906 (2022)
12. Chen, W., Samuel, R., Krishnamoorthy, S.: Computer vision for dynamic student data management in higher education platform. J. Multiple-Valued Logic and Soft Computing **36**(1/3), 5–23 (2021)
13. Kassawat, M., Cervera, E., Pobil, A.P.D.: An omnidirectional platform for education and research in cooperative robotics. Electronics **11**(3), 499 (2022)

Online Teaching Platform of Career Guidance Course Based on Virtual Reality

Weiwei Zhang(✉)

Changchun University of Finance and Economics, Changchun 130000, China
zhangww6565@163.com

Abstract. The current online teaching platform function module of career guidance course is generally one-way structure, and the efficiency of teaching resources acquisition is low, which leads to the extension of the response time of the platform. Therefore, the design and verification analysis of online teaching platform of career guidance course based on virtual reality is proposed. According to the actual platform application requirements and standards, build an ARM data processor, access the S3C4510B information storage chip, and complete the design of system hardware. Next, we will first analyze the teaching needs of career guidance courses, improve the efficiency of obtaining career guidance teaching resources in a multi-level way, realize the design of virtual reality multi-level functional teaching modules, associate intelligent virtual databases, and complete the design of platform software. The final test results show that the response time of the teaching platform finally obtained through the test and comparison of five classes has been controlled within 2s, indicating that with the assistance and support of virtual reality technology, the teaching efficiency of the platform has been significantly improved, with stronger pertinence and stability, and great practical application value.

Keywords: Virtual Reality Technology · Career Guidance Courses · Online Teaching Platform

1 Introduction

Today, with the continuous strengthening of information technology, online teaching plaorm has become an important part of people's education industry. To some extent, the form of online teaching platform can increase the number and scale of information carriers, expand the actual teaching scope, and form a more complete and detailed teaching structure. This time, take the employment guidance course as an example for research and analysis [1]. In fact, with the stimulation and support of the modern network platform, the proportion of online education in employment guidance courses has gradually expanded [2]. Especially in recent years, with the deepening of informatization in the education industry, the establishment of an efficient and convenient online course teaching platform has become the main trend of modern education. The online teaching

L. Yun et al. (Eds.): ADHIP 2023, LNICST 549, pp. 439–453, 2024.
https://doi.org/10.1007/978-3-031-50549-2_30

platform with modern network technology, information technology and communication technology as the main support has become a new teaching model and has been widely used [3]. Therefore, this kind of career guidance teaching curriculum platform not only promotes the innovation of the education industry, but also provides great convenience for daily teaching. Generally, a variety of teaching forms and resource forms are used to realize online teaching of career guidance courses, which provides great convenience for current online teaching [4]. However, there are still some problems in the practical application of the online course teaching platform for career guidance at this stage, which are mainly reflected in the processing of teaching resources on the platform. The processing efficiency is low, leading to the extension of the waiting time for users to process business; The data integration environment is not fixed, resulting in errors in teaching design, which to some extent affects the daily teaching effect of career guidance courses. Although, in the face of these problems, relevant personnel have made some targeted optimization, it still does not meet the current user requirements [5]. Therefore, this paper proposes the design and verification of online teaching platform for employment guidance courses based on virtual reality. The so-called Virtual Reality, abbreviated as VR), also known as Virtual reality or spiritual realm technology is a brand new Practical technology. Virtual reality technology includes computer Electronic information, Simulation technology, the basic implementation method is to computer technology give priority to, utilize and integrate3D graphics Technology multimedia technology, simulation technology display technique, servo technology and other high-tech latest development achievements, with the help of computers and other equipment to produce a realistic three-dimensional vision, touch, smell and other sensory experience virtual world So that people in the virtual world can have a feeling of immersive [6]. With the society productivity With the continuous development of science and technology, all walks of life are increasingly demanding VR technology [7]. Integrate this technology with the online teaching platform of career guidance courses to further expand the actual teaching scope and gradually form a more stable and diversified teaching form. Although the VR platform requires additional equipment purchase, it still has many advantages, which are as follows:

1) The VR platform can provide students with an immersive learning experience, enabling them to participate in the virtual environment and enhance the feeling and experience of learning.
2) Interaction and participation: Through the VR platform, students can actively participate in the teaching process, interact with the virtual environment, conduct practical operations and simulation experiments, so as to deepen their understanding and memory of knowledge.
3) Across time and space constraints: VR platforms can break the constraints of time and space, enabling students to learn at any time and anywhere. This is very useful for distance education, remote teaching and other scenarios.
4) Innovation and stimulate interest: VR platform can provide teachers with innovative teaching tools and methods to stimulate students' learning interest and improve learning motivation.

Therefore, while a VR platform requires additional equipment, the teaching advantages and experience it brings may offset the cost to some extent. In addition, as the technology continues to evolve, the price of VR devices is expected to gradually decrease and become more widespread and affordable. Different from the traditional teaching platform of career guidance courses, this platform combines the actual teaching needs to carry out hierarchical or targeted management of dynamic teaching resources, so as to design an efficient teaching resource management process, build the practical application of online teaching platform of career guidance courses under the network environment, in order to reduce the waiting time of users and improve user satisfaction. In addition, with the help and support of virtual reality technology, the online teaching platform for career guidance courses has also added a corresponding interactive processing device, which involves many interactive operations and conversion teaching processing in daily use, which can specifically enhance the creativity and learning of students in career guidance courses, which is conducive to their long-term growth and development, while strengthening the teaching characteristics of the platform, Improve teaching efficiency and quality [8].

2 Design Online Virtual Reality Teaching Platform Hardware for Employment Guidance

2.1 Design of ARM Data Processor

Before the basic platform construction, hardware design should be carried out in combination with the actual measurement requirements and standards. In view of the changes in the teaching needs and standards of employment guidance, it is necessary to first design an ARM data processor [9]. The online teaching interactive platform contains many types of data and a large amount of data, so there are many instructions issued during the operation of the platform [10]. In order to ensure the processing efficiency of instructions and improve the operating performance of the platform, the online teaching interactive platform for career guidance courses designed in this paper adds ARM microprocessors with more sources, which can achieve coverage control and guidance. In fact, ARM microprocessor is a special RISC processor, which has the advantages of small size, low power consumption and supports Thumb instruction set. A special register is added inside the processor to minimize the difficulty of addressing and improve the efficiency of instruction execution on the platform.

Different processors have different internal structures and different application fields [11]. This paper selects ARM9 processor as the processor of the design platform based on actual needs, and selects LCD (Liquid Crystal Display) expansion board for data expansion. Then, based on this, we use SSH framework to design the processing structure of the teaching platform hardware for career guidance courses. Apache and Tommcat servers are selected as the application server and database server of the online teaching platform [12]. The Apache server contains 67 MHz and 105 MHz external frequency processors, as well as hard disks of more than 10 GB. Its random access memory can ensure the efficient transport of the server's electrical signals. Tommcat server adopts Intel 8028678 6 MHz central processing unit (CPU), which can ensure stable transmission of detection signals.

Add a file processor of Java Server Pages (JSP) in the server, which can mark the dynamic data generated in real time in the activity of employment guidance for teachers and students to the static page [13]. The designed platform can still process access data in real time when facing large-scale access. In order to ensure that the platform can run smoothly in view, model, and controller (MVC) modes, a controller needs to be added. The controller model used is Philips LPC3180, which is a 16 bit/32 bit Advanced Reduced Instruction Set Computer Machines (ARM). In addition, the floating point coprocessor of the processor access part also needs to be adjusted. The synchronous dynamic random access memory (SDRAM) memory interface can run at the 200 MHz CPU frequency [14]. By starting the integrated control command of the instruction register of the host computer, the platform gradually completes the summary and processing of the data and information of the career guidance course, providing convenient conditions for subsequent processing.

2.2 Design of S3C4510B Information Storage Chip

After completing the design of ARM data processor, the next step is to build and apply the S3C4510B information storage chip. There is a huge amount of data information in the online teaching interactive platform for career guidance courses. In order to ensure the normal operation of the platform, S3C4510B memory chip is added inside the designed platform [15]. The S3C4510B memory chip has two cache descriptors, including two Universal Asynchronous Receiver Transmitter (UART) channels and two Code Division Multiple Access (CDMA) channels. At this time, set the control range and value of the multiple access channel, as shown in Table 1 below:

Table 1. Multiple Access Channel Control Range and Value Setting Table

Multiple Access Channel Number	Directional control range ratio	Controllable ratio
Multiple Access Channel 1	2.05	1.3
Multiple Access Channel 2	2.11	1.2
Multiple Access Channel 3	2.36	1.2
Multiple Access Channel 4	2.16	1.8
Multiple Access Channel 5	2.54	1.6

According to Table 1, complete the setting and adjustment of the control range and value of the multiple access channel. Next, based on this, set a DMA engine inside the memory chip, which can simultaneously send or receive 256 bytes of data, and the internal data calibration logic matches the 10 Mb/s working conversion rate. Because the interactive platform for career guidance teaching designed in this paper includes the High level Data Link Control (HDLC) protocol, in order to increase compatibility, the S3C4510B memory chip provides a Media Independent Interface (MII) and a 10 Mb/s interface, which are perfectly compatible with IEEE802.3. The memory chip can be expanded to a maximum of 4 bytes, When in use, different Cyclic Redundancy Check

(CRC) modules can be selected to connect with the digital clock. The transmission and reception of stored data support the NRZNRZIFM format, and each HDLC has two buffer channels for the transmission and reception of stored data. After using the Modem interface to connect, an 8-bit HDLC will be generated inside the S3C4510B memory chip. When the signal is XCLK, the output clock data source is low level; When the signal is MCLKO/SDCLK, the output clock source is CLKSEN high level, and the resulting reset signal can be processed by nRESET to improve the overall performance of the memory chip.

However, in this part, it should be noted that the chip settings should be related and overlapped with the ARM data processor as much as possible, and many practical modules can be integrated, such as the excellent performance and low-power 8051 microcontroller core with code prefetching function, the total number of three sets of universal IO interfaces reaches 21, five channel DMA, and four timer modules, of which T1 is a 16 bit timer, T2 is a dedicated timer for MAC, T3T4 is an 8-bit timer, with two powerful USART and RF modules supporting multiple serial communication protocols. At the same time, in order to further ensure the stability and reliability of the platform operation, it is also necessary to connect the photoresist, LM393 voltage comparator, sliding rheostat, power supply circuit and other devices at specific locations.

Connect the photosensitive resistor in series with a 10k Ω resistor between the power supply and the ground. The positive terminal of LM393 is connected to the upper end of the photosensitive resistor. The sliding resistor is also connected in series between the power supply and the ground. The sliding piece terminal is connected to the negative terminal of LM393. According to the principle of voltage comparison during platform operation, when the positive terminal voltage is greater than the negative terminal voltage, the positive voltage (logic high level) will be output, otherwise, the 0 voltage (logic low level) will be output. According to the changing state of the level, the basic teaching data and information are collected directionally through the chip. At the same time, the built-in equipment is used to screen the data, which is convenient for the near level design and innovative processing of the platform in the later stage.

3 Design Online Virtual Reality Teaching Platform Software for Employment Guidance

3.1 Analysis of Teaching Needs of Career Guidance Courses

After completing the design of the above system hardware, the next step is to analyze the teaching needs of the career guidance course by integrating the virtual reality technology and the changes in the daily teaching needs and standards of the career guidance course. Requirements formulation of platform design generally refers to functional requirements, that is, some capabilities that the system must complete or have. The online teaching platform for career guidance courses is an intelligent teaching platform developed for students, teachers, and system administrators. Its main purpose is to provide a platform for students to learn online and communicate, so as to improve students' learning efficiency, and to assist teachers in offline teaching, and comprehensively improve the quality of school teaching. The traditional online teaching platform has a single function and is not attractive to students, which cannot meet the diversified learning needs of students, resulting in a weak willingness of students to use it. Therefore, this paper will expand the functions on the basis of the traditional online teaching platform, and the functional requirements will be analyzed below. The first is the registration and login requirements.

As an online teaching platform for employment courses in colleges and universities, the security of the system needs to be ensured first, and the resources inside the system can only be viewed by students of the university, so it is necessary to set targeted login permissions to improve the encryption and stability of the platform. In order to achieve unified authentication of multiple services and avoid multiple logins, the platform's login function must be implemented using single sign on technology. See Fig. 1 below for details:

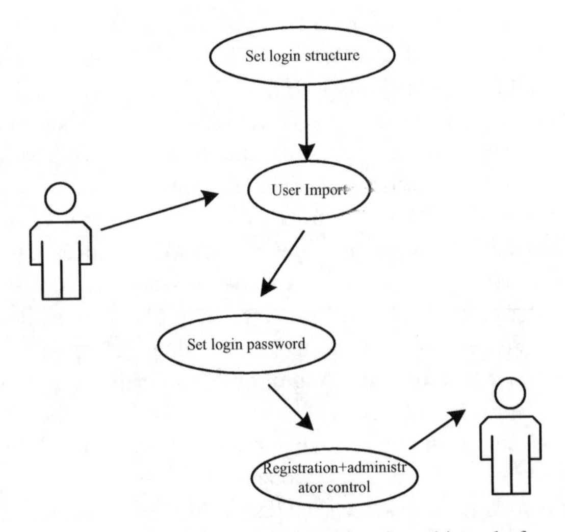

Fig. 1. Diagram of login relationship of teaching platform

According to Fig. 1, complete the setting and adjustment of the login relationship of the teaching platform. Then, based on this, we set the requirements for online teaching course permission management. This part mainly designs a complete set of permission management scheme for different types of users. Permission management in this paper is divided into user, permission and role management. A user can be assigned a role, and a role can have multiple permissions. These operations can only be performed by the system administrator. Next is the demand for curriculum management. After entering the platform, students can search for career guidance courses according to their own interests, and collect or add courses at the same time. Career guidance courses are uploaded by teachers, including different chapters and sections. During the uploading process, you can edit the basic information of the course, including the cover, course introduction, course keywords, and departments displayed on the front desk. In addition, teachers can also modify or delete the published career guidance courses. Combined with virtual reality technology, build a multi-level and multi-objective demand structure, import the above set requirements uniformly, and achieve the set analysis of renewal.

In addition, it also includes the demand for mutual help question and answer. In order to improve the interaction between students, it is necessary to set up a mutual help question and answer module. In case the teacher can't answer the questions in time, students can ask other students questions and seek answers in the question and answer module, or they can answer others' questions to consolidate their knowledge again, so as to improve the learning efficiency. Not only that, this module can also be operated as a forum model. Students can discuss some things they are interested in outside the course.

The level inspection requirement is a key link, which is equivalent to the capability test. The existing online teaching platform for career guidance courses lacks the function of evaluation. On the one hand, teachers can not understand students' learning process and learning dynamics by uploading course videos; On the other hand, students do not know whether their understanding of the content is in place or to what extent they have mastered the knowledge after learning the course. Therefore, the platform needs to set the test function. The teacher can set up multiple sets of test papers for the career guidance course, and the questions of the test paper are mainly multiple-choice questions. After learning the course, students can choose any set of test papers to check their learning level. After submitting the answers, the system needs to return the test scores and wrong questions to students. In this way, students can review in a targeted way to improve their learning achievements, and teachers can also evaluate their teaching achievements through students' test results.

Finally, recommend course requirements. With the increase of the number of courses, students will have no way to start when choosing courses, so a good recommendation function is indispensable. The current recommendation modes are divided into manual recommendation and intelligent recommendation. Manual recommendation is mainly used by the administrator to set recommendation resources in the background system and display them on the front page in the form of slides. The recommended resources of this platform include courses and notes, such as some national famous teacher courses, popular courses, popular notes, etc. Automatic recommendation is to recommend content to users through algorithms. The resources recommended by this platform are mainly for courses. By analyzing the course selection records of users, they recommend courses of interest to users according to their course selection records.

3.2 Design of Multi-Level Functional Teaching Module of Virtual Reality

After completing the analysis of the teaching needs of the career guidance course, the design of the multi-level functional teaching module of virtual reality is carried out immediately. The practical homework function and online examination of the online teaching interaction platform of career guidance courses are the core of their online teaching interaction. Therefore, the functional modules can be designed in combination with the basic needs and changes in standards in the teaching process of career guidance courses and virtual reality technology.

(1) Job module. In this module, virtual reality technology is combined to build a multi-level teaching virtual structure, so that students can continuously upload their own employment guidance assignments, which are then corrected by teachers. To form a complete teaching model has strong application value.

(2) Q&A module. This module mainly guarantees the online question answering function of teachers. Combined with virtual reality technology, students can switch specific employment scenes and network environment in the learning process. In the face of various problems encountered in the virtual scene, students can independently grow, and then teachers can supplement. This module is very helpful for students' subsequent learning.
(3) Chat module. In this module, teachers and students can log in at the same time. In this module, teachers and students can query chat content, send chat information, and delete chat records, which is very helpful for interaction between teachers and students.
(4) Online examination module. In this module, combined with virtual reality technology, students can choose their own exam type for the exam, but each student's exam content and direction are also different, mainly because virtual display technology will design the best test scheme for students to improve the effectiveness of the test. After receiving the students' test papers, teachers can make online comments and record the students' scores. Through virtual reality technology, the test papers can be converted into virtual intelligent format and distributed to students. In addition, teachers can design examination papers that meet the needs according to the types of learning knowledge points, and students can answer them to master the learning status of students in a timely manner.
(5) Online examinee account management module. Students can register and log in through this module, and then be managed by the platform. Combined with virtual reality technology, users can be managed to build a large "teaching tree". In the built-in virtual structure, users can not only view their status, but also add new accounts, delete ordinary accounts, and browse user information. Users who log in for the first time need to register to log in to the platform. So far, the design of the focal function teaching module of the system has been completed.

3.3 Design of Intelligent Virtual Database

After completing the design of the multi-level functional teaching module of virtual reality, the next step is to design an intelligent virtual database. Different from the traditional database, this time because of the higher requirements of the teaching chapters and practice of the career guidance course, it is necessary to change the collection and collection form of the database to expand the actual data storage capacity and space. In order to ensure the performance of the interactive platform for career guidance courses, this paper uses MySQL to design the platform database. Because the teaching interactive platform involves a large amount of data and a complex data table structure, the platform designed in this paper selects SQL2020 as the database management platform and designs the database. According to the actual needs of the platform, this paper lists some data structures and data items, as shown in Table 2 below:

Set the database structure and data according to Table 2. Next, use big data technology and virtual reality technology to build a multi-dimensional Python crawler teaching transformation scheme, crawl the online teaching resources and related extracurricular resources of career guidance courses in the teaching network, and transfer the resources to the MySQL database structure of the platform, so as to establish the teaching resource

Table 2. Database structure and data setting table

Field name	data type	role
Id	digit	Complete automatic numbering settings
detail	-	Tag Keywords
page	-	Tag Target Page
title	text	Implement announcement title design
Fiag-shoe	-	Display Tag Data
Fiag-answer	-	Returning to data questions

database of career guidance courses. Before storing data resources, it is necessary to clean and convert the crawled data to ensure the accuracy and reliability of teaching resources.

Add the curriculum retrieval function with Hadoop as the core in the database, and design it around employment direction, employment trend, job selection and other aspects. While meeting the resource storage needs of the teaching platform, ensure the real-time retrieval and acquisition of teaching resources, so that the career guidance curriculum teaching platform has the characteristics of flexible, efficient and convenient teaching resource management, It lays a data foundation for classroom teaching of career guidance courses. Further expand the storage capacity of the database and complete the system software design.

4 Platform Test

This time is mainly to analyze and verify the actual application effect of the online teaching platform for employment guidance courses based on virtual reality. Considering the authenticity and reliability of the final test results, the analysis is carried out by comparison, and the H online teaching platform is selected as the main target of the test, Use professional equipment and devices to collect the basic teaching data and information of the platform, and then summarize and integrate them for future use. Next, according to the actual teaching needs and standard changes of the career guidance course, the final test results are compared, verified and studied. Next, combined with the real-time test requirements, the basic test environment is built.

4.1 Test Preparation

Combined with virtual reality technology, this paper analyzes and studies the practical application effect of H online teaching platform. Next, build a basic test environment. The front page of the teaching platform is produced by the Bootstrap framework, and the back end development uses the Spring Boot framework. The MySQL database is used to connect My Batis. The running environment is Windows 10, the background writing language is Java, and the front page label is a Hyper Text Markup Language (HTML) label. The setting platform can be directly accessed during operation, forming a more controllable teaching platform environment. Then, based on this, the basic teaching environment was built, as shown in Table 3 below:

Table 3. Hardware Environment Settings of Basic Teaching Platform

Set indicator name	Controllable basic parameter standards	Standard for measured control parameters
CPU model	11th Gen Intel Core i7-1165G7 @ 2.80 GHz GHz	GHz
memory size	12 GB	24 GB
Graphics card	NVIDIA GeForce MX4502 GB/ associate	NVIDIA GeForce MX4504 GB/ associate
operating system	Windows10	Windows10
browser	Google Chrome browser	Yes
Server Memory	2 GB	8 GB
personal computer	8 GB	16 GB

According to Table 3, realize the setting and research of the basic teaching platform operating environment. After setting the hardware environment of the system, the software environment is designed by integrating virtual reality technology, as shown in Table 4 below:

Table 4. Software Test Environment Setting Table

Software Name	version	Access Platform
FFmpeg	4.2.5	CentOS
Safari	15.1	iOS
Chrome	95.0.4638.69	Mac OS
Safari	15.1	Mac OS
Openssl	1.1.1 g	CentOS
MySql	5.7	CentOS
Spring Boot	2.4	CentOS

The software environment of the platform was set up and analyzed according to Table 4. So far, the basic operating environment of the platform has been basically built. Next comes the design of the online teaching module of the employment guidance course, and the module functions and uses, which are shown in Table 5.

So far, the basic test environment of the platform has been built. Next, the virtual reality technology will be combined with specific verification and analysis research.

Table 5. Functional module test point setting

Function module	Test function point
Live online class, review and real-time instruction modules	Whether the teacher can push real-time audio and video stream normally, whether the student can pull real-time audio and video stream normally, instant messaging function, microphone, online question and other interactive functions
Service function module	Registration and login function, personal information maintenance function, student side subscription function, teacher side course management function, subscription SMS notification function, learning statistics function
Platform operation and maintenance management module	Overview of system monitoring, CPU usage, disk I/O data, network I/O data, and memory consumption ratio Detailed viewing and abnormal consumption Automatically save time-stamped information

4.2 Test Process and Result Analysis

Combined with virtual reality technology, the design of H career guidance teaching platform is analyzed and studied. Describe the test cases and test requirements of each teaching function module of the platform, and test the concurrency performance of the platform. First, five classes in a school are selected as the main target objects of the test. Combining the acquired data and information and integrating virtual reality technology, the teaching weight value of the platform at this time is calculated, as shown in Formula 1 below:

$$A = \frac{W_2 \times \phi}{W_1 + (1 - m)} - (m\chi + W_2)^2 \quad (1)$$

In Formula 1: A Indicates the teaching weight value of the platform, W_1 It indicates controllable teaching coverage, W_2 It indicates the controllable teaching coverage of actual measurement, m Represents the conversion deviation, ϕ Indicates that the directional response takes time, χ Represents a course chapter. According to the above measurement, complete the calculation of the teaching weight value of the platform, and according to the data and information obtained, delimit the chapters and teaching contents of the career guidance course in the platform. First, the basic teaching service function was tested.

Generally, the basic employment guidance service platform mainly includes user management, authority management, teacher management, department management, and category management. User management mainly tests the operations of adding, deleting, modifying and querying users, registering and logging in. Permission management tests the modification and granting of user permissions, while teacher management, category management and department management tests the basic addition, deletion, modification and check. The details are as follows:

(1) Test purpose: check whether the functions of the basic service can run normally.
(2) Test steps: a. Click the registration button in the foreground to fill in the information to register a new user, and use the user to log in to the foreground. The administrator logs in to the background to assign roles to the new user, and finally deletes the user.
   a. The administrator logs in to the background to test the operation of adding, deleting, modifying and querying the departments and categories.
   b. The administrator tests the addition, deletion, modification and query of the teacher, and checks whether the object storage is normal when uploading the teacher's avatar.
(3) Expected results: users can successfully register and log in; Administrators can successfully add, delete, modify and query users; Administrators can modify user permissions; Administrators can maintain department and category information; The administrator can successfully operate the teacher's information. According to the daily teaching needs and standards of the career guidance course, we have realized the basic measurement of teaching services. This time, we can know that the teaching platform is in a relatively stable and real operating state. Then, based on this, we designed the test process of the teaching platform, as shown in Fig. 2 below:

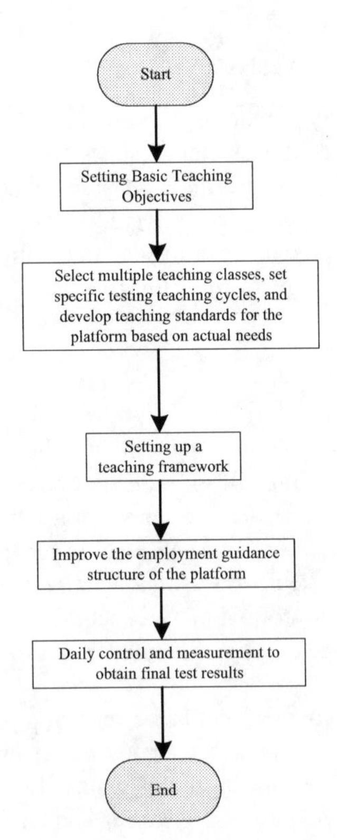

Fig. 2. Structure Diagram of Platform Test Process

According to Fig. 2, complete the design and verification analysis of the platform test process structure. According to the above process, and in combination with the actual test needs, the platform's basic test indicators and parameter standards should be reasonably adjusted to ensure that the platform is in the best operating state. Then, on this basis, the corresponding employment guidance teaching test instructions were set up and introduced into the internal control structure to form a complete program based on the requirements of the platform measurement at this time. According to the changes of career guidance teaching content and daily increase, set the test indicators and parameters of the adjustment platform, as shown in Table 6 below:

Table 6. Adjustment Platform Test Index and Parameter Setting Table

Platform testing metrics	Initial test standard value	Actual test standard value
Reads/time	12	18
Access Relationship Value	16.35	18.11
Interactive control ratio	2.5	2.6
Learning behavior	Refers to the ability to replicate the content of employment guidance courses	On the basis of the ability to replicate employment guidance courses, add a shared transmission function module
Operating average	6.34	7.22
Unit running time/s	20	24
Interference frequency/time	7	11

According to Table 6, realize the secondary setting of the test. Issue the corresponding control commands on the platform, and test according to the initial set order. The five classes with 22, 35, 30, 47, and 45 people are preset for the test. Finally, the response time of the teaching platform was selected as an evaluation indicator to verify the performance of the design platform, for the following reasons:

1) Fast response time can improve user satisfaction and make users more willing to continue using the platform.
2) Rapid response time can ensure real-time communication and interaction between teachers and students, promoting the improvement of learning outcomes.
3) Quick response time can help teachers conduct teaching activities more efficiently and improve teaching efficiency.

The calculation formula for the teaching response time of this platform under different test number states is shown in Eq. 2:

$$M = \lambda^2 + \sqrt{(\pi\lambda - e)} + \aleph\lambda \tag{2}$$

Equation 2: M Indicates the teaching reaction time, λ Indicates the directional test range, π Represents the reaction time of the teaching unit, e Represents the number of conversions, ℵ Indicates the waiting time. According to the above settings, the analysis of test results is achieved, as shown in Table 7 below:

Table 7. Comparison and Analysis of Test Results

Test Class and Number of Students	Preset teaching response time/s	Teaching response time/s
Classes 1–22	2.1	1.8
Classes 2–35	1.6	1.3
Classes 3–30	1.9	1.5
Classes 4–47	1.8	1.4
Classes 5–45	1.6	1.3

According to Table 7, the analysis of the test results is completed: after the test and comparison of five classes, the response time of the teaching platform is finally controlled within 2 s, which indicates that with the assistance and support of virtual reality technology, the teaching efficiency of the platform has been significantly improved, with stronger pertinence and stability, and great practical application value.

5 Conclusion

To sum up, it is the design and verification analysis of the online teaching platform for employment guidance courses based on virtual reality. With the help and support of virtual reality technology, in view of the changes in the teaching needs and standards of career guidance courses, the problems of insufficient educational resources and uneven distribution have been gradually solved. After analyzing the teaching needs of functional and non functional courses, the overall platform framework has been proposed, and thus the various teaching function modules of this platform for career guidance courses have been divided, It includes real-time teaching interaction module, online classroom live broadcast and review module, personal information (including subscription, statistics, etc.) module and system operation and maintenance management module. In the process of teaching, help students learn and understand each business function module in turn, improve the concurrency of the teaching platform, and realize online interactive teaching. In addition, on this basis, it is also necessary to add targeted teaching algorithms to the platform, further improve and optimize the overall teaching structure, expand the actual scope of employment guidance, ensure stable operation under the high load and high traffic demand in the actual teaching scene, and provide reference and theoretical reference for the design and innovation of subsequent related platforms. Of course, there are still some shortcomings in this study, and the relevant course assessment function is not set in the platform, which will be further improved in this aspect in the future.

References

1. Phattanawasin, P., Toyama, O., Rojanarata, T., et al.: Students' perspectives and achievements toward online teaching of medicinal chemistry courses at pharmacy school in Thailand During the COVID-19 pandemic. J. Chem. Educ. **10**, 98 (2021)
2. Deniz, S., Mueller, U.C., Steiner, I., et al.: Online (Remote) teaching for laboratory based courses using "digital twins" of the experiments. J. Eng. gas Turbines and Power: Transactions of the ASME **5**, 144 (2022)
3. Yalagi, P.S., Dixit, R.K., Nirgude, M.A.: Effective use of online teaching-learning platform and MOOC for virtual learning. J. Physics: Conference Series **1854**(1), 012019 (8 pp) (2021)
4. Ye, L., Zhong, J.: Study on blended teaching in principles of chemical engineering based on cloud platform. IOP Conference Series: Earth and Environmental Science **693**(1), 012027 (6pp) (2021)
5. Wang, S., Sun, T., Qu, X., et al.: Online education of atomic physics based on MOOC platform. J. Physics: Conference Series **1881**(3), 032009 (5pp) (2021)
6. Sun, N., Wang, Y., Liu, Y.: The application of cloud class in the teaching of biochemistry. IOP Conference Series: Earth and Environmental Science **692**(3), 032033 (6pp) (2021)
7. Wei, Y.: Exploration and research on the mixed teaching mode of basic Japanese course under the background of information technology. J. Phys. Conf. Ser. **1852**(4), 042032 (2021)
8. Narayanan, R., Mathew, P.: Teaching international english language testing system (IELTS) academic writing and exam strategies online to develop omani students' writing proficiency. Arab World English J. **2**(2), 49–63 (2021)
9. Liu, X., Gao, F., Jiao, Q.: Massive open online course fast adaptable computer engineering education model. Complexity **2021**(1), 1–11 (2021)
10. Wang, H., Yang, Z.: Research on digital virtual reproduction simulation of local micro damage in venues. Computer Simulation **39**(11), 5 (2022)
11. Cui, H., Cheng, M.: Digital media art teaching strategy based on virtual reality technology. J. Shanxi University of Finance and Economics **44**(S2), 125–127 (2022)
12. Wang, S., Jiang, Z.: A multiplayer online teaching system based on virtual reality live streaming. Computer Applications and Software **39**(10), 132–140 (2022)
13. Wang, H., Wang, Z.: Design of simulation teaching system based on virtual reality. Computer Simulation **39**(04), 205–209 (2022)
14. Li, X., Wang, Y., Wang, G.: Implementation of a digital teaching factory based on digital twins and virtual reality. Automation Technology Appl. **40**(09), 113–115 (2021)
15. Zhang, J., Wang, H., Ban, J.: Optimization design of vocal music teaching platform based on virtual reality system. Computer Simulation **38**(06), 160–164 (2021)

Intelligent Control Method of Indoor Physical Environment in Atrium Under Social Information Network

Hai Huang[1], Linmei Shi[1](✉), and Xian Zhou[2]

[1] Shanghai Dong Hai Vocational Technical College, Shanghai 200241, China
007001007001@163.com

[2] Institute of Economics and Management, Zhi Xing College of Hubei University, Wuhan 430011, China

Abstract. In order to meet the needs of human body for indoor environment, an intelligent control method of indoor physical environment in atrium under social information network is proposed. Considering the fluid dynamic characteristics of the building atrium, the physical environment model of the building atrium is constructed. Use social information network to collect physical environment parameters such as temperature and humidity. Set the environment intelligent control target according to the thermal comfort of human body. The intelligent controller of indoor physical environment is designed by combining fuzzy control and PID control. With the support of control commands, the intelligent control task of indoor physical environment in the atrium of the building is realized. Through the performance test experiment, it is concluded that compared with the traditional control method, the control errors of indoor ambient temperature, humidity and daylighting coefficient are reduced by 0.61°C, 0.51% and 0.145 respectively, while the control error of sound pressure is reduced.

Keywords: Social Information Network · Building Atrium · Indoor Physical Environment · Intelligent Control

1 Introduction

Stalls space architectural space A form of "outdoor space" refers to the courtyard space inside the building, which is characterized by the formation of "outdoor space" located inside the building. At first, it was the main hall or internal middle hall in a Roman house. The center was open to the sky, and there was usually a pool to collect rainwater. It is a shared space formed by the upper and lower floors in the building. In recent years, with the development of large-scale and comprehensive buildings, large atrium space buildings with several or even dozens of floors have appeared. The atrium building space is relatively large and often connects with the openings of the surrounding buildings and plays the role of vertical traffic. In order to create a better indoor light environment, a large area daylight opening is usually set on the top or side wall of the atrium [1]. The

L. Yun et al. (Eds.): ADHIP 2023, LNICST 549, pp. 454–469, 2024.
https://doi.org/10.1007/978-3-031-50549-2_31

solar short wave radiation enters the room through the top lighting glass, and transfers the heat to the indoor furniture, walls and ground through convection or thermal radiation. However, the long wave radiation emitted from the surface of these objects cannot pass through ordinary glass to reach the outside, resulting in heat accumulation indoors, thus forming the greenhouse effect. Another part of the heat of the solar radiation is absorbed by the glass, so that more heat is gathered at the top of the large space in the atrium. In addition, the large use of refrigeration equipment at the height of personnel flow at the bottom of the atrium, there is a large thermal pressure difference in the scale of the atrium perpendicular to the ground, which forms the indoor chimney effect, and the gradient distribution of indoor temperature in the vertical direction is large. In the past, the indoor thermal and humid environment of commercial complexes was mainly regulated by air conditioning equipment, and passive cooling technology was basically not used. The continuous operation of equipment throughout the day caused indoor users to feel too cold or overheated, and a large amount of unnecessary building energy consumption was generated.

The indoor physical environment in building atrium often includes indoor thermal environment, indoor light environment, indoor acoustic environment and indoor air quality. The impact of indoor environmental quality on people can be divided into direct impact and indirect impact. Direct impact refers to the direct effect of the direct factors of the environment on human health and comfort, such as good indoor lighting, especially the use of natural light can promote people's health; The indoor layout and color that people like can ease the tension caused by the pressure of work and life; Appropriate indoor temperature, humidity and fresh air can improve people's work efficiency. Indirect influence refers to the factors that promote the alleviation of the positive or negative effects on people, such as the suitable environment when the mood is stable makes people excited, and the unsuitable environment when the mood is depressed makes people more depressed Fidgety etc. This shows that improving the quality of indoor environment can increase the comfort and health protection of indoor personnel, and meet the requirements of people for indoor environment from both physical and mental health. In order to meet the requirements of human life, an intelligent control method of indoor physical environment in atrium was proposed.

Intelligent control is a control mode with intelligent information processing, intelligent information feedback and intelligent control decision-making. It is an advanced stage of control theory development, mainly used to solve the control problems of complex systems that are difficult to solve with traditional methods. The main characteristics of intelligent control research objects are uncertain mathematical models, high nonlinearity and complex task requirements. At present, the more mature research results of intelligent control methods for indoor physical environment specifically include: intelligent control based on sensors and fuzzy rules, intelligent environment control based on network communication, and environmental intelligent control based on PMV indicators. However, the above intelligent control methods have obvious control accuracy problems in the actual operation process, which is mainly reflected in the large gap between the indoor physical environment parameters of the building atrium and the control objectives. Therefore, the concept of social information network is introduced.

Social information network refers toSocial individual Relatively stable relationship system formed due to interaction between members, social information Network attention. It is the interaction and connection between people. Social interaction will affect people social behavior. A social network is a social structure composed of many nodes. Nodes usually refer to individuals or organizations. Social networks represent various social relationships. Through these social relationships, social networks transform casual acquaintances into closely connected ones Family relations. All kinds of people or organizations are linked together. The social information network is used to optimize the intelligent control method of the indoor physical environment of the building atrium in order to improve the control effect of the environment.

2 Design of Intelligent Control Method for Indoor Physical Environment of Building Atrium

The indoor physical ambient intelligence control of building atrium can be automatically adjusted and controlled according to the real-time environmental data and user's needs, without manual intervention. This can improve work efficiency and reduce manpower burden. Real-time monitoring of indoor humidity, air quality, light, sound and other environmental parameters through various sensors. Compared with the traditional fixed set-point control mode, the intelligent control method can more accurately perceive environmental changes and make corresponding adjustments.

2.1 Construction of Indoor Physical Environment Model of Building Atrium

The intelligent control method of the indoor physical environment in the atrium of the optimized design building mainly implements the intelligent control program from the aspects of indoor thermal environment, air quality, acoustic environment, light environment, etc. Before environmental control, it is necessary to simulate the flow characteristics of the indoor physical environment in the atrium under natural conditions, and establish the corresponding dynamic environment model [2]. Figure 1 shows the dynamic simulation results of the indoor thermal environment of the atrium.

The state space method discretizes the space of the room envelope, indoor furniture, etc., establishes the heat balance equation of each discrete node, and keeps the temperature of each node continuous in time; Then, the heat balance equations of all nodes in the room are solved, and a series of coefficients representing the thermal characteristics of the room are obtained; On this basis, the room is expressed as a function of thermal response coefficient and various thermal disturbances, and the heat balance equation of air nodes in all rooms is solved simultaneously. The expression of the heat balance equation of the inner wall of the atrium is:

$$-\kappa_{\text{heat}}\frac{\partial t}{\partial x}\bigg|_{x=L} = \kappa_{\text{transfer}}(T_{room} - T) + R_{\text{absorb}} + \sum_{j}\kappa_{\text{radiation}}T + R \quad (1)$$

Variables in Formula (1), κ_{heat}, κ_{transfer} and $\kappa_{\text{radiation}}$ They are the thermal conductivity of the inner wall of the atrium, the convective heat transfer coefficient with the air, and the

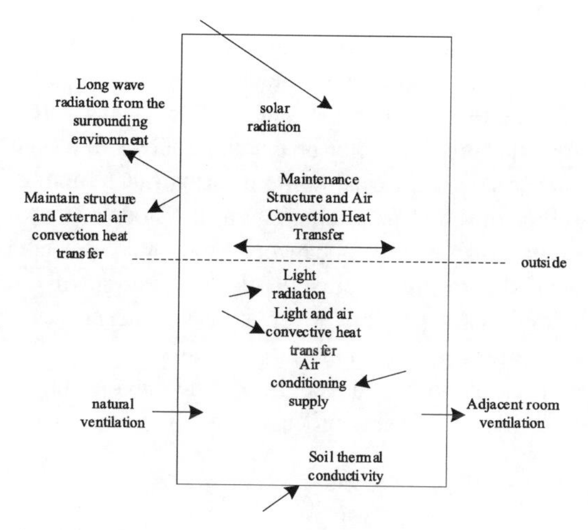

Fig. 1. Dynamic model of indoor thermal environment in atrium

long wave radiation heat transfer coefficient, x and L corresponding to the wall thickness, T_{room} and T represents indoor temperature and internal wall temperature respectively, R_{absorb} and R are respectively the solar radiation heat absorbed by the inner wall of the atrium through the window and the heat transferred to the surface by other indoor heat sources in the form of radiation. Similarly, it can be concluded that the circulation characteristics of indoor ambient air in the atrium of the building are:

$$\frac{d\left(\text{H}_{\text{air}} A \rho_{\text{pollute}} T_{\text{pollute}} C_P\right)}{dt} = \kappa_h + mC_\rho T_0 \tag{2}$$

Among H_{air} is the thickness of the air layer, ρ_{pollute} and T_{pollute} corresponds to the density and temperature of pollutants in the indoor air environment of the atrium, C_ρ is the specific heat of polluted air, κ_h represents the convective heat transfer rate of the pollution plume, T_0 is the temperature value of ordinary air, m is polluted air quality [3]. Similarly, the simulation results of the indoor physical environment characteristics of the atrium can be obtained, and the corresponding model construction results can be obtained.

2.2 Use Social Information Network to Collect Physical Environment Parameters

In the process of information transmission in a multi-layer social network, the information can only be transmitted from the node with high information trust in the active state to the node with low information trust, rather than back propagation. In the process of communication, the information reliability decreases with the relationship resistance in the direction of communication. For the same information, the greater the network relationship resistance, the more the information reliability decreases, as follows:

$$\delta_x - \delta_y = \kappa_{\text{propagate}} Z_{x,y} \tag{3}$$

Among δ_x and δ_y represent nodes in social information network respectively x and y Information reliability, $Z_{x,y}$ represents a node x and y total relationship resistance between, $\kappa_{\text{propagate}}$ is the propagation coefficient of social network. In the process of network information transmission, after any node receives a certain information, the node will have a certain degree of trust in the information. When it receives the same information from other different neighbors again, the node's trust in the information will be strengthened once [4]. As the node receives the same information from more and more other neighbor nodes, the node's trust in this information will continue to increase, but in any case the node's trust in this information will never exceed 100%, but may exceed the trust of one of the propagation sources in this information. The superposition principle of information reliability can use the Wayne diagram in set theory to remove the overlapping effect. The processing process is as follows:

$$\delta(x \cup y) = \delta(x) + \delta(y) - T(x \cap y) \tag{4}$$

In the above formula $\delta(x)$ and $\delta(y)$ value range of is [0,1]. In the social information network, multiple sensor devices are used to collect the physical parameters in the indoor environment of the building atrium, and then the initial value of the control method is obtained. Install temperature and humidity sensors, illumination sensors and other equipment in the indoor physical environment of the building atrium. The structure of temperature and humidity sensors is shown in Fig. 2.

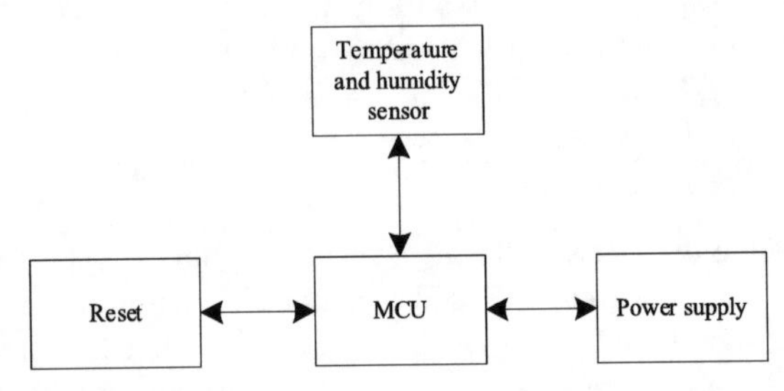

Fig. 2. Structure diagram of indoor ambient temperature and humidity sensor in atrium

The temperature and humidity sensor shown in Fig. 2 transmits data with MCU through a serial interface, which is different from the I2C interface. A set of "start transmission" timing is used to indicate the initialization of data transmission, followed by 8-bit data commands for temperature measurement, humidity measurement, reading and writing of status registers, etc.; After sending the measurement command, wait for the "data ready" signal to read out the data [5]. The upper computer of the system sends control commands to the control node based on the data collected by various sensors, so the effective reading of data by sensors is an important prerequisite for the intelligent control system to achieve control functions. The specific process of the sensor information reading program is: after receiving the query command, send the query command to the lower computer, monitor the data through the serial port. For the received data saved in the database and the number of times of no received commands cleared to zero, the system will report errors and alarm when no data is received and the number of times of no received commands is 3. In this process, the upper computer sends the No. 03 command in the Modbus protocol to the sensor node through the Zigbee communication network. Install illuminance sensor, carbon monoxide sensor, formaldehyde sensor, etc. in the indoor physical environment of the building atrium according to the above method, and complete the collection of physical environment parameters through the process shown in Fig. 3, where the collection result of ambient temperature is:

$$T = \frac{\kappa_{\text{radiation}} T_{air} + \kappa_{\text{Commutation}} T_{\text{radiation}}}{\kappa_{\text{radiation}} + \kappa_{\text{Commutation}}} \tag{5}$$

Among $\kappa_{\text{radiation}}$ and $\kappa_{\text{Commutation}}$ are radiation heat transfer coefficient and convection heat transfer coefficient, T_{air} and $T_{\text{radiation}}$ corresponding is indoor air temperature and average radiation temperature [6]. In addition, the collection results of light environment parameters include: illuminance, contrast sensitivity, etc. The collection results of illuminance parameters are:

$$\psi = \frac{d\Phi}{dS} \tag{6}$$

Among $d\Phi$ and dS are respectively the luminous flux and the area of the panel at a point on the surface, and the illuminance can be added directly. When several light sources illuminate the illuminated surface at the same time, the illuminance on them is the algebraic sum of the illuminance formed when a single light source exists separately. The collection results of contrast sensitivity parameters are as follows:

$$\sigma = \frac{B_{\text{background}}}{\left| B_{\text{target}} - B_{\text{background}} \right|} \tag{7}$$

among $B_{\text{background}}$ and B_{target} It corresponds to the background brightness and target brightness. The collection results of physical environment parameters can be obtained according to the above method, and transmitted to the control terminal through the social information network.

2.3 Set the Intelligent Control Target of the Indoor Physical Environment of the Atrium

According to the human body's demand for the comfort of the indoor physical environment in the atrium, the control target of the physical environment is set. Thermal comfort refers to the subjective satisfaction expression of people with the thermal environment. The intuitive feeling of human body in the environment is influenced by objective environment and subjective factors. According to the indexes of various building materials, the influence on indoor environment is studied, and the influence degree of various factors on indoor environment is analyzed. The heat exchange between building materials and human body in the air will also have an impact on human thermal comfort [7]. When the human body feels comfortable in the steady heat and humidity environment, it needs to meet three conditions at the same time: the gain and loss of heat between the human body and the environment are equal, that is, the heat storage of the human body in the heat balance equation is equal to zero; The average skin temperature of human body shall be kept within the range corresponding to comfort; The actual sweat evaporation heat loss of human body should be kept in a small range. Under the above conditions, the thermal comfort equation is:

$$M - W = P_a + M(34 - T_{air}) - 8\vartheta\left[(T_w + 273)^4 - (T_h + 273)^4\right] + \vartheta(T_w - T_{air}) \tag{8}$$

Among M and W are the metabolic rate of the human body and the mechanical work done by the human body, P_a is the partial pressure of water vapor in the air, T_w and T_h corresponds to the average temperature of human body surface or clothing surface and the average radiation temperature of the environment, ϑ is the convective heat transfer coefficient. According to the construction result of human thermal comfort equation, determine the intelligent control target of indoor physical environment temperature, air quality, humidity, light and other parameters in the atrium, and mark it as U_{target}.

2.4 Design of Intelligent Controller for Indoor Physical Environment

The intelligent controller of indoor physical environment is the executive element of the control function of the intelligent control method of the indoor physical environment of the building atrium under the social information network. The optimized intelligent controller consists of window opening controller, R&D switch, transponder and other equipment. The internal logic principle of the intelligent controller is shown in Fig. 3.

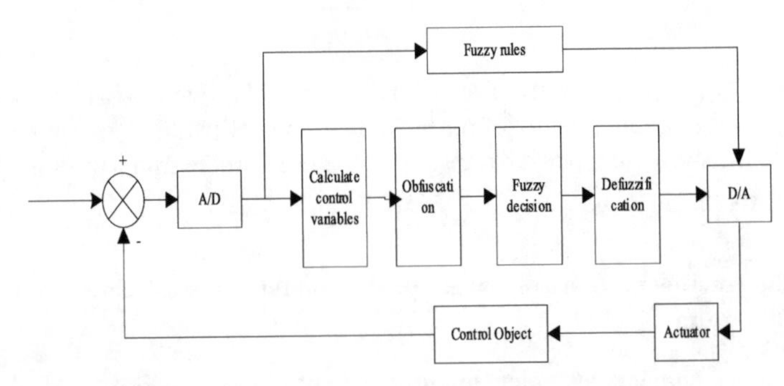

Fig. 3. Logic schematic diagram of intelligent controller

The working logic of the intelligent controller shown in Fig. 3 consists of fuzzy control and PID control. Fuzzy control is an imitative intelligent control method, which mainly imitates the thinking mode of human learning, judgment and selection. Fuzzy set theory, fuzzy language variables and fuzzy logic reasoning are the basic elements of fuzzy control [8]. The working principle of fuzzy control is to fuzzy process the signal of parameter measuring equipment according to the fuzzy rules formulated by relevant personnel, then analyze the signal in the fuzzy rules, and finally add the conclusion output to the execution area of the system. During the implementation of fuzzy control, first of all, the actual value of the controlled variable measured by the measuring equipment is transmitted to the system, and the difference signal is obtained by comparing the value with the standard value as the input port value of the control method. Then, the difference signal is fuzzily processed, and the fuzzy algorithm is used to present its form, and the fuzzy set of difference signals is obtained. Finally, the final output can be obtained by making a decision between the set and the fuzzy relationship. The deviation and deviation change rate of the current building atrium indoor physical environment parameters from the control target are defined as the input items of the controller. The calculation formula of the deviation is as follows:

$$e_U = U_{\text{target}} - U_{\text{gather}} \tag{9}$$

Among U_{gather} is the collection result of the indoor physical environment parameters of the building atrium, including temperature, humidity, illumination, air quality, etc. According to the structure of the intelligent controller, the precise quantity can be used to control the output only after being fuzzed. The fuzzification formula generally uses:

$$\xi = 2\zeta \frac{\text{x} - 0.5(\text{a} + \text{b})}{\text{b} - \text{a}} \tag{10}$$

The input precision is in the range $[a, b]$, which is converted to the interval through the above formula $[-\zeta, \zeta]$ triangular distribution is selected as the membership function of the four fuzzy control variables. According to the basis for formulating the control rules described above, a suitable fuzzy rule table is established, and the fuzzy sets obtained by fuzzy reasoning are converted into accurate control quantities after anti fuzzy processing [9]. The area center of gravity method is selected, and the center of gravity of several continuous points within the output range is obtained through Formula 11, or discrete values within several output ranges are selected and calculated.

$$\Delta = \frac{\int U e_U(U) dU}{\int e_U(U) dU} \tag{11}$$

In the PID part of the intelligent controller, the transfer relationship of the control function is shown as follows:

$$G(s) = K_P \left(1 + \frac{1}{K_i e} + K_d e\right) \tag{12}$$

among K_P, K_i and K_d respectively corresponding to proportional gain and product/differential time constant, PID discrete control law is as follows:

$$g = K_p \left\{ e(i) + K_i \sum e(i) + \frac{K_d}{t_{\text{sampling}}} [e(i) - e(i-1)] \right\} \tag{13}$$

Variables in Eq. (13) t_{sampling} is the sampling period. Put the calculated deviation into the fuzzy control module and the PID control module, complete the design of the intelligent controller for the indoor physical environment, and generate the intelligent control instructions for the indoor physical environment of the building atrium.

2.5 Realize Intelligent Control of Indoor Physical Environment of Building Atrium

From the aspects of temperature and humidity, air, light, sound, etc., the intelligent controller with optimized design is used to complete the intelligent control of the indoor physical environment in the atrium of the building. Taking indoor ambient temperature parameters as an example, its control principle is mainly to control the air valve, adjust the temperature and amount of air supply. The humidity of air supply is adjusted by humidifier, which depends on solenoid valve control. The coordination between various components plays a good role in control. According to the law of conservation of energy, the change rate of energy storage in the air conditioning environment is equal to the heat flowing into the air conditioning environment per unit time minus the heat flowing out, and the equation is expressed as:

$$\begin{cases} Q_0 - Q_1 + Q_2 + Q_3 = c_{air} dT \\ Q_0 - Q_1 = Wc_p(T_0 - T) \\ Q_3 = \kappa_c S_{\text{exterior wall}} \left(T_{\text{outdoor}} - T + \dfrac{\kappa_{\text{surround to guard}}}{\kappa_c} \Delta T \right) \end{cases} \tag{14}$$

Among Q_0、Q_1、Q_2 and Q_3 heat brought into the room by the supply air, the heat that can be taken away by the return air, the heat generated by indoor personnel, mechanical and electrical equipment, and the heat transferred from outside the room, c_{air} and c_p are air heat capacity and air specific constant pressure heat capacity respectively, W supply air volume for air-conditioned rooms, $S_{\text{exterior wall}}$ represents the area of the exterior wall, T_{outdoor} represents the average comprehensive outdoor temperature, $\kappa_{\text{surround to guard}}$ and κ_c corresponds to the heat transfer coefficient of the inner surface of the enclosure and the heat transfer coefficient of the enclosure, ΔT is the temperature fluctuation on the inner surface of the enclosure [10, 11]. In the actual intelligent control process, the air conditioning amplification factor and air supply temperature input to the controller are:

$$\begin{cases} \gamma = \dfrac{Wc_p}{Wc_p + \kappa_{c\,\text{exterior wall}}} \\ T_{\text{Air supply}} = \dfrac{\kappa_c S_{\text{exterior wall}} T + \dfrac{\kappa_{\text{surround to guard}}}{\kappa_c} \Delta T + Q_2}{Wc_p + \kappa_c S_{\text{exterior wall}}} \end{cases} \tag{15}$$

With the support of the controller, control instructions are generated and applied to the corresponding control end to complete the intelligent control of the indoor physical environment temperature in the atrium of the building [12, 13]. In the same way, with the support of intelligent controller optimization design, it can realize the intelligent control of environmental parameters such as humidity, air quality, light, sound, etc. in the indoor physical environment of the building atrium, can automatically adjust and control the physical environment in the atrium of the building according to real-time environmental data and user needs, so as to provide a more comfortable and healthy living and working environment.

3 Experimental Analysis of Control Performance Test

The control performance design test experiment of the intelligent control method for the indoor physical environment of the building atrium under the social information network of the test optimization design is mainly aimed at the control accuracy performance test, that is, to observe the error between the indoor physical environment parameters of the building atrium and the control target under the effect of the optimization design method. Through the statistics of relevant data and the comparison with traditional methods, the advantages of optimal design methods in control performance are reflected.

3.1 Selection of Indoor Environment of Building Atrium

In this experiment, the atrium space of several buildings in a commercial building is selected as the experimental research environment. The commercial building integrates the functions of catering, sports and leisure, culture and entertainment, office and banking, which plays an important role in improving the quality of the city in the location. The landscaping and humanized environment of the commercial building atrium space is an excellent shared space for shopping, leisure, entertainment and communication. Its artistry is changeable in form and rich in content. It often creates a gorgeous and dazzling business atmosphere to attract customers. The comfortable environment atmosphere affects customers' consumption psychology and affects their shopping behavior. The commercial atrium has the characteristics of facility. The atrium space is a space for people to rest and communicate, improve public service facilities such as rest, transportation, health and information provision in the space, and increase consumers' comfort and sense of belonging. It enables people to carry out shopping behavior and social interaction activities in high-quality indoor leisure and shared atrium space. The indoor environment of the atrium of the selected building adopts the typical polymerization enclosed atrium space, surrounded by the store space on all sides, and the glass roof is the main natural lighting interface. The roof at the top elevation of 42m adopts the structural form of I-shaped steel beam with a height of 1.2m, and the composite floor is laid on it, which solves the problem of formwork erection in large-span space construction. The light colored stone floor and stone wrapped columns in the atrium space make them an integral whole. The escalator wrapped with white aluminum plate and the glass handrail together look very light. The natural light through the glass roof, combined with the artificial lighting to set off the goods, makes the space more lively and rich, and creates a comfortable light environment.

3.2 Setting Initial Values and Control Objectives of Indoor Physical Environment

Through the operation control of fans, air conditioners and other equipment, the initial value of the indoor physical environment in the atrium of the building is set, and the environmental control objectives are determined in combination with human needs. Table 1 shows the setting of initial values and control objectives of indoor physical environment.

The measuring points in Table 1 are located on the floor, wall and ceiling of the indoor physical environment of the building atrium, and the distance between any two measuring points shall not be less than 2 m.

Table 1. Initial Values and Control Objectives of Indoor Physical Environment

No. of measuring point	Initial value of indoor physical environment				Indoor physical environment control objectives			
	Temperature (°C)	Humidity (%)	Dayligh ting coefficient	Sound pressure (Pa)	Temperature (°C)	Humidity (%)	Daylighting coefficient	Sound pressure (Pa)
1	28.3	22	0.2	35	25	50	0.5	12
2	30.2	18	0 .3	28	25	50	0.5	12
3	29.5	20	0.3	36	25	50	0.5	12
4	21.4	16	0.8	47	25	50	0.5	12
5	18.6	75	0.7	39	25	50	0.5	12
6	9.4	68	0.6	40	25	50	0.5	12
7	10.5	63	0.2	42	25	50	0.5	12
8	14.7	32	0.2	45	25	50	0.5	12

3.3 Installation and Commissioning of Environmental Intelligent Control Equipment

The indoor environment of the atrium of the selected building is equipped with European sun shading louvers and FSS ceiling curtains as the control objects of the light environment. The European sun shading louvers have a leaf width of 68 mm, a wall thickness of 0.8 mm, a stainless steel silver color, and an arc shaped section with folded edges. It is resistant to moisture, heat and bumps. The FSS ceiling curtain is made of 30% polyester fiber 60% PVC, with an opening rate of about 5%, a weight of 500 g/m^2$\pm$ 5%, a thickness of 0.8 MM $\pm$ 5%, and a UV shading rate of about 96%. It is characterized by tensile strength, tear resistance, fire resistance, light resistance, anti-bacterial mildew resistance, environmental protection and no odor. In order to adjust the extremely hot and humid indoor environment, the atrium is equipped with several air conditioners and dehumidifiers. The actual layout is shown in Fig. 4.

Three split air conditioners are arranged between the wall mounted fans, each with a cooling capacity of about 2.0 horsepower. Although this cooling capacity is not enough to fully meet the cooling demand of this large space, it alleviates the problem of indoor temperature overheating to a certain extent, and the layout position of the air conditioners also simulates the layout of the air conditioning outlets in the mall, which can play a role in cooling pedestrians walking in the corridor to a certain extent. In addition to the control object, it is also necessary to install the intelligent controller and sensor devices at the designated locations. Before the experiment starts, it is necessary to debug all hardware devices in the physical environment of the building atrium to determine whether the control object, controller and sensor are in normal operation. If the debugging shows that the device is operating abnormally. Relevant equipment shall be replaced in time.

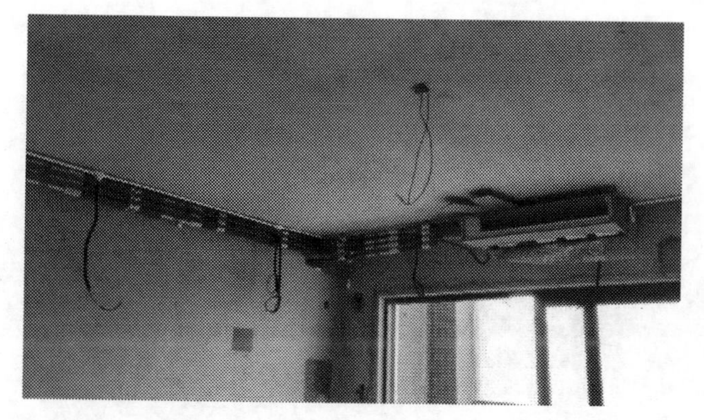

Fig. 4. Realistic view of air conditioning equipment layout

3.4 Describe the Test and Experiment Process

The experiment was conducted from four aspects: ambient temperature, humidity, light and sound. Simulink was used as the operating platform for optimizing the design of intelligent control methods for the indoor physical environment of the building atrium under the social information network. Simulink, as a software integration, was based on the block diagram design environment of MATLAB, and could support multi rate systems at the same time. It plays a good role in digital signal processing. At the same time, the unique GUI interface of the platform simplifies the development process of the method. It does not require input language, but only selects the properties on the interface, which can be completed by simple clicking and moving. The running program corresponding to the intelligent control method of the indoor physical environment of the atrium under the social information network is substituted into the main testing computer to complete the development of the optimization design method. Connect the hardware equipment successfully debugged into the control program, input the environmental control target, and get the intelligent control results of the indoor physical environment of the building atrium, including the control results of temperature, light environment and sound environment, as shown in Fig. 5.

According to the above method, the intelligent control results of various environmental parameters on all measuring points in the indoor physical environment of the building atrium can be obtained. In order to reflect the advantages of the optimization design method in the control performance, the traditional intelligent control based on sensors and fuzzy rules and the intelligent environment control based on network communication are set as the comparison method of the experiment, and the development of the comparison method is realized according to the above way, and the corresponding intelligent control results are output.

3.5 Set Quantitative Test Indicators of Control Performance

In this experiment, temperature control error, humidity control error, daylighting coefficient control error and sound pressure control error are respectively set as quantitative

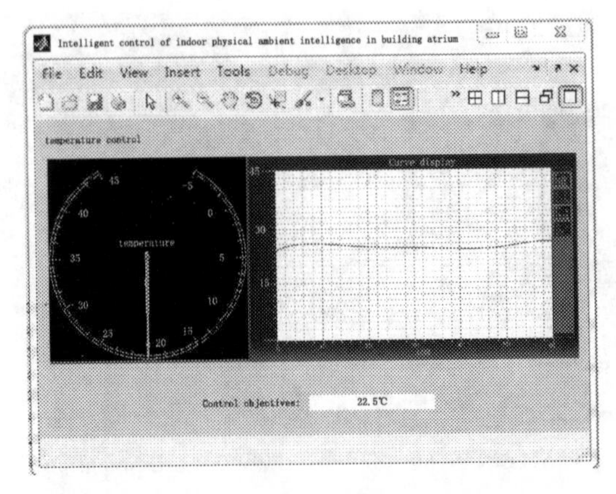

(a) Physical ambient temperature

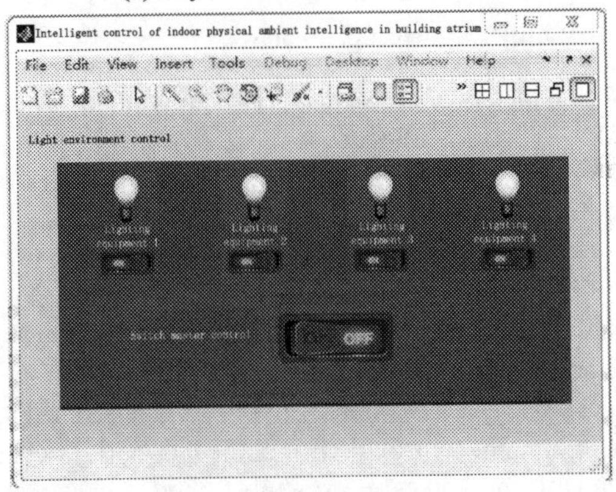

(b) Physical light environment

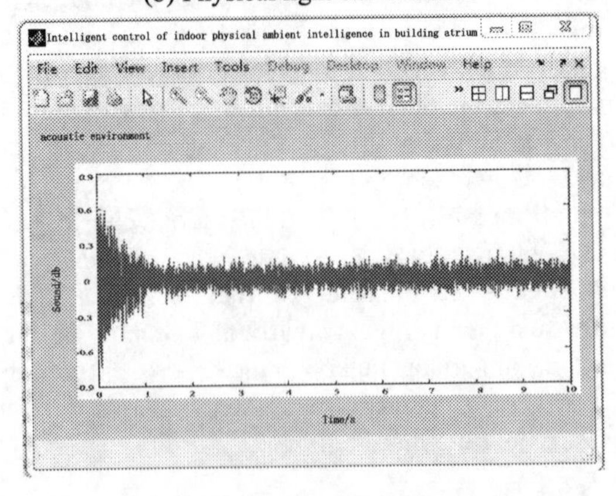

(c) Physical acoustic environment

Fig. 5. Intelligent Control Results of Indoor Physical Environment in Atrium

test indicators of control performance. The test results of temperature control error, humidity control error and daylighting coefficient control error are as follows:

$$\begin{cases} \varepsilon_T = |T_{\text{control}} - T_{\text{target}}| \\ \varepsilon_\chi = |\chi_{\text{control}} - \chi_{\text{target}}| \\ \varepsilon_\upsilon = |\upsilon_{\text{control}} - \upsilon_{\text{target}}| \end{cases} \tag{16}$$

Among T_{control} and T_{target} are the control value and control target value of temperature, χ_{control} and χ_{target} corresponding to the control value and control target value of humidity, $\upsilon_{\text{control}}$ and υ_{target} respectively represent the control value and control target value of the daylighting coefficient. In addition, the numerical results of sound pressure control error are as follows:

$$\varepsilon P_{\text{sound}} = |P_{\text{sound - control}} - P_{\text{sound - target}}| \tag{17}$$

Variables in the above formula $P_{\text{sound - control}}$ and $P_{\text{sound - target}}$ are respectively the control value and control target value of sound pressure. Finally, the smaller the control error is, the better the control performance of the corresponding method is proved.

3.6 Control Performance Test Results and Analysis

The test results of temperature control error, humidity control error and daylighting coefficient control error are obtained through the statistics of relevant data, as shown in Table 2.

Table 2. Test Results of Temperature, Humidity and Daylight Coefficient Control

No. of measuring point	Intelligent control based on sensors and fuzzy rules			Intelligent environment control method based on network communication			Intelligent control method of indoor physical environment in atrium under social information network		
	Temperature (°C)	Humidity (%)	Daylighting coefficient	Temperature (°C)	Humidity (%)	Daylighting coefficient	Temperature (°C)	Humidity (%)	Daylighting coefficient
1	25.6	50.7	0.5	25.5	50.7	0.6	25.1	50.3	0.5
2	25.8	50.9	0.8	25.4	50.3	0.6	25.0	50.2	0.4
3	25.7	50.8	0.8	24.7	49.5	0.7	24.9	50.1	0.5
4	26.0	50.7	0.6	24.5	49.6	0.7	25.0	50.3	0.5
5	26.2	49.1	0.9	24.4	50.4	0.6	25.0	49.8	0.6
6	24.1	49.2	0.7	25.4	50.6	0.6	24.9	49.9	0.5
7	24.3	51.1	0.4	25.6	50.5	0.4	25.1	50.1	0.4
8	24.0	51.4	0.9	25.5	50.5	0.3	24.9	50.2	0.5

The data in Table 1 and Table 2 are substituted into Formula 16 to obtain the calculation results of the control error of temperature, humidity and daylighting coefficient. The average temperature control error of the two comparison methods is 0.86°C and 0.48°C, the average humidity control error is 0.91% and 0.49%, the average value of the control error of daylighting coefficient is 0.23 and 0.14, and the average value of the

temperature, humidity. The average values of daylighting coefficient control errors are 0.06°C, 0.19% and 0.04 respectively. In addition, the test comparison results of sound pressure control error are obtained through the calculation of Formula 17, as shown in Fig. 6.

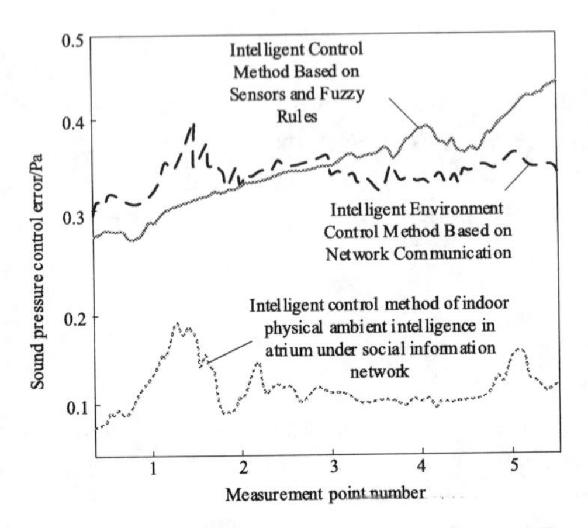

Fig. 6. Comparison Curve of Sound Pressure Control Error Test

It can be seen intuitively from Fig. 6 that the control error of the optimal design method is significantly lower than that of the two comparison methods, that is, the optimal design method has obvious advantages in control performance.

4 Conclusion

In this paper, a physical ambient intelligence control method of building atrium under social information network is proposed. This method considers the dynamic characteristics of fluid in building atrium and constructs the indoor physical environment model of building atrium. Using social information network to collect physical environment parameters such as temperature and humidity; According to the thermal comfort of human body, the ambient intelligence control target is set, and the indoor physical ambient intelligence controller is designed. With the support of control instructions, the indoor physical ambient intelligence control task of building atrium under the social information network is realized. The intelligent control method of the indoor physical environment of the building atrium can display the indoor environmental parameters in real time, and can control the work of household appliances intelligently through the terminal control device, effectively ensuring the safety and comfort of the indoor environment.

In the future, the indoor physical ambient intelligence control method of building atrium will continue to develop and innovate, creating a more intelligent, sustainable and healthy indoor environment for people. The following is the prospect of future work:

(1) With the continuous progress of artificial intelligence and big data technology, the indoor physical ambient intelligence control method of building atrium will be more

accurate and intelligent. By analyzing huge data sets, intelligent control systems can learn and predict users' behaviors and preferences, thus providing personalized environmental adjustment and services.

(2) Future work will pay more attention to energy management and sustainability. Intelligent control system will combine energy monitoring and analysis technology, optimize energy utilization and management strategy, realize energy efficient utilization and saving, and reduce the impact on the environment.

References

1. Hu, C., Liu, C., Hu, N., et al.: Government environmental control measures on CO2 emission during the 2014 youth olympic games in Nanjing: perspectives from a top-down approach. J. Environ. Sci. **113**(1), 165–178 (2022)
2. He, B.: Quantifying the effect of index-based operation logic for building environmental control system—taking shading as example. Buildings **12**(12), 2043–2052 (2022)
3. He, D., Teng, X., Chen, Y., et al.: Piston wind characteristic and energy saving of metro station environmental control system. J. Building Eng. **44**(6), 102–114 (2021)
4. Zhang, Z., Jin, T., Wu, H., et al.: Experimental investigation on environmental control of a 50-person mine refuge chamber. Build. Environ. **210**(2), 108–116 (2022)
5. Pedersen, E., Gao, C., Wierzbicka, A.: Tenant perceptions of post-renovation indoor environmental quality in rental housing: Improved for some, but not for those reporting health-related symptoms. Build. Environ. **189**(2), 1–10 (2021)
6. Pereira, J., Rivero, C.C., Gomes, M.G., et al.: Energy, environmental and economic analysis of windows' retrofit with solar control films: a case study in Mediterranean climate. Energy **233**(15), 1–14 (2021)
7. Park, S.: Zero-energy building integrated planning methodology for office building considering passive and active environmental control method. Appl. Sci. **11**(8), 3686–3695 (2021)
8. Pedersen, E., Borell, J., Li, Y., et al.: Good indoor environmental quality (IEQ) and high energy efficiency in multifamily dwellings: how do tenants view the conditions needed to achieve both? Build. Environ. **191**(3), 1–9 (2021)
9. Lee, J., Woo, D.O., Jang, J., et al.: Collection and utilization of indoor environmental quality information using affordable image sensing technology. Energies **15**(3), 1–10 (2022)
10. Moon, H.J.: Novel integrated and optimal control of indoor environmental devices for thermal comfort using double deep Q-network. Atmosphere **12**(5), 629–635 (2021)
11. Liu, S., Xu, X., Zhang, Y., Muhammad, K., Fu, W.: A reliable sample selection strategy for weakly-supervised visual tracking. IEEE Trans. Reliab. **72**(1), 15–26 (2023)
12. Liu, S., et al.: Human inertial thinking strategy: a novel fuzzy reasoning mechanism for IoT-assisted visual monitoring. IEEE Internet of Things J. **10**(5), 3735–3748 (2023)
13. Fan, J., Zhang, L.: Home energy intelligent control system simulation based on NB-IOT. Computer Simulation **38**(9), 336–339, 365 (2021)

Author Index

L. Yun et al. (Eds.): ADHIP 2023, LNICST 549, pp. 471–472, 2024.
https://doi.org/10.1007/978-3-031-50549-2

Printed in the United States
by Baker & Taylor Publisher Services